T0135333

Lecture Notes in Electrical Engineering

Volume 653

The book series *Lecture Notes in Electrical Engineering* (LNEE) publishes the latest developments in Electrical Engineering - quickly, informally and in high quality. While original research reported in proceedings and monographs has traditionally formed the core of LNEE, we also encourage authors to submit books devoted to supporting student education and professional training in the various fields and applications areas of electrical engineering. The series cover classical and emerging topics concerning:

- Communication Engineering, Information Theory and Networks
- Electronics Engineering and Microelectronics
- Signal, Image and Speech Processing
- Wireless and Mobile Communication
- Circuits and Systems
- Energy Systems, Power Electronics and Electrical Machines
- Electro-optical Engineering
- Instrumentation Engineering
- Avionics Engineering
- Control Systems
- Internet-of-Things and Cybersecurity
- Biomedical Devices, MEMS and NEMS

For general information about this book series, comments or suggestions, please contact leontina.dicecco@springer.com.

To submit a proposal or request further information, please contact the Publishing Editor in your country:

China

Jasmine Dou, Editor (jasmine.dou@springer.com)

India, Japan, Rest of Asia

Swati Meherishi, Editorial Director (Swati.Meherishi@springer.com)

Southeast Asia, Australia, New Zealand

Ramesh Nath Premnath, Editor (ramesh.premnath@springernature.com)

USA, Canada:

Michael Luby, Senior Editor (michael.luby@springer.com)

All other Countries:

Leontina Di Cecco, Senior Editor (leontina.dicecco@springer.com)

**** This series is indexed by EI Compendex and Scopus databases. ****

More information about this series at http://www.springer.com/series/7818

Qilian Liang · Wei Wang · Jiasong Mu ·
Xin Liu · Zhenyu Na · Xiantao Cai
Editors

Artificial Intelligence in China

Proceedings of the 2nd International Conference on Artificial Intelligence in China

Editors
Qilian Liang
Department of Electrical Engineering
University of Texas at Arlington
Arlington, TX, USA

Jiasong Mu
Tianjin Normal University
Tianjin, China

Zhenyu Na
School of Information Science and Technology
Dalian Maritime University
Dalian, China

Wei Wang
Tianjin Normal University
Tianjin, China

Xin Liu
Dalian University of Technology
Dalian, China

Xiantao Cai
Wuhan University
Wuhan, Hubei, China

ISSN 1876-1100 ISSN 1876-1119 (electronic)
Lecture Notes in Electrical Engineering
ISBN 978-981-15-8601-9 ISBN 978-981-15-8599-9 (eBook)
https://doi.org/10.1007/978-981-15-8599-9

This Springer imprint is published by the registered company Springer Nature Singapore Pte Ltd.
The registered company address is: 152 Beach Road, #21-01/04 Gateway East, Singapore 189721, Singapore

Contents

Research on Improved K-Means Clustering Algorithm for Smart Energy Meter Based on Climatic Features 1
Hongyu Cao, Huiying Liu, Xin Yin, Ruxin Wen, and Yue Chen

A Method to Solve the Measurement Error of Track Safety Control Based on Weighted Fusion 9
Jianping Pan, Shengdong Ji, Pengfei Hao, Zhiguo Ren, and Quansheng He

Improved Spectral Efficiency Based on Double Spatially Sparse in Millimeter Wave MIMO System 18
Yanping Zhu, Sicheng Guo, Yinxuan Sun, and Jing Liu

Electromagnetic Compatibility Test Analysis of Space Launch Field Based on Wavelet Analysis 26
Bo Zhou, Yao lou Jiang, Zong ju Xiong, Chun guang Shi, Zhen Chen, Mei Yang, and Gai qin Li

Design of a Biometric Access Control System Based on Fingerprint Identification Technology 34
Hai Wang, Zhihong Wang, Guiling Sun, Limin Zhang, Yi Gao, Ying Zhang, and Chaoran Bi

Aeronautical Meteorological Decision Supporting Technology Based on 4D Trajectory Prediction 43
Yi Mao, Yuxin Hu, and Jiajing Zhang

Evaluation Index System and Case Study for Smart ATM Based on ASBU 52
Chao Jiang

Research on the Development Trend of Artificial Intelligence Based on Papers and Patent Analysis—Data Comparison Between China and the United States 60
Jinyi Feng and Xiangyang Li

Analysis of 5G and Beyond: Opportunities and Challenges 68
Liedong Wang

Prediction of PM2.5 Concentration Based on Support Vector Machine and Ridge 76
Haocun Zhang

Helmet Detection Based on an Enhanced YOLO Method 84
Weizhou Zheng and Jiayi Chang

Feature Extraction and Classification of Unknown Types of Communication Emitter 93
Xu Zhang, Zhuo Sun, Suyu Huang, Shaolin Ma, and Anhao Ye

Research on Escape Strategy Based on Intelligent Firefighting Internet of Things Virtual Simulation System 102
Hai Wang, Guiling Sun, Yi Gao, and Xiaochen Li

Weather Identification-Based Multi-level Visual Feature Combination 111
Ziheng Li, Anliang Zhou, and Yilong Geng

Railway Tracks Defects Detection Based on Deep Convolution Neural Networks 119
Zhong-Jun Wan and Song-Qi Chen

Efficiency Evaluation of Deep Model for Person Re-identification 130
Haijia Zhang, Sen Wang, Nuoran Wang, Shuang Liu, and Zhong Zhang

Cloud Recognition Using Multimodal Information: A Review 137
Linlin Duan, Jingrui Zhang, Yaxiu Zhang, Zhong Zhang, Shuang Liu, and Xiaozhong Cao

Graph Convolution Network for Person Re-identification 145
Wenmin Huang, Yilin Xu, Zhong Zhang, and Shuang Liu

Cross-Domain Person Re-identification: A Review 153
Yanan Wang, Shuzhen Yang, Shuang Liu, and Zhong Zhang

A Weighted Least Square Support Vector Regression Method with MPP-GGP Based Sequential Sampling for Efficient Reliability Analysis 161
Yang Guo, Nan-nan Wang, and Gen-shen Kai

Design of Quadrotor Automatic Tracking UAV Based on OpenMV . . . 171
Hai Wang, Ying Zhang, Guiling Sun, Zhihong Wang, Binghao Tian, and Penghui Li

A Reliability Evaluation Model of Intelligent Energy Meter in Typical Environment 179
You Gong, Huiying Liu, Xin Yin, Heng Hu, and Guorui Wu

Analysis and Prediction of the Resettlement for Climate Refugees in the Maldives 186
Jiasong Mu and Hao Ma

Smart Electricity Meters Test Data Management Service System 194
Liu Huiying, Yin Xin, Wang Xiaoyu, Wen Ruxin, Zhang Qiuyue, and Du Bo

Power Equipment Identification Based on Single Shot Detector 202
Hanwu Luo, Wenzhen Li, Qirui Wu, Hailong Zhang, and Zhonghan Peng

RETRACTED CHAPTER: Power Equipment Defect Detection Algorithm Based on Deep Learning 211
Hanwu Luo, Qirui Wu, Kai Chen, Zhonghan Peng, Peng Fan, and Jingliang Hu

Research on Active Learning Method Based on Domain Adaptation and Collaborative Training 220
Wenzhen Li, Qirui Wu, Hanwu Luo, Guoli Zhang, Zhonghan Peng, and Kai Chen

Evolution Analysis of Research Hotspots in the Field of Machine Learning Based on Complex Network 228
Tala, Cui Yimin, Li Junmei, and Su Xiaoyan

Research on Index Network Construction and Effectiveness Evaluation Method of Air Traffic Control System Based on CDM Data 242
Zheng Li, Yinfeng Li, Xiaowen Wang, Meng Xu, and Shenghao Fu

Rerouting Path Planning Based on MAKLINK Diagram and MS-Genetic Algorithm 251
Tong Wei, Manzhen Duan, Bin Dong, Yinfeng Li, and Shenghao Fu

Data-Driven Fault-Aware Multi-objective Optimization for Flexible Job-Shop Scheduling Problem 261
Zhibo Sui, Xiaoxia Li, Jie Yang, and Jianxing Liu

Research on Airspace Conflict Resolution Algorithm Based on Dempster–Shafer Theory 270
Zelin Li, Tianhao Tan, and Shiming Zhu

A Learning Based Automated Algorithm Selection for Flexible Job-Shop Scheduling 277
Xinyu Wang, Xiaoxia Li, Rongyin Zhu, and Zetao Lv

Controller's Workload and Sector Capacity Assessment Based on 4D Track 285
Changcheng Li and Yuxin Hu

Classification of Tea Pests Based on Automatic Machine Learning 296
Heng Zhou, Fuchuan Ni, Ziyan Wang, Fang Zheng, and Na Yao

Thunderstorm Service and Decision Support Technology Based on Composite Reflectivity Information 307
Yao Shan, Yuxin Hu, and Yungang Tian

Research on Intrusion Detection Method Based on PGoogLeNet-IDS Model 315
Min Sun, Xue Hao, and Wenbin Li

Research on ADS-B Interference Principle and Suppression Method 324
Yi Yang, Ruheng Xie, and Yang Ding

A Multispectral Image Enhancement Algorithm Based on Frame Accumulation and LOG Detection Operator 334
FengJuan Wang, BaoJu Zhang, CuiPing Zhang, ChengCheng Zhang, and Man Wang

Research on Intelligent Release Strategy of Air Traffic Control Automation System Based on Control Experience 344
Liu Yan

A Method Based on Deep Reinforcement Learning to Generate Control Strategy for Aircrafts in Terminal Sector 356
Qiucheng Xu, Jinglei Huang, Zeyuan Liu, and Hui Ding

Sand Table Design of Virtual Reality Psychotherapy 364
Chaoran Bi, Hai Wang, and Rui Dong

Massive Flights Real-Time Rerouting Planning Based on Parallel Discrete Potential Field Method 372
Yang Ding, Zelin Li, Bingyu Li, and Ruheng Xie

Remote Sensing Image Detection Based on FasterRCNN 380
Shunmin Liu, Zhiming Ma, and Bingcai Chen

Research on Airport Surface Simulation Method Based on Dynamic Path Planning 387
Shenghao Fu, Xiaowen Wang, Bin Dong, and Yan Liu

Triple-Channel Feature Mixed Sentiment Analysis Model Based on Attention Mechanism 396
DeGang Chen, Azragul, and Bingcai Chen

National Food Safety Standard Graph and Its Correlation Research 405
Li Qin and ZhiGang Hao

Research on the Hydraulic System of the Expandable Shelter Improved by the Servo Motor Pump 412
Fang Bai and Shucheng Wang

Recognition of Grape Species with Small Samples Based on Attention Mechanism 424
Yanuo Lu and Bingcai Chen

F-Measure Optimization of Forest Flame Salient Object Detection Based on Boundary Perception 436
Tiantian Tang and Bingcai Chen

Cluster Analysis of Student Scores Based on Global K-Means Algorithm 443
Jiashan Cui, Mei Nian, Jun Zhang, and Bingcai Chen

Microblog Rumors Detection Based on Bert-GRU 450
Lianjin Han, Weimin Pan, and Haijun Zhang

Intelligent Ocean Governance—Deep Learning-Based Ship Behavior Detection and Application 458
Peng Qin and Yang Cao

The Research on Disruptive Technology Identification Based on Scientific and Technological Information Mining and Expert Consultation: A Case Study on the Energy Field 469
Lucheng Lyu, Xuezhao Wang, Wei Chen, Xin Zhang, Xiaoli Chen, and Xiwen Liu

Research on Transfer Learning Technology in Natural Language Processing 483
Ruilin Shen and Weimin Pan

Identification of Key Audience Groups Based on Maximizing Influence 489
Jie Zhou, Weimin Pan, and Haijun Zhang

Technical Theme Analysis of WeChat Graphic Based on Domain Science and Technology Information 497
Min Zhang, Rui Yang, Wei Chen, Jun Chen, Jinglin Xu, and Yanli Zhou

Coronavirus Disease (COVID-19) X-Ray Film Classification Based on Convolutional Neural Network 508
Shixiang Yan and Bingcai Chen

The Shortest Path Network Rumor Source Identification Method Based on SIR Model 516
Zhongyue Zhou, Hai-Jun Zhang, Weimin Pan, Bingcai Chen, and Yanjun Li

Microblog Rumor Detection Based on Bert-DPCNN 524
Yan-Jun Li, Hai-Jun Zhang, Wei-Min Pan, Ru-Jia Feng,
and Zhong-Yue Zhou

A New Method of Microblog Rumor Detection Based on Transformer Model 531
Ru-Jia Feng, Hai-Jun Zhang, Wei-Min Pan, Zhong-Yue Zhou,
and Yan-Jun Li

Research on Image Classification Method Based on Improved Xception Model 538
Shuping Chen and Bingcai Chen

Research on Real-Time Expression Recognition of Complex Environment Based on Attention Mechanism 548
Shunping Li, Cheng Peng, and Bingcai Chen

Identification Model of Crop Diseases and Insect Pests Based on Convolutional Neural Network 557
Yong Ai, Chong Sun, Anran Liu, Feng Ding, and Jun Tie

UBHIC: Top-Down Semi-supervised Hierarchical Image Classification Algorithm 564
Jiang Qing Wang, Jian Quan Bi, Lei Zhang, Chong Sun, and Jun Tie

Topic Mining and Effectiveness Evaluation of China's Coal-Related Policy Based on LDA Model 573
Fang Yue, Kaimo Guo, Mingliang Yue, and Wei Chen

Feature Extraction and Selection in Hidden Layer of Deep Learning Based on Graph Compressive Sensing 582
Yifei Yuan, Lei Xu, Yiman Ma, and Wei Wang

Vehicle Detection in Aerial Images Based on YOLOv3 588
Ruiheng Hu, Bingcai Chen, and Tiantian Tang

A New Node Optimization Algorithm of Wireless Sensor Network Based on Graph Signal 594
Lei Xu, Yifei Yuan, and Wei Wang

Research and Practice of Intelligent Water Conservancy Integration Management Platform in Xinjiang 600
Yumeng Lin, Bingcai Chen, Zhiming Ma, Qian Ning, Yanting Xiao,
Lun Shao, Qiang Luo, and Xinzhi Zhou

Retraction Note to: Power Equipment Defect Detection Algorithm Based on Deep Learning C1
Hanwu Luo, Qirui Wu, Kai Chen, Zhonghan Peng, Peng Fan,
and Jingliang Hu

Research on Improved K-Means Clustering Algorithm for Smart Energy Meter Based on Climatic Features

Hongyu Cao[1], Huiying Liu[1], Xin Yin[1], Ruxin Wen[1], and Yue Chen[2](✉)

[1] State Grid Heilongjiang Electric Power Co., Ltd., Electric Power Research Institute, 150000 Harbin, Heilongjiang, China

[2] Heilongjiang Electrical Instrument Engineering Research Center Co., Ltd., 150000 Harbin, Heilongjiang, China

736057854@qq.com

Abstract. In the existing clustering algorithm research of smart meters, it is based on the operation data or load curve of smart meters, and there is no research on the influence of climatic factors. Therefore, this paper proposes an improved K-means algorithm based on regional features to cluster smart energy meters. In view of the problem that the traditional K-means algorithm is dependent on the initial value and easy to fall into the local optimum, the initial value selection method and the movement rule of the cluster center are improved, so as to improve the accuracy of the algorithm. Through the simulation of the data, the accuracy and calculation speed of this method are improved compared with the traditional K-means algorithm method. It is more suitable for the solution of this problem to cluster the smart meters accurately according to the climate characteristics.

Keywords: Smart meter · Climatic features · Improved K-means algorithm

1 Introduction

In order to study the reliability of smart meters in typical environment, State Grid Corporation of China established operation bases of smart meters in Heilongjiang, Fujian, Xinjiang, and Tibet[1]. The operation data and local environmental information of are uploaded to the main station and used to study the impact of environmental characteristics on the operation error of smart meters. The premise of studying the above problems is to divide the smart meters according to the environmental features and then study the influence of different environmental features on the error of the smart meters.

The existing classification of smart energy meter is based on the operation parameters, the type of power consumption, or the load conditions. In reference [1], the clustering method of power load curve is studied. By reducing the dimension of data, the

[1] This paper is supported by the science and technology project of the headquarters of State Grid Corporation of China (project no.: 5230HQ19000F).

Q. Liang et al. (eds.), *Artificial Intelligence in China*, Lecture Notes in Electrical Engineering 653, https://doi.org/10.1007/978-981-15-8599-9_1

classical integrated clustering algorithm is used to cluster the load curve. In reference [2], a distributed clustering algorithm is proposed, which uses the adaptive K-means algorithm to cluster the power consumption data stored in the smart energy meter. At present, there is no research on the classification of smart energy meter according to the climate characteristics.

In order to meet the needs of large-scale smart energy meters to be classified in a specific climate, this paper proposed a classification method of smart energy meters based on climate characteristics. The principle of climate feature extraction is put forward. Then, according to the characteristics of extracted climateal data, an improved K-means algorithm is proposed, which improves the selection of initial clustering center to improve the dependence of classification results on initial values. This improved K-means algorithm based on climate characteristics can effectively classify the smart meters running in multiple places and provide effective guidance for further mining the impact of climateal parameters on the reliability of smart energy meters.

2 Selection of Climatic Features

In order to further study the operation of smart meters in complex climate, the measurement center of State Grid Corporation of China has carried out the reliability verification of smart meters in four areas, namely Heilongjiang Mohe, Xinjiang chatkale, Tibet Yangbajing, and Meizhou Island of Fujian Province, respectively, under the conditions of severe cold, dry heat, high altitude, humid heat.

Mohe County, Heilongjiang Province, is located in the northernmost part of China, with an annual average temperature of—5.5 °C. Turpan is a unique warm temperate continental arid desert climate. The annual average temperature is 14 °C. Yangbajing is located in Tibet, 4300 m above sea level, an annual average atmospheric pressure of 0.06 MPa. Meizhou Island is located in the south of Putian City. The average annual rainfall is about 1000 mm. The change of temperature in every day shows the following regularity: the peak value appears at about 14 o'clock, and the valley value generally appears at about 4 o'clock before sunrise.

The principle of selecting climatic information is to maximize the difference. The climatic data to distinguish these areas are the lowest temperature, the highest temperature, the humidity, and the altitude. Therefore, it can select the temperature data at 14:00 in a day, the temperature at 4:00 in a day, and then increase one of the humidities or air pressure data to create a three-dimensional data vector for each smart energy meter $X_i(a_i, b_i, c_i)$.

$$y_i^{'} = [y_i - \min(y_i)]/[\max(y_i) - \min(y_i)] \tag{2.1}$$

$\max(y_i)$ and $\min(y_i)$ represent the maximum and minimum values in the corresponding element item. Calculated results are $x_i(a_i^{'}, b_i^{'}, c_i^{'})$

3 The Clustering Method of Smart Energy Meter Based on Improved K-Means Algorithm

K-means algorithm is a kind of hierarchy method [3], which usually uses Euclidean distance as the evaluation index of similarity degree of two samples. The principle of the algorithm is: firstly, randomly select points in the sample set as the initial clustering center, compare the Euclidean distance from each point to each clustering center, and divide them into the category with the smallest distance. After all the points are classified, the average value of all kinds of data is calculated, and the new clustering center is the average value, until the square error criterion function meets the requirements.

Sample set is $M = \{x_i | x_i \in R^m, i = 1, 2, \ldots n\}$, where m is the dimension of data and n is the size of the data set. The Euclidean distance between x_i and x_j is calculated as follows [4]:

$$\mathrm{Dist}(x_i, x_j) = \sqrt{(x_i - x_j)^{\mathrm{T}}(x_i - x_j)} \tag{3.1}$$

Category set to which the sample belongs is $N = \{c_t | c_t \in R^m, t = 1, 2, \ldots k\}$, where k is the number of clusters. Each cluster center is calculated by the following formula [5]:

$$z_t = \frac{1}{n_t} \sum_{x_i \in c_t} x_i \tag{3.2}$$

K-means algorithm has the advantages of simple and fast convergence, but the selection of initial clustering center has a great influence on the clustering results. Therefore, when choosing the initial value, we choose the point as far away as possible as the initial clustering center. This method of selecting initial value meets the principle of clustering algorithm that the gap between categories is as large as possible. The specific steps are as follows:

(1) Randomly select a data point in data set as the first initial cluster center.
(2) Calculate the distance between each point and the selected cluster center.
(3) Set a threshold value ε. If the distance between the data point and any selected clustering center is less than the threshold value, the data point will be removed.
(4) Calculate the sum of the distance from the remaining data points to the selected cluster center.

$$\mathrm{Dist}(x_i, z_j) = \sum_{j=1}^{j} \sqrt{(x_i - z_j)^{\mathrm{T}}(x_i - z_j)} \tag{3.3}$$

where $z_1, z_2, \ldots z_j$ represents the selected cluster centers. Repeat (2) to (4) until all cluster centers are selected. Then, K-means clustering was carried out. The calculation flow of the improved K-means algorithm is shown in Fig. 1.

After all the initial clustering centers are selected, the distance between each sample point and each initial clustering center is calculated. The data points and the nearest center points are grouped into a cluster, and then, the cluster center is updated. It should be noted that in the traditional K-means clustering, the way of taking the average value of data in each cluster as a new clustering center may result in a local optimal result. Therefore, this paper proposes to select the median of data samples as the moving direction of clustering center.

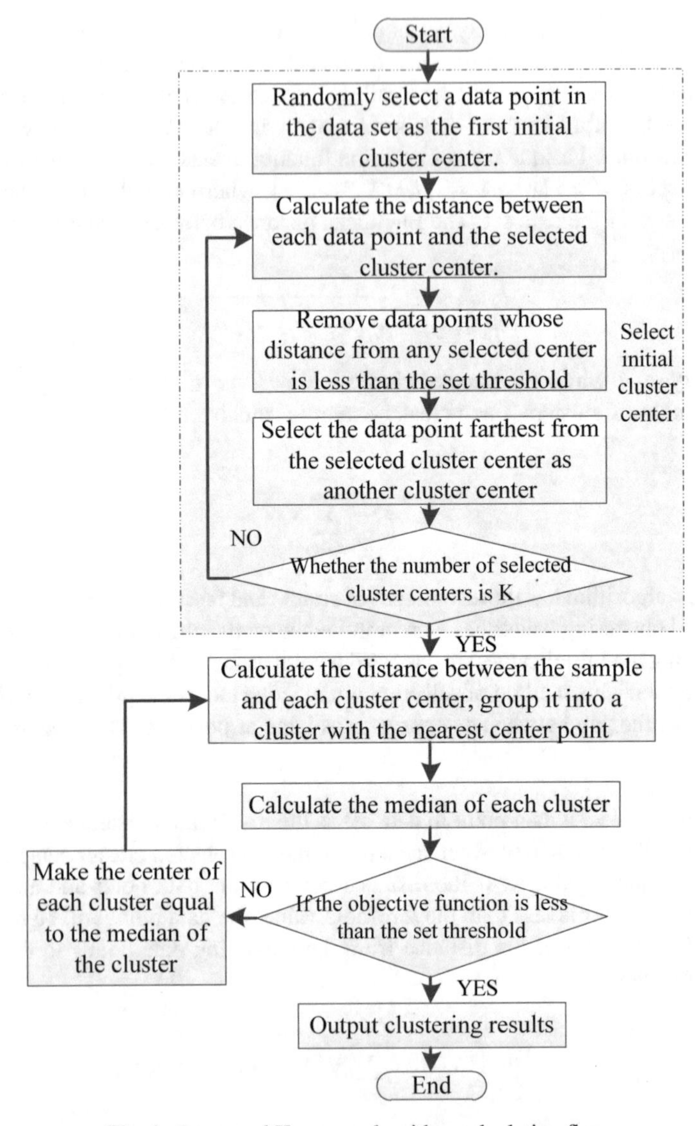

Fig. 1. Improved K-means algorithm calculation flow

4 Simulation and Result Analysis

The case data are from the typical environmental laboratory data of Heilongjiang, Fujian, Tibet, and Xinjiang. Select 100 smart energy meters in each of the four provinces. Extract the temperature at 4:00 on February 3, 2018, the temperature at 14:00 on August 1, 2018, and the air pressure at 0:00 on February 3, 2018. The data statistics of 400 smart meters are shown in Table 1.

The distribution of the raw data is shown in Fig. 2. It can be seen that the climate data of smart meters selected according to the climate feature extraction principle proposed

Table 1. Statistics of raw sample data

	Average minimum temperature (°C)	Average maximum temperature (°C)	Average pressure(kPa)
Heilongjiang	−35	28	95
Fujian	9	38	101
Tibet	−5	13	61
Xinjiang	−12	41	110

in this paper makes the difference between the four types of meters larger, and the difference between the four clusters of data is more obvious. This kind of data distribution can facilitate clustering calculation and effectively improve the accuracy of clustering results.

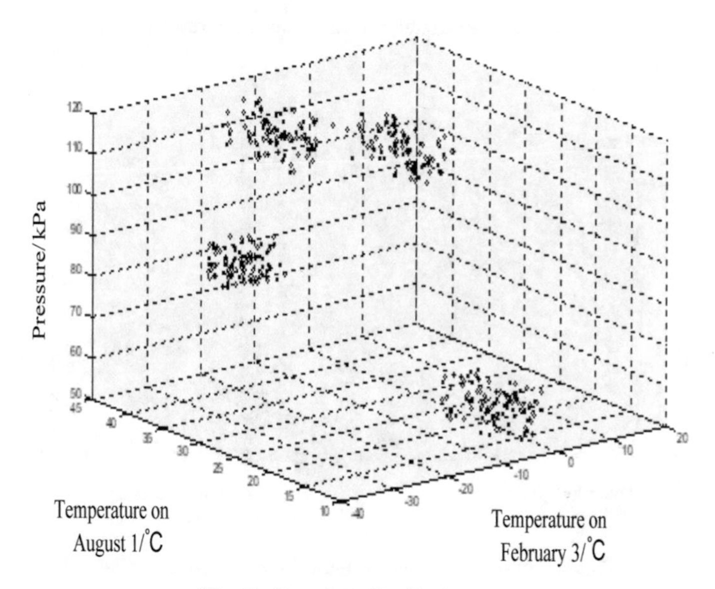

Fig. 2. Raw data distribution map

For the original data, the traditional K-means algorithm and the improved K-means algorithm are used for clustering. Each algorithm repeatedly performs 10 calculations, recording the accuracy of the results of 10 calculations and the time taken for each calculation. The experimental results are shown in Figs. 3, 4, 5 and 6.

In Figs. 3 and 4, red circle, green square, blue triangle, and black cross are used to represent the four clusters. Figure 3 shows the clustering results using the traditional K-means algorithm, which classifies Heilongjiang Province into one category, Xinjiang Province and Fujian Province into one category, and Tibet province into two categories. The result is obviously wrong

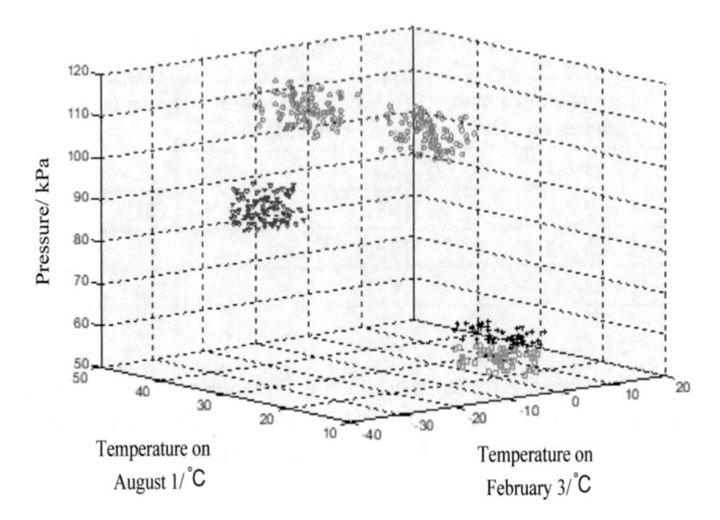

Fig. 3. Traditional K-means clustering results

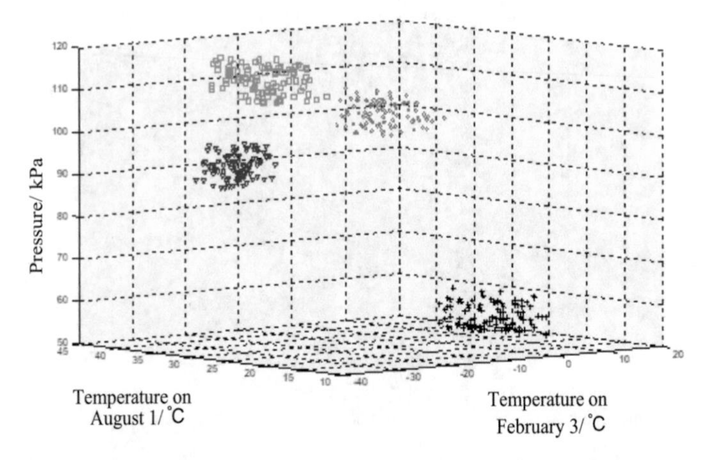

Fig. 4. Improved K-means algorithm clustering results

Figure 4 shows the clustering results using the improved K-means algorithm proposed in this paper. This algorithm can distinguish the intelligent energy meters of four provinces accurately. Among them, the red circle represents Fujian provincial, the blue triangle represents Heilongjiang provincial, the black cross represents Tibet provincial, and the green square represents Xinjiang provincial.

Figure 5 is a comparison of the accuracy of the two algorithms. The clustering results of the traditional K-means algorithm are inconsistent each time, and the accuracy is not high. The average accuracy of 10 times is 74.4%. The main reason for this result is that the initial clustering centers are randomly selected, and different initial clustering centers may produce different clustering results. The accuracy of the improved K-means

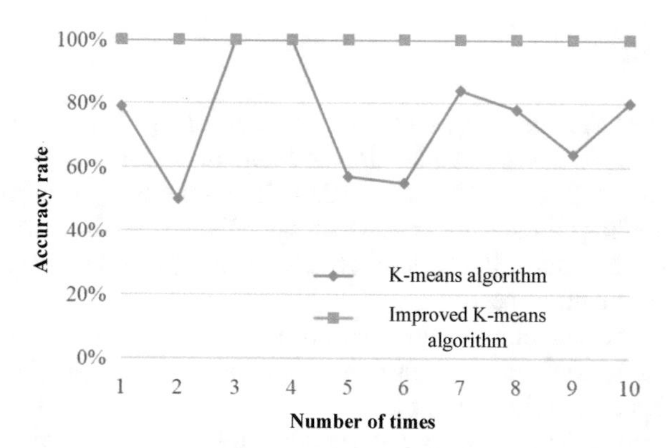

Fig. 5. Comparison of accuracy

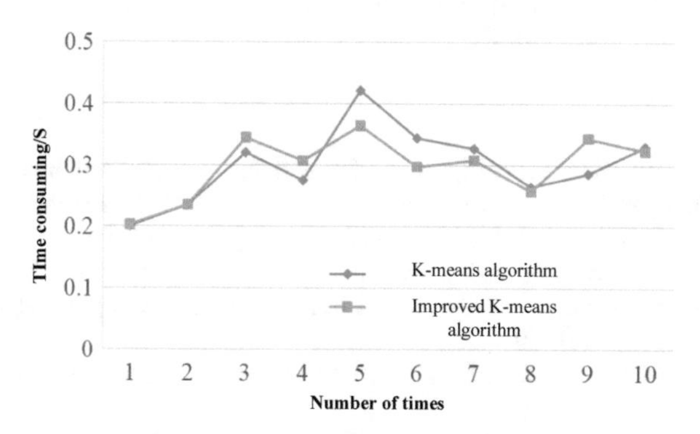

Fig. 6. Time-consuming comparison

algorithm is 100% in 10 times. The reason for the improvement of accuracy is the accuracy of initial cluster center selection. The improved selection method of initial clustering centers in this paper can make the four initial clustering centers in four clusters, so the accuracy of clustering results is greatly improved.

Figure 6 is a comparison of the time-consuming of the two algorithms. The average time of the traditional K-means algorithm 10 times is 0.334 s. The average time of the improved K-means algorithm 10 times is 0.308 s. The time-consuming of the two algorithms is almost the same. The time-consuming of the improved K-means algorithm is slightly lower than that of the traditional K-means algorithm. The traditional algorithm takes less time in the selection of initial clustering center, but the selection of initial clustering center will affect the number of subsequent iterations, which will lead to long time. The improved K-means algorithm takes a certain time in the selection of initial clustering center, but greatly reduces the number of subsequent iterations, so it relatively reduces the time-consuming of the algorithm.

5 Conclusion

This paper studies the clustering algorithm of smart energy meter based on climatic features. According to the climatic features of Mohe in Heilongjiang, Chatkale in Xinjiang, Yangbajing in Tibet, and Meizhou Island in Fujian, this paper puts forward the principle of extracting climatic features, which can effectively reduce the data dimension without affecting the calculation accuracy, thus reducing the calculation complexity and improving the calculation speed.

The selection method of initial clustering center and the moving principle of clustering center of traditional K-means algorithm are improved. The distant points are selected as the initial clustering center. The median of cluster is selected as the moving direction of cluster center. The accuracy and calculation speed of clustering algorithm are improved. Finally, MATLAB simulation is used to verify the accuracy and effectiveness of the theory and algorithm proposed in this paper.

References

1. Bin Z, Chijie Z, Jun HU et al (2015) Ensemble clustering algorithm combined with dimension reduction techniques for power load profiles. Proc Csee 35(15):3741–3749
2. Wenjun Z, Yi W, Min L et al (2016) Distributed clustering algorithm for awareness of electricity consumption characteristics of massive consumers. Autom Electr Power Syst
3. Zhong C, Malinen M, Miao D et al (2015) A fast minimum spanning tree algorithm based on K-means. Inf Sci 295(3):1–19
4. Smith J, Jones M Jr, Houghton L et al (1999) Future of health insurance. N Engl J Med 965:325–329
5. Li ZHAO, Xingzhe HOU, Jun HU, Hong FU, Hongliang SUN (2014) Improved K-means algorithm based analysis on massive data of intelligent power utilization. Power Syst Tech 38(10):2715–2720

A Method to Solve the Measurement Error of Track Safety Control Based on Weighted Fusion

Jianping Pan(✉), Shengdong Ji, Pengfei Hao, Zhiguo Ren, and Quansheng He

Technology Department of Taiyuan Satellite Launch Center, Shanxi Taiyuan 030027, China
2468954081@qq.com

Abstract. In this paper, a method of data fusion measurement for error solution based on weight factor is proposed. Firstly, the mathematical model of orbit control in data fusion is given, the main factors affecting the accuracy of track control are analyzed, a method of calculating weight factor based on correlation number is proposed, the rationality of this method compared with the average measurement error method is theoretically analyzed. Finally, an example of weight measurement errors orbit controlling is given.

Keywords: Track control · Weight factor · Measurement error · Data fusion

1 Introduction

Safe flight control is an indispensable part of spacecraft test; its main purpose is to carry out the optimal blow up when the spacecraft fail in flight, so as to protect the safety of important targets and ground facility personnel under the flight path as such as possible. Safety control can be generally divided into independent safety control and ground safety control. Ground safety control mainly calculates space position alarm in advance, warning of landing point, platform frame angle attitude angle deviation and other safety control orbit. In the real-time flight process, the trajectory calculated before the flight curve of the measured spacecraft is compared; once the measured curve flight line exceeds the calculated sailing path trajectory, the corresponding safety control is carried.

Ground safety control orbit as standard to judge whether the real-time flight process is normal or not, which is the key to the success of spacecraft flight. Safety control track mainly includes theoretical term and error term. The error term has measurement error, interference error, display error, method error, delay error, etc. Measurement error generally refers to the real-time ballistic accuracy of measurement and control equipment. The current calculation method is to make a difference between the real-time trajectory

Q. Liang et al. (eds.), *Artificial Intelligence in China*, Lecture Notes in Electrical Engineering 653, https://doi.org/10.1007/978-981-15-8599-9_2

results of historical missions and the high-precision trajectory results after the event. Calculate the mean and variance of the residuals; this is equivalent to multiple averages. Because there are differences in the theoretical trajectory of each mission, difference theoretical ballistics will lead to difference measurement errors of measurement and control equipment. Therefore, the average of multiple tasks can not accurately reflect the measurement error. It is necessary to analyze the weight factor [1–5].

In this paper, a method of ground measurement error solution based on weight factor is proposed; this method calculates the correlation coefficient between the theoretical trajectory of multiple mission and the theoretical trajectory of present missions. Weight factor is obtained by normalization. Finally, the weight measurement error is calculated. This method considers the difference of theoretical trajectory of multiple missions. Therefore, it can better reflect the actual situation of measurement error; it can improve the calculation accuracy of safety control track.

2 Calculation Model of Ground Safety Control Track

Generally, the ground safety control orbit can be divided into position orbit, velocity orbit and landing point prediction. Their mathematical models are given below.

(1) **Position orbit**

The calculation formula of position is as follow:

$$P_i(t) = P_0(t) \pm \sigma_p^i \tag{1}$$

where i is x or y or z; $P_0(t)$ is the theoretical position trajectory at time t; $P_i(t)$ position orbit indicating direction i at time t; σ_p^i is position orbit error, the calculation formula is as follows:

$$\sigma_p^i = \sqrt{\left(\sigma_p^{i1}\right)^2 + \left(\sigma_p^{i0}\right)^2} \tag{2}$$

where σ_p^{i1} is the measurement error; σ_p^{i0} indicates others errors other than measurement errors.

(2) **Speed track**

The calculation formula for the velocity orbit is as follows.

$$V_i(t) = V_0(t) \pm \sigma_v^i \tag{3}$$

where $V_0(t)$ is the theoretical velocity v at time t; where $V_i(t)$ the velocity orbit in direction i at time t; σ_v^i is position orbit error, the calculation formula is as follow.

$$\sigma_v^i = \sqrt{(\sigma_v^{i1})^2 + (\sigma_v^{i0})^2} \tag{4}$$

σ_v^{i1} is the speed measurement error; σ_v^{i0} is the error other than the measurement error..

(3) **Fall point indicates orbit**

The calculation formula of the predicted trajectory of the landing point is as follows:

$$L_j(t) = L_{j0}(t) \pm \Delta L_j \ j = x, y \tag{5}$$

where $L_{j0}(t)$ is the theoretical prediction of time t, the prediction of theoretical landing point is based on the theoretical position and theoretical velocity of trajectory, consider air resistance, Coriolis force, acceleration of gravity, etc., the solution is obtained by Runge–Kutta integral method; where ΔL_j is prediction deviation of falling point in the direction of j, the calculation method is shown in Eq. (6).

$$\Delta L_j = f(\sigma_p^{j1}, \sigma_p^{j0}, \sigma_v^{j1}, \sigma_v^{j0}) \ j = x, z \tag{6}$$

In summary, Position orbit velocity orbit and fall point prediction orbit are all related to measurement error, to some extent, the accuracy of orbit is determined by the accuracy of measurement error calculation. In this paper, the measurement error is divided into position measurement error and velocity error and velocity measurement, calculation of measurement error by weight factor method.

3 Calculation Method of Weight Measurement Error

The solution of weight measurement error mainly includes the calculation of correlation number, normalization of correlation number, calculation of non weight measurement error, four steps of weight measurement error calculation.

3.1 Calculation of Correlation Number

Assuming there are samples $X = (X_1, X_2, \ldots X_n)$ and $Y = (Y_1, Y_2, \ldots Y_n)$ the mean value of the sample is as follows:

$$\bar{X} = \frac{1}{n}\sum_{i=1}^{n} X_i, \bar{Y} = \frac{1}{n}\sum_{i=1}^{n} Y_i \tag{7}$$

The covariance between samples X and Y is

$$S_{12} = \frac{1}{n}\sum_{i=1}^{n} (X_i - \bar{X})(Y_i - \bar{Y}) \tag{8}$$

The number of correlation between samples X and Y is

$$\rho_{12} = \frac{n \cdot S_{12}}{\sqrt{\sum_{i=1}^{n} (X_i - \bar{X})^2 \cdot \sum_{i=1}^{n} (Y_i - \bar{Y})^2}} \tag{9}$$

3.2 Weight Factor Calculation

The weight factor calculation is carried out in two steps: (1) Calculate the correlation coefficient of theoretical ballistic position and velocity of different missions: (2) The correlation number is normalized, and the position weight factor and speed weight factor are obtained. Suppose there are three different historical task data, *A*, *B* and *C* respectively, datum data is represented by *D*. The positions and speeds of *A*, *B* and *C* are expressed as follows:

$$P = \begin{bmatrix} X_A & Y_A & Z_A \\ X_B & Y_B & Z_B \\ X_C & Y_C & Z_C \end{bmatrix}, V = \begin{bmatrix} V_{XA} & V_{YA} & V_{ZA} \\ V_{XB} & V_{YB} & V_{ZB} \\ V_{XC} & V_{YC} & V_{ZC} \end{bmatrix} \tag{10}$$

where *P* is the position data and *V* is the speed data, then the correlation number of *A*, *B*, *C* and *D* is:

$$\rho_P = \begin{bmatrix} \rho_{XAD} & \rho_{YAD} & \rho_{ZAD} \\ \rho_{XBD} & \rho_{YBD} & \rho_{ZBD} \\ \rho_{XCD} & \rho_{YCD} & \rho_{ZCD} \end{bmatrix}, \rho_V = \begin{bmatrix} \rho_{VXAD} & \rho_{VYAD} & \rho_{VZAD} \\ \rho_{VXBD} & \rho_{VYBD} & \rho_{VZBD} \\ \rho_{VXCD} & \rho_{VYCD} & \rho_{VZCD} \end{bmatrix} \tag{11}$$

where ρ_p is position cross-correlation coefficient and ρ_V is the velocity cross-correlation coefficient, the calculation results of its normalized weight factor are as follows.

$$\overline{\overline{\rho_P}} = \begin{bmatrix} \frac{\rho_{XAD}}{\sum_i^{A,B,C} \rho_{XiD}} & \frac{\rho_{YAD}}{\sum_i^{A,B,C} \rho_{YiD}} & \frac{\rho_{ZAD}}{\sum_i^{A,B,C} \rho_{ZiD}} \\ \frac{\rho_{XBD}}{\sum_i^{A,B,C} \rho_{XiD}} & \frac{\rho_{YBD}}{\sum_i^{A,B,C} \rho_{YiD}} & \frac{\rho_{ZBD}}{\sum_i^{A,B,C} \rho_{ZiD}} \\ \frac{\rho_{XCD}}{\sum_i^{A,B,C} \rho_{XiD}} & \frac{\rho_{YCD}}{\sum_i^{A,B,C} \rho_{YiD}} & \frac{\rho_{ZCD}}{\sum_i^{A,B,C} \rho_{ZiD}} \end{bmatrix}, \overline{\overline{\rho_V}} = \begin{bmatrix} \frac{\rho_{VXAD}}{\sum_i^{A,B,C} \rho_{VXiD}} & \frac{\rho_{VYAD}}{\sum_i^{A,B,C} \rho_{VYiD}} & \frac{\rho_{VZAD}}{\sum_i^{A,B,C} \rho_{VZiD}} \\ \frac{\rho_{VXBD}}{\sum_i^{A,B,C} \rho_{VXiD}} & \frac{\rho_{VYBD}}{\sum_i^{A,B,C} \rho_{VYiD}} & \frac{\rho_{VZBD}}{\sum_i^{A,B,C} \rho_{VZiD}} \\ \frac{\rho_{VXCD}}{\sum_i^{A,B,C} \rho_{VXiD}} & \frac{\rho_{VYCD}}{\sum_i^{A,B,C} \rho_{VYiD}} & \frac{\rho_{VZCD}}{\sum_i^{A,B,C} \rho_{VZiD}} \end{bmatrix} \tag{12}$$

where is $\overline{\overline{\rho_P}}$ and $\overline{\overline{\rho_V}}$ represent the position weight factor and speed weight factor calculated by normalization.

3.3 Weight Measurement Error Calculation

The measurement error is divided into position measurement error and speed measurement error. The calculation steps of weight measurement error are as follows: (1) Calculate the non weight measurement error of each historical task. Using real-time mission trajectory data to make error with high-precision trajectory data, take the mean and variance of the residual as the result; (2) Calculation of weight measurement error. Multiply the non weight measurement error of each historical task by the corresponding weight factor, it is the result of weight measurement error.

(1) Calculate the non weight measurement error of each historical task. The position vector and speed vector of the measurement and control equipment in the real-time task error are as follows.

$$E^s_{p(t)} = \begin{bmatrix} X^s_A(t) & Y^s_A(t) & Z^s_A(t) \\ X^s_B(t) & Y^s_B(t) & Z^s_B(t) \\ X^s_C(t) & Y^s_C(t) & Y^s_C(t) \end{bmatrix}, E^s_{v(t)} = \begin{bmatrix} V^s_{XA}(t) & V^s_{YA}(t) & V^s_{ZA}(t) \\ V^s_{XB}(t) & V^s_{YB}(t) & V^s_{ZB}(t) \\ V^s_{XC}(t) & V^s_{YC}(t) & V^s_{ZC}(t) \end{bmatrix} \tag{13}$$

$X_A^S(t)$ and $V_{XA}^s(t)$ represent the position and speed in X direction of A task, B and C represent B and C tasks, X, Y and Z are position in X, Y and Z directions, V_x, V_y and V_z represent the speed in X, Y and Z directions.

Where $\mathrm{E}_{p(t)}^s$ and $\mathrm{E}_{v(t)}^s$ are position and speed, respectively, the corresponding post event high-precision results are expressed as:

$$E_{p(t)}^h = \begin{bmatrix} X_A^h(t) & Y_A^h(t) & Z_A^h(t) \\ X_B^h(t) & Y_B^h(t) & Z_B^h(t) \\ X_C^h(t) & Y_C^h(t) & Y_C^h(t) \end{bmatrix}, E_{v(t)}^h = \begin{bmatrix} V_{XA}^h(t) & V_{YA}^h(t) & V_{ZA}^h(t) \\ V_{XB}^h(t) & V_{YB}^h(t) & V_{ZB}^h(t) \\ V_{XC}^h(t) & V_{YC}^h(t) & V_{ZC}^h(t) \end{bmatrix} \tag{14}$$

$X_A^H(t)$ and $V_{XA}^H(t)$ represent the position and speed in X direction of A task, B and C represent B and C tasks, X, Y and Z are position in X, Y and Z directions, V_x, V_y and V_z represent the speed in X, Y and Z directions.

Use Eq. (14) to subtract Eq. (13) to get the residual

$$\Delta P(t) = E_{p(t)}^h - E_{p(t)}^s = \begin{bmatrix} X_A^h(t) - X_A^s(t) & Y_A^h(t) - Y_A^s(t) & Z_A^h(t) - Z_A^s(t) \\ X_B^h(t) - X_B^s(t) & Y_B^h(t) - Y_B^s(t) & Z_B^h(t) - Z_B^s(t) \\ X_C^h(t) - X_C^s(t) & Y_C^h(t) - Y_C^s(t) & Z_C^h(t) - Z_C^s(t) \end{bmatrix},$$

$$\Delta V(t) = E_{V(t)}^h - E_{V(t)}^s = \begin{bmatrix} V_{XA}^h(t) - V_{XA}^s(t) & V_{YA}^h(t) - V_{YA}^s(t) & V_{ZA}^h(t) - V_{ZA}^s(t) \\ V_{XB}^h(t) - V_{XB}^s(t) & V_{YB}^h(t) - V_{YB}^s(t) & V_{ZB}^h(t) - Z_{ZB}^s(t) \\ V_{XC}^h(t) - V_{XC}^s(t) & V_{YC}^h(t) - V_{YC}^s(t) & V_{ZC}^h(t) - Z_{ZC}^s(t) \end{bmatrix} \tag{15}$$

Calculate the mean value and variance of position and velocity residuals:

$$E[\Delta P(t)] = \begin{bmatrix} E(\Delta X_A) & E(\Delta Y_A) & E(\Delta Z_A) \\ E(\Delta X_B) & E(\Delta Y_B) & E(\Delta Z_B) \\ E(\Delta X_C) & E(\Delta Y_C) & E(\Delta Z_C) \end{bmatrix},$$

$$\sigma[\Delta P(t)] = \begin{bmatrix} \sigma(\Delta X_A) & \sigma(\Delta Y_A) & \sigma(\Delta Z_A) \\ \sigma(\Delta X_B) & \sigma(\Delta Y_B) & \sigma(\Delta Z_B) \\ \sigma(\Delta X_C) & \sigma(\Delta Y_C) & \sigma(\Delta Z_C) \end{bmatrix} \tag{16}$$

$$E[\Delta V(t)] = \begin{bmatrix} E(\Delta V_{XA}) & E(\Delta V_{YA}) & E(\Delta V_{ZA}) \\ E(\Delta V_{XB}) & E(\Delta V_{YB}) & E(\Delta V_{ZB}) \\ E(\Delta V_{XC}) & E(\Delta V_{YC}) & E(\Delta V_{ZC}) \end{bmatrix},$$

$$\sigma[\Delta V(t)] = \begin{bmatrix} \sigma(\Delta V_{XA}) & \sigma(\Delta V_{YA}) & \sigma(\Delta V_{ZA}) \\ \sigma(\Delta V_{XB}) & \sigma(\Delta V_{YB}) & \sigma(\Delta V_{ZB}) \\ \sigma(\Delta V_{XC}) & \sigma(\Delta V_{YC}) & \sigma(\Delta V_{ZC}) \end{bmatrix} \tag{17}$$

Use formula (16) to point multiply formula (12) to get

$$\begin{bmatrix} E(\Delta X_D) \\ E(\Delta Y_D) \\ E(\Delta Z_D) \end{bmatrix} = U_P \begin{bmatrix} (1,1) \\ (2,2) \\ (3,3) \end{bmatrix} = \begin{bmatrix} E(\Delta X_A) & E(\Delta Y_A) & E(\Delta Z_A) \\ E(\Delta X_B) & E(\Delta Y_B) & E(\Delta Z_B) \\ E(\Delta X_C) & E(\Delta Y_C) & E(\Delta Z_C) \end{bmatrix}^{\mathrm{T}} \cdot \overline{\overline{\rho_P}}$$

$$\begin{bmatrix} \sigma(\Delta X_D) \\ \sigma(\Delta Y_D) \\ \sigma(\Delta Z_D) \end{bmatrix} = M_P \begin{bmatrix} (1,1) \\ (2,2) \\ (3,3) \end{bmatrix} = \begin{bmatrix} \sigma(\Delta X_A)\ \sigma(\Delta Y_A)\ \sigma(\Delta Z_A) \\ \sigma(\Delta X_B)\ \sigma(\Delta Y_B)\ \sigma(\Delta Z_B) \\ \sigma(\Delta X_C)\ \sigma(\Delta Y_C)\ \sigma(\Delta Z_C) \end{bmatrix}^T \cdot \overline{\overline{\rho_P}} \tag{18}$$

where $\left[E(\Delta X_D)\ E(\Delta Y_D)\ E(\Delta Z_D)\right]$ and $\left[\sigma(\Delta X_D)\ \sigma(\Delta Y_D)\ \sigma(\Delta Z_D)\right]$ are the mean and variance vectors of weight position measurement errors respectively; $U\left[(1,1)\ (2,2)\ (3,3)\right]$, $M\left[(1,1)\ (2,2)\ (3,3)\right]$ is the diagonal element after matrix multiplication, Eq. (18) is the calculation formula of weight position measurement error.

In the same way, the point multiplication of Eqs. (17) and (13) is used to obtain.

$$\begin{bmatrix} E(\Delta V_{XD}) \\ E(\Delta V_{YD}) \\ E(\Delta V_{ZD}) \end{bmatrix} = U_V \begin{bmatrix} (1,1) \\ (2,2) \\ (3,3) \end{bmatrix} = \begin{bmatrix} E(\Delta V_{XA})\ E(\Delta V_{YA})\ E(\Delta V_{ZA}) \\ E(\Delta V_{XB})\ E(\Delta V_{YB})\ E(\Delta V_{ZB}) \\ E(\Delta V_{XC})\ E(\Delta V_{YC})\ E(\Delta V_{ZC}) \end{bmatrix}^T \cdot \overline{\overline{\rho_V}},$$

$$\begin{bmatrix} \sigma(\Delta V_{XD}) \\ \sigma(\Delta V_{YD}) \\ \sigma(\Delta V_{ZD}) \end{bmatrix} = M_V \begin{bmatrix} (1,1) \\ (2,2) \\ (3,3) \end{bmatrix} = \begin{bmatrix} \sigma(\Delta V_{XA})\ \sigma(\Delta V_{YA})\ \sigma(\Delta V_{ZA}) \\ \sigma(\Delta V_{XB})\ \sigma(\Delta V_{YB})\ \sigma(\Delta V_{ZB}) \\ \sigma(\Delta V_{XC})\ \sigma(\Delta V_{YC})\ \sigma(\Delta V_{ZC}) \end{bmatrix}^T \cdot \overline{\overline{\rho_V}} \tag{19}$$

where $\left[E(\Delta V_{XD})\ E(\Delta V_{YD})\ E(\Delta V_{ZD})\right]$ and $\left[\sigma(\Delta V_{XD})\ \sigma(\Delta V_{YD})\ \sigma(\Delta V_{ZD})\right]$ are the mean and variance vectors of the weighted speed measurement errors respectively; $M_V\left[(1,1)\ (2,2)\ (3,3)\right]$ is the diagonal element after matrix multiplication, Eq. (19) is the calculation formula of weight speed measurement error.

4 Calculation Example

Take three historical task data, record as *A*, *B* and *C* respectively, and record the reference data as *D*. First, the cross-correlation coefficient of the theoretical data of each task is calculated, then the weight factor is solved by normalization, secondly, the weight measurement error is calculated, and finally the safety control trajectory based on the weight measurement error is calculated.

4.1 Calculation of Correlation Number

The correlation coefficient matrix is calculated by applying the theoretical position and velocity data in formula (9), as shown in formula (20). The left side is the position correlation coefficient and the right side is the velocity correlation.

$$\rho_P = \begin{bmatrix} 1.0000\ 1.0000\ 0.9994 \\ 0.9999\ 0.9923\ 0.9965 \\ 1.0000\ 1.0000\ 0.9987 \end{bmatrix}, \rho_v = \begin{bmatrix} 0.9999\ 0.9996\ 0.9993 \\ 0.9975\ 0.7434\ 0.9944 \\ 0.9998\ 0.9994\ 0.9986 \end{bmatrix} \tag{20}$$

After normalization, the weight factors are as follows

$$\overline{\overline{\rho_P}} = \begin{bmatrix} 0.3333\ 0.3342\ 0.3337 \\ 0.3333\ 0.3316\ 0.3328 \\ 0.3333\ 0.3342\ 0.3335 \end{bmatrix}, \overline{\overline{\rho_v}} = \begin{bmatrix} 0.3336\ 0.3645\ 0.3340 \\ 0.3328\ 0.2711\ 0.3323 \\ 0.3336\ 0.3644\ 0.3337 \end{bmatrix} \tag{21}$$

4.2 Weight Measurement Errors Calculation

The non weight measurement error curve of historical tasks *A*, *B* and *C* is obtained by real-time and post event high-precision ballistic error, as shown in Fig. 1, Where Fig. 1a–c are non weighted position measurement errors, Fig. 1d–f are non weighted speed measurement errors.

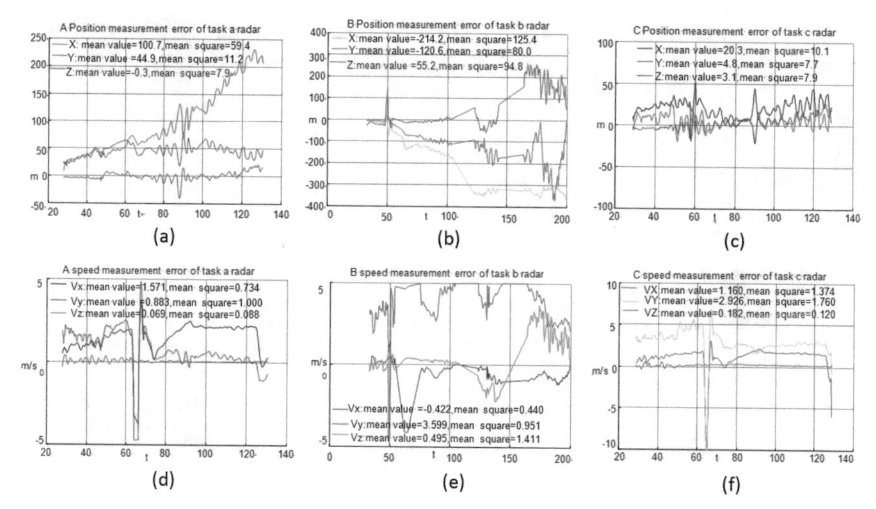

Fig. 1 a, b, c, d, e, f Non weight measurement error curve

In the figure, the non weighted measurement position and velocity error curves in *X*, *Y* and *Z* directions were calculated from tasks *A*, *B* and *C*. According to the data indicated in the curves in the figure, the position and velocity error values of the rockets flight orbit are obtained, which can be used as the input data of non weight measurement in flight safety orbit control to calculate next test mission.

Weight measurement error is the product of non weight measurement error and weight factor, the calculation results are as follows.

$$\begin{bmatrix} E(\Delta X_D) \\ E(\Delta Y_D) \\ E(\Delta Z_D) \end{bmatrix} = \left\{ E[\Delta P(t)]^{\mathrm{T}} \cdot \overline{\overline{\rho_P}} \right\} \begin{bmatrix} (1,1) \\ (2,2) \\ (3,3) \end{bmatrix} = \begin{bmatrix} -31.1 \\ -23.4 \\ 19.3 \end{bmatrix}, \begin{bmatrix} \sigma(\Delta X_D) \\ \sigma(\Delta Y_D) \\ \sigma(\Delta Z_D) \end{bmatrix}$$

$$= \left\{ \sigma[\Delta P(t)]^{\mathrm{T}} \cdot \overline{\overline{\rho_P}} \right\} \begin{bmatrix} (1,1) \\ (2,2) \\ (3,3) \end{bmatrix} = \begin{bmatrix} 65.0 \\ 32.8 \\ 53.4 \end{bmatrix}$$

$$\begin{bmatrix} E(\Delta V_{XD}) \\ E(\Delta V_{YD}) \\ E(\Delta V_{ZD}) \end{bmatrix} = E[\Delta V(t)]^{\mathrm{T}} \cdot \rho V \begin{bmatrix} (1,1) \\ (2,2) \\ (3,3) \end{bmatrix} = \begin{bmatrix} 0.7706 \\ 2.3638 \\ 0.2483 \end{bmatrix}, \begin{bmatrix} \sigma(\Delta V_{XD}) \\ \sigma(\Delta V_{YD}) \\ \omega(\Delta V_{ZD}) \end{bmatrix}$$

$$= \sigma[\Delta V(t)]^{\mathrm{T}} \cdot \rho V \begin{bmatrix} (1,1) \\ (2,2) \\ (3,3) \end{bmatrix} = \begin{bmatrix} 0.8497 \\ 1.2637 \\ 0.5383 \end{bmatrix} \quad (22)$$

4.3 Track Safety Control Calculation

The orbit safety control of weight measurement errors is based on the calculated weight measurement error as the input, the result curve is shown in Fig. 2, where Fig. 2a–c are weighted position safety control tracks, Fig. 2d–f are weighted speed safety control tracks.

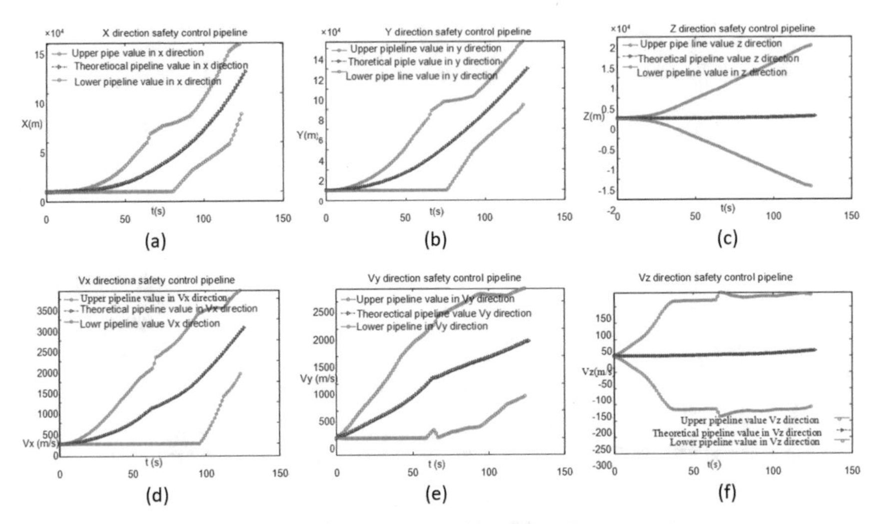

Fig. 2 The weight measurement error safety control track curve

According to the calculation results, if its position and speed exceed the red upper and lower limits in the figure, the technicians should implement safety control to ensure flight safety during the rocket flight. The method is verified to be effective by subsequent tasks.

5 Conclusion

This paper starts from the mathematical model of track safety control, the factors affecting the accuracy of track safety control were analyzed, it is concluded that the measurement error is one of the factors affecting the safety accuracy of track. The rationality of the measurement error of the weight factor compared with the average measurement error was analyzed theoretically, and the method of calculating the weight factor based on the correlation of theoretical data is proposed. Finally, the real inversion of the measurement error of weight factor is given, and it was applied to the calculation of track safety controlling. The correctness and validity of this method are verified by theoretical analysis and simulation calculation.

This method has the following advantages: (a) it can improve the calculation accuracy of track safety control; (b) It has a good adaptability to new tasks, it is more reasonable to calculate measurement errors through the correlation between historical tasks and new tasks. There are mainly the following limitations: (1) Time variation of weight factor is ignored. The tracking state of the measurement and control equipment changes with time, and the measurement error changes with time. The time-varying weight factor should be calculated. (2) The relevant factors of weight factor are only qualitative analysis and the theoretical basis is insufficient. (3) Assume all weighting factors are uncouple.

References

1. Wei YM, Dong GG, Fan TJ (1995) An investigation on neural network based method for computing the multi-object weight. J Wuhan Inst Chem Tech 4(17):37–41 (in Chinese)
2. Wang H (2002) The Bayesian network of being used forecasting. J Northeast Normal Univ (Natural Science Edition) 34(1):9–14 (in Chinese)
3. Zhang AH, Jing HF, Wang B, Xu Y (2010) Research on the function of feature weight factor in text classification. Chin Inf J 24(3):97–103
4. Dong YH, Guo SC (2015) Improved text clustering algorithm based on weight factor and eigenvector. Comput Eng Des 36(4):1051–1056
5. Sun B, Wang LJ (2006) Research on weight determination method based on rough set theory. Comput Eng Appl 42(29):216–217

Improved Spectral Efficiency Based on Double Spatially Sparse in Millimeter Wave MIMO System

Yanping Zhu[1(✉)], Sicheng Guo[1], Yinxuan Sun[1], and Jing Liu[2]

[1] School of Electronic & Information Engineering, Nanjing University of Information Science and Technology, Nanjing 210044, China
001520@nuist.edu.cn
[2] Jinling Institute of Technology, Jiangsu, China

Abstract. Existing millimeter wave (mmWave) systems must take advantage of large antenna arrays to make wavelength reduction possible against path losses and beamforming gains. To solve the above problem and improve the spectral efficiency, double spatially sparse processing is proposed in this paper. One sparse is the spatially sparse precoding with orthogonal matching pursuit (OMP) reconstruction method in mmWave hybrid beamforming, which can reduce the data rate; The other sparse is the sparse array generated by augmented nested array at transmitter to greatly decrease the array elements. Simulation results show that this double sparse processing has better spectral efficiency than linear array MIMO hybrid beamforming.

Keyword: Hybrid beamforming · Augmented nested arrays · Spectral efficiency · OMP

1 Introduction

The 5th generation (5G) mobile communication system is a new generation of cellular mobile communication technology with the highest data transmission rate up to 10Gbit/s. Its network delay is less than 1 ms and spectrum efficiency is ten times higher than LTE [1, 2]. One of its key technologies—Massive MIMO technology, is equipped with a massive array of antenna elements far more than the number of users, which enables the system to generate a larger array gain, bringing an order of magnitude increase in spectral efficiency [3].

In order to avoid the interference and further improve the spectral efficiency, the modern mmWave MIMO systems adopts precoding technology to beamforming the signals transmitted by the base station and optimizes the network performance by eliminating interference in advance. Considering the limitations of traditional precoding technologies like analog or RF processing which have low performance and high cost [4, 5], Omar El Ayach et al. proposed an orthogonal of spatial sparse hybrid precoding to enable the mmWave Massive MIMO system to reach the limit of the optimal unconstrained precoder in low-cost hardware design [6].

Q. Liang et al. (eds.), *Artificial Intelligence in China*, Lecture Notes
in Electrical Engineering 653, https://doi.org/10.1007/978-981-15-8599-9_3

In addition, the 5G system deployed large-scale antenna arrays [7] to alleviate severe transmission losses in the mm Wave band. Literature [8] proposed an augmented nested array concept and compared to the (super) nested array [9] with the same element number, the newly formed augmented nested array possesses higher degree-of-freedom capacity and less mutual coupling.

Therefore, on the basis of hybrid beamforming, this paper proposes to further optimize MIMO system by constructing sparse array with augmented nested matrix, so as to improve spectral efficiency while reducing array element number.

2 System Model

2.1 System Model

This paper considers a large-scale single-user MIMO system shown in Fig. 1, which transmits N_s data streams using N_{ts} antennas to a receiver equipped with N_r antennas. To reduce the number of antennas, N_t linear antenna array connected to N_t phase shifters are optimized to N_{ts} sparse array. In this hardware architecture, N_{RF}^t and N_{RF}^r denote the number of RF chains at the transmitter and receiver respectively and the constrains are satisfied with $\boldsymbol{N}_s < \boldsymbol{N}_{\text{RF}}^t < \boldsymbol{N}_t, \boldsymbol{N}_s < \boldsymbol{N}_{\text{RF}}^r < \boldsymbol{N}_r$.

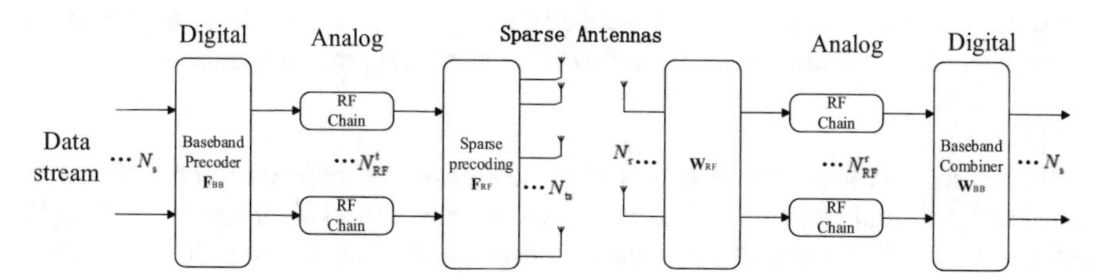

Fig. 1 The double sparse mmWave MIMO model

This downline precoding enables the transmitter to apply an $N_{\text{RF}}^t \times N_s$ baseband precoder $\mathcal{F}_{\text{BB}}$ followed by an $N_t \times N_{\text{RF}}^t$ RF precoder $\mathcal{F}_{\text{RF}}$, and the transmitter's total power constraint is $\|\mathcal{F}_{\text{RF}}\mathbf{F}_{\text{BB}}\|_F^2 = \boldsymbol{N}_s$, besides, the receiver applies an $N_r \times N_{\text{RF}}^r$ RF combining matrix $\mathcal{W}_{\text{RF}}$ and an $\boldsymbol{N}_{\text{RF}}^r \times \boldsymbol{N}_s$ baseband combining matrix $\mathcal{W}_{\text{BB}}$.

Given the ordered singular value decomposition (SVD) of the channel matrix:$\mathcal{H} = U\sum\mathcal{V}^*$, the optimal unconstrained unitary precoder $\mathcal{F}_{\text{opt}}$ for $\mathcal{H}$ can be found that $\mathcal{F}_{\text{opt}} = \mathcal{V}_{(:,1:N_s)}$.

Reference [6] proposed the algorithm of spatially sparse precoding via orthogonal matching pursuit to obtain precoder $\mathcal{F}_{\text{RF}}$ and $\mathcal{F}_{\text{BB}}$. The algorithm starts by finding the basis vectors $a_t(\phi_{il}^t, \theta_{il}^t)$ of channel $\mathcal{H}$ along which $\mathcal{F}_{\text{opt}}$ has the maximum projection and then appends it to $\mathcal{F}_{\text{RF}}$. The algorithm then calculates the least squares solution to $\mathcal{F}_{\text{BB}}$ and the contribution of the selected vectors is removed in the following step. Afterward, the column will be found to make the residual precoding matrix $\mathcal{F}_{res}$ have the largest projection. It finally stops when all $\boldsymbol{N}_{\text{RF}}^t$ beamforming vectors have been selected. Moreover, the combiner $\mathcal{W}_{\text{RF}}$ and $\mathcal{W}_{\text{BB}}$ can also be calculated. The maximization problem can be translated into (1) to get the optimal precoders.

$$(\mathcal{F}_{\text{RF}}^{\text{opt}} \mathcal{F}_{\text{BB}}^{\text{opt}}) = \arg\min||\mathcal{F}_{\text{opt}} - \mathcal{F}_{\text{RF}}\mathcal{F}_{\text{BB}}||_F,$$
$$s.t. \mathcal{F}_{\text{RF}}^{(i)} \in \{a_t(\phi_{il}^t, \theta_{il}^t), \forall i, \boldsymbol{l}\},$$
$$||\mathcal{F}_{\text{RF}} \mathcal{F}_{\text{BB}}||_F^2 = N_s \tag{1}$$

And the spectral efficiency achieved can be written as

$$\boldsymbol{R} = \log_2\left(|\mathcal{I}_{N_s} + \frac{\rho}{\boldsymbol{N}_s} \mathcal{R}_n^{-1} \mathcal{W}_{\text{BB}}^* \mathcal{W}_{\text{RF}}^* \mathcal{H} \mathcal{F}_{\text{RF}} \mathcal{F}_{\text{BB}} \times \mathcal{F}_{\text{BB}}^* \mathcal{F}_{\text{RF}}^* \mathcal{H}^* \mathcal{W}_{\text{RF}} \mathcal{W}_{\text{BB}}| \right) \tag{2}$$

where $\mathcal{R}_n = \sigma_n^2 \mathcal{W}_{\text{BB}}^* \mathcal{W}_{\text{RF}}^* \mathcal{W}_{\text{RF}} \mathcal{W}_{\text{BB}}$ is the noise covariance matrix after combining, (2) shows that the spectral efficiency will be improved by an optimized channel matrix and precoders with better performance.

As shown in Fig. 1, our system adopts the mentioned sparse precoding algorithm just before the sparse transmitting antenna array. We focus on the design of sparse array in the following.

2.2 Sparse Array Model

We want to design a sparse array generated by augmented nested array to replace the square array at transmitter terminal to form the channel matrix of sparse array.

ANAI-2

In [8], based on these properties of ANAI-1 and known that there are still two dense part in ANAI-1, we could find a better element distribution of two sub-arrays by splitting the dense sub-array into odd part and even part respectively, thus another kind of two level ANA is generated as ANAI-2.

The geometry is shown in Fig. 2 and its sub-arrays can be expressed as:

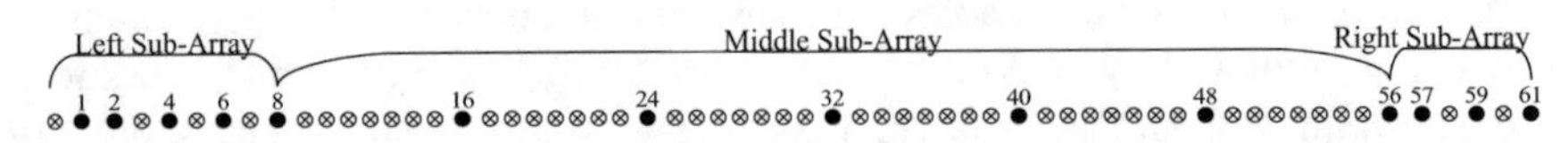

Fig. 2 The two level ANAI-2 defined in (2), where $M = 14$, $(N_1 + 1)$ is an odd integer, $N_1 = 7$

$$\begin{cases} \text{V}_{21} = \{N_1, N_1 - 1, N_1 - 3, \ldots, N_1 - 2k_1 - 1, \ldots, 0\} \\ \text{V}_{22} = \{N_1 - 2, N_1 - 4, \ldots, N_1 - 2k_2 \ldots, 0\} \end{cases} \tag{3}$$

where V_{21} and V_{22} are two sub-sets split from the non-negative reminder V_2 as mentioned in [1], $k_1 \in [0, R\lfloor (N_1 - 1)/2 \rfloor]$ and $k_2 \in [1, R\lfloor N_1/2 \rfloor]$.

ANAII-2

Motivated by ANAI-2 configuration, by splitting all the elements into odd/even parts respectively, one kind of four level ANA is designed, as shown in Fig. 3.

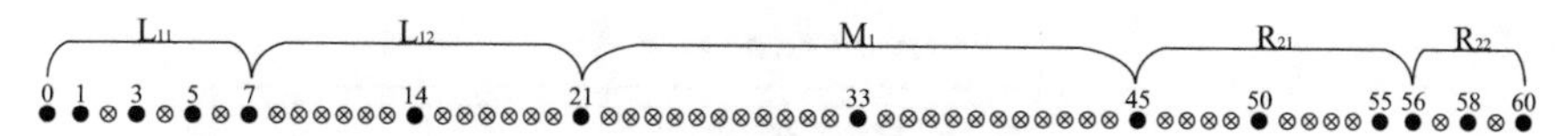

Fig. 3 The four level ANAII-2 defined in (3), where $M = 14, d_{12} = 7, d_{21} = 5$

The corresponding configuration can be expressed as:

$$\begin{cases} V_{11} = \{0, 2, 4, 6, \ldots, 2L_1 - 2, 2L_1, 2L_1 + 1\} \\ V_{22} = \{0, 1, 3, 5, \ldots, 2L_1 - 3, 2L_1 - 1\} \\ V_{12} = V_{21} = [0, L_3] \\ d_{12} = 2L_1 + 1 \\ d_{21} = 2L_1 - 1 \end{cases} \tag{4}$$

where $(L_1 + 2)$ and $(L_1 + 1)$ denote the element number of leftmost/right most dense sub-array (i.e., L_{11} and R_{22}), respectively. L_3 denotes that if $2L_1 - 1 = d_{21}$, there will be $L_3 + 1$ pairs of elements in the right and a pair of elements in the left. For ANAII-2 geometry, $N_1 + 1 = d_{12} + d_{21} = 4L_1$, which means that the inter-element spacing in N_1 can only be an even number.

Two experiments in [2] considering the DOF ratio $\gamma(M) = \frac{M^2}{L_u(M)}$ and mutual coupling $L(M) = \left\| C - \tilde{C} \right\|_F / \|C\|_F$ showed that when element number M is 14, ANAII-2 and ANAII-2 possess better performance in increased DOF and reduced mutual coupling: $L(14)$ of ANAI-2 is 0.221 and that of ANAII-2 is 0.248, $\gamma(14)$ of ANAI-2 is 3.267 and that of ANAII-2 is 3.063. Therefore, we choose these two ANAs to make the linear antenna a sparse one.

3 Simulation Results

This section simulates a MIMO hybrid beamforming system, with a sparse planar antenna array processed from a 64-element square array with 4 RF chains on the transmitter side and a 16-element square array with 4 RF chains on the receiver side.

Considering an ANAI-2 with $M = 14$, $(N_1 + 1)$ is an odd integer, $N_1 = 7$ and an ANAII-2 with $M = 14$, $d_{12} = 7$, $d_{21} = 5, L_1 = 3, L_3 = 2$. The ANAs generated according to this configuration finely match the original ULA transmitting antenna array. We will replace the position of the original transmission array with the ANAs after supplementing three vacancy values. We could therefore reduce the 64 transmitting antennas in the MIMO system transmitter part to the actual 14 by taking advantage of the fact that the sparse array can generate a large number of virtual array elements.

Figure 4 shows the spectral efficiency of achieved against various SNR values for different processing methods. The figures in Figs. 4 and 6 show that when $N_s = 1$, the spectral efficiency has been increased by 10% at least using proposed double sparse processing method compared to the single spatially sparse processing method. Figures in Figs. 5 and 7 show that when $N_s = 2$, the spectral efficiency could also be obviously increased at low SNR while that at high SNR is sacrificed.

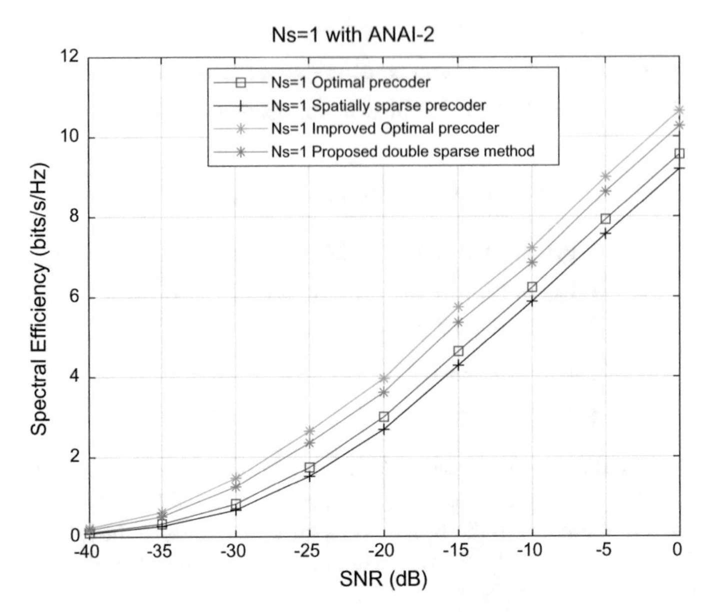

Fig. 4 $N_s = 1$ using ANAI-2

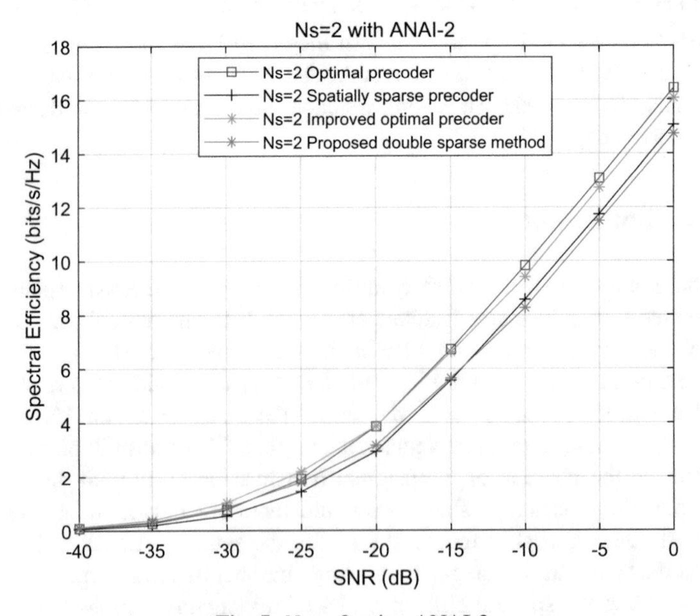

Fig. 5 $N_s = 2$ using ANAI-2

Considering that there may be differences in the performance of the two ANAs, Table 1 shows the spectral efficiency values of them under different SNRs and N_s. The

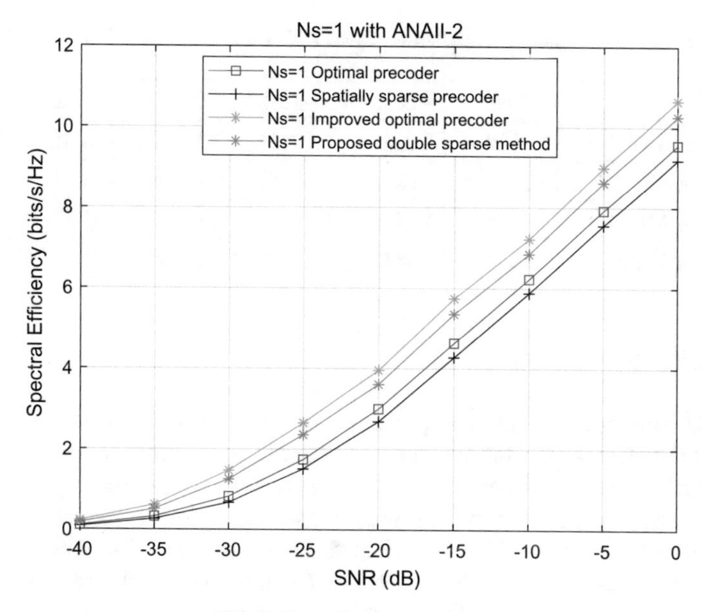

Fig. 6 $N_S = 1$ using ANAII-2

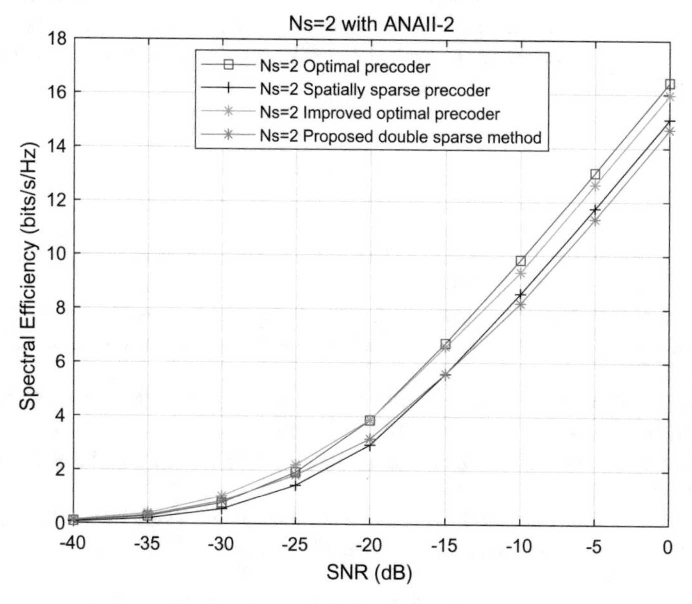

Fig. 7 Ns = 2 using ANAII-2

comparison found that the results of the two ANAs under the same conditions have similar results in improving the spectral efficiency.

Table 1 Performance comparisons of two ANAs under different SNRs and N_S

SNR	−30 dB	−20 dB	−10 dB	0 dB	10 dB	20 dB
$N_S = 1$ with ANAII-2	1.2528	3.6061	6.8481	10.259	13.613	16.837
$N_S = 1$ with ANAI-2	1.2473	3.6076	6.8461	10.261	13.618	16.832
$N_S = 2$ with ANAII-2	0.8396	3.1709	8.2160	14.684	21.303	27.792
$N_S = 2$ with ANAI-2	0.8373	3.1645	8.2643	14.729	21.348	27.773

4 Conclusion

Due to the hardware cost and computational complexity of the linear uniform array antenna of the MIMO system, we combined spatially sparse precoding with augmented nested sparse array and further proposed the method of using sparse algorithm to deal with sparse array in this paper. We first analyzed several different types of ANAs performance, and then chose two ANAs mentioned that are most compatible with our system to do the sparse processing. In the end, numerical simulation results shown that the spectral efficiency is improved, which validates the effectiveness of the proposed method.

Acknowledgements. The Project was supported by the National Natural Science Foundation (Grant No. 61801231) of China, the Natural Science Foundation of Higher Education of Jiangsu Province (Grant No. 17KJB510020) and the Natural Science Foundation of Jiangsu Province (Grant No. BK2019 1399).

References

1. Boccardi F, Heath RW, Lozano A et al (2014) Five disruptive technology directions for 5G. IEEE Commun Mag 52(2):74–80
2. Dierks S, Zirwas W, Jager M, Panzner B, Kramer G (2015) MIMO and massive MIMO-analysis for a local area scenario. In: 2015 23rd European signal processing conference (EUSIPCO), pp 2451–2455
3. Andrews JG, Buzzi S, Choi W, Hanly SV, Lozano A, Soong AC, Zhang JC (2014) What will 5G be? IEEE J Sel Areas Commun 32(6):1065–1082
4. Pi Z, Khan F (2011) An introduction to millimeter-wave mobile broadband systems. IEEE Commun Mag 49(6):101–107
5. El Ayach Omar, Rajagopal Sridhar, Abu-Surra Shadi, Pi Zhouyue, Heath Robert W (2014) Spatially sparse precoding in millimeter wave MIMO systems. IEEE Trans Acoust Speech Sig Process 13(3):1499–1513
6. Rahimian A, Jilani SF, Abbasi QH, Alomainy A, Alfadhl Y (2019) A millimetre-wave two-dimensional 64-element array for large-scale 5G antenna subsystems. IEEE Xplore. June 2019
7. Liu Jianyan, Zhang Yanmei, Yilong Lu, Ren Shiwei, Cao Shan (2017) Augmented nested arrays with enhanced DOF and reduced mutual coupling. IEEE Trans Acoust Speech Signal Process 65(21):5549–5563
8. Liu CL, Vaidyanathan PP (2016) Super nested arrays: linear sparse arrays with reduced mutual coupling—part I: fundamentals. IEEE Trans Signal Process 64(15):3997–4012

9. Xia P, Yong SK, Oh J, Ngo C (2008) A practical SDMA protocol for 60 GHz millimeter wave communications. In: Proceedings of 2008 asilomar conference signals, systems computer, pp 2019–2023

Electromagnetic Compatibility Test Analysis of Space Launch Field Based on Wavelet Analysis

Bo Zhou[1(✉)], Yao lou Jiang[1], Zong ju Xiong[1], Chun guang Shi[2], Zhen Chen[1], Mei Yang[1], and Gai qin Li[1]

[1] XiChang Satellite Launch Center of China, XiChang, SiChuan 615000, China
koof123@163.com
[2] TaiYuan Satellite Launch Center of China, TaiYuan, ShanXi 030027, China
flyhingshicg@126.com

Abstract. Before launch of a spacecraft, an electromagnetic compatibility test is required at the launch site. This paper discusses a method for testing and analyzing electromagnetic compatibility of a space launch field based on wavelet analysis. Through the theoretical analysis of wavelet analysis and the multiresolution characteristics of wavelet analysis, the electrostatic discharge of electrical equipment in the electromagnetic compatibility test of launch site is taken as an example to analyze the electromagnetic interference.

Keywords: Space launch site · Wavelet analysis · Electromagnetic compatibility

1 Introduction

The space launch site is responsible for the testing and launching of carrier rockets and spacecraft. In order to ensure the success launch of the spacecraft, electromagnetic compatibility tests must be performed before launch. In recent years, the number of satellites launched by China's launch sites has increased, and the launch cycle has become shorter and shorter, and the emission frequency becomes higher and higher, how to effectively implement the electromagnetic compatibility test of the emission field is a very important task. What kind of analysis and research method to use is also a step by step discussion and improve. This article therefore proposes an analysis method for electromagnetic compatibility test of space launch field based on wavelet analysis. At present, the data analysis of electromagnetic compatibility (EMC) test is mainly Fourier analysis (EMC) test is mainly Fourier analysis [1] but Fourier analysis is not applicable to sudden signals, and electromagnetic interference is often a transient with rich frequency components. The use of multi-resolution characteristics of wavelet analysis can focus on the analysis of transient signals and non-stationary signals when analyzing signals, which provides a new theoretical basis for electromagnetic [2].

Q. Liang et al. (eds.), *Artificial Intelligence in China*, Lecture Notes
in Electrical Engineering 653, https://doi.org/10.1007/978-981-15-8599-9_4

2 Necessity of Electromagnetic Compatibility Test of Space Launch Field

After the spacecraft enters the launch site, it is necessary to perform electromagnetic compatibility tests between multiple systems such as launch vehicles, satellites, and ground measurement and control equipment. The main consideration is the following two points:

1. The frequency band of spacecraft electronic equipment is getting wider and wider, and their start and stop processes are prone to electromagnetic interference, which makes the electromagnetic environment of the launch field more and more complicated; radars, radios, etc. on the rockets and launch site areas emit electromagnetic waves. Energy devices will cause electromagnetic interference (EMI) to other electronic devices; the bandwidth resources of each electronic device are limited, and a large number of devices reuse certain frequency bands, causing serious mutual interference. Based on the above factors, the requirements for electromagnetic compatibility is getting higher and higher [3]. Electromagnetic compatibility is achieved by controlling electromagnetic interference, and how to test and suppress interference sources is an important part of electromagnetic compatibility research.
2. Emission field electromagnetic compatibility tests will also be conducted during the development stage of satellites. Some telemetry carrier frequencies on spacecraft are generated by local oscillator frequency multiplication or frequency conversion, and are affected by factors such as the non-linearity of the transmitter power amplifier and receiver amplifier [4]. It is easy to generate harmonic interference, combined frequency interference, cross-modulation, and image frequency interference. If the electromagnetic compatibility test is performed before the launch site is launched, it will be costly to take anti-interference measures. Therefore in the early stage of satellite development, in order to avoid the satellite's under-design on electromagnetic compatibility, the electromagnetic compatibility test of the transmitting field is also necessary.

3 Multi-resolution Characteristics of Wavelet Analysis

3.1 Wavelet and Wavelet Transform

The mathematical definition of wavelet is: for $\Psi_{a\tau}(t) \in L^2(R)$, If

$$\Psi_{a\tau}(t) = |a|^{\frac{-1}{2}} \Psi \left| \frac{t-\tau}{a} \right| \tag{1}$$

And meet the allowable conditions

$$C_{\Psi} = \int_{0}^{+\infty} \frac{|\Psi(\omega)|^2}{\omega} d\omega < +\infty \tag{2}$$

Then said $\Psi_{a\tau}(t)$ is the base wavelet. From the formula, we see that the wavelet transform has two variables, scale a(scale) and translation amount. τ (translation) [5].

Scale a controls the scaling and translation of the wavelet function τ. To control the translation of the wavelet function, the scale corresponds to the frequency and the amount of translation corresponds to time.

3.2 Multi-resolution Characteristics of Wavelet Analysis

Wavelet analysis has a variable time window, so it is suitable for analyzing interference signals. Using the multi-resolution characteristics of wavelet analysis, the interference signals are decomposed, and the wavelets on multiple frequency bands are analyzed. Decomposing signals can decompose electromagnetic signals into multiple frequency bands, and extract the time-frequency characteristics of a certain frequency range for spectrum analysis. According to the results of spectrum analysis, electromagnetic interference from different interference sources can be obtained.

For functions $\Psi_{a\tau}(t) \in L^2(R)$ by $\{\Psi_{a\tau}, \tau \in Z\}$ generate the closed subspace as V_a,which is [6]:

$$V_a = \overline{\text{span}\{\Psi_{a\tau}, \tau \in Z\}}, \; a \in Z \tag{3}$$

In order to build a multi-resolution wavelet model of the function, the following definitions are given [7]:

Space $L^2(R)$ multi-resolution analysis (mra) in $L^2(R)$ a spatial sequence in $\{V_a\}$, $a \in Z$:

1. consistent monotonicity: $\ldots V_{-1} \subset V_0 \subset V_1 \ldots$
2. Progressive completeness:

$$\bigcap_{a \in Z} V_a = \{0\} \tag{4}$$

$$\text{clos}_{L^2(R)}\left\{\bigcap_{a \in Z} V_a\right\} = L^2(R) \tag{5}$$

3. Scalability rules:

$$f(t) \in V_a \Leftrightarrow f(2t) \in V_{a+1}, a \in Z \tag{6}$$

4. Riesz basis existence: existence $\emptyset \in V_0$ so that $\{\emptyset(t-k)\}$, $k \in Z$, with Riesz bounds A and B V_a Riesz basis in

$$\emptyset_{a,k}(x) = 2^{\frac{a}{2}}\emptyset\left(2^a t - \tau\right) \tag{7}$$

Then $\forall_a \in Z$, $\left\{\emptyset_{a,k}, \tau \in Z\right\}$, which also has the same Riesz bounds A and B V_a a Riesz base called $\emptyset$ is a scaling function.

In case $\emptyset$ generate an mra, then because $\emptyset \in V_0 \subset V_1$ and because $\left\{\emptyset_{a,\tau}, \tau \in Z\right\}$ Yes V_1 a Riesz basis, so there is only one ι^2 two-scale sequence $\{p_\tau\}$ describe the scaling function $\emptyset$ the two-scale relationship:

$$\emptyset(t) = \sum_{\tau=-\infty}^{\infty} p_\tau \emptyset(2t - \tau) \tag{8}$$

Assuming that the wavelet here satisfies the wavelet admissibility condition, then V_1 the wavelets in can be generated as follows:

$$\Psi(t) = \sum_{\tau=-\infty}^{\infty} p_\tau \emptyset(2t - \tau) \tag{9}$$

Function family $\{\Psi_{0,k}\}$ also generates a closed subspace W_0, which is

$$\Psi_0 = \overline{\text{span}\{\Psi_{0,\tau}, \tau \in Z\}} \tag{10}$$

Therefore, when constructing a wavelet, it is generally at least guaranteed V_1 yes W_0 with V_0 the direct sum, that is:

$$V_1 = W_0 + V_0 \tag{11}$$

In practical systems, since the resolution of the measuring element is always limited, the information obtained can be considered $f(t(\tau - 1)) \in V_0$ and break it down into:

$$\begin{aligned} f(t(\tau - 1)) &= f_L(t(\tau - 1)) + g_L(t(\tau - 1)) \\ &\quad + \cdots + g_2(t(\tau - 1)) + g_1(t(\tau - 1)) \\ &= \sum_{\tau \in z} d_{L,\tau} \emptyset_{L,\tau}(t) + \sum_{a \geq L} \sum_{\tau \geq z} \emptyset_{L,\tau} \Psi_{a,\tau}(t) \end{aligned} \tag{12}$$

When orthogonal wavelet is selected, then:

$$d_{L,\tau} = f, \emptyset_{L,\tau}, C_{a,\tau} = f, \Psi_{a,\tau} \tag{13}$$

The above formula is the signal $f \in L^2(R)$ multi-resolution wavelet model can be rewritten as

$$f(t) = \sum_{a,\tau} \omega_{a,\tau} g_{a,\tau}(t) \tag{14}$$

4 Electromagnetic Compatibility Test Analysis of Space Launch Field Based on Wavelet Analysis

4.1 Transmit Field Electromagnetic Compatibility Test

The systems participating in the test include carrier rocket systems, satellite systems and ground systems. The technical status of all the test equipment is set according to the launch status. Each system is turned on in accordance with the program, and the working conditions and related data of each system equipment are recorded. At the same time, during the launch An electromagnetic spectrum monitoring point is set up on the tower. Before the test starts, and the equipment has not powered on, the environmental background noise is tested and recorded; during the test, the frequency spectrum near the operating frequency of the carrier rocket and satellite is mainly recorded; After the test, the background is re-tested to further confirm the key environmental impacts. In order to achieve the purpose of electromagnetic compatibility of the system, it is necessary to reduce the interference source as much as possible, suppress the interference propagation path, and reduce the sensitivity of each device.

4.2 Wavelet Analysis of Typical Electromagnetic Interference in Electromagnetic Compatibility Test

Electromagnetic compatibility refers to the coexistence of devices or systems that can perform their respective functions together in a common electromagnetic environment. Therefore, it includes two requirements: On the one hand, it means that the electromagnetic disturbance of the environment during the operation of the device cannot exceed a certain limit; On the other hand, it means that the device has a certain degree of immunity to the electromagnetic disturbance existing in the electromagnetic environment, that is, electromagnetic sensitivity. The so-called electromagnetic disturbance refers to any electromagnetic phenomenon that can reduce the performance of the device or system. Refers to the degradation of the equipment or system caused by electromagnetic disturbance.

Multi-resolution wavelet analysis can be used as a mathematical analysis tool for electromagnetic compatibility diagnostic testing. Through the theoretical analysis of multi-resolution wavelets in the previous section, the signal can be gradually subdivided by multi-resolution wavelets to find the frequency of interest and its time. Characteristics, this kind of search is a process from coarse to fine to browse the subdivided signals with a variety of different resolutions. An electronic device is composed of many subsystems, components and components. When the system has severe interference, The interference source needs to be diagnosed [8]. The operating frequencies of different subsystems, components and components are not the same, and the working timing may also be different. For the objects that are suspected, you can use multi-resolution wavelet analysis to determine whether it is causing serious electromagnetic interference. The reason for this is to provide a basis for electromagnetic compatibility improvement.

In the electromagnetic compatibility test of the transmitting field, the electrostatic discharge caused by electrical equipment is the most important electromagnetic interference [9]. The multi-resolution characteristics of wavelet analysis are used to conduct time-frequency analysis of this electromagnetic interference.

Electrostatic discharge is a harmful source of interference. As the charge on the device continues to accumulate, voltages of thousands to tens of thousands of volts may be generated. When a device with static electricity is discharged, it will cause a short strong electromagnetic field. The strong electromagnetic field can directly Breakdown equipment, or coupling to sensitive circuits, damage electrical equipment. As can be seen from Fig. 1, because the duration of electrostatic discharge is extremely short and the frequency spectrum is extremely wide, it is difficult to distinguish it from other interferences in spectrum analysis. Using traditional The Fourier spectrum analysis method is difficult to distinguish it from other interferences. The good local time-frequency analysis ability of wavelet analysis enables it to process transient and sudden electrostatic discharge signals. The current pulse of electrostatic discharge is an oscillating attenuation wave, so the decaying oscillation function can be used as the discharge simulation signal:

$$f(t) = 0 \quad t < t_0 \tag{15}$$

$$f(t) = A_e^{-\frac{t-t_0}{\tau}} \cos\left[2\pi f_0(t - t_0)\right] t > t_0 \tag{16}$$

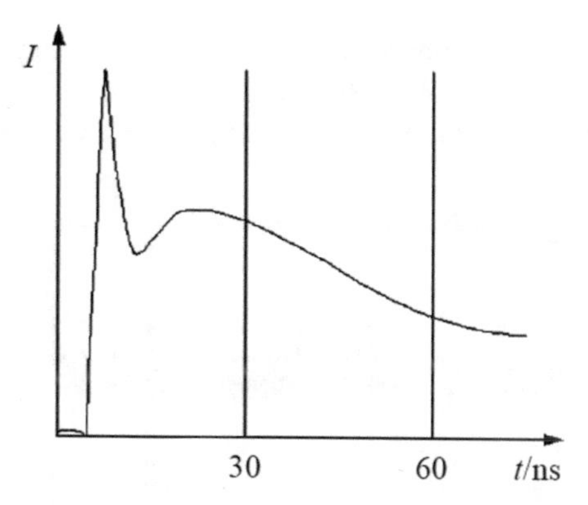

Fig. 1 Typical waveform of electrostatic discharge

Among them, t_0, A, τ with f_0 Represent the start time (i.e., the delay time), amplitude, attenuation coefficient, and oscillation frequency of the discharge pulse. We chose the parameters $a = 0.5$, $t_0 = 800\ \mu s$ $\tau = 5\ \mu s$, $f_0 = 200$ kHz Superimpose multiple background noises on the discharge signal, and then sample the discharge signal at a sampling frequency of 1000 kHz, as shown in Fig. 2. At this time, the discharge signal is submerged in the noise, and the moment when the discharge signal occurs cannot be distinguished from the picture.

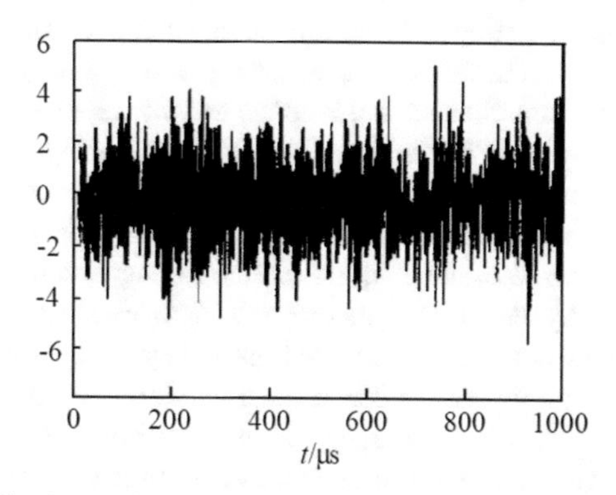

Fig. 2 Interference waveform after superimposed noise

Using the previous multi-resolution analysis method, a convolution operation is performed on this discharge signal to perform 3 layers of wavelet decomposition. The sample points are 1200 points, and the sampling interval is 1 μs. The original signal is decomposed into a low-frequency approximation signal and 3 sets of high-frequency detail signals, which can be clearly seen in the second layer of wavelet details 800 μs. The decomposed results of the discharge signal are shown in Fig. 3. This shows that the wavelet analysis can correctly extract the oscillating electrostatic discharge signal from the narrow-band interference, make a measurement diagnosis of the electrostatic discharge in electromagnetic interference, and if you can know the electrical equipment Or the working sequence of the system, you can further diagnose and locate the electromagnetic interference in time, so that the equipment can be better rectified.

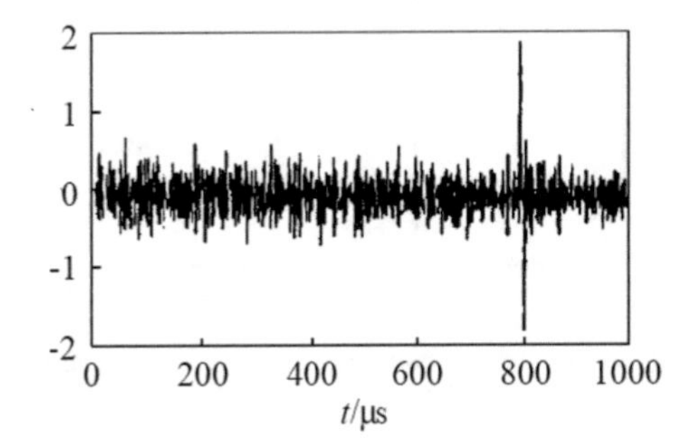

Fig. 3 Schematic diagram of the second layer signal after wavelet transform

5 Concluding Remarks

Spectrum analysis using Fourier transform is the basic theory and method of electromagnetic interference measurement and analysis at present. It is widely used in electromagnetic compatibility testing and electromagnetic compatibility prediction analysis. However, electromagnetic interference signal is a signal with a wide frequency band, and the composition is very complicated and belongs to non-stationary random signals, so Fourier analysis is not suitable for sudden signals. Wavelet analysis is a time-frequency analysis tool. Through the change and translation of the wavelet window, you can analyze the signal at the same time and frequency, and give the frequency these time characteristics play an important role in the analysis of electromagnetic interference. The electromagnetic compatibility test of the launching field is an electromagnetic compatibility test between the launch vehicle, satellite, and ground system. At present, there are no specific national standard requirements and specifications. Based on the actual test, an analysis mode is discussed, which can speed up the interference finding and elimination when ensuring the accuracy of the test conclusion. With the deepening of the test and the accumulation of experience, the test analysis method will be continuously improved.

References

1. Klein AF (2006) Conventional and wavelet coherence applied to sensory-evoked electrical brain activity. IEEE Trans Biomed Eng
2. Surajit Midya F (2007) An overview of electromagnetic compatibility challenges in European rail traffic management system. J Transp Res Part C 05:36–38
3. Sheng L, Zhang Lanyong FS (2010) Research on measurement technology of electromagnetic interference based on wavelet analysis. J Electron Inf Tech 05:34–36
4. Lu ZF (2012) A novel method of ambient interferences suppressing for In situ electromagnetic radiated emission test. IEEE Trans Electro-magnetic Compact

5. Deng HF (2013) Design of EMC test system based on wavelet analysis. Sichuan Ordnance Eng J 4:41–42
6. Berg TF (2019) Calibration of potential drop measuring and damage extent prediction by Bayesian filtering and smoothing. Int J Fatigue 05:38–39
7. Xu JF (2017) A modified LOT model for image denoising. Multim Tools Appl 6:47–49
8. Yin Qingsong F (2018) Research on image denoising algorithm based on improved wavelet threshold. J Softw Guide 1:24–26
9. Tao Yuan F (2019) A comparative study on the reduction of BDS multipath errors by wavelet analysis and empirical mode decomposition. Global Position Syst 3:46–48

Design of a Biometric Access Control System Based on Fingerprint Identification Technology

Hai Wang[1], Zhihong Wang[1](✉), Guiling Sun[1], Limin Zhang[1], Yi Gao[1], Ying Zhang[1], and Chaoran Bi[2]

[1] Teaching Center for Experimental Electronic Information, College of Electronic Information and Optical Engineering, Nankai University, Tianjin 300350, China
wanghao801226@nankai.edu.cn

[2] College of Media Design, Tianjin Modern Vocational Technology College, Tianjin 300350, China

Abstract. Biometric technology is a significant constituent of pattern recognition technique in artificial intelligence (AI). In this paper, an access control system is designed and implemented based on fingerprint identification, a typical and extensively utilized biometric technology in various fields recently. The hardware modules of the system include microcontroller unit (MCU) C8051F020, semiconductor fingerprint sensor chip FPS200, network interface chip RTL8019AS, liquid crystal display chip HY12864, keyboard, loudspeaker, audible and visual alarm, etc. The software architecture adopts client/server (C/S) model, with online and independent operative modes. Test results indicate that the system achieves the expected requirements and has the characteristic of high reliability, rapid response and expandable function.

Keywords: Artificial intelligence · Fingerprint identification · MCU · C/S model

1 Introduction

Artificial intelligence is a new technology science which researches and develops the theory, method, technique and application system for simulating and extending human intelligence [1, 2]. With the development and maturity of this technology, AI has been widespreadly applied in computer science, agriculture, medical treatment, transportation, commerce, security and other fields in recent years [3–5]. AI research includes three aspects: calculation, perception and cognition. Biometrics is an important method of perceptual intelligence, which is be able to constantly improve cognitive intelligence. It can provide a stable, unique and convenient authentication solution by perceiving some inherent biological characteristics of human body that cannot be copied, stolen or forgotten easily. This biometric information consists of voice, fingerprint, iris, face, vein and so on [6–9].

Fingerprints are lines formed by concave and convex skin at the end of human fingers, which have been formed before they birth. As an individual grows, the finger marks

Q. Liang et al. (eds.), *Artificial Intelligence in China*, Lecture Notes in Electrical Engineering 653, https://doi.org/10.1007/978-981-15-8599-9_5

deepen but the shape do not change. Sometimes two fingerprints have similar general characteristic, however their detailed characteristics cannot be identical. Fingerprint lines are not continuous, smooth and straight, but often break, bifurcate or turn. These breakpoints, bifurcate points, and turning points have a collective name, "feature points" [10, 11]. Different people have different finger marks and feature points, thus it provides an appropriate method of confirming identity uniqueness [12–15].

The fingerprint identification access control system replaces the traditional key with the finger marks. To unlock this system, the only action of that users need to do, is to place their fingers on the fingerprint acquisition chip. Not only is this biometric entrance guard system convenient and time-efficient, it but also avoids the risk of forgery, embezzlement, oblivion and destruction of the other access control methods, such as traditional mechanical lock, password lock and identification card [16–18]. A fingerprint identification access control system based on MCU technique and C/S model is designed in this paper, which has the characteristics of high reliability, rapid response and expandable function.

2 Hardware Design of the System

The microcontroller unit of fingerprint identification access control system is C8051F020, as the core control chip. Besides, the system is composed of semiconductor fingerprint sensor chip FPS200, network interface chip RTL8019AS, liquid crystal display chip HY12864, keyboard, loudspeaker, audible and visual alarm and other components. The hardware block diagram is illustrated in Fig. 1. This client terminal provides the completions like the fingerprint image acquisition, data exchange with the upper computer and the other corresponding control functions. The efficiency of executing program is significantly improved due to the introduction of C8051F020, the MCU chip. Meanwhile, the response time of this system is also shortened. The application of low-power devices and chip package components is capable of effectively reducing power consumption, decreasing PCB area, improving the anti-interference performance of the circuit, and enhancing the stability and reliability of the system.

2.1 MCU Chip C8051F020

C8051F020, which is a fully integrated mixed signal system level MCU chip, is applied in this design as the core component. It is provided by CIP-51 microcontroller kernel patented by CYGNAL, ADC with 12-bit and eight channels, DAC with programmable data update mode, 64 K byte flash memory, five universal 16-bit timers, on-chip watchdog timers and other components. When the system clock frequency is 25 MHz, its peak speed can be 25 MIPS because of the pipeline structure. So the sending time of fingerprint packets is reduced to half a second. In addition, C8051F020 has 8-byte I/O port, which meets the needs of I/O interface expansion in the system. All analog and digital peripherals can be enabled, disabled, and configured by the user firmware.

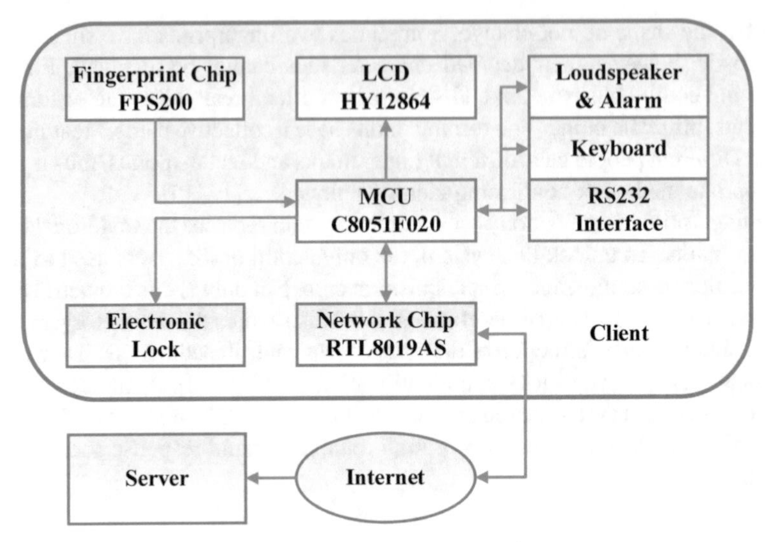

Fig. 1 Block diagram of the hardware

2.2 Fingerprint Sensor FPS200

FPS200, an ideal contact fingerprint authentication device, is a kind of complementary metal oxide semiconductor (CMOS) sensors based on the principle of capacitor charging and discharging. It consists of 256 × 300 capacitance sensor arrays, as well as the resolution of 500 dpi and the working voltage range of 3.3–5 V. The outside is an insulating surface. Each point of the sensor array is a metal electrode, and the finger acts as the other electrode of the capacitor. When an individual touches the fingerprint acquisition area of FPS200, a dielectric layer between the two electrodes is formed. Due to the different distance between the ridge and valley of fingerprint relative to the other pole, the capacitance values of silicon surface capacitance array varies. According to this way, the capacitance array values can represent a fingerprint image.

Each column of FPS200 has two sets of sample and hold circuits. The pattern acquisition is realized by lines. Select a line, charge all the capacitors in this line, acquire and save voltage values by the circuits. Then the residual voltage is obtained by another set of circuit. Two groups of data are calculated by the built-in ADC to get the fingerprint image with corresponding gray level, eventually.

2.3 Gate Array Logic (GAL) Device and External SRAM Circuit

GAL22LV10D logic device is applied in this design, and HY628100B with TSSOP-1 chip package is adopted as the external SRAM. Circuit schematic diagram is shown in Fig. 2. All I/O ports remain high from MCU power on to completion of initialization configuration. Therefore, the electronic lock will receive a pulse of rising edge, causing the electronic lock to open. In order to prevent this action, a NOT gate of 74HC04 is utilized in the circuit to reverse the PWRCTRL signal, which controls the electronic door lock. Both/INFPS1 and/INFPS2 are optoelectronic detection analog signals generated when fingers are pressed. Sharp rising or falling edges can be shaped and transformed by two other NOT gates when converting from analog signals to digital signals. This procedure is propitious to precisely and smoothly read the data, which is obtained by fingerprint acquisition chip in the system.

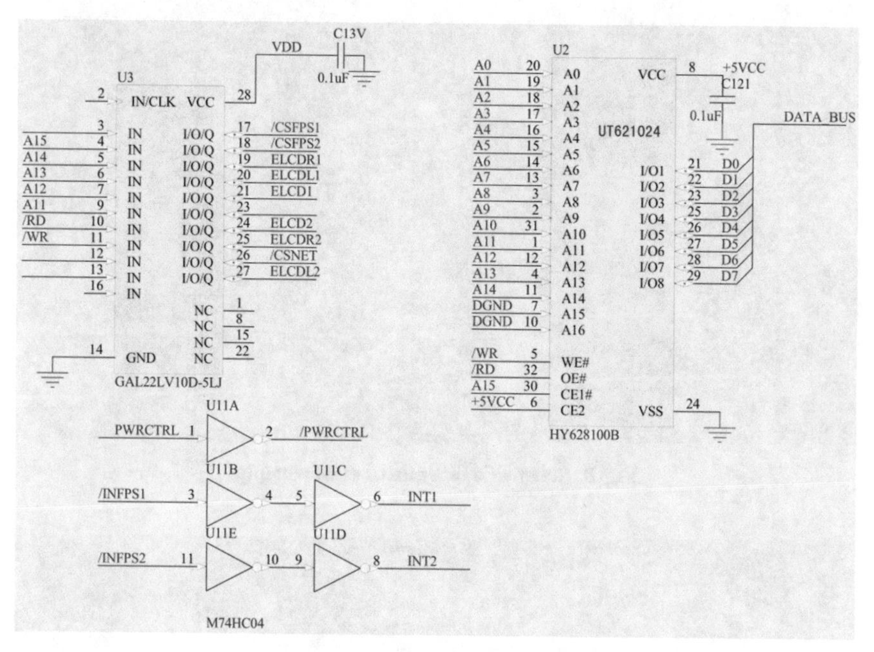

Fig. 2 Schematic diagram of GAL and external SRAM

2.4 Other Module Circuits

The whole access control solution requires three different voltage inputs, which are respectively +12 VDC, +5 VDC and +3.3 VDC. The main power supply of the whole system is +12 VDC, directly introduced from the outside. It is connected to electronic door lock as the working power supply and shunted to the input of power supply regulator chip LM7805, which generates +5 VDC. The output client terminal is supplied to the input terminal of every +5 V device, of course included ALS1117, on the main control board. The voltage stabilizing chip ALS1117 generates +3.3 VDC, as illustrated in Fig. 3, supplied to MCU and GAL devices.

The circuit diagram of network card chip RTL8019AS is shown in Fig. 4. The schematics including door lock control interface, keyboard, LED indicator, buzzer, RS232 interface, expansion port and other parts is demonstrated in Fig. 5.

3 Software Design of the System

Most of the software program of the client terminal, with the front-and-back program structure, is written in C51 language. After the initialization, it remains the query state

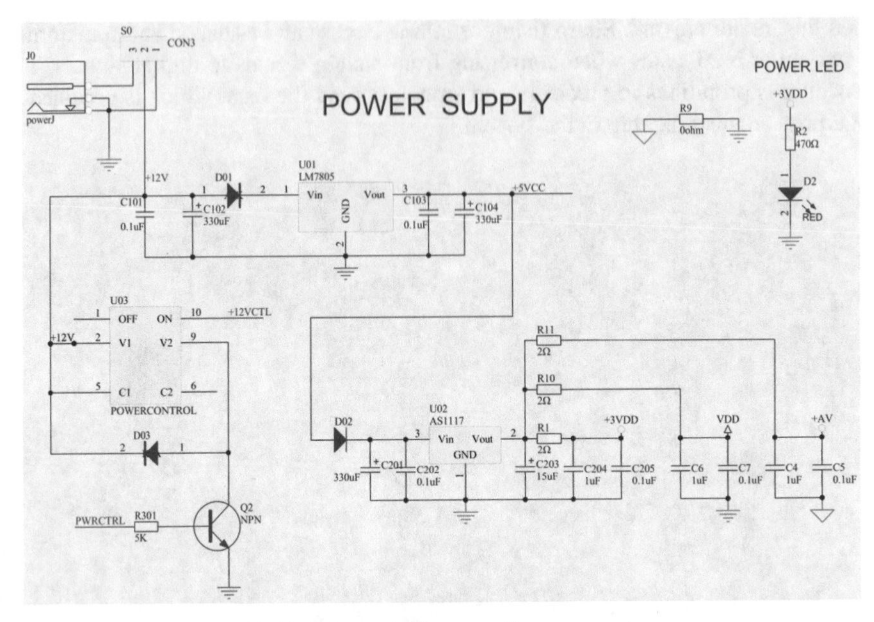

Fig. 3 Schematic diagram of power supply

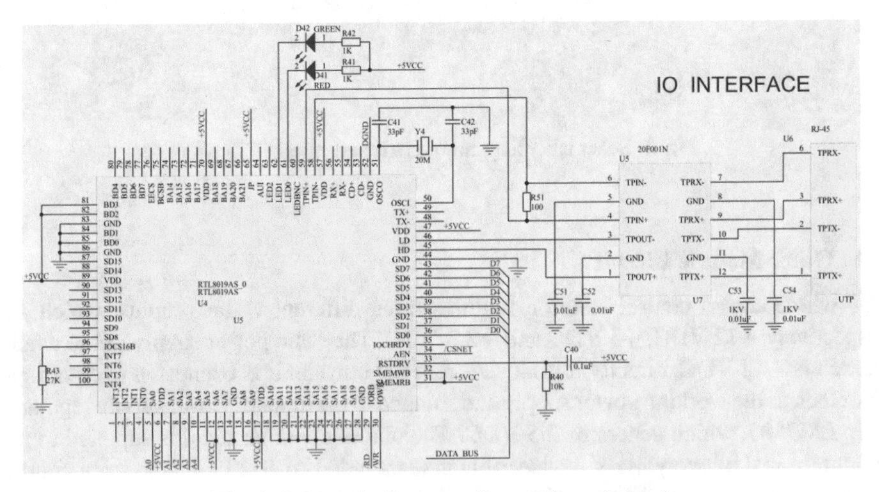

Fig. 4 Schematic diagram of network card circuit

and waits for the interrupt to trigger. There are two operation modes in the fingerprint identification access control system: online and independent. After the system starts, check the network connection to enter the online mode. If the connection is invalid, switch to the independent working mode. The client terminal software flow chart is illustrated in Fig. 6.

When the server starts and runs the service program, the client terminal automatically works in online mode. The data is acquired, packaged and sent to the server directly while

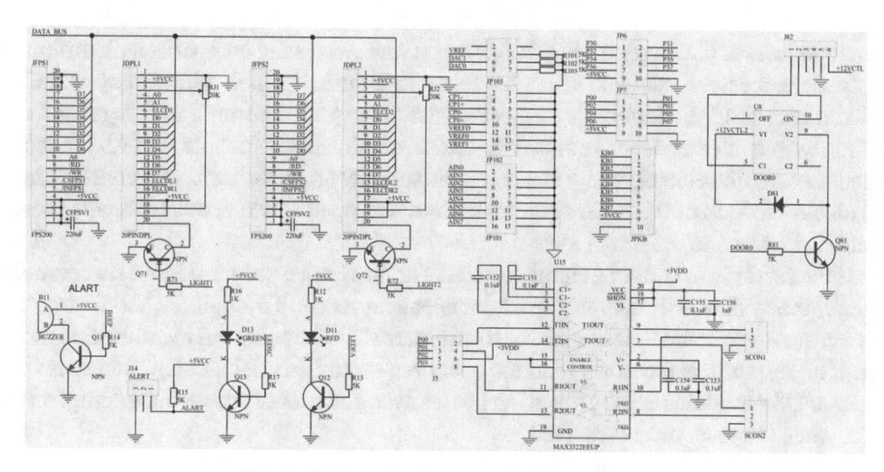

Fig. 5 Schematic diagram of other modules

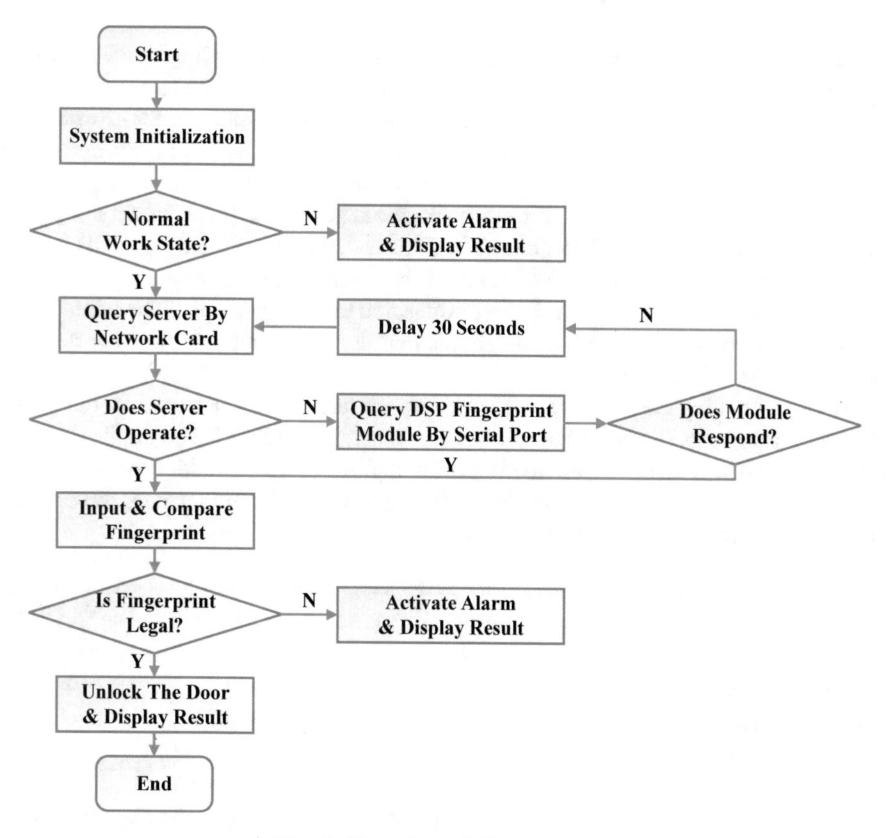

Fig. 6 Flow chart of client software

someone presses the finger on FPS200. The server invokes the comparison algorithm to analyze the received data and sends the result to terminal, which will display on LCD. If the match result is correct, it means that the fingerprint information is legal and the authority to enter the access control system is obtained. Meanwhile, the server sends signal to make the electric relay close for approximately two seconds. As a consequence, the electronic lock of the access control system will open. Then, it automatically closes again and returns to the initial state.

If the server is in case of a shutdown status or the service program is not successfully started, the system will operates in an independent mode. Through the serial port, the system queries whether it is connected to an external DSP fingerprint identification module. If no response is received, send the query request again after a delay. Otherwise the external DSP module replaces the server to analyze and process the acquired fingerprint data, when the response is received.

4 Debugging and Implementation of the System

Printed Circuit Board (PCB) is designed and manufactured according to the hardware circuit schematic diagram of the part of the system. The function test is carried out in the order of power module, MCU and its reset circuit, fingerprint module, SRAM module, network card chip and its peripheral circuit, LCD display module and other parts. The hardware performance is dramatically affected by many factors, such as the quality of electrical components, the anti-interference performance of PCB, the stability of power supply. Since the crystal oscillator circuit in C8051F020 is very sensitive to PCB layout, the crystal should be as close to XTAL pin of the device as possible when PCB wiring. In addition, the load capacitance shall be connected to the crystal pin. The wire should be as short as possible, and link to the ground in order to prevent noise and interference caused by other circuit leads.

When debugging the software, it is necessary to check the efficiency and redundancy of the program code. Optimize the source code generated by Keil C51 compiler system. A STARTUP.A51 assembler, provided by the compiling system, is added to the program for port configuring and clock switching. The following section is part of the code of the clock switching delay program.

```
        mov    0b1H, #67H      ; make external crystal work
        mov    R0, #06H        ; run delay program
loop3:  mov    R1, #0ffH
loop2:  mov    R2, #0ffH
loop1:  djnz   R2, loop1
        djnz   R1, loop2
        djnz   R0, loop3
loopa:  mov    A, 0b1H         ; detect the effective flag of crystal oscillator XTLVLD
        anl    A, #80H
        cjne   A, #80H, loopa
        mov    0b2H, #98H      ; switch to external clock
```

After final debugging and operation test, the expected function of fingerprint identification access control system has been achieved. The legal fingerprints can open the access control, while the illegal fingerprints will trigger the alarm device. In the process of fingerprint acquisition and recognition, the system has the characteristics of simple operation, high accuracy and short average response time, which is less than 1 s. The successful operation of the system is achieved by utilizing a lower development cost, which is particularly worthy to mention.

5 Conclusion

A type of fingerprint identification access control system with core processor C8051F020 and C/S model software architecture is designed and manufactured in this paper based on biometric technology in this paper. The system has sorts of positive features, such as online and independent operate modes, simple structure, rapid response, expandable function, stable performance and high reliability. Nevertheless, no single biometric technology can guarantee 100% security in practical applications. How to recognize identity information more accurately, conveniently, quickly and cheaply combined with other biometrics, such as voice, iris, vein and so on, therefore, will be the most critical and inevitable aspect to develop fingerprint identification technology in the future.

Acknowledgements. This work is supported by the Self-made Experimental Teaching Instrument and Equipment Project Fund of Nankai University; the 2020 Undergraduate Education Reform Project Fund of Nankai University (NKJG2020004) and Teaching Center for Experimental Electronic Information.

References

1. Vercauteren T, Unberath M, Padoy N et al (2020) CAI4CAI: The Rise of Contextual Artificial Intelligence in Computer-Assisted Interventions. Proc IEEE 108(1):198–214
2. Humood K, Mohammad B, Abunahla H et al (2020) On-chip tunable memristor-based flash-ADC converter for artificial intelligence applications. IEEE Trans Industr Inf 14(1):107–114
3. Iqbal K, Odetayo MO, James A (2014) Face detection of ubiquitous surveillance images for biometric security from an image enhancement perspective. J Ambient Intell Human Comput 5(1):133–146
4. Liu Y, Ling J, Liu ZS et al (2018) Finger vein secure biometric template generation based on deep learning. Soft Comput 22(7):2257–2265
5. Al-Hmouz R, Pedrycz W, Daqrouq K et al (2018) Development of multimodal biometric systems with three-way and fuzzy set-based decision mechanisms. Int J Fuzzy Syst 20(1):128–140
6. Srivastava V, Tripathi BK, Pathak VK (2014) Biometric recognition by hybridization of evolutionary fuzzy clustering with functional neural networks. J Ambient Intell Human Comput 5(4):525–537
7. Kumar N, Singh S, Kumar A (2018) Random permutation principal component analysis for cancelable biometric recognition. Appl Intell 48(9):2824–2836
8. Zhu Q, Xu NY, Huang SJ et al (2020) Adaptive feature weighting for robust Lp-norm sparse representation with application to biometric image classification. Int J Mach Learn Cybern 11(2):463–474

9. Tistarelli M, Schouten B (2011) Biometrics in ambient intelligence. J Ambient Intell Human Comput 2(2):113–126
10. Jang HU, Kim D, Mun SM et al (2017) DeepPore: fingerprint pore extraction using deep convolutional neural networks. IEEE Sig Process Lett 24(12):1808–1812
11. Jiang A, Yuan YG, Liu N et al (2019) Transparent capacitive-type fingerprint sensing based on zinc oxide thin-film transistors. IEEE Electr Dev Lett 40(3):403–406
12. Engelsma JJ, Arora SS, Jain AK et al (2018) Universal 3D wearable fingerprint targets: advancing fingerprint reader evaluations. IEEE Trans Inf Foren Secur 13(6):1564–1578
13. Labati RD, Genovese A, Piuri V et al (2016) Toward unconstrained fingerprint recognition: a fully touchless 3-D system based on two views on the move. IEEE Trans Syst Man Cybern Syst 46(2):202–219
14. Yuan CS, Xia ZH, Jiang LQ et al (2019) Fingerprint liveness detection using an improved CNN with image scale equalization. IEEE Access 7:26953–26966
15. Ivanov VI, Baras JS (2017) Authentication of swipe fingerprint scanners. IEEE Trans Inf Forens Secur 12(9):2212–2226
16. Shu YC, Gu YJ, Chen JM (2014) Dynamic authentication with sensory information for the access control systems. IEEE Trans Parall Distrib Syst 25(2):427–436
17. Sichkar VN (2018) Fingerprint identification as access control system. In: International conference on industrial engineering, applications and manufacturing (ICIEAM)
18. Geralde DD, Manaloto MM, Loresca DED et al (2017) Microcontroller-based room access control system with professor attendance monitoring using fingerprint biometrics technology with backup keypad access system. In: IEEE 9th international conference on humanoid, nanotechnology, information technology, communication and control, environment and management (HNICEM)

Aeronautical Meteorological Decision Supporting Technology Based on 4D Trajectory Prediction

Yi Mao, Yuxin Hu, and Jiajing Zhang(✉)

State Key Laboratory of Air Traffic Management System and Technology, Nanjing 210007, China
40414663@qq.com

Abstract. Aeronautical meteorological technology is essential for supporting a safe air space and efficient air traffic management. This paper proposed a 4D trajectory prediction-based aeronautical meteorological decision supporting technology, which is based on the concepts of "next-generation" ATC system. This methodology analyzes both dynamic radar track data and weather data around the airways, then generates combined weather images for a certain air space. Empirically, this technology provides an efficient way to decrease weather restriction time, which makes a good effort on reducing flight delays, flight plan coordination and flight routes decision-making to reduce return and alternate.

Keywords: Aeronautical meteorology · 4D trajectory prediction · Assistant decision

1 Introduction

With the rapid economic development, time-saving and highly efficient air transportation has become the first choice for business travelers. The air transportation industry at home and abroad is developing rapidly. Flight safety is the most concern of the public, and it is the critical task and responsibility of all aviation service personnel. Since air transportation is operated in the air, the aircraft is affected by meteorological conditions all the time in flight.

For example, the cloud is the most common meteorological factor affecting flight. Low visibility in the cloud will affect visual flight, supercooled water droplets in the cloud will make the aircraft accumulate ice, turbulence in the cloud will cause the aircraft to bump, uneven lightness in the clouds interferes with the pilot's judgment and lightning in the cloud can damage the aircraft, etc. [1]. During the flight of a civil aviation aircraft, if a cloud layer affecting the safety is found in front of the route through ground notification or airborne weather radar, it can generally take a change in altitude or deviate from the original flight route to avoid the dangerous cloud; however, the routes of civil aviation aircraft have strict regulations on certain height, width and route. If deviation from its safe route happens, there will be a risk of losing contact, getting drift off course and

Q. Liang et al. (eds.), *Artificial Intelligence in China*, Lecture Notes in Electrical Engineering 653, https://doi.org/10.1007/978-981-15-8599-9_6

intruding into a low altitude danger zone. Therefore, once the corresponding measures are taken within the safe range and still the dangerous cloud cannot be avoided, the original flight plan can only be canceled, alternate to other airports or return flight. If the meteorological department predicts in advance that the weather on the route affects flight safety and cannot be avoided, the aircraft will wait for the weather to improve before performing the corresponding flight. In addition, under overcast conditions, even if the ground visibility is good, the low cloud layer above the airport runway and near the landing path will prevent the pilot from seeing the runway during the landing process, and the aircraft is too close to the ground once descended through the low cloud layer to deal with possible unexpected situations, which will directly affect the landing of the aircraft and even cause the airport to close.

In the process of air traffic management, one of the main reasons for flight accidents and delays is the restrictions of the aircraft under certain severe weather conditions. In recent years, China's civil aviation has developed rapidly. It is estimated that in the next few years, China's civil aviation traffic will grow at an average annual rate of 10%. Although the increasingly advanced airport navigation facilities and aircraft performance have reduced the restrictive effect of adverse meteorological conditions on flight, with the increase of volumes of flights, the numbers of passenger detention and flight delays caused by airports and airways under adverse meteorological conditions and climate environment did not decrease significantly. The continuous development of civil aviation has put forward higher requirements for aviation meteorological services. The demand for accurate forecasting of meteorological information affecting flight has been transformed into the need to obtain auxiliary decision support to complete the flight under adverse weather conditions. By accurately predicting the weather along the route, flight delay time can be effectively shortened, and unnecessary alternation and return flights can be avoided.

2 Overview of Aviation Meteorology

Current sources of aeronautical meteorological information mainly include meteorological reports, Doppler weather radars and meteorological satellites. The analysis of the weather report mainly includes wind direction, wind speed, visibility, temperature, QNH, etc.; weather radar detection data directly reflect the intensity, direction of movement, speed of movement, precipitation distribution, precipitation intensity, cloud volume, cloud status and cloud height of cloud rain echoes in the form of echo maps; satellite cloud images can intuitively express the cloud general categories, characteristics, distribution and changes of clouds [2].

Aeronautical meteorological services are mainly oriented to airlines, airports, air traffic control departments, air traffic flow management departments and airspace management departments [3]. Airlines need to formulate and revise flight plans according to the weather conditions at take-off and landing airports, alternate landing airports and air routes, and can pre-assess the fuel consumption and total flight hours with reference to the high-altitude wind field and temperature field information. The airport needs to be prepared for protection of surface facilities and operational capabilities based on weather forecasts. When it is threatened by severe weather such as typhoons, heavy snow, hail, thunderstorms, etc., it is prepared to take emergency measures in advance to ensure the normal operation of the airport and reduce the number of flight delays. The controller of the air traffic control department needs to implement dispatch and command according to the weather conditions (cloud, wind, QNH, visibility, etc.) to guide the aircraft to avoid the wind shear area and the strong thunderstorm area to avoid flight accidents. The air traffic flow management department needs to timely grasp the weather forecast and weather conditions to formulate flow management measures, make the most possible use of the available capacity of airports and airspace under the constraints of adverse meteorological conditions, and reasonably schedule the arrival and departure. The airspace management department needs to fully consider local meteorological statistics for site selection of new airports, determination of direction of runways and planning of new routes.

3 A New Generation of Aviation Meteorological Technology Research

The US Joint Planning and Development Office formed the Meteorological Function Demand Research Group. Based on the development strategy of the "New Generation Air Transport System," NextGen Air Traffic Management released version 0.1 of the 4D weather function requirements [4] and proposed three prominent concepts: (1) Provide common weather scenarios for all decision makers and aviation system users. (2) Meteorological information is directly integrated into a complex decision support system to help decision makers. (3) Use Internet transmission to achieve flexible and effective access to all necessary meteorological information.

Based on the analysis and reference of the research on the new generation of air transportation system in the USA, China cooperates with the construction concept of the integration, high efficiency and seamlessness of China's new generation air traffic management system [5], combining with the development trend of China's civil aviation; in recent years, the application of aviation meteorological technology in air traffic management has been expanded, comprehensively improved the level of weather observation and forecasting, and enabled meteorology to participate in the cooperative decision-making of the entire process of aviation operations to adapt to the future air transportation's pursuit of safety, flexibility and efficiency.

3.1 Meteorological Information Sharing Technology

Existing aeronautical meteorological technology collects real-time aeronautical meteorological information required for air traffic control, processes or merges weather information in the same geographic area into a unique forecast, transmits it via the Internet and displays it in integrated meteorological information display system in the form of graphics, text, etc. The system builds an aeronautical meteorological data sharing platform, which enables various departments to flexibly and effectively access the required meteorological information, and ensures that all decisions made by every air traffic management departments are based on the same meteorological information. The disadvantage is that this system and the air traffic control system radar track situation are displayed on separate screens so when the controller checks the weather conditions of the route, he needs to locate on the display interface of the meteorological information system through latitude and longitude, and it will increase workload and distract the controller's attention.

3.2 Meteorological Fusion Display Technology

In order to achieve the same screen display of meteorological information and route and track information, the ATC automation system converts the weather radar echo value into data conforming to the ASTERIX radar data format standard and simulates it as an air traffic radar with weather information output function, and then integrated it as input to the automation system. The original information represents the cloud coverage area with vector lines and the automation system processes the data and expands each vector line into a quadrangle. After superimposition, the bottom layer is displayed on the ATC automation system radar track situation display interface, where a clouded area is represented in the form of a filled map. This method achieves the same screen display, but can only roughly express the range and level of the cloud image, and it cannot accurately express the reference data of the control decision such as cloud bottom height and cloud top height. Similarly, the automated system receives meteorological METAR reports and GRIB reports, analyzes QNH and high-altitude wind information and pops up new windows in texts or tables on the man–machine interfaces of the air traffic controller positions, which is difficult to play a direct reference role for collaborative decision-making.

3.3 Five-Dimensional Visualization Meteorological Information Technology of Routes

The five-dimensional graphical visualization system of flight meteorological information provides intuitive and three-dimensional visualization of route information. This technology provides an assessment of the fuel consumption of high-altitude winds and high-altitude airways, and generates 3D dynamic display of real-time field and forecasts field along the route, in order to realize the display of wind field, temperature field, vertical profile of important weather area, horizontal profile of different flight heights and route profile on the route area [6]. Compared with the first two technologies mentioned before, the application of this technology has its advantages. The three-dimensional display of track information and meteorological information on the same screen is realized, and the current weather situation at the current time is expressed in real time [7], which is in line with the research and development direction of China's new generation of aviation meteorological system [8].

4 Aeronautical Meteorological Assistant Decision-Making Technology Based on 4D Trajectory Prediction

Based on the requirements of civil aviation, this paper further improves the new generation of aeronautical meteorological technology and proposes an aeronautical meteorological assistant decision-making technology based on 4D trajectory prediction. The technology uses the time t as the axis to give the weather conditions along the route $W(t)$, where t is the time when the aircraft is estimated to reach a certain point R on the route based on the 4D trajectory, and $W(t)$ is the weather condition at time t at the point R calculated based on the weather forecast.

4D trajectory prediction is based on extracting route points, take-off/arrival airport, take-off/arrival airport altitude, requested altitude, cruise altitude, requested speed, offline defined aircraft performance, high-altitude wind, real-time radar correction information and offline settings and the empirical data of the flight plan waypoint time, etc., after calculating the flight trajectory, and the estimated flying height and the estimated flying time for each reporting point along the route are obtained as shown in Fig. 1.

4D model calculation timing:

(1) When creating a flight plan, build a 4D model;
(2) When the route or departure airport or destination airport changes, re-establish the 4D model;
(3) When manually modifying the height of the midpoint of the route, re-establish the 4D model;
(4) When the height change of the position report is received, re-establish the 4D model;
(5) Receive the real target height, speed, climb/descent rate detected by the monitoring data source in real time, and modify the existing 4D model according to the real track to make the subsequent prediction more accurate.

The 4D model of the technology is divided into three stages: take-off stage, level flight stage and descent stage, as shown in Fig. 2. In different flight stages, the influence index of meteorological elements that affect the flight of the aircraft is different. In order to enhance the practicability, the information of key meteorological elements and the trajectory should be extracted and synthesized. Climbing phase: Provide the ground temperature, wind direction, QNH, visibility and other factors of the airport at the take-off time of the flight. Descent phase: Provide the ground temperature, wind direction and speed, QNH, visibility and other factors of the landing airport at the time the flight landed. Level flight phase: Provide information on wind direction, wind speed, temperature, cloud shape, etc. on the flight level of the flight path during the entire process of take-off, approach, level flight, approach again and landing.

The aeronautical meteorological assistant decision-making technology based on 4D trajectory prediction, combined with the weather forecast information, carries out a comprehensive route-to-point weather synthetic forecast. According to the actual needs, it provides the most critical and valuable meteorological information in air traffic management at this stage in the form of graphs or tables. The effect is shown in Fig. 3. With reference to the prediction model provided by this technology, weather flow control can be lifted in advance, which helps air traffic management personnel to reasonably select the route and flight release time, and effectively shortens the delay caused by the weather. At the same time, it can also refer to the prediction model provided by this technology to predict the weather conditions that will be encountered on the route, assist airlines in making flight plan changing decisions and reduce return flights and alternate landings.

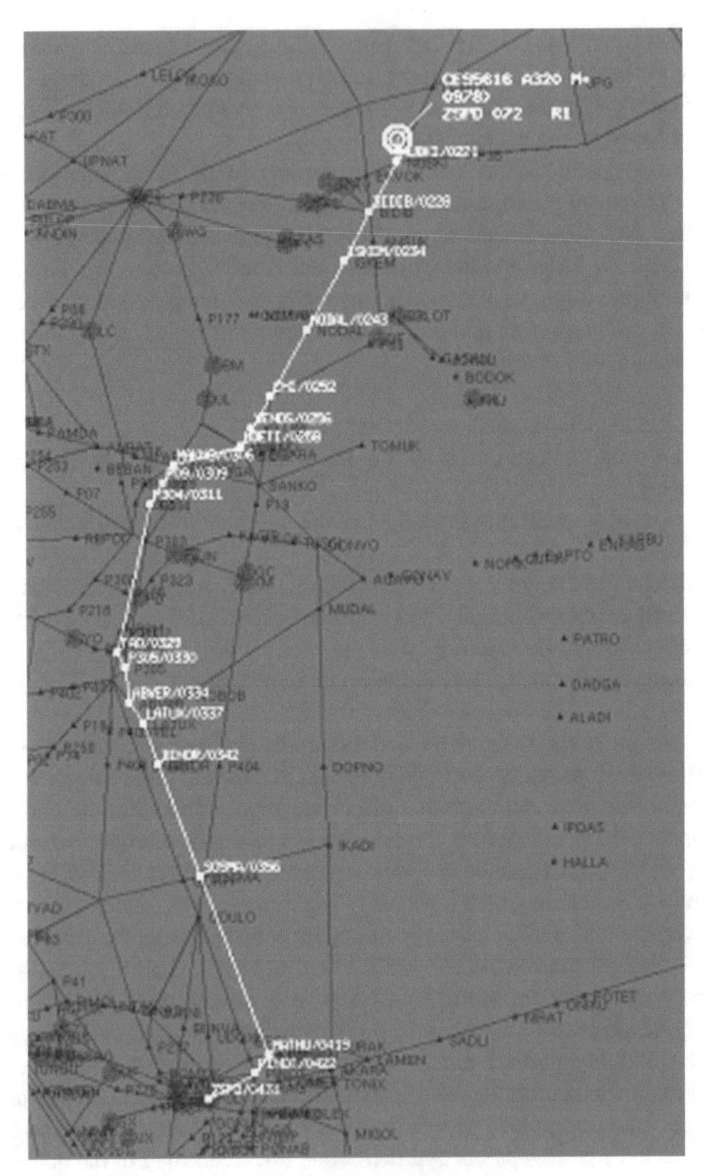

Fig. 1. Future route forecast

The meteorological information that can be accessed by the system established based on this technology includes meteorological reports, Doppler weather radar echoes, meteorological satellite cloud images, weather forecast information of the meteorological information sharing platform, etc. Various types of meteorological information are analyzed and normalized on the meteorological data preprocessing server, combined into an internal format input system. The information to be extracted includes wind direction,

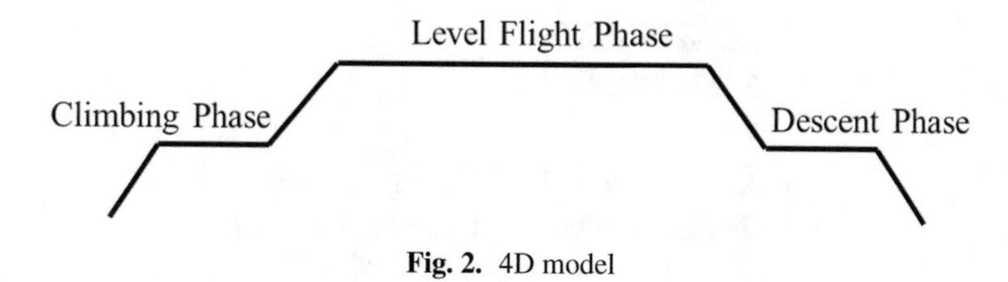

Fig. 2. 4D model

Fig. 3. Aeronautical meteorological information synthesis model based on 4D trajectory prediction

wind speed, visibility, temperature, QNH, cloud and rain intensity, cloud and rain movement direction, cloud and rain movement speed, precipitation distribution, precipitation intensity and the general category of clouds, characteristics, distribution and changes, etc. At the same time, in order to make the prediction of the 4D trajectory of the executed plan more accurate, the comprehensive trajectory information is connected, and the passing time and passing height of the future waypoint are corrected accordingly. The data processing server performs 4D trajectory calculations and meteorological prediction calculations with the input data, and finally gives an integration of dynamic track information and dynamic weather information to predict the future time and space, in the form of images or tables in the man–machine interface. It can also create a flight plan through the man–machine interface, plan the route and the estimated take-off time, and the system will perform the prediction and synthesis of the track trajectory and meteorological information based on this. The main processing flow is shown in Fig. 4.

The aeronautical meteorological assistant decision-making system based on 4D trajectory prediction is composed of an information docking subsystem (*meteorological information docking* and *integrated trajectory information docking*), a *meteorological data preprocessing server*, a *data processing server*, and a *man-machine interface client*. The overall system architecture is shown in Fig. 5.

5 Conclusion

The aeronautical meteorological assistant decision-making technology based on 4D trajectory is a new type of aeronautical meteorological technology developed based on the research and application of a new generation of aeronautical meteorological technology, combined with the actual work needs of air traffic management personnel, airlines and

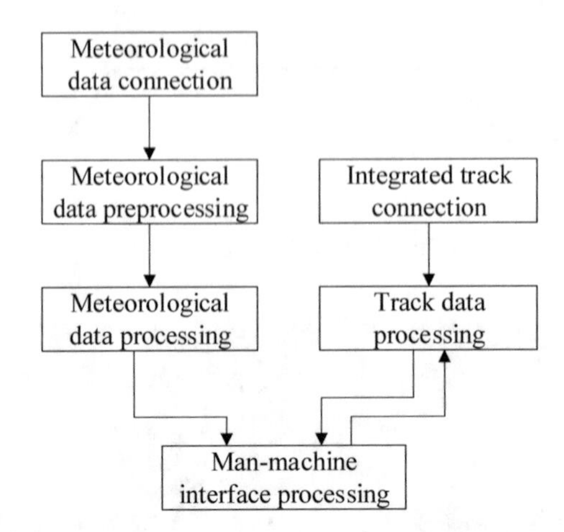

Fig. 4. Schematic diagram of the software information interface

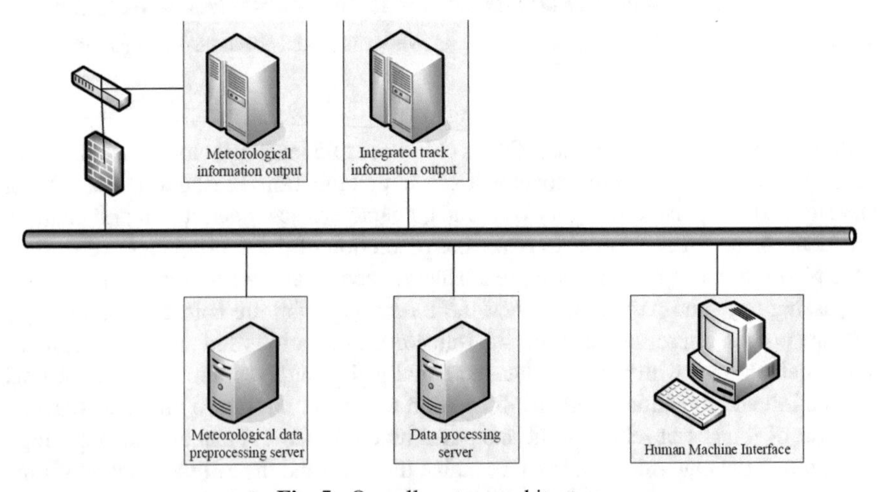

Fig. 5. Overall system architecture

other units. The technology simultaneously predicts, synthesizes and displays dynamic track information and dynamic weather information to achieve the early release of restrictions on unfavorable meteorological conditions, effectively reducing flight delay time. Meanwhile, it can also predict the weather conditions along the route based on this to re-plan, select reasonable air routes to avoid dangerous weather and effectively reduce return and alternate landing. The technology is currently limited to play a role in assisting decision-making and has not yet involved automatic route planning under adverse weather conditions. This will be the next step in this research direction.

References

1. ARINC specification 620-6, Air/ground character-oriented protocol specification [S]. Annapolis: ARINC, 2007, pp 11–39
2. Dietz SJ, Kneringer P (2019) Forecasting low-visibility procedure states with tree-based statistical methods. Pure appl Geophys 176(6):2631–2644
3. Chan PW, Yeung WL (2019) A comparison between laser ceilometers at Hong Kong International Airport. Weather 74:6
4. Blickensderfer B, Lanicci J, Guinn TA, King K, Ortiz Y, Thomas R (2017) Assessing general aviation pilots' understanding of aviation weather products. Int J Aerosp Psychol 27:3–4
5. Wu T-C, Hon K-K (2017) Application of spectral decomposition of LIDAR-based headwind profiles in windshear detection at the Hong Kong International Airport. Meteorologische Zeitschrif 27(1):33–42
6. Meteorology; Recent findings from national center for atmospheric research provides new insights into meteorology (data fusion enables better recognition of ceiling and visibility hazards in aviation). Sci Lett (2015)
7. Cao Y, Wu Z, Xu Z (2014) Effects of rainfall on aircraft aerodynamics. Prog Aerosp Sci p 71
8. Hon KK, Chan PW (2014) Application of LIDAR-derived eddy dissipation rate profiles in low-level wind shear and turbulence alerts at Hong Kong International Airport. Meteorolog Appl 21(1)

Evaluation Index System and Case Study for Smart ATM Based on ASBU

Chao Jiang[1,2(✉)]

[1] The 28th Research Institute of China Electronics Technology Group Corporation, Nanjing 210000, China
capjiangchao@163.com

[2] State Key Laboratory of Air Traffic Management System and Technology, Nanjing 210016, China

Abstract. Air Traffic Management Bureau of Civil Aviation Administration of China (ATMB) has steadily promoted capacity building of smart ATM under the guidance of the *Action Plan of Four Enhancement of ATM*. However, at present, the evaluation method for the effectiveness of capacity building of smart ATM is not systematic, there is also a lack of statistical analysis of capacity building of smart ATM. In 2012, the International Civil Aviation Organization (ICAO) proposed an authoritative and instructive *Aviation System Block Upgrade* plan, which provides guidance for the capacity building of smart ATM. Therefore, based on the analysis of the threads of capacity improvement proposed in *ASBU* and the current situation of ATMB, this paper constructs an evaluation index system of capacity building of smart ATM, at the same time, the current situation of ATMB in the implementation of various threads is obtained according to the survey. And the above-mentioned evaluation index system is used to implement a case study for the effectiveness analysis of the capacity building of smart ATM, which verifies the feasibility of the evaluation index system.

Keywords: Smart ATM · ASBU · Index system · Case study

1 Introduction

Air Traffic Management Bureau of Civil Aviation Administration of China (hereinafter referred to as ATMB) issued the *Action Plan of Four Enhancements of Air Traffic Management* [1] in July 2018, it indicates that the four enhancements of safety, efficiency, smartness and coordination have become the general development direction of ATMB in the future. In the *Action Plan of Four Enhancements of Air Traffic Management*, smart air traffic management (ATM) is clearly proposed as a means to achieve the four enhancements of ATM, and specific capacity building contents such as smart operation, smart service, smart facilities and smart management are proposed. Based on the above requirements, ATMB is also vigorously promoting the capacity building of smart ATM, and some results have achieved. However, at present, the evaluation method for the effectiveness of capacity building of smart ATM is relatively single and does not form a

Q. Liang et al. (eds.), *Artificial Intelligence in China*, Lecture Notes
in Electrical Engineering 653, https://doi.org/10.1007/978-981-15-8599-9_7

system, and there is a lack of statistical analysis on the effectiveness of capacity building of smart ATM nationwide, it will have a negative impact on the current assessment of capacity building of smart ATM and subsequent planning and building.

In 2012, ICAO put forward the authoritative and guiding *Aviation System Block Upgrade (ASBU)* [2], it provides guidance for the development of global air navigation system in the next 15 years through a set of system engineering methods. *ASBU* can be said to be a high-level guidance document for the development planning of ATM. Its new technology application and capacity upgrading plan are basically consistent with the content of capacity building of smart ATM, therefore, the relevant index requirements proposed by *ASBU* can be used as an important guide for the development of smart ATM. Based on the above reasons, this paper constructs an evaluation index system for smart ATM based on *ASBU* and the development status of smart ATM in China, and evaluate the effectiveness of capacity building of smart ATM by using the index system, so as to provide a reference for the effectiveness of capacity building of smart ATM of airports in China.

2 Research Status

The research results of smart ATM index system are still relatively scarce, but there are some achievements in the evaluation index system of civil aviation informatization. Based on the principles of integrity, comparability, wide data coverage, relativity, data characteristics, completeness, time limit, applicability and adaptability, a two-class evaluation index system of civil aviation informatization was constructed [3]. From the perspective of evaluation of ATM information system and equipment, a set of scientific and reasonable evaluation index system guided by the evaluation problem was constructed [4]. Guanzhong et al. [5] thought that it is of great significance to construct a set of relatively complete evaluation index system of ATM operation efficiency to provide scientific decision-making foundation for improving ATM operation efficiency, which has a good reference for the construction of index system of the paper. Yufeng and Qifeng [6] focused on constructing an index system for the control center of smart airport, which provided the basis for the quantitative management of the operation of smart airport. It can be seen from the existing research results that the index system can provide a better reference for the effectiveness evaluation of capacity building of smart ATM, at the same time, the research results of index system for civil aviation informatization, ATM operation efficiency and control center of smart airport can be used as a good reference.

3 Construction of Index System

3.1 Target and Principle

The evaluation index system of smart ATM should be statistical investigation indexes that can reflect the basic situation of new technology application and capacity building of smart ATM in airports across the country, and after quantitative and qualitative analysis, index system can give conclusions on objective phenomena. The selection of evaluation indexes shall conform to the characteristics of the civil aviation industry, the development status of each airport and national conditions in China. The evaluation index system and evaluation results shall conform to the requirements of the four enhancements of ATM and other strategies proposed by CAAC and ATMB.

3.2 Analysis of Hierarchy of ASBU

The task of *ASBU* in the application of new ATM technology and capacity building mainly includes four parts: module, thread, block and performance improvement areas, the overall hierarchy of *ASBU* is shown in Fig. 1.

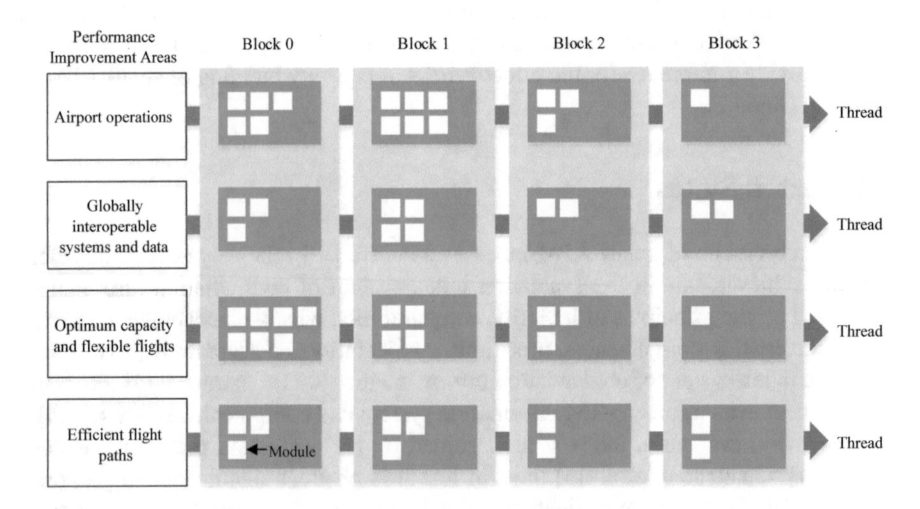

Fig. 1. Hierarchy of *ASBU*

Module: Each small diamond in Fig. 1 is a module, it represents a certain degree of operation improvement. It is a deployable package based on performance improvement and each package provides clear performance improvement index which is supported by procedures, technologies, regulations, standards, etc.

Thread: The horizontal arrow in Fig. 1 runs through a series of interrelated modules, representing the evolution from basic to more advanced performance improvement and across time. Threads directly reflect key aspects of the concept of global air traffic management operations.

The grey rectangular part in Fig. 1 is called the block. The block is composed of several modules, which can promote significant improvement when combined. Based on the five-year interval, the concept of the block is divided into four blocks: Block 0, Block 1, Block 2 and Block 3.

Performance Improvement Areas (PIA): The leftmost column in Fig. 1 shows four performance improvement areas: airport operations, globally interoperable systems and data, optimum capacity and flexible flights, efficient flight paths. Each module in *ASBU* can be classified into a specific PIA.

ASBU, as a systematic capacity improvement plan, can provide important reference for the evaluation index system of smart ATM. Modules, threads, modules and PIAs can also be used as the basic framework of the evaluation index system of smart ATM to evaluate the effectiveness of capacity building of airports in China. However, it is difficult to comprehensively evaluate the capability improvement in the short term due to the fact that the indexes of blocks and PIAs in *ASBU* are long-term continuity index. Based on the current situation of capacity building of smart ATM of airports in China, the relevant indexes in the modules and threads in *ASBU* can be used as the main indexes of evaluation index system of smart ATM. Besides, some of the threads have not been applied in China for many reasons, so the indexes with the actual application or implementation plan are mainly selected.

3.3 Index System

According to the results of above chapters, 17 of the 21 threads of *ASBU* that are closely related to the capacity building of smart ATM in China at the current stage are selected as the first-class index of evaluation index system of smart ATM (four threads that are not included in the indexes are mainly long-term or lack of implementation conditions). Meanwhile, in order to combine the modules of each thread with the capacity building of smart ATM, and in combination with national conditions in China and current construction situation, the technical products represented by the modules with implementation conditions of capacity building under each thread are selected as the second-class index, so as to comprehensively reflect the situation of capacity building of smart ATM of airport in China. The index system constructed according to this idea includes 17 first-class indexes and 32 second-class indexes, as shown in Tables 1 and 2.

4 Case Study

To verify the feasibility and accuracy of the index system proposed in Chap. Improved Spectral Efficiency Based on Double Spatially Sparse in Millimeter Wave MIMO System, 44 airports which are directly provided control services by ATMB in China are selected as research objects, and grading 32 second-class indexes (first-class indexes are selected for the lack of second-class indexes) for 44 airports by means of telephone and field survey. There are three grading standards: 3 points for the airport who have the capacity of the index above mentioned, 1 point for the airport who plan to carry out capacity building of the index above mentioned, 0 point for the airport who do not have and do not plan to carry out capacity building of the index above mentioned temporarily.

Define the parameters as follows:

i, number of airports, 1–44;j, number of indexes, 1–32;T, total grade of airport;S, total grade of indexes;

The total grade of airport i shall be calculated as shown in Formula 1.

$$T(i) = \sum_{j=1}^{32} S(j) \tag{1}$$

Table 1. Index system of smart ATM based on *ASBU*

No.	First-class index	Second-class index
1.	Thread APTA: airport accessibility	PBN operation of airport
2.		Vertical guidance
3.		HUD
4.	Thread RSEQ: runway sequencing	AMAN
5.		DMAN
6.		SMAN
7.	Thread: surface operations	A-SMGCS
8.	Thread ACDM: airport collaborative decision-making	Collaborative decision-making mechanism of ATM and Airport
9.		A-CDM
10.	Thread WAKE: wake turbulence separation	RECAT-CN
11.	Thread FICE: FF-ICE	AIDC
12.	Thread AMET: advanced MET information	Doppler weather radar
13.		Wind profile radar
14.		Numerical weather prediction system
15.		Lidar
16.	Thread DAIM: digital ATM information	Stage1-3
17.	Thread SWIM: system-wide information management	-
18.	Thread FRTO: free-route operations	PBN operation of en-route
19.	Thread NOPS: network operations	Flow management system
20.	Thread ASUR: alternative surveillance	ADS-B
21.		Satellite-based ADS-B
22.		Multilateration systems
23.		Mode-S radar and airborne downlink data link

1408 grades of indexes are confirmed by survey and the total grade of each airport is calculated through the method of Formula 1. It can be concluded that the total grade of each airport represents the level of effectiveness of capacity building of smart ATM in the airport.

According to the above survey and grading statistics, the top 10 airports in total grades are Beijing Capital International Airport (ZBAA, ICAO Code of airport, similarly hereinafter), Guangzhou Baiyun International Airport (ZGGG), Xi'an Xianyang International Airport (ZLXY), Urumqi Diwobao International Airport (ZWWW), Changsha Huanghua International Airport (ZGHA), Chengdu Shuangliu International Airport

Table 2. Index system of smart ATM based on *ASBU*

No.	First-class index	Second-class index
1.	Thread ASEP: airborne separation	Visual approach and visual interval control operation
2.		Equivalent visual separation capability based on airborne ads-b in technology
3.	Thread OPFL: optimum flight levels	ITP operation based on ADS-B IN
4.		CDP operation based on ADS-C
5.	Thread SNET: safety nets	CPAR
6.	Thread TBO: trajectory-based operations	ATC VHF data link
7.		CPDLC and ADS-C
8.		DCL and D-ATIS
9.	Thread CCO/CDO: continuous climb/descent operations	–

(ZUUU), Shanghai Hongqiao International Airport (ZSSS), Kunming Changshui International Airport (ZPPP), Shanghai Pudong International Airport (ZSPD), Zhengzhou Xinzheng International Airport (ZHCC), the total grade of each airport is shown in Table 3.

Table 3. Total grades of airports

No.	Airports	Total grade	No.	Airports	Total grade	No.	Airports	Total grade	No.	Airports	Total grade
1	ZBAA	61	***12***	ZUCK	46	***23***	ZJHK	38	***34***	ZYCC	28
2	ZGGG	61	***13***	ZGNN	45	***24***	ZLLL	38	***35***	ZSWZ	28
3	ZLXY	55	***14***	ZHHH	45	***25***	ZYTX	37	***36***	ZLIC	28
4	ZWWW	54	***15***	ZUGY	44	***26***	ZJSY	37	***37***	ZBSJ	27
5	ZGHA	52	***16***	ZGKL	43	***27***	ZGSD	37	***38***	ZBYN	27
6	ZUUU	52	***17***	ZSQD	41	***28***	ZSHC	36	***39***	ZSNB	26
7	ZSSS	51	***18***	ZYTL	40	***29***	ZSOF	35	***40***	ZLXN	26
8	ZPPP	50	***19***	ZBTJ	39	***30***	ZYHB	34	***41***	ZBHH	24
9	ZSPD	49	***20***	ZGZJ	39	***31***	ZSJN	34	***42***	ZSFZ	24
10	ZHCC	48	***21***	ZSCN	38	***32***	ZSNJ	33	***43***	ZBLA	23
11	ZGSZ	47	***22***	ZSAM	38	***33***	ZGOW	30	***44***	ZWAK	21

It can be seen from Table 3 that Beijing Capital International Airport (ZBAA), Guangzhou Baiyun International Airport (ZGGG) have the most remarkable achievements in capacity building of smart ATM, and Chengdu Shuangliu International Airport (ZUUU), Shanghai Hongqiao International Airport (ZSSS) and other hub airports of China are also in the top position of the effectiveness of capacity building of smart ATM. By qualitative analyzing the relationship between the capacity building of smart ATM and the number of flights in one airport, it can be easily deduced that the relationship should be positively related, that is, the level of effectiveness of the capacity building of smart ATM of each airport should be basically the same as the level of flight support capacity of each airport. According to the *2019 Production Statistics Bulletin of Civil Aviation Airport* issued by CAAC, The passenger throughput (positively related to flight support capacity) ranking of the above airports has a high coincidence with the ranking in Table 3. So it is easy to infer that the total grade ranking of each airport basically keeps the same trend with the flight numbers ranking it undertakes, which conforms to the conclusion of qualitative analysis.

Therefore, the evaluation index system of smart ATM based on *ASBU* proposed in this paper is feasible and relatively accurate, which can provide some evaluation reference for the capacity building of smart ATM of airports in China. In the meantime, according to the comparative analysis of the ranking of specific airports, Hangzhou Xiaoshan International Airport (ZSHC) ranks 10th in terms of passenger throughput nationwide, but only 28th in Table 3, the big difference between the two rankings shows that under the condition of large demand of flight support, the level of effectiveness of capacity building of smart ATM of ZSHC needs to be improved urgently.

5 Conclusion

The objective of the study on the evaluation index system of smart ATM is to explore the evaluation method for the effectiveness of capacity building of smart ATM. Based on the survey and statistical analysis of the development status of smart ATM in China, the paper makes a qualitative and quantitative analysis of the effectiveness of capacity building of smart ATM in airports across the country, so as to provide evaluation and follow-up development reference for the next stage of capacity building of smart ATM in airports. How to optimize the evaluation index system for capacity building path of smart ATM in China, and further improve the evaluation index system based on other domestic and foreign high-level guidance documents, are topics that need further in-depth study in the future.

References

1. Air Traffic Management Bureau of Civil Aviation Administration of China (2018) Action Plan of Four Enhancements of Air Traffic Management. Beijing
2. International Civil Aviation Organization (2012) Aviation System Block Upgrade. Montréal
3. Liyuan S, Yaru D (2014) Study on evaluation index system for informatization construction of china's civil aviation industry. Tianjin Sci Technol 41(08):16–18
4. Ping C (2011) Equipment evaluation and index systems for ATC systems. Command Inf Syst Technol 2(03):10–13

5. Guanzhong L, Minghua H, Qiqian Z (2012) Effectiveness assessment of the ATM operation. J Transp Inf Saf 30(02):99–102
6. Yufeng S, Qingfeng S (2019) Construction of control center index system of smart airport. Sci Tech Inf 01:71–72
7. Haichao G, Jiahui W (2017) Comparative analysis and enlightenment of next generation air traffic management system planning in Euramerica. Command Inf Syst Tech 8(04):83–87
8. Civil Aviation Administration of China (2020) Production Statistics Bulletin of Civil Aviation Airport. http://www.caac.gov.cn/XXGK/XXGK/TJSJ/202003/t20200309_201358.html

Research on the Development Trend of Artificial Intelligence Based on Papers and Patent Analysis—Data Comparison Between China and the United States

Jinyi Feng and Xiangyang Li(✉)

Information Research Center of Military Science, 100142 Beijing, China
lixyljx@163.com

Abstract. With the development of artificial intelligence ushered in the third wave, how to develop artificial intelligence in the future has become the focus of attention. This article first introduces the basic situation of the development of artificial intelligence, and secondly analyzes the development context of artificial intelligence based on the data of the paper. It then analyzes the hotspots of artificial intelligence application according to the patent data, and finally, it summarizes and forecasts the development trend of artificial intelligence related fields.

Keywords: Artificial intelligence · Development trends · Text mining

1 Introduction

Since the term "Artificial Intelligence (AI)" was proposed in 1956, it has experienced two low ebbs [1]. Artificial intelligence has entered a new round of rapid development, ushering in the third wave. Countries around the world have introduced policies related to artificial intelligence, actively launched the field of artificial intelligence, and led and promoted the development of artificial intelligence. China has included the strategy of artificial intelligence in reports on the work of the government for three consecutive years from 2017 to 2019. The US government also attaches great importance to it and has issued multiple documents to maintain the "overall leadership" of US artificial intelligence. The United Kingdom, Japan, France, Canada, and the European Union have also released multiple special strategies or plans for artificial intelligence in the past five years, which has gradually formed a continuous driving force to promote the development of artificial intelligence.

The field of artificial intelligence not only involves basic research but also focuses on technology applications. Papers and patents, as the national technology direct measurement system, can effectively reflect the development status of national scientific research [2]. Therefore, the combined analysis of papers and patent achievements is necessary to fully grasp the development trend in this field. This paper will start with the influence of papers and patent research and development in various countries (mainly China and the United States) and analyze the scientific performance of China and the United States.

Q. Liang et al. (eds.), *Artificial Intelligence in China*, Lecture Notes
in Electrical Engineering 653, https://doi.org/10.1007/978-981-15-8599-9_8

2 Paper Analysis

2.1 Retrieval Ideas and Strategies

The number of papers is the direct expression of the scientific research volume and the strength of basic research, which reflects the size of the research scale to some extent. Therefore, the analysis of the paper is quite necessary. Because artificial intelligence involves a wide range, and there is currently no recognized or relatively mature retrieval mode in the field of artificial intelligence, there are some limitations in using only "artificial intelligence" for retrieval.

Therefore, a total of two Searches have been conducted in determining the relevant subject areas of artificial intelligence. The retrieval range of the second retrieval is determined by extracting the keywords of the literature of the first retrieval results and the word frequency statistics. The retrieval ideas and strategies are shown in Fig. 1.

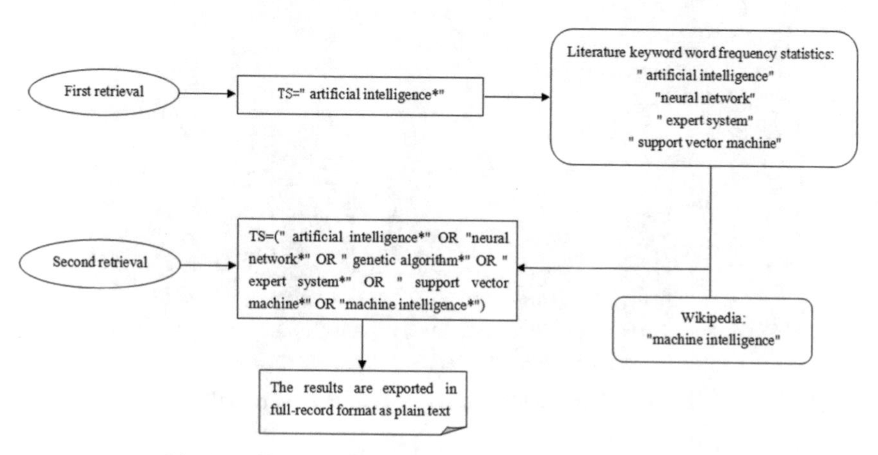

Fig. 1. Schematic diagram of retrieval ideas and strategies

First, the Web of Science Core Collection was selected and the theme retrieved from 2015 to 2019 was "artificial intelligence". After a preliminary search, 19,956 documents were obtained. After excluding conference abstracts, news, and other data, 18,119 documents were obtained. After counting the word frequency of these documents and removing some common words, the words with higher word frequency are: "artificial intelligence", "neural network", "genetic algorithm", "expert system", "support vector machine", etc.

At the same time, the definition of "artificial intelligence" on Wikipedia gives another way to name "artificial intelligence", that is, "machine intelligence". Machine intelligence refers to the intelligence expressed by machines made by people [3, 4].

Combining the high-frequency words from a single search with Wikipedia's interpretation of "artificial intelligence," another name for "machine intelligence," I chose to use "TS = ("artificial intelligence*" OR "neural network*" OR "genetic algorithm*" OR "expert system*" OR "support vector machine*" OR "machine intelligence*")" as a search type in Web of Science Core collection for the second search of documents from 2015 to 2019, Select Artificial for full record data analysis.

2.2 Annual Trends and Analysis of Publication Volume

Paper, as the basic manifestation of scientific research activities, is an important indicator of scientific and technological level in the field of the response. The more papers, the more active the national scientific and technological research. From 2015 to 2019, the total number of papers published globally on artificial intelligence is 117,384, as shown in Fig. 2. From 2015 to 2019, the number of papers published on artificial intelligence around the world increases every year. It shows that the global research interest in artificial intelligence is increasing year by year, and the policies issued by various countries have certain guiding and promoting effects on scientific research.

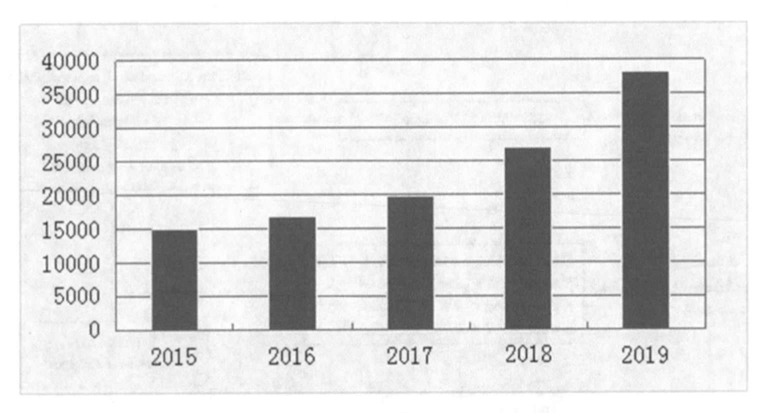

Fig. 2. Total publications global from 2015 to 2019

From the statistics of the number of documents issued by various countries, it can be seen that the total number of documents issued by China in the past five years is much higher than that of other countries, accounting for about 37.7% of the total, and 2.31 times that of the United States (Fig. 3). From 2015 to 2019, China has launched a number of top-level strategies in the field of artificial intelligence. The great importance attached by the government has led to a high investment of research forces. Various universities and research institutions have carried out research on artificial intelligence.

The policy orientation of the country is also influencing the research direction of papers. The above introduced how this search strategy is determined, combined with the analysis of the number of papers can generally draw a conclusion: because the national policy orientation will guide the research direction of the papers, so China's related areas of the volume of papers grew fast, at the same time, the increase in the number of papers will also affect the keyword frequency so that China in the relevant fields of the volume of papers far higher than other countries.

Since China and the United States account for half of the world's papers on artificial intelligence, this paper will analyze the number of papers and highly cited papers in China and the United States.

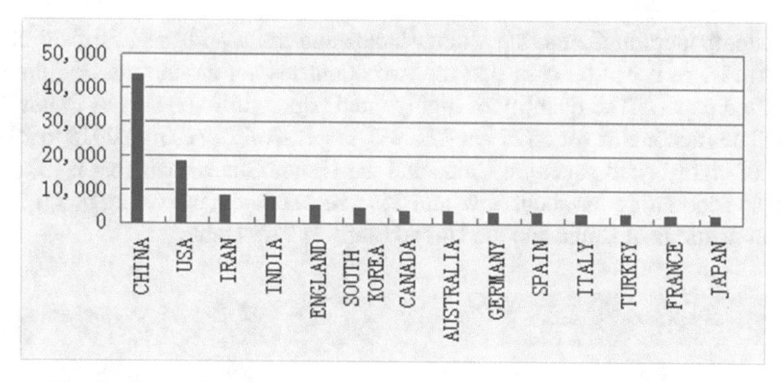

Fig. 3. Comparison of global countries' publications from 2015 to 2019

Figure 4 is a comparison chart of the number of annual publications of China and the United States from 2015 to 2019. It can be seen from the figure that from 2015 to 2017, the number of articles published in both countries increased steadily, but from 2018 to 2019, the number of articles published in both countries, especially in China, surged, which should be related to the strong support of China's policies (China wrote the development of artificial intelligence into its government report for three consecutive years from 2017 to 2019).

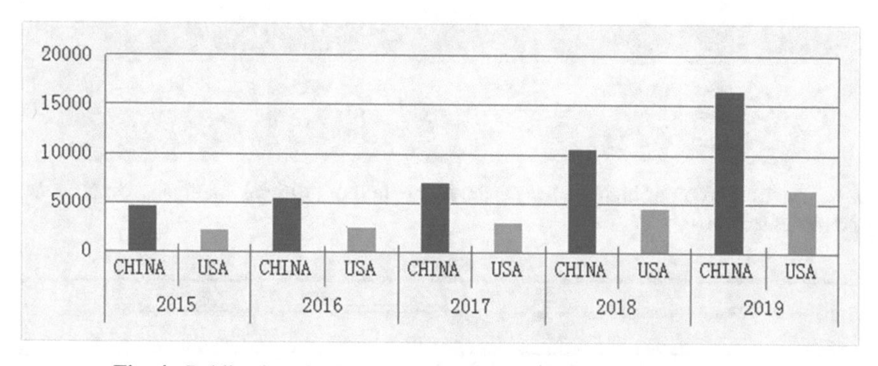

Fig. 4. Publications in China and the United States from 2015 to 2019

At the same time, the number of highly cited papers per year has a similar trend to the total number of papers, and the number of highly cited papers in China is much higher than that in the United States. The high number of cited papers in the top 1% of the number of citations by field and publication year, usually used to measure a country's high level of scientific research capabilities and high-quality scientific research output.

2.3 Analysis of Highly Cited Papers

The statistics of the proportion of top-cited papers in the field of artificial intelligence (top 10) in the past 5 years are shown in Fig. 5, and the statistics of the number of highly cited papers in China, the United States and the world from 2015 to 2019 are shown in Fig. 6, it can be seen that the two countries are among the best amount of highly cited papers. The quantity of highly cited papers in five years in China and the United States accounted for 55.6% and 25.9%, respectively, and from 2015 to 2019, the number of highly cited papers in China and the United States each year is of a similar proportion accounting for about 55% and 25%, respectively, indicating that the quality of the publications in China and the United States is very stable.

Field: Countries/Regions	Record Count	% of 2,090	Bar Chart
CHINA	1,161	55.550 %	
USA	541	25.885 %	
IRAN	201	9.617 %	
AUSTRALIA	163	7.799 %	
ENGLAND	142	6.794 %	
CANADA	117	5.598 %	
INDIA	102	4.880 %	
GERMANY	100	4.785 %	
SOUTH KOREA	98	4.689 %	
MALAYSIA	89	4.258 %	

Fig. 5. The proportion of highly cited papers in the field of artificial intelligence in 2015–2019 by countries (top 10)

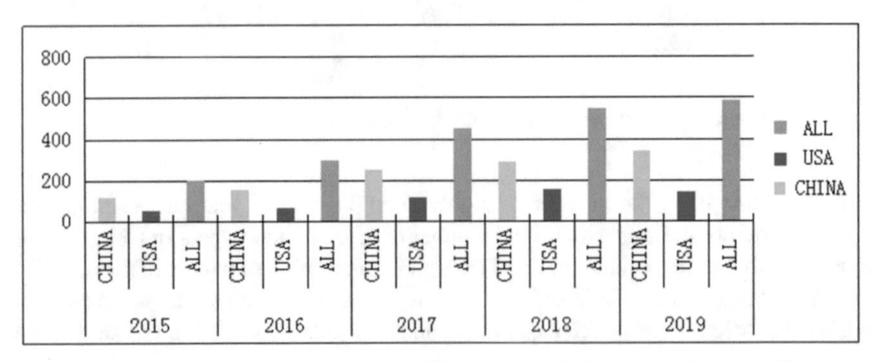

Fig. 6. The number of highly cited papers in China, the United States and the world each year from 2015 to 2019

It is worth mentioning that previous studies have shown that the quality of papers in the United States is high, while the quantity of Chinese papers still lags behind that of the United States to some extent. However, the results of this search in the field related to artificial intelligence show that the quality of Chinese papers is also high, which has a tendency to catch up with the United States.

To sum up, the national policy orientation will play a guiding role in the research direction of papers, which should be one of the reasons why the quantity of papers in related fields in China is much higher than that in other countries. At the same time, the number of cited papers in China and the United States is among the highest, and the quality of published papers in China and the United States is very stable.

3 Patent Analysis

Patent is a kind of monopoly right with exclusive effect, which can monopolize technology in a specific time, space and region. Obtaining exclusive benefits through the exclusive rights of patents is the core and essence of the value of patent rights. Therefore, all countries apply for patents to maintain their dominant position in the technical field. Therefore, the analysis of patent data is quite necessary to judge the advantages of various countries in this field.

This article uses the Derwent Innovations Index for the patent search database, and the patent search scope is consistent with the paper search scope. The search terms are related to "artificial intelligence", "neural network", "genetic algorithm", "expert system", "support vector machine", "machine intelligence", etc. The data analysis is directed to the number of patent authorizations rather than the number of patent applications.

3.1 Basic Trends of Patents

From the basic situation of related patents in the field of artificial intelligence, China and the United States are the two countries that account for about 89% of all patents in the world. In recent years, the number of patents in China has been increasing, and its share of the global patent market has also been increasing. At present, it has surpassed the United States to occupy the top position in the global list (Fig. 7).

3.2 Patent Data Analysis

The number of applications for artificial intelligence patents in China has increased particularly rapidly, mainly for the following reasons: first, the country strongly supports the advancement of artificial intelligence; Second, the enterprises led by Huawei and Ali vigorously develop artificial intelligence technology; Third, universities are constantly increasing their research and development efforts. In contrast to China, the number of patents in the United States declined after 2018, showing a downward trend (Fig. 8).

From the perspective of patent authorization, the application direction is not unitary, but crossed with each other [5]. The top five patents are engineering 99.52%, computer science 95.03%, instrumentation 34.71%, telecommunication 19.75%, general medical medicine 6.66%, from which we can see the main direction of the patent application (Fig. 9).

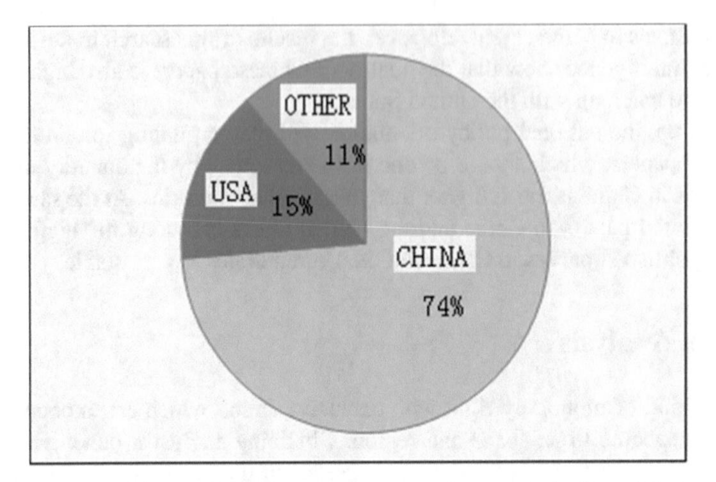

Fig. 7. Comparison of the number of patents between China and the United States and other countries in the world from 2015 to 2019

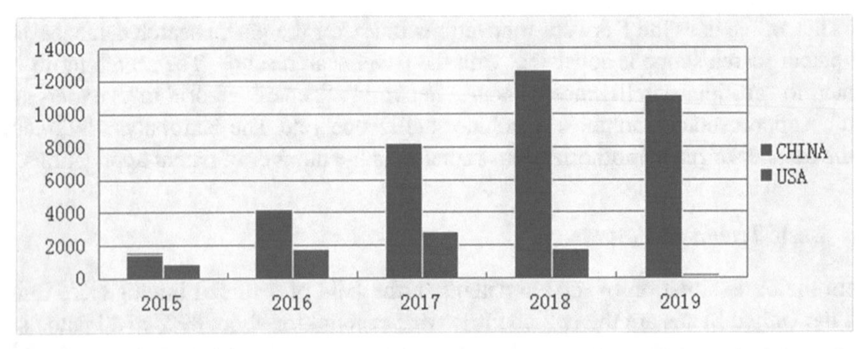

Fig. 8. Comparison of the number of relevant patents between China and the United States from 2015 to 2019

4 Summary and Prospect

This article searches and analyzes the papers and patent data in the field of artificial intelligence. Since the data of China and the United States are at the forefront, a comparative analysis of the publication of papers and patent data in China and the United States is conducted.

After comparison, it is found that the number of articles published globally increased from 2015 to 2019. Among them, the number of papers published in China is far ahead, much higher than other countries, including the United States, which ranks second in the number of articles published. At the same time, the number of highly cited papers in China also exceeded 50% of the total number in the world in five years, and the quality of Chinese papers is on a trend to match that of the United States. The trend of patents is basically synchronized with the development of the paper, which reflects the promotion of national policies on artificial intelligence research.

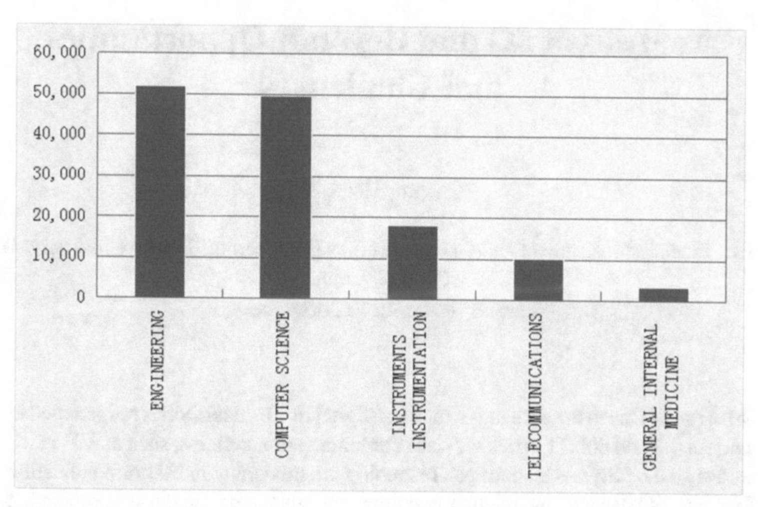

Fig. 9. Statistics of the top five application fields and directions of global artificial intelligence patents from 2015 to 2019

Now countries around the world are trying to seize the opportunity of the third wave of artificial intelligence, therefore, the popularity of artificial intelligence-related research will not subside in a short period of time, and the same of the number of papers issued, The number of patent applications and the amount of authorization should continue to show a trend of increasing in the coming years.

References

1. Bench-Capon TJM, Dunne PE (2007) Argumentation in artificial intelligence. Artif Intell 171(10):619–641
2. Lai KK, Lin CY, Chang YH, Yang MC, Yang WG (2017) A structured approach to explore technological competencies through R&D portfolio of photovoltaic companies by patent statistics. Scientometrics 111(3):1327–1351
3. Aletras N, Baldwin T, Lau JH et al (2017) Evaluating topic representations for exploring document collections. J Assoc Inf Sci Tech 68(1):154–167
4. Wang H, Wu F, Lu W et al (2018) Identifying objective and subjective words via topic modeling. IEEE Trans Neur Netw Learn Syst 29(3):718–730
5. Shin HC, Lu L, Kim L et al (2017) Interleaved text/image deep mining on a large-scale radiology database for automated image interpretation. J Mach Learn Res 17(1):3729–3759

Analysis of 5G and Beyond: Opportunities and Challenges

Liedong Wang(✉)

School of Precision Instrument and Optoelectronics Engineering, Tianjin University, Tianjin 300072, China
wld_0201@tju.edu.cn

Abstract. Due to the demand of modern digital life, critical QoS is not satisfied in current 4G and 802.11 series wireless communication systems, such as 4 K video and virtual reality. As an outbreak technology in modern wireless communication systems, 5G has become popular and come into its commercial use as scheduled according to 3GPP specifications. In this paper, we analyze the state of art of 5G in its three typical scenarios, listed as EMBB, URLLC and MMTC, discuss the main advantages in the current version of specification and then provide the general discussion in challenges of future 5G, for the risk of security, deployment and spectrum resources.

Keywords: 5G · Progress · Opportunities · Challenges

1 Introduction

5G technology is currently the most advanced cellular mobile communication technology. It is based on 4G, 3G and 2G systems. As the main development direction of the new generation of information and communication technology, 5G has become a new type of infrastructure for digital transformation. It has changed the traditional mobile Internet field and developed it into the field of mobile Internet of Things. The service target extends from person-to-person communication to person-to-thing, thing-to-thing communication; its impact will cover all areas of the economy and society and cause profound changes in people's production and lifestyle. The International Telecommunication Union (ITU) defines three major application scenarios for 5G, including enhanced mobile broadband (EMBB), massive machine-type communication (MMTC), and ultra-high-reliability low-latency communication (URLLC). Enhanced mobile broadband is mainly aimed at the current large demand for mobile Internet traffic and hopes to provide a more comfortable and smooth application experience for a large number of mobile Internet users; Massive machine communications is mainly targeted at specific applications. These applications usually require a lot of machine work, such as smart home and environmental monitoring applications. The goals of these applications are sensing and data collection; Ultra-high-reliability and low-latency communications are targeted at applications with low time-latency and high-reliability requirements, such as telemedicine,

Q. Liang et al. (eds.), *Artificial Intelligence in China*, Lecture Notes in Electrical Engineering 653, https://doi.org/10.1007/978-981-15-8599-9_9

autonomous driving, and mission-critical applications. 5G focuses on adopting a flexible system design. In the aspect of architecture, in order to support the three application scenarios, flexible new system design is adopted, such as flexible parameters and framework structure. In the aspect of core technology, in order to meet the needs of high-speed transmission and make the coverage wider, 5G uses a new scheme to complete channel coding, and improves the large-scale antenna technology. In order to ensure low delay and high reliability, 5G adopts high-speed feedback, short frame structure and multilayer data return. In the aspect of frequency band, 5G adopts a new technical scheme to meet the design requirements of medium- and low-frequency band and high-frequency band. In the aspect of core network architecture, 5G is totally different from 4G. 5G core network uses a new service-oriented architecture and a comprehensive service architecture. It can not only realize more flexible and differentiated business deployment, but also supports modular network functions, on-demand calls and required function reconfiguration. At the same time, network slicing technology is a feature of 5G. It can customize multiple logical networks with different functions and features, and flexibly provide services for users with different needs in the industry. Through the construction and deployment of 5G, a new mobile network infrastructure will be built, which will not only drive the growth of the entire industrial chain of the information and communications manufacturing and service industries, but more importantly, it will also accelerate 5G technology and industry, agriculture, transportation, energy, etc. The integration of vertical industries, comprehensive transformation, and upgrading of traditional industries, at the same time, accelerate the development of new technologies, new forms, and new industries, win the initiative in the latest technological revolution and industrial reform, and create great economic value.

In [1], the authors outline the current state of 5G research, describes the standardized trials that have been carried out, and the challenges faced by current deployments. Moreover, the focus of its description is mainly user-centric, not device-centric. In [2], Nokia has successfully established the world's first connection excuse, which is based on Verizon's fifth-generation (V5G) Technology Forum (TF) interface and is considered to be the industry's recognized 5G application specification. In [3], the authors describe in detail some large-scale field trials conducted in China and analyzes their results. These test results are of great significance for selecting 5G key technology components and their combinations, including filtered OFDM (f-OFDM), Sparse code multiple access (SCMA), and massive MIMO for uplink and downlink cellular networks. In [4], the authors outline the features of a fifth-generation (5G) wireless communication system currently being developed for the millimeter-wave band. In [5], the authors provide a comprehensive review of emerging technologies related to 5G that are designed to support the exponential traffic growth of the Internet of Things. In [6], the authors reviewed the latest technology of 5G network slicing and creatively proposed an overall framework for researchers to discuss existing work and carry out subsequent research. In [7], the authors analyzed and discussed how flexible wireless access networks can be implemented. The result is to rely on network slicing and its impact on 5G mobile network design. In [8], the authors discuss several current mobile network design options, features, and technical challenges, and analyze potential research topics and challenges in the future. In [9], the authors describe in detail the data consistency issues that arise when people use mobile applications for cloud-based collaborative work. In [10], the authors conducted a comprehensive investigation of the birth of NOMA, recent developments and future research directions. In [11], the authors outline the basic requirements for 5G channel model, and conduct an extensive review of their channel measurements and theoretical models.

In this article, I discussed the advantages of 5G in the first part, described the opportunities and challenges of 5G in the second part, and finally summarized and prospected the paper.

2 Typical Advantages of 5G

5G includes three typical advantages: enhanced mobile broadband (EMBB), ultra-high-reliability low-latency communication (URLLC), and massive machine-type communication (MMTC). EMBB targets mobile multimedia, interactive high-definition video, cloud office, AR/VR applications, etc. URLLC targets smart grid, environmental monitoring, smart home, vehicle management and other businesses. MMTC targets smart banking, smart securities, smart transportation, mobile healthcare and other businesses. A brief description follows (Fig. 1).

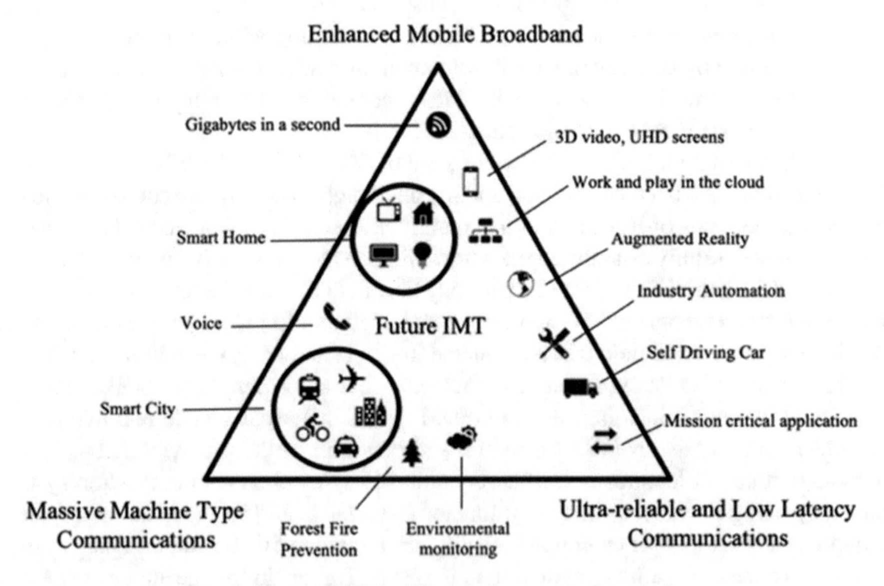

Fig. 1 Future IMT for 5G with three typical scenarios

2.1 EMBB

EMBB mainly addresses the shortcomings of the large bandwidth and high-speed requirements that cannot be met by 4G and achieves a single user 10 Gbps connection rate, which can bring users a more extreme experience and meet requirements such as AR and online HD conferences. EMBB should meet three requirements: high speed and large bandwidth, ultra-high clock accuracy, and flattening, virtualization, and programmable network architecture. 5G can use available spectrum in the centimeter- and millimeter-wave bands (28–300 GHz) to meet the needs of large bandwidth and high speed. This application scenario covers multiple situations, including hotspots and wide-area coverage. For hotspots, 5G needs to be able to support higher user density, higher traffic, and lowest latency. Under the assumption that 3.4G–3.5G spectrum resources are obtained, the bandwidth of 5G low-frequency base stations can reach 100 MHz, and high-frequency base stations can reach 20 Gbps. At the same time, the EMBB delay requirement is 10 ms, and the bearer network requires a two-way delay less than 4 ms. These applications can solve many social problems, for example, in concerts, student dormitories and other densely populated places, to ensure that all users can get connect to the Internet, reduce the situation of unreachable in emergency situations; to ensure that users can access the network at any time to achieve the lowest delay; help the rapid development of AR/VR applications. For wide-area situations, it is necessary to meet seamless seams and high mobility. Compared with the data rates provided today, under normal circumstances, the user data upload and download rates must be greatly increased, saving users time.

2.2 URLLC

URLLC mainly reflects the communication needs between things and is characterized by high reliability and low latency, extremely high availability, and has strict requirements on it. There are two main technical characteristics of URLLC. The first one is in terms of delay and reliability. Compared with the previous cellular mobile communication technology, 5G URLLC has been greatly improved. The second is to achieve a user plane delay of 0.5 ms between the base station and the terminal. At present, URLLC's reliability index is: within 1 ms of user plane delay, the reliability of transmitting a 32-byte packet at a time is 99.999%. It is reported that when the 5G standard is finally formulated, the above reliability index is expected to achieve further improvement from 99.999%. In addition, if the transmission delay is allowed, 5G URLLC can also adopt the retransmission mechanism to further improve the success rate. It can meet the applications of many scenarios: intelligent transportation systems, smart grids, industrial control applications, remote manufacturing, telemedicine surgery, etc. For example, there are a limited number of outstanding surgeons in the world, unable to go to various places for surgery, which requires the application of remote surgery, which requires extremely low delay to ensure smooth operation. URLLC can provide latency as low as 10 ms. In terms of driverless driving, the transmission delay is even lower, which can reach 1 ms to ensure that the vehicle can smoothly avoid obstacles and avoid traffic accidents.

2.3 MMTC

MMTC is mainly an information exchange between people and things. The 5G low-power, large-connection, low-latency, and high-reliability scenarios are mainly geared toward IoT services. Its application goals should have three characteristics, large-scale connection, insensitive latency and small bandwidth. As a new extended application scenario of 5G, the focus is on solving the current problems of mobile communications and enabling it to support the Internet of Things and vertical industry applications. Its application scenarios are mainly smart cities, environmental monitoring, intelligent agriculture, forest fire prevention, etc. that target sensing and data collection. With features such as power consumption and massive connections, MMTC deployments consist of a large number of devices with relatively few (or relatively high) non-latency sensitive data, and typically require lower cost and longer battery life. Such terminals have a wide distribution range and a large number. They not only require the network to support more than 100 billion connections, but also to meet the density requirements of 1 million/km^2 connections, while also ensuring low power consumption and low cost of terminals. For example, millions of shared bicycles are deployed around the city, but only 2G networks are used for communication. 5G can achieve the purpose of mass communication and increase the monitoring of individual bicycles.

3 Opportunities and Challenges

With the acceleration of the 5G standardization process, the basic enhancements brought by 3GPP Release 17 can improve network capacity, signal coverage, lower latency, device power and mobility. Globally, the number of devices and network connections are growing faster than the population and Internet users, which means that the average number of devices and connections per household and per capita has increased significantly. Therefore, the most critical research area is further research and promotion of 5G NR massive MIMO technology, support 5G systems to provide better performance and efficiency (for example, higher spectral efficiency), will also focus on beam management in the millimeter-wave band, and multicast operation, higher mobility and other improvements. Another key research area is technology to improve the coverage of low-frequency and millimeter-wave frequency bands. The current Cisco Visual Network Index (VNI) predicts that from 2017 to 2022, global IP traffic will increase nearly three times. The improvement is essential; the technology closely related to coverage is integrated access and backhaul (IAB) technology, which may make the mesh network topology more dynamic. Release 17 will also explore more mechanisms, such as improving the energy-saving effect of equipment, enhancing the dual connection combination of NR + LTE and NR + NR.

These improvements bring new opportunities and challenges.

3.1 Opportunities

There are two main opportunities for 5G.

First, a wider application space. Compared with 3G and 4G, 5G will greatly increase the data transmission rate and reduce the delay. It is reported that 5G is expected to significantly reduce latency to less than 1 ms, which is suitable for time-sensitive data services. At the same time, thanks to its high transmission speed, 5G networks can provide many high-speed broadband services. EMBB will enable high-speed mobile broadband in congested areas, enable consumers to enjoy high-speed streaming capabilities indoors, on screens, and mobile devices, and will promote corporate collaboration services and development. At the same time, to address the current lack of fiber connectivity in some homes, operators are considering EMBB as a last-mile solution.

It is expected that in order to promote the development of smart cities and the Internet of Things, 5G will also form a sensor network by deploying a large number of ultra-low-power sensors in urban and rural areas. 5G's built-in security makes it suitable for public safety events and some mission-critical tasks, such as smart grids, security services, utilities and healthcare. The low-latency and safety features of 5G will promote the development of intelligent transportation systems to achieve the purpose of autonomous driving by enabling intelligent vehicles to communicate with each other. For example, a self-driving car running through a cloud-based autonomous driving system must be able to make timely instructions when receiving a message, control the car to stop, accelerate, or turn. Any network delay or signal loss due to signal coverage will prevent the exchange of messages. This can have disastrous consequences.

Second, support the informatization of national defense and the army. First, 5G communications technology will likely enable the military to have dedicated frequencies. As 5G communication technology not only makes full use of existing communication resources, it is also expanding to millimeter-wave communication resources, thereby making it possible for the military to have special frequencies. This will effectively solve the problems of overlapping and sharing of frequency bands and mutual interference between military mobile communication systems and civil mobile communication systems. Secondly, 5G communication technology will promote the interconnection of battlefield global weapon platforms. 5G communication technology can improve the frequency application efficiency without increasing the base station density. The application of this technology will be able to interconnect all weapon platforms equipped with 5G communication modules in the battlefield area. Based on this, various types of troops on the battlefield can not only achieve interconnection and interoperability of battlefield information terminals, but also realize barrier-free communication. Finally, 5G communication technology will achieve deep integration of battlefield information networks. 5G communication technology can deeply integrate various heterogeneous information communication networks applied to the battlefield to form a compatible high-speed information network to achieve more efficient combat operations. Therefore, 5G communication technology will provide the military with a wide-area coverage, high-speed transmission, and strong compatibility of air-ground integrated information and communication networks, thereby greatly improving the information support and support capabilities of the battlefield.

3.2 Challenges

The main challenges of 5G are three points.

First, the issue of network security. When 5G combines the characteristics of "low latency + high reliability" with the key vertical industries such as connected cars, telemedicine, industrial automation, and smart grids, the target of cyber-attacks and rights will further expand; when 5G supports "big connection services," When more critical infrastructure and important application architectures are placed on them, these high-value targets may attract greater attack power-national hackers enter; when 5G breaks the network boundary, the network world and the physical world are further integrated At times, attacks against virtual worlds will all become physical damage, and the impact of cyber-attacks will increase exponentially.

Second, 5G network deployment issues. In some countries, the administrative and financial obligations imposed on operators by the policies of relevant departments and local authorities have hampered investment and slowed down the development of micro base stations. Due to the poor availability of fiber networks in many cities, the biggest problem that operators need to solve is to deploy fiber backhaul networks for micro base stations to support transmission at high data rates and lower time delays. If deploying fiber backhaul is extremely cost-effective, operators should consider wireless backhaul technology. In this case, in addition to fiber optics, other combinations of wireless technologies should be considered, including PMP application models, millimeter-wave RF systems, and satellite communications.

Third, 5G spectrum planning challenges. Coordination between international organizations, regional telecommunication organizations and national regulators is required to allocate and determine globally unified spectrum. This is one of the biggest challenges NRA faces when successfully deploying 5G networks. Traditionally, NRA only allocates spectrum to mobile operators. However, as demand continues to grow, sharing can be a way to improve the effective use of existing spectrum. In addition, NRA also needs to further consider the licensing and usage models of 5G spectrum, especially above 24 GHz. The spectrum above 24 GHz is relatively easier to obtain, so it does not have a strong scarcity, which will also affect business models and spectrum auctions.

4 Conclusion

In conclusion, this article describes the development levels of three typical cases of 5G: enhanced mobile broadband (EMBB), massive machine-type communication (MMTC), and ultra-high-reliability low-latency communication (URLLC), and a thorough investigation of their strengths in the application area was conducted. In addition, this article also investigates the current opportunities and challenges of 5G, states the important opportunities of 5G in the application space and the construction of national defense and military information, and also discusses the risks of security, deployment and spectrum resources of 5G.

References

1. Shafi M, Molisch AF, Smith PJ, Haustein T, Zhu P, De Silva P et al (2017) 5G: a tutorial overview of standards, trials, challenges, deployment and practice. IEEE J Sel Areas Commun 1–1
2. Patzold M (2017) 5G developments are in full swing [mobile radio]. IEEE Veh Technol Mag 12(2):4–12
3. Dong L, Zhao H, Chen Y, Chen D, Wang T, Lu L et al (2017). Introduction on imt-2020 5G trials in China. IEEE J Sel Areas Commun 1–1
4. Rappaport TS, Xing Y, Maccartney GR, Molisch AF, Mellios E, Zhang J (2017) Overview of millimeter wave communications for fifth-generation (5G) wireless networks-with a focus on propagation models. IEEE Trans Antenn Propag 1–1

5. Fan W, Carton I, Kyosti P, Karstensen A, Jamsa T, Gustafsson M et al (2017) A step toward 5G in 2020: low-cost ota performance evaluation of massive mimo base stations. IEEE Antennas Propag Mag 59(1):38–47
6. Foukas Xenofon, Patounas Georgios, Elmokashfi Ahmed, Marina Mahesh K (2017) Network slicing in 5G: survey and challenges. IEEE Commun Mag 55(5):94–100
7. Rost P, Mannweiler C, Michalopoulos DS, Sartori C, Sciancalepore V, Sastry N et al (2017) Network slicing to enable scalability and flexibility in 5G mobile networks. IEEE Commun Mag 55(5):72–79
8. Alsharif Mohammed H, Nordin Rosdiadee (2017) Evolution towards fifth generation (5G) wireless networks: current trends and challenges in the deployment of millimetre wave, massive mimo, and small cells. Telecommun Syst 64(4):617–637
9. Mavromoustakis CX, Mastorakis G, Dobre C (2017) Advances in mobile cloud computing and big data in the 5G era
10. Dai L, Wang B, Ding Z, Wang Z, Chen S, Hanzo L (2018) A survey of non-orthogonal multiple access for 5G. IEEE Commun Surv Tutor 1–1
11. Wang CX, Bian J, Sun J, Zhang W, Zhang M (2018) A survey of 5G channel measurements and models. IEEE Commun Surv Tut 1–1

Prediction of PM2.5 Concentration Based on Support Vector Machine and Ridge

Haocun Zhang(✉)

Nanjing University of Information Science & Technology, No. 219 Ningliu Road, Nanjing City, Jiangsu Province, China
zhang58_00@qq.com

Abstract. Air Pollution has a great impact on people's life and health. The main reason is that industrial development causes a great deal of burning of fossil fuels, which causes serious pollution to the environment and releases chemicals into the atmosphere, to human life and production and physical health have a serious impact. Serious air pollution has also occurred in beijing-tianjin-hebei and its surrounding areas [1]. If we don't deal with the environmental deterioration caused by the economic development in time, the air pollution problem will further restrict the economic development, so in order to create a more suitable environment for people to live in, it is particularly urgent to study the factors affecting air pollution and to propose feasible measures to improve air quality [2]. In most Chinese cities, PM2.5 is the main pollutant, causing smog, reducing visibility and affecting traffic safety. It also enters the Alveoli through the airways, endangering human health. High levels of PM2.5 can also lead to serious social problems that affect people's lives, health is also seriously affected. Therefore, the research on air pollution prediction can help the government and the public to take timely measures to prevent and control air pollution. Based on the correlation between PM2.5 and other components in the atmosphere, the concentration prediction model of PM2.5 was established.

Keywords: PM2.5 · Concentration prediction · SVM

1 Research Background and Significance

Environmental Quality plays a vital role in the sustainable development of society. Haze weather is more and more frequent and large-scale occurrence, resulting in more and more serious human health and environmental pollution [3]. PM2.5 generally refers to fine particulate matter. It refers to pollutant particles in ambient air with an equivalent diameter of less than 2.5 μm, which can be suspended in the air for a long time. It has an important impact on air quality and visibility, etc., easy to attach toxic and harmful substances (for example, heavy metals, microorganisms, etc.), and in the atmosphere of the long residence time, transport distance, so the impact on Human Health and atmospheric environmental quality more. Long term exposure to particles can cause cardiovascular disease and respiratory disease, as well as lung cancer.

Q. Liang et al. (eds.), *Artificial Intelligence in China*, Lecture Notes
in Electrical Engineering 653, https://doi.org/10.1007/978-981-15-8599-9_10

The WHO's latest air quality guidelines, published in 2005, set strict limits, particularly for concentrations of particulates in the atmosphere, with an annual average PM2.5 concentration of 10 micrograms per cubic metre ($\mu g/m^3$), the 24 h average concentration was 25 $\mu g/m^3$.

Beijing's air quality is an important environmental issue that society is concerned about nowadays. An objective evaluation of its changing trend and influencing factors is not only conducive to a correct understanding of Beijing's air quality, it can also provide the basis for the effective forecast, control and treatment of air pollution. In this paper, PM2.5 and some other historical atmospheric data from 2013 to 2017 in Beijing, China are selected for analysis and prediction.

In the past 30 years, the history of air pollution in Beijing has experienced the following stages: The overall air quality declined from 1983 to 1998, the air quality improved from 1999 to 2008, and the air quality has once again shown a downward trend since 2009. The continuous smog weather in Beijing in autumn 2011 resulted in reduced visibility and serious air pollution, further arousing the public's great attention to the air quality monitoring, especially the PM2.5 pollution situation.

If early warning systems are established in advance and effectively, their impact on the human health situation will be significantly reduced.

2 Related Work

The atmospheric environment is very complex. The concentration of pollutants in the atmosphere is not only related to the source of emission, but also affected by local meteorological conditions. Many scholars use the regression model of pollutant concentration and related meteorological factors to predict and analyze the air quality. In recent years, using air pollution monitoring data, researchers have developed a number of different methods to estimate and predict the levels of different air pollutants, which are very useful for air pollution forecasting and early warning systems. However, despite the importance of research and experience in the analysis and prediction of atmospheric pollutants, they remain incomplete.

Therefore, this paper studies and puts forward a new type of air pollutant prediction system, which can not only effectively meet the requirements of air quality regulation to reduce and control the emission of air pollutants, it can also give the public effective warning and warning before the coming of dangerous weather such as heavy smog.

3 Experiment

The training set is used to train the model, which contains 27,000 lines of data. The test set is used to see the performance of the model, which contains 8,000 lines of data.

The data correlation was then analyzed, including year, month, day, hour, PM2.5, PM10, SO_2, O_3, NO_2, Co, TEMP, PRES, DEWP. The correlation between PM 2.5 and "month", "day" and "hour" is not high. Other variables, such as "RAIN", "o 3" and "WSPM", were not significantly correlated with PM2.5, so "month", "day", "hour", "RAIN", "o 3" and "WSPM" were rounded off.

In addition, other data (“PM10”, “SO_2”, “NO_2”, “Co”, “TEMP” and so on) and PM2.5 have some relevance.

The previous understanding of the data type can be found, SO_2, NO_2 equivalence is missing, here extract feature engineering, remove missing values and unwanted columns, the data into quantitative data.

The next step is to use standard scaler normalization (also known as deviation normalization, which is a linear transformation of the original data so that the result maps to the [0, 1] interval).

Transform the sequence $x_1, x_2, x_3 \ldots, x_n$:

$$y_i = \frac{x_i - \min_{1 \le j \le n}\{x_j\}}{\max_{1 \le j \le n}\{x_j\} - \min_{1 \le j \le n}\{x_j\}}$$

Then the new sequence is dimensionless.

Import the linear model below:

Suppose the function is as follows:

$$h(x) = \theta_0 + \theta_1 x_1 + \theta_2 x_2 = \sum_{i=0}^{2} \theta_i x_i = h_\theta(x)$$

where, represents a parameter. For General Problems, the formula is as follows:

$$h_\theta(x) = \sum_{i=0}^{n} \theta_i x_i = \theta^{\mathrm{T}} x$$

where X is the vector and N is the length of X. Thus, we can define the function to be optimized:

$$J(\theta) = \frac{1}{2} \sum_{i=1}^{m} \left(h_\theta\left(x^{(i)}\right) - y^{(i)} \right)^2$$

The best parameter for the fitting training is the parameter value that minimizes the function.

Here’s how to use the gradient descent:

$$\theta_j := \theta_j - \alpha \frac{\partial}{\partial \theta_j} J(\theta)$$

When there is only one training example, the partial derivative is calculated as follows:

$$\begin{aligned} \frac{\partial}{\partial \theta_j} J(\theta) &= \frac{\partial}{\partial \theta_j} \frac{1}{2} (h_\theta(x) - y)^2 = (h_\theta(x) - y) \frac{\partial}{\partial \theta_j} (h_\theta(x) - y) \\ &= (h_\theta(x) - y) \frac{\partial}{\partial \theta_j} \sum_{i=0}^{n} \theta_i x_i = (h_\theta(x) - y) x_j \end{aligned}$$

$$\theta_j := \theta_j - \alpha (h_\theta(x) - y) x_j$$

This formula is for cases where there is only one training instance, also known as a least squares.

Taking into account all M training instances, the updated change rule is:

$$\theta_j := \theta_j - \alpha \sum_{i=0}^{m} \left(h_\theta\left(x^{(i)}\right) - y^{(i)} \right) x_j^{(i)}$$

Running this rule until it converges is the batch gradient descent algorithm [15] (the "a" in the rule is the learning rate, which needs to be adjusted in practice. Too Small a value will lead to many iterations before it can converge, and too large a value will lead to oscillation over the best point [5].

For the solution of the above formula, when the amount of data is large, each iteration of the data to be traversed once, this will make the operation become slow. The solution is as follows:

Repeat Until Converge{
For i=1 to m{

$$\theta_j := \theta_j - \alpha \left(h_\theta\left(x^{(i)}\right) - y^{(i)} \right) x_j^{(i)} \quad \text{(for every j)}$$

This method is called incremental gradient descent or Stochastic gradient descent, which means that when updating a parameter, instead of traversing the entire data set, one instance is sufficient.

print(linear.score(train_data_X,train_data_Y))

print(linear.score(test_data_X,test_data_Y))

Here, the predicted Coefficient R^2 is returned by score (self, x, y, sample_weight = None) [15]

$$R^2 = (1 - u/v)$$

u = ((y_true − y_pred) ** 2).sum()

v = ((y_true − y_true.mean()) ** 2).sum()

The best possible score is 1, which returns 0 when a model always outputs the expected Y, regardless of the characteristic value entered

The output of the training set and the test set is used to evaluate the prediction effect and the result of the reaction training.

The next step is to find the mean square error (the average of the sum of the squares of distances from the true values, the smaller the mean square error is, the better the model is at fitting the experimental data). The calculation formula is as follows:

$$\text{MSE} = \sum_{i=1}^{n} \frac{1}{n} (f(x_i) - y_i)^2$$

And drew the test set forecast value and the actual value contrast chart, draws the graph as follows (Fig. 1):

The main idea of this study is summarized as follows: By comparing several different models including linear regression, KNNRegressor, SVR support vector regression, Ridge, Lasso, mlpressor, Multilayer perceptron, decision tree, XGB, AdaBoost, Gradient, Bagging, the model with better prediction performance was chosen.

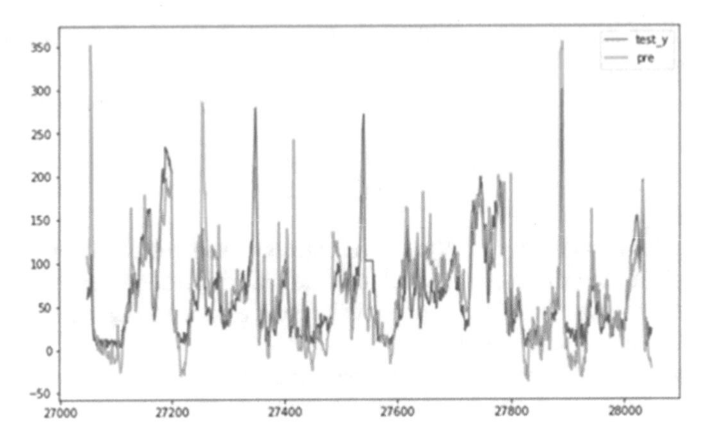

Fig. 1. Test set prediction and actual values

To train and evaluate the MODEL, the accuracy of prediction is an important standard of prediction model evaluation. According to the output of "training set score" and "test set score" to judge the prediction accuracy of the model, here choose, "training set score" and "test set score" are very high and not much difference.

Ridge, SVR meets the evaluation requirements, so they are chosen as the prediction model.

4 Ridge

For some matrices, a small change in an element of a matrix will cause a large error in the final calculation. This matrix is called a "sick Matrix". Sometimes incorrect calculation method can also make a normal matrix in the operation of the ill-conditioned. In the case of Gaussian elimination, if the elements on the main element (that is, the elements on the diagonal) are very small, the calculation will show ill-conditioned features.

The least squares commonly used in regression analysis is an unbiased estimate. For A well-posed problem, x is usually rank-full

$$X\theta = y$$

using least squares, the loss function is defined as the square of the residuals to minimize the loss function

$$\|X\theta - y\|^2$$

The above optimization problem can be solved by gradient descent or directly by the following formula

$$\theta = \left(X^{\mathrm{T}}X\right)^{-1}X^{\mathrm{T}}y$$

When X is not a full rank, or when the Linear independence between certain columns is large, $X^{\mathrm{T}}X$, The determinant of x is close to zero, $X^{\mathrm{T}}X$, Approaching singularity, the above problem becomes an ill-posed problem, at which point the computation

$$\left(X^{\mathrm{T}}X\right)^{-1}$$

The time error can be very large, traditional least squares lack of stability and reliability.

In order to solve the above problem, we need to transform ill-posed problem into well-posed problem: We add a regularization term to the loss function and change it to

$$\|X\theta - y\|^2 + \|\Gamma\theta\|^2$$

Among them, we define [3]

$$\Gamma = \alpha I$$

And so:

$$\theta(\alpha) = \left(X^{\mathrm{T}}X + \alpha I\right)^{-1} X^{\mathrm{T}}y$$

In the upper form, I is the identity Matrix.

The absolute values of each element tend to decrease as they increase, and their deviations from the correct values become larger and larger. When it goes to infinity, it goes to zero. Among them, with the change and change of track, called Ridge track. The actual calculation can choose a lot of values, make a ridge trace map, to see which value of this map when the time to take a stable, that will determine the value.

Ridge regression is a complement to least squares regression, which loses its unbiasedness in exchange for high numerical stability and thus higher computational accuracy.

5 SVM

The use of hard margin SVM IN LINEAR unclassifiable problems will produce classification error, so a new optimization problem can be constructed by introducing loss function based on maximizing margin. SVM uses the hinge loss function and adopts the optimization problem of SVM with hard boundary. The optimization problem of soft margin SVM is as follows:

$$\min_{\boldsymbol{\omega},\boldsymbol{b}} \frac{1}{2}\omega^2 + C\sum_{i=1}^{N} L_i, L_i = \left[0, 1 - y_i\left(\omega^{\mathrm{T}}X_i + b\right)\right]$$

$$\text{s.t.}\quad y_i\left(\omega^{\mathrm{T}}X_i + b\right) \geq 1 - L_i, L_i \geq 0$$

The above expression shows that the soft margin SVM is an L_2 regularized classifier, in which the hinge loss function is expressed. Using the relaxation variable: After treating the value of the loss function of the hinge in segments, the upper formula can be reduced to:

$$\min_{\omega,b} \frac{1}{2}\|\omega\|^2 + c\sum_{i=1}^{N}\xi_i$$

$$y_i\left(\omega^{\mathrm{T}}X_i + b\right) \geq 1 - \xi_i, \xi_i \geq 0$$

The duality of the optimization problem is usually used to solve the above-mentioned soft-margin SVM:

The optimization problem of SVM with soft margin is defined as primal problem by Lagrange multiplier:

$$\alpha = \{\alpha_1, \ldots, \alpha_N\}, \mu = \{\mu_1, \ldots, \mu_N\}$$

To obtain their Lagrangian:

$$\mathcal{L}(\omega, b, \xi, \alpha, \mu) = \frac{1}{2}\omega^2 + C\sum_{i=1}^{N}\xi_i + \sum_{i=1}^{N}\alpha_i\left[1 - \xi_i - y_{i(\omega^{\mathrm{T}}X_i+b)} - \sum_{i=1}^{N}\mu_i\xi_i\right]$$

Let the partial derivative of the Lagrangian to the optimization objective be 0, and a series of expressions containing the Lagrange multiplier can be obtained:

$$\frac{\partial\mathcal{L}}{\partial\omega} = 0 \Rightarrow \omega = \sum_{i=1}^{N}\alpha_i y_i X_i$$

$$\frac{\partial\mathcal{L}}{\partial b} = 0 \Rightarrow \sum_{i=1}^{N}\alpha_i y_i = 0.$$

Bring it into the dual problem of the Lagrangian problem:

$$\max_{\alpha}\sum_{i=1}^{N}\alpha_i - \frac{1}{2}\sum_{i=1}^{N}\sum_{j=1}^{N}\left[\alpha_i y_i (X_i)^{\mathrm{T}}\left(X_j\right)y_j\alpha_j\right]$$

$$\text{s.t.}\sum_{i=1}^{N}\alpha_i y_i = 0, 0 \leq \alpha_i \leq C$$

The constraint condition of the dual problem contains inequality, so the local optimal condition of the dual problem is that the Lagrange multiplier satisfies the Karush-Kuhn-Tucker condition (KKT):

$$\alpha_i \geq 0, \mu_i \geq 0$$

$$\xi_i \geq 0, \mu_i\xi_i = 0$$

$$y_i\left(\omega^{\mathrm{T}}X_i + b\right) - 1 + L_i \geq 0$$

$$\alpha_i\left[y_i\left(\omega^{\mathrm{T}}X_i + b\right) - 1 + L_i\right] = 0$$

From the above KKT conditions, for any sample (X_i, y_i), there is always $\alpha_i = 0$ OR $y_i(\omega^{\mathrm{T}}X_i + b) = 1 - \xi_i$, for the former, the sample does not affect the decision boundary $\omega^{\mathrm{T}}X_i + b = 0$, for the latter, the sample satisfies the

$$y_i\left(\omega^{\mathrm{T}}X_i + b\right) = 1 - \xi_i$$

Which means that it is on the edge of the interval $(\alpha_i < C)$, within the interval $(\alpha_i = C)$, or misclassified $(\alpha_i > C)$, that is, the sample is a support vector. It can be seen that the decision boundary of SVM with soft margin is only related to support vector, and the SVM has sparsity by using hinge loss function [1].

6 Conclusion

Prediction of atmospheric pollutants is one of the fastest growing research fields in environmental and meteorological sciences in recent years. It is not only of great significance in demonstrating the theoretical problem of how Human Activities Affect Air quality, it also has important practical value for urban environmental management, pollution control, environmental planning, urban construction and public health [2]. This paper presents a prediction model based on machine learning to analyze and predict PM2.5 in Beijing. The empirical results show that SVR and Ridge have higher prediction accuracy and stability than other models.

References

1. Friedman J, Hastie T, Tibshirani R (2001) The elements of statistical learning (Chap. 12). vol 1, No 10. Springer, New York, NY, pp 417–438
2. Dong Y, Li Y, Du T, Tang X, Ma Z (2016) Comparative analysis of air environmental quality during 2014–2015 in Guanzhong Area. J Environ Eng Tech 6(6):532–538
3. Li J, Jiang W, Wen Y (2018) Analysis of annual water use time-series and influencing factors in China. In: IOP conference series: materials science and engineering
4. Wang L, Yan PB (2014) Research on prediction of air quality index based on NARX and SVM. Appl Mech Mater
5. Mariyam S, Osman A, Ramadhe C (2013) Weight changes for learning mechanisms in two-term back-propagation network (Chap. 3). IntechOpen

Helmet Detection Based on an Enhanced YOLO Method

Weizhou Zheng[1(✉)] and Jiayi Chang[2,3]

[1] Beijing-Dublin International College, Beijing University of Technology, Beijing, China
zhengweizhou0321@163.com

[2] Department of Information, Beijing City University, Beijing, China
changjiayi1225@163.com

[3] Institute of Automation, Chinese Academy of Science, Beijing, China

Abstract. Wearing a safety helmet is one of the most important requirements of the construction site and is essential to the safety of workers. Computer vision can be applied to identifying the helmet worn by the workers as external supervision. In this paper, helmet detection algorithms based on YOLO models with a special data set where the training set consists of simple helmet pictures but the test set holds complicated real construction sites are studied. In view of actual situations of the construction site, some pretreatment methods for the training set are tested to enhance the performance. The result shows that with proper pretreatment, the YOLOv3 model with a simple training set can have good performance in detecting helmet in complicated construction sites.

Keywords: Helmet · YOLO · Object detector · Convolutional neural network

1 Introduction

Safety is always an important topic in construction sites since construction sites are often very complex and full of dangers. Workers may get injured by falling objects on the construction site at any time. To prevent it, wearing a safety helmet is a very effective safety measure. However, many workers would choose to not wear one for multiple reasons [1]. Therefore, the external supervision of helmet wearing is very necessary [2]. In the early days, most of the construction sites were equipped with full-time safety personnel to supervise whether the workers wore safety equipment. However, this method is time-consuming as well as laborious, and the effectiveness of supervision cannot be guaranteed.

In recent years, scholars have done a lot of innovative research work in helmet automatic recognition methods. Dong [3] tried to put the pressure sensor in the helmet and uses the positioning system and pressure information to determine whether a worker wears the helmet. Khairullah et al. [4] combined Bluetooth technology with sensor networks. But Bluetooth devices need to be recharged regularly, so it cannot be used for a long time. For these methods, it is very costly to build such a system, since every helmet should be customized. Besides, their accuracy is not satisfied.

Q. Liang et al. (eds.), *Artificial Intelligence in China*, Lecture Notes
in Electrical Engineering 653, https://doi.org/10.1007/978-981-15-8599-9_11

As for methods based on traditional image processing ways, Wen [5] applied an improved Hough transform arc recognition method on the ATM monitoring system. However, this method only works when people are close to the monitor. Shrestha [6] proposed an edge detection algorithm to recognize the edge of the object in the head area to identify the helmet. But the facial features of construction workers are required for this method. These methods cannot adapt to the extremely complex scene of the construction site, and usually need to obtain the face features of workers in advance, so the robustness is poor.

When it comes to deep learning, object detection based on the convolutional neural networks (CNNs) become very popular. Many successful CNN architectures have been proposed and developed, such as RCNN [7], SSD [8], R-FCN [9] and YOLO [10]. Now algorithms based on CNN have been widely used in various fields, including face recognition [11], vehicle recognition [12], cancer cell recognition [13], etc. For construction sites, Fang et al. [14] used a fast RCNN based computer vision method to identify the remote monitoring image of the staff without a safety helmet. Fang [15] also tried a fast RCNN algorithm on supervising workers working at heights but without a safety belt.

It can be seen that deep learning methods based on convolutional neural networks can have good performance and fewer limits for target recognition, which can be widely applied to different industries.

At present, deep learning models require huge data set to reach object detection. To improve accuracy, a training set with similar pictures to practical application scenarios is highly appreciated, since even for the same objects, different background environments will widely influence the result. How to gain such huge data often becomes a big problem to apply deep learning. This paper tries YOLO models with simple and general training sets to realize the safety helmet detection on the complicated construction sites. Also, some pretreatment methods according to difficulties in real construction site situations are used to enhance the performance.

In the next section, a brief introduction to the YOLO models is given. The details of the data composition, label method, and data augment methods are discussed in Sect. 3. Evaluation methods are introduced in Sect. 4. Result and analysis are presented in Sect. 5. Section 6 is about summary and prospect.

2 Yolo Models

The YOLO model first proposed by Joe Redmon in 2016 is an end-to-end learning model with real-time speed for target detection.

YOLO [10] only uses one single convolutional network, including 24 convolutional layers and two fully connected layers. The convolutional layers are in charge of feature extraction and the fully connected layers are used to generate the bounding boxes. YOLO will divide the input image into $S \times S$ grid cells and each cell will have several bounding boxes for prediction. Every bounding boxes consist of five parameters: x, y, w, h, c, where x and y stand for the location of the center point of the bounding boxes, w and h are used for determining the size of the bounding box, and c stands for the confidence. The confidence value is related to the intersection-over-union (IOU) of the bounding box and the ground truth box. If the box doesn't hold any object, the confidence value will be zero. The model of YOLO is presented in Fig. 1.

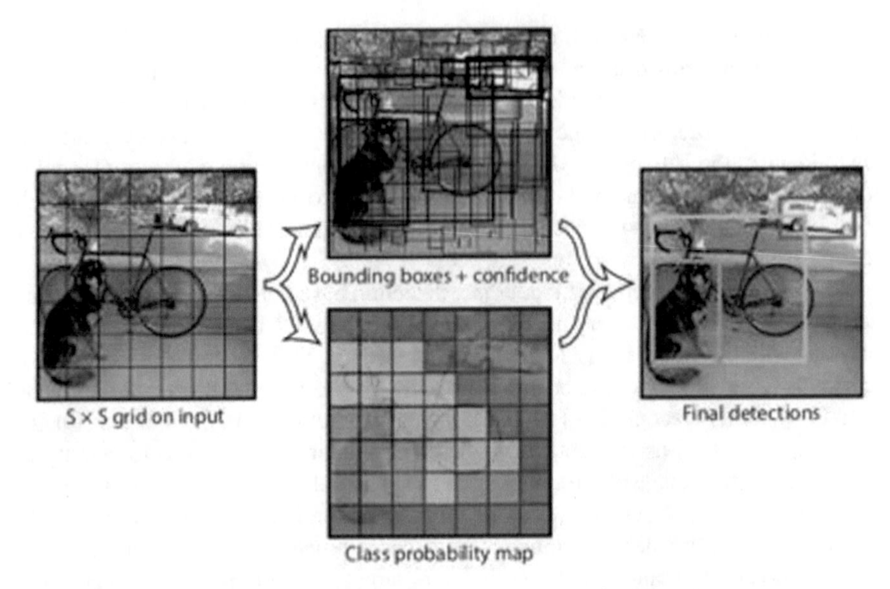

Fig. 1 The detection pipeline of YOLO [10]

In YOLOv2 [16], 5 convolutional layers are substituted by max-pooling layers. Besides, the way of generating bounding box proposals is changed. Originally, 2 fully connected layers are used to predict the bounding boxes, but in YoloV2 fully connected layers are removed and predefined anchor boxes are used. Then, the k-means clustering method is added to better define the anchor boxes.

In YOLOv3 [17], a deeper network including 53 layers of 3 × 3 and 1 × 1 filters with skip connections is introduced to substitute the remaining 19 convolutional layers. Besides, the bounding box prediction method is optimized. Now, YOLOv3 will generate 3 sets of bounding boxes prediction based on features on different scales. Also, multilabel classification is realized in YOLOv3, by using logistic regression instead of the softmax.

3 Data Engineering

3.1 Data Set

The helmet data set used in this paper consists of two kinds of pictures: pictures with simple scenes and usually helmet only, and pictures from the complicated real construction sites. All pictures in the data set are from search engines such as Baidu and Google.

The idea of this paper is to use relatively simple pictures to train and have a good result in complicated scenes. Therefore, 3200 simple pictures containing diverse helmets are selected to form the training set and all 513 pictures in the test set are complicated real constructions sites. Most of the pictures in the train set contain only one helmet and only a few pictures may contain more. However, in the test set, a large proportion of pictures hold more than one helmet. Figure 2 shows some pictures from the training set and the test set.

(a) Some pictures from the training set

(b) Some pictures from the test set

Fig. 2 Some pictures from the data set

3.2 Data Annotation

In real life, there are many kinds of helmets and some of them have different accessory parts to reach various functions. For example, the goggle may be added to protect the eyes. Therefore, there is no existing uniform standard for labeling. We formulated a set of specific labeling rules and manually labeled the positions of helmets in each picture.

For the helmet with accessories, we only border the main part of the helmet instead of labeling the whole helmets. For helmets with straps hanging over there, the straps would not be included to decrease the influence of the background environment. For the helmets stacking on the floor, only the helmet on the top will be labeled. Figure 3 shows some examples of our annotation.

3.3 Data Enhancement

For target detection algorithms based on deep learning, it is always important to have a large amount of data as a training set. The data augment is often used to enrich the data set. There are several methods for data augment, such as flip, rotation, contrast rise and fall, scale, etc. For practical use, we carefully consider the characteristics of real construction sites, where lighting changes over time, helmets are often blocked by the obstacles, and the distance between people and monitors is not constant. Therefore, the

Fig. 3 Some examples of the annotations

main influence factors are lighting and scale, so we pretreat our training set with contrast and scale to meet the requirement of the application.

For contrast, we use the following formula to process every pixel in one picture:

$$v' = 127 + (v - 127) * \sigma \quad (1)$$

where v is the pixel of the point and σ is a random number between 0.5 and 2 for different pictures. As for the scale, the length and width of the picture are randomly multiplied by a factor between 0.6 and 1.4 separately. The results are presented in Part 4.

4 Experiment Evaluation

4.1 Evaluation Setup

All the experiments were carried out on a personal computer with GeForceGTX 980 Ti.

This paper uses the Precision-Recall (PR) curve and the Average Precision (AP) to evaluate the performance of the model. The prediction results are compared with the ground truth boxes, and their IoU is calculated. The IoU is defined as the ratio of the overlapping area of prediction results and the ground truth to the total area of two regions. This paper takes IoU = 0.5 as a threshold value, which means if the IoU of a prediction is higher than or equal to 0.5, it is regarded as a true positive, otherwise, it is a false positive.

The formula for calculating precision and recall are as follows.

$$\mathrm{pre} = \frac{\mathrm{TP}}{\mathrm{TP} + \mathrm{FP}} \tag{2}$$

$$\mathrm{recall} = \frac{\mathrm{TP}}{\mathrm{TP} + \mathrm{FN}} \tag{3}$$

where TP represents the number of true positives, FP represents the number of false negatives, and FN is the number of false negatives, i.e., the number of targets objects that are not be predicted. In other words, the precision is the ratio of the number of true positives to the number of all prediction results and recall is the ratio of the number of true positives to the number of all target objects in the test set.

By changing the confidence levels, a set of (pre, recall) point can be got. Based on these points, we can draw a PR recurve on a rectangular coordinate system, where the horizontal axis is related to recall and the vertical axis is related to precision. AP is the area bounded by curves, the horizontal axis, and the vertical axis, which is the integral of the PR curve and can be calculated as follow:

$$\mathrm{AP} = \int_{0}^{1} p(r)\mathrm{d}r \tag{4}$$

For multiple classes, the mean of AP will be calculated as the evaluation index. However, in this paper, only the helmet class is considered.

5 Result

In this paper, different YOLO model with the same simple training set and complicated test set are compared. From Table 1 and Fig. 4, it can be seen that YOLO can hardly handle such a special data set. For YOLOv2 and YOLOv3_tiny, they have relatively good precision when the recall value is at a low level. However, with the gradual rise of the recall value, their precision values begin to fall rapidly and their AP values are quite low. For YOLOv3, the result is very good. The precision value of YOLOv3 can keep stable for a wide range and AP value reaches 0.736. Figure 5 shows some results of different models.

The results for different pretreatment methods are shown in Fig. 6 and Table 2. From the table, it is clear that both methods improve the value of AP by about 3 respectively. With the joint effect of two methods, the AP can reach 0.7750. From the curve, we can

Table 1 AP results on helmets test set under different models

Model	AP
YOLO	0.0012
YOLOv2	0.4276
YOLOv3_tiny	0.2958
YOLOv3	**0.7368**

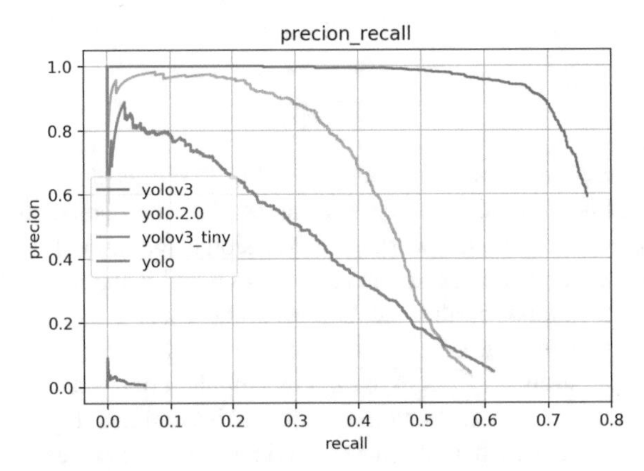

Fig. 4 PR curves on helmet test set for different models

(a) YOLOv2 (b) YOLOv3_tiny (c) YOLOv3

Fig. 5 Some results of detection in the test set for different models

see the curve of YOLOv3 with the joint effect of two pretreatment methods start to fall rapidly later than the single YOLOv3 model.

The above results imply that the data annotation and preprocessing methods presented in this paper are suitable for the helmet detection task. It also shows that the YOLOv3 model has the potential to achieve detecting objects in complex scenes by learning simple pictures.

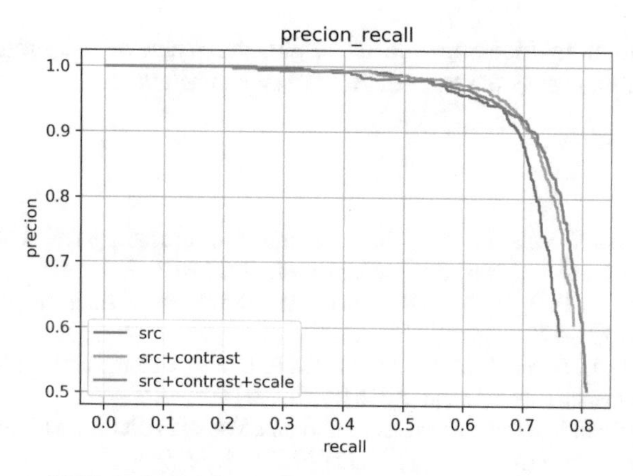

Fig. 6 PR curves of YOLOv3 model on helmet test set under different and processing modes

Table 2 AP of YOLOv3 model on helmet test set under different and processing modes

Pretreatment	Map
Null	0.7362
Contrast	0.7605
Scale	0.7679
OUR (Contrast + Scale)	**0.7750**

6 Conclusion

This essay examines the common deep learning model for the detection effect of common deep learning models with the simple training set but the complex test set. The task is trying to let the machine learn the characteristics of the helmet and then be able to tell the existence of the helmet under serious conditions. It turns out that YOLOv3 is well adapted to the special training and testing sets, while YOLOv3_tiny and YOLOv2 models are not. As for YOLO, the oldest model, it can hardly recognize helmet in complicated scenarios. It shows that technology has indeed achieved the development of machine learning from simply remembering things to preliminarily understanding things. We also pretreat the training set with different methods according to the real application scenarios to improve the performance and get a good result, which shows that the correct choice of the pretreatment method according to the practical application can increase the accuracy.

YOLO is an excellent end-to-end model for target detection. Its real-time speed characteristic can meet the requirement of many real scenarios. The result of the experiments can be used as guidance to build surveillance systems for not only construction sites. The results verify that simple training sets can also lead to high accuracy in special scenarios, making the build of the training set less costly. The proposed YOLOv3 model is able to recognize the helmet in a variety of harsh environments, including but not limited to insufficient lighting, blocking, unusual angle, etc. When we use simple pictures like those in the training set as the test set, AP can even reach 96%.

References

1. Li H, Li X, Luo X, Siebert J (2017) Investigation of the causality patterns of non-helmet use behavior of construction workers. Autom Constr 80:95–103
2. Hale AR, Heming BHJ, Carthey J, Kirwan B (1997) Modelling of safety management systems. Saf Sci 26(1):121–140
3. Dong S, He Q, Li H, Yin Q (2015) Automated PPE misuse identification and assessment for safety performance enhancement. ICCREM 204–214
4. Khairullah M, Habibur Rahman M, Hasanul Banna M (2012) BlueAd: a location based service using bluetooth. Int J Comput Appl 43(15):19–22
5. Wen C (2004) The safety helmet detection technology and its application to the surveillance system. J Forens Sci 49(4):770–780
6. Shrestha K, Shrestha PP, Bajracharya D, Yfantis EA (2015) Hard-hat detection for construction safety visualization. J Constr Eng 2015:1–8
7. Girshick R (2015) Fast R-CNN. IEEE, pp 1440
8. Liu W, Anguelov D, Erhan D, Szegedy C, Reed S, Fu C, Berg AC (2016) 2015 SSD: single shot multibox detector, pp 21
9. Dai J, Li Y, He K, Sun J (2016) R-FCN: object detection via region-based fully convolutional networks
10. Redmon J, Divvala S, Girshick R, Farhadi A (2016) You only look once: unified, real-time object detection. IEEE, pp 779
11. Peng Q, Luo W, Hong G, Feng M, Xia Y, Yu L, Hao X, Wang X, Li M (2016) Pedestrian detection for transformer substation based on gaussian mixture model and YOLO. IEEE, pp 562
12. Kim H, Lee Y, Yim B, Park E, Kim H (2016) On-road object detection using deep neural network. IEEE, pp 1
13. Zhang J, Hu H, Chen S, Huang Y, Guan Q (2016) Cancer cells detection in phase-contrast microscopy images based on faster R-CNN IEEE pp 363
14. Fang Q, Li H, Luo X, Ding L, Luo H, Rose TM, An W (2018) Detecting non-hardhat-use by a deep learning method from far-field surveillance videos. Autom Constr 85:1–9
15. Fang W, Ding L, Luo H, Love PED (2018) Falls from heights: a computer vision-based approach for safety harness detection. Autom Constr 91:53–61
16. Redmon J, Farhadi A (2016) YOLO9000: better, faster, stronger
17. Redmon J, Farhadi A (2018) YOLOv3: an incremental improvement

Feature Extraction and Classification of Unknown Types of Communication Emitter

Xu Zhang(✉), Zhuo Sun(✉), Suyu Huang, Shaolin Ma, and Anhao Ye

Wireless Signal Processing and Network Laboratory, Beijing University of Posts and Telecommunications, Beijing 100876, China
{Xu_Zhang,zhuosun}@bupt.edu.cn

Abstract. In the field of cognitive radio, communication emitter recognition plays an important role. The traditional methods need to determine expert feature in advance, which leads to inefficiency in some case. One is that the emitters may not be observed in the training and we need to classify much larger new classes examples without knowledge of the label. Therefore, this paper proposes a semi-supervised method for extracting RF fingerprint based on supervised Convolutional Neural Networks and unsupervised clustering. Experimental results demonstrate that the features extracted from CNN can distinguish emitters and the proposed method has a great performance in classifying unknown types of emitters.

Keywords: RF fingerprinting · CNN · K-Means · Machine learning · Deep neural network

1 Introduction

With the progress of modern science and technology, electronic warfare will become an important part of future battlefield. In order to gain advantage in the modern digital information war, it is necessary to carry out electronic reconnaissance on the enemy communication equipment and emitters. In addition, due to the rapid development of 5G technology, more and more users and devices can access to the wireless system. Although these wireless devices provide huge convenience for people,the security of wireless communication can not be ignored. Effective identification of wireless network intrusion is an essential part of enhancing wireless security, in which the communication emitter identification plays an important role.

The traditional communication emitter classification methods firstly extract the expert features including high-order cumulant [1], multifractal mature [2], wavelet analysis [3], Fourier transform [4] that need to be determined in advance,

Q. Liang et al. (eds.), *Artificial Intelligence in China*, Lecture Notes in Electrical Engineering 653, https://doi.org/10.1007/978-981-15-8599-9_12

and then classifiers such as support vector machine [5], bagging [6] are used to classify the extracted features. However, traditional methods have its limitations and they can not apply to some particular situations. One such case is that the prior knowledge of test signal is unknown and the emitter may not observed in the training. To solve the problem, this paper explores how to automatically extract and fuse the RF fingerprinting that can distinguish the unknown type emitters and classify the extracted feature without any prior knowledge.

In recent years, with the development of artificial intelligence technology, deep learning has been used for identifying emitters [7], but their work only focused on the classification, not on the extraction of RF fingerprint features.This paper proposes a semi-supervised method based on CNN to learn the unique high dimensional features that can be used as RF fingerprinting. As shown in Fig.1, CNN is trained to extract features of the emitter signals and implement classification with Softmax. This step is a supervised learning process. Once trained, as long as the last layer of network is removed, it can be considered as feature extractor, and the output of the first full connection layer is high dimensional features. Then, we can cluster the features of test signal captured by feature extractor to classify emitters without any prior knowledge, which is an unsupervised process. In particular, this method does not need to pre-process the signal, and does not need know any prior knowledge of test signal, so it can meet the needs of non-cooperative scenarios.

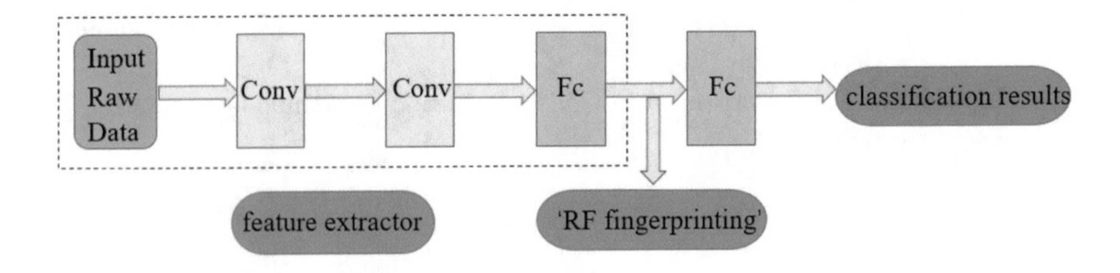

Fig. 1. Workflow for the approach proposed in this paper

The structure of the paper is as follows: Sect. 2 gives a description of the methodology details of the proposed approach. Section 3 provides the experimental results from the models and the related analysis. Finally, Sect. 4 concludes the paper.

2 Feature Extraction and Classification

2.1 Feature Extraction with Convolutional Neural Network

The fingerprinting of communication emitter comes from the uncontrollable or unconscious error factors in the process of transmitter design, manufacturing and transmitting signal. We consider choosing the appropriate network layer to extract the RF fingerprint features. In recent years, convolutional neural network has made a great breakthrough in the field of image recognition [8,9], which is formed by multi-layer convolution layers, each layer with multiple convolution kernel, and often uses pooling layer to reduce dimension. And some people have successfully applied CNN network to modulation recognition [10]. Convolution neural network has the ability of pattern matching and each convolution kernel defines a pattern. Convolution operation can be regarded as signal conversion from time domain to frequency domain. Some of the RF fingerprint features are more obvious in the frequency domain, which is the important reason why we choose convolutional neural network for feature extraction.

In this paper, we try to use CNN network to classify communication emitters. We hope the model can automatically learn the characteristics of communication emitter. Each filter can match different characteristics. Finally, combine multiple characteristics to form stable and effective RF fingerprinting. Using convolutional neural network to match the micro characteristics of signal, there is no need to delve into the bottom of the signal. By setting up experiments to compare different convolution layers and convolution kernel sizes, the final network structure with the best performance is shown in Table 1. The input of our network is 1024 raw samples, which contains about 30 modulation symbols. The network structure includes three convolution layers and two full connection layers. Except for the last layer, the activation function is ReLU. In addition, dropout is used to avoid overfitting.

Table 1. Layers of the network for emitter identification

Layer	Dimension	Parameters	Activation
Input	1 * 1024	–	–
Convolution1	1 * 5	300	ReLU
Maxpooling	2	–	–
Convolution2	1 * 5	12550	ReLU
Maxpooling	2	–	–
Convolution3	1 * 5	12550	ReLU
Maxpooling	2	–	–
Flatten	–	–	–
Dense1	256	768256	ReLU
Dropout(0.5)	–	–	–
Dense2	10	2570	Softmax

Once the network is trained, all parameters are fixed and the output of Dense1 layer can be used as RF fingerprinting to classify emitters. The feature extracted by convolution neural network is 256 dimensions, and it is reduced to two dimensions by dimension reduction algorithm to complete visualization.

Figure 2 shows the distribution of signal features of ten different emitters, in which different colors represent different emitters. It can be seen that the signals of the same emitter will gather together, and the signals of different emitters will disperse in the feature space. Therefore, our proposed feature extraction method is effective and distinguishable.

2.2 Clustering with K-Means

Using the feature extractor, we can get the high dimensional features of test signals. After that, we will exploit the extracted features to classify the emitters by clustering. Because this work does not assume the emitters in the test set are included in the training set, it can not implement by supervised classification method. In this paper, we consider using K-Means algorithm to group a large number of unlabeled test signals, resulting in identifying the class number of emitters and the signals belonging to the same emitters.

K-means is one of the most widely used clustering algorithms, which has the advantages of simple implementation and high efficiency. The application of clustering algorithm is also very extensive, including document classification, music classification, classification based on user purchase behavior [11]. The core idea of K-Means is to select k samples randomly from the sample set as the cluster center, and calculate the distance between all samples and the cluster center. For each sample, it is divided into the nearest cluster. For new clusters, calculate the new cluster center of each cluster. Repeat the above process until the cluster center no longer changes, and finally determine the category of each sample and the centroid of each category.

There is a challenging problem that the number of emitters is unknown. We need to determine the appropriate K value first, that is the number of emitters. We try to use the silhouette coefficient method to determine. The key index of this method is the silhouette coefficient. The definition of the silhouette coefficient of a sample point X is as follows:

$$S = \frac{b - a}{max(a, b)} \tag{1}$$

where a is the average distance between X and other samples in the same cluster, which is called cohesion, and b is the average distance between X and all samples in the nearest cluster, which is called separation. By calculating the silhouette coefficient of all samples, we can get the average silhouette coefficient.The closer the distance between samples in the same cluster is, the larger the average silhouette coefficient is, the better the clustering performance is. So, the K with the largest average silhouette coefficient is the best clustering number, that is the number of communication emitters.

We choose the feature extractor trained by 10 transmitters to carry out experiments to verify the above method. When the class number of emitters in test set is 15, Fig. 3 shows how the average Silhouette Coefficient changing with increasing K value. It can be seen that when $K = 15$, the average Silhouette

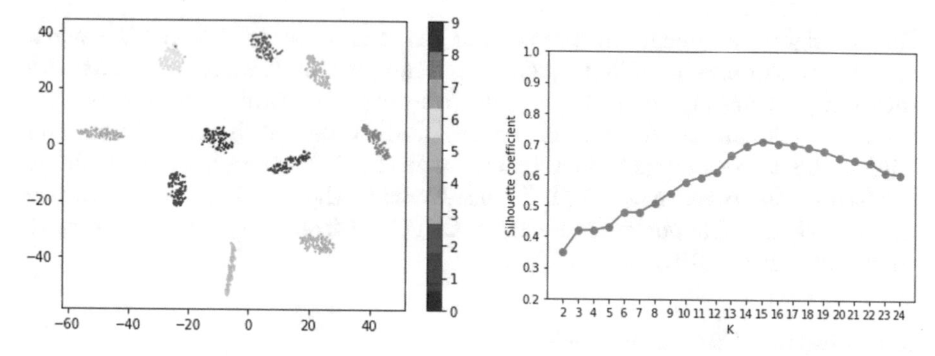

Fig. 2. The feature distribution of signals transmitted by ten different emitters

Fig. 3. The Silhouette Coefficient with different K

Coefficient is the largest, so taking K as 15 is the best clustering number, which is consistent with the truth. Therefore, this method correctly identifies the class number of emitters. In addition, it also shows that the method proposed in this paper can identify the emitter unseen in the training, because the number of emitters in the training set is only 10.

The output of clustering is a series of labels, in which the samples with the same labels are transmitted from the same transmitter. Unlike classification, clustering cannot measure the performance of the algorithm with accuracy. In this paper, the adjusted Rand index (ARI) is used to measure the similarity between the predicted labels and the real labels. ARI measures the degree of agreement between the two data distributions, the larger the value is, the more consistent the clustering result is with the real situation. The features extracted from CNN is a 256 dimension vector, and we can use the t-SNE dimension reduction algorithm to reduce the feature dimension to two-dimensional space [12]. It can be seen intuitively that the signals of the same emitters will gather together, and the signals of different emitters will disperse.

3 Experiments and Results

In this section, we verify the feasibility of the method in Sect. 2. Firstly, data collection and experimental setup are described. Then through a series of experiments, the performance of feature extractor and classification is analyzed.

3.1 Data Collection and Experimental Setup

We collected 30 continuous signals generated by communication emitter sources as samples. The parameters of the signals are as follows: the carrier frequency is 1000 MHz, the modulation mode is QPSK, the modulation rate is 1M Baud/s,

70M IF signal sampling, and the sampling rate is 30 MSps. We cut the signal into several samples by sliding window, each sample is 1024 sampling points, and each sample contains about 30 symbols. We use an improved version of Adam. Adam remembers the direction of the last gradient descent by momentum, and adjusts the descent range of each parameter to get better results. Each training is based on the batch size of 100. We use Keras as the frontend and Tensorflow as the backend. The bottom is NVIDIA CUDA environment, and the hardware condition is 1080 GPU.

3.2 Feature Extractor Performance

The number of samples of each emitter is 30,000. At the same time, the samples are randomly divided into training set, test set and verification set, with the proportion of 0.8, 0.1 and 0.1 respectively. We have conducted experiments on the number of emitters from 5 to 30. In order to prove the effectiveness of our proposed feature extraction network, we use accuracy as the performance metric, that is the ratio of the number of samples classified correctly to the total number of samples in the dataset.

Figures 4 and 5 show the performance when the training set contains 15 and 20 emitters, from which we can see that the accuracy is increasing with the number of epochs. When the epoch reaches a certain value, the accuracy tends to be constant, indicating that the network begins to converge. The test accuracy of different number of emitters is shown in Table 2. It can be seen that when the number of emitters is 10, the accuracy is as high as 99.25%, which is a good classification performance. With the increase of the number of emitters, the test accuracy of classification gradually decreased. The result is understandable, because with the increase of the number of emitters, more samples are needed to be provided to the network for training. Meanwhile, as classes in the classifier becomes more, the neural network needs to fit a more complex function, which is a difficult challenge. From the above results, we can see that the effective features are captured by our proposed feature extractor, which can represent the emitter.

Table 2. Test accuracy of convolutional neural network

Number of emitters	Accuracy (%)
10	99.25
15	97.63
20	96.36
25	95.79
30	94.44

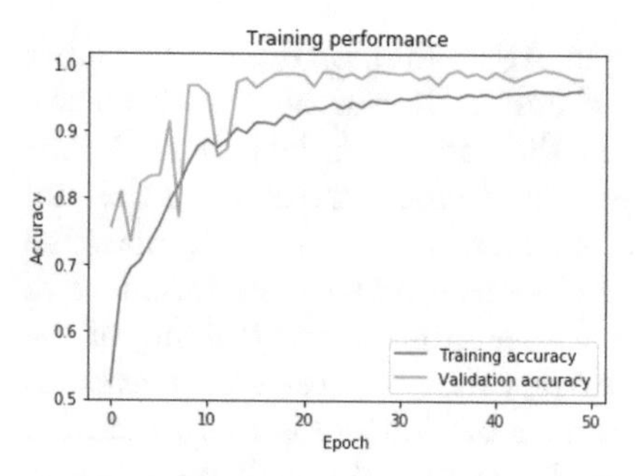

Fig. 4. Accuracy for emitter classification with 15 emitters

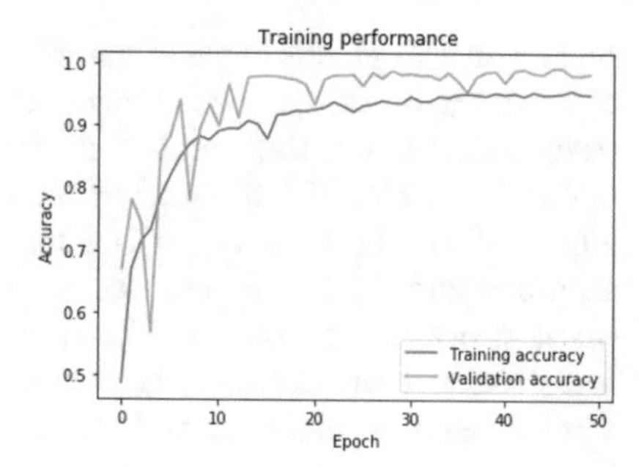

Fig. 5. Accuracy for emitter classification with 20 emitters

3.3 The Performance of Classification

Once we get the high-dimensional features through the feature extractor based on CNN, we first use PCA algorithm to reduce the high-dimensional features to 50 dimensions, and then use k-means to cluster the features. We use two metrics to evaluate the performance of classification. First, the accuracy is used to determine the ability to identify the number of unknown type emitters. Accuracy refers to the ratio of the number of identified emitters to the number of emitters in test set. The method of identifying the number of emitters is discussed in Sect. 2.2. After getting the K value, that is, the number of emitters, the output of K-means is a series of label. We use ARI to evaluate the similarity between the predicted labels and the real labels.

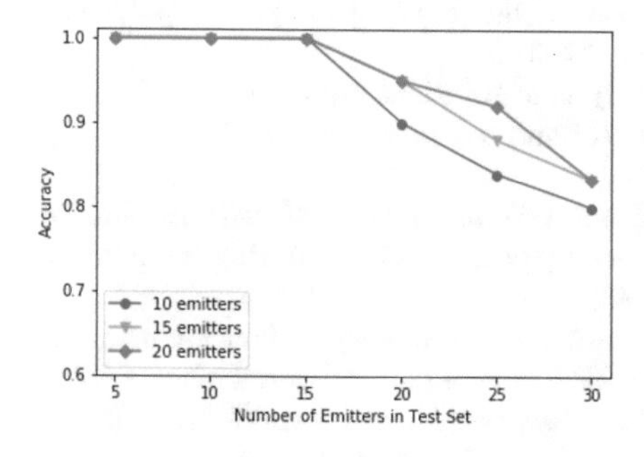

Fig. 6. The accuracy with different emitters in test set

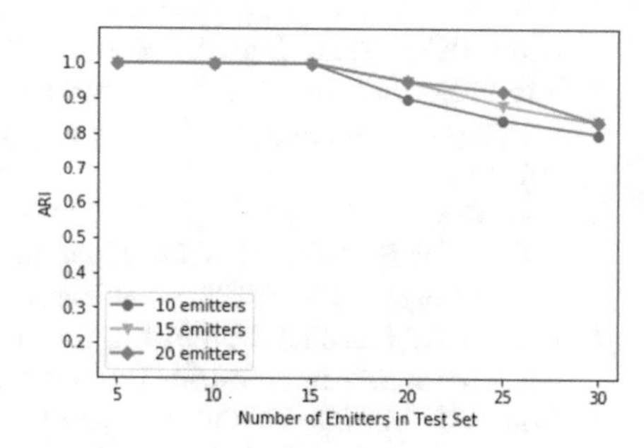

Fig. 7. The adjusted rand index with different emitters in test set

We performed experiments on three feature extractors, which is trained with 10,15 and 20 emitters. When the accuracy is 1, it represents that the network identifies all emitters in test set. The closer the ARI value is to 1, the more successful the clustering is. The accuracy and ARI value for different number of emitters in test set are shown in Figs. 6 and 7. It can be seen from the results that when the number of emitters in the test set is less than 15, our method can identify all emitters and the predicted labels are close to the real labels. With the increase of the number of emitters in the test set, both the accuracy and ARI value are decreasing. It is obviously that the performance of the network will be better if more emitters are seen during the training of the feature extractor. Because the network can extract more representative features if more samples are observed. When the number of emitters is more than 20, the clustering performance decreases, which shows that the ability of feature extraction network is limited.

4 Conclusion

In this paper, we propose a semi-supervised method for extracting and classifying RF fingerprint based on CNN and K-Means. The experimental results show that this method is simple and effective compared with the traditional method using expert features to identify. Moreover, the feature extractor also performs well for emitters unseen in training, which indicates that this method is suitable for many non-cooperative scenes. However, when the number of emitters in the test set exceeds a certain range, our method is challenged greatly with a drop in the performance of classification. In the future work, we will do further study on this issue.

References

1. Tugnait JK (1994) Detection of non-Gaussian signals using integrated polyspectrum. IEEE Trans Signal Process 42(11):3137–3149
2. Sun L, Serinken N (2000) Multifractal analysis of transient power system. In: Proceedings of Canadian conference on electrical and computer engineering, vol 1, pp 307–311
3. Toonstra J, Kinsner W (1995) Transient analysis and genetic algorithms classification. In: WESCANEX 95. Communications, power and computing. conference proceedings, vol 2. IEEE, New York, pp 432–437
4. Kwok HK, Jones DL (2000) Improved instantaneous frequency estimation using an adaptive short-time Fourier transform. IEEE Trans Signal Process 48:2964–2972
5. Comes C, Vapnik V (1995) Support vector networks. Mach Learn 20:273–297
6. Breiman L (1996) Bagging predictors. Mach Learn 24:123–140
7. Wu Q, Feres C, Kuzmenko D (2018) Deep learning based RF fingerprinting for device identification and wireless security. Electron Lett 54:1405–1407
8. Krizhevsky A, Hinton G (2012) Imagenet classification with deep convolutional neural networks. In: Advances in neural information processing systems, vol 25, pp 1106–1114
9. Szegedy C, Liu W, Jia Y, Sermanet P, Reed S, Anguelov D, Erhan D, Vanhoucke V, Rabinovish A (2015) Going deeper with convolutions. In: CVPR

10. O'Shea TJ, Corgan J, Clancy TC (2016) Convolutional radio modulation recognition networks. Engineering applications of neural networks. Springer International Publishing, Berlin
11. Xu JH, Liu H (2010) Web user clustering analysis based on K-means algorithm. In: 2010 international conference on information, networking and automation
12. van der Maaten L (2014) Accelerating t-SNE using tree-based algorithms. J Mach Learn Res 15:3221–3245

Research on Escape Strategy Based on Intelligent Firefighting Internet of Things Virtual Simulation System

Hai Wang, Guiling Sun(✉), Yi Gao, and Xiaochen Li

Teaching Center for Experimental Electronic Information, College of Electronic Information and Optical Engineering, Nankai University, Tianjin 300350, China
sungl@nankai.edu.cn

Abstract. Intelligent firefighting accomplishes the intellectualization of city fire protection by utilized the latest technologies such as the Internet of Things (IOT), artificial intelligence, virtual reality, and mobile Internet plus. Besides, it cooperates with the professional application of big data cloud computing platform, fire alarm intelligence judgment, and so on. Intelligent fire control is not only the digital foundation of fire information service, but also an important part of smart perception, interconnection, and intelligent application architecture in smart city. In this paper, the virtual simulation technology is used to simulate the real fire scene. According to the specific scene, the appropriate network framework and communication protocol are selected. Different types of sensors are utilized to collect the real-time fire data. Through the simulation calculation and mathematical manipulation, the escape strategy and escape route are determined for the trapped people.

Keywords: Intelligent firefighting · Internet of Things · Artificial intelligence · Virtual simulation · Escape strategy

1 Introduction

With the rapid development of urban construction in China, there are more and more high-rise buildings and super high-rise buildings, which are developing toward modernization, large scale, and multi-functional. Once a fire occurs, it rapidly spreads and easily forms three-dimensional combustion, dense smoke, and toxic gas, which brings significant difficulties to the firefighting work and personnel evacuation [1, 2]. Thus, great economic losses and casualties are caused. How to prevent fire and ensure fire safety has become a key and focus issue in urban governance. There is an urgent need for firefighting and rescue technology, which is necessary to improve the social fire prevention and control ability and realize the coordinated development of firefighting work and economic society [3, 4].

Using emerging information technologies, such as sensors, the Internet of Things, cloud computing, big data, mobile Internet, etc., to build an "Intelligent Firefighting"

Q. Liang et al. (eds.), *Artificial Intelligence in China*, Lecture Notes in Electrical Engineering 653, https://doi.org/10.1007/978-981-15-8599-9_13

firewall with artificial intelligence characteristics can play a significant role to ensure urban public security and fire prevention and control system [5–7]. It assists users to figure out the number of fires and master the initiative of firefighting and disaster relief, to carry out dynamic supervision and dispatching of fire resources, to improve the utilization rate of firefighting equipment and equipment, and to standardize fire safety process management and improve people's fire safety awareness. Deep reinforcement learning technology is very important in intelligent firefighting. It combines the perceptual ability of deep learning with the decision-making ability of reinforcement learning and can be controlled directly according to the input image. It is an artificial intelligence method closer to human thinking mode [8, 9]. In the field of intelligent fire protection, deep reinforcement learning technology combines the perception ability of deep learning with the decision-making ability of reinforcement learning, which can be controlled directly according to the input image. As a result, Deep Q Network (DQN) algorithm and deep reinforcement learning based on convolutional neural network or recurrent neural network are widely used [10–13].

This paper realizes the application of the IOT technology in the field of intelligent fire protection. The virtual simulation technology is used to simulate the real fire scenes. And the appropriate network architecture and communication protocol can be selected according to the specific scene. The system calculates and processes according to the relevant data collected by different types of sensors, so as to help trapped people make decisions and provides them with escape strategy design and optimization.

2 Design of Virtual Simulation System

The architecture of the Internet of Things can be divided into three layers: perception layer, network layer, and application layer. The perception layer, data source of IOT, consists of various sensors and sensor gateways, which act as the nerve endings of human's eyes, ears, nose, throat, and skin. The network layer is composed of various private networks, the Internet, wired and wireless communication networks, network management systems, and cloud computing platforms, which is equivalent to the human nerve center and brain, responsible for transmitting and processing the information acquired by the perception layer. The application layer is the interface between IOT and users (including people, organizations, and other systems). It combines with the acquirements of the industry to realize the intelligent application of IOT [14–16]. In this paper, the intelligent firefighting Internet of Things virtual simulation system is designed based on these three layers.

2.1 Perception Layer Design

As the main way to obtain information, sensors have long been applied to such extensive fields as industrial production, space development, ocean exploration, environmental protection, resource investigation, medical diagnosis, bioengineering, and even cultural relics protection. It is no exaggeration to say that almost every modern project, from the vast space and ocean to all kinds of complex engineering systems, is inseparable from all kinds of sensors.

The system includes temperature sensor, humidity sensor, smoke sensor, harmful gas sensor, etc., which play a very important role in monitoring the occurrence and spread of fire, and provides basic data supported for subsequent escape strategy decision. MiCS-4514 (illustrated in Fig. 1), a kind of MEMS gas sensor integrates two independent heating resistors and two sensitive layers, one for detecting oxidizing gas and the other for detecting reducing gas. The working voltage is DC 4.9–5.1 V, and the current is 30 mA. It has the characteristics with rapid response time and the wide temperature and humidity range of working environment. MiCS-4514 can detect carbon monoxide and hydrogen in the concentration range of 1–1000 parts per million (ppm). In addition, it also can detect carbon dioxide, ethanol, ammonia, methane, and other gases.

Fig. 1. MiCS-4514 gas sensor

2.2 Network Layer Design

The main function of network layer is "transmission," that is, information transmission through communication network. The network layer, as a link between network layer and application layer, is responsible for processing the information obtained by the sensors according to different application requirements and finally transmitting it safely and reliably to the application layer. The network layer of the Internet of Things basically integrates all the existing network forms to build a broader "interconnection." Each kind of network has its own characteristics and application scenarios. Only by combining with each other can it play a maximum role. Therefore, in practical applications, information is often transmitted through a single kind of network or combination of several networks.

This system mainly involves three network frameworks: Narrow band Internet of Things (NB-IOT), ZigBee, and 1Long range radio (LORA). NB-IOT has the advantages of wide coverage, strong support and connection ability, low power consumption, and low module cost. ZigBee is a kind of wireless network protocol with low-speed and short-distance transmission. The bottom layer is the media access layer and physical layer based on IEEE 802.15.4 standard. It is suitable for a series of electronic devices with short transmission range and low-data transmission rate. Lora is a low-power local area network wireless standard created by Semtech. Its biggest feature is that under the same power consumption condition, Lora can travel 3–5 times longer than other wireless modes. It realizes the unification of low power consumption and long distance.

2.3 Application Layer Design

The application layer is located at the top level of the Internet of Things, which process information by the cloud computing platform. The application layer and the lowest level, the perception layer, are the significant characteristics and core of IOT. The former can calculate and process the data collected by the latter, so as to realize the real-time control, precise management and scientific decision-making of the physical world. The application layer, whose main technologies are software technology and computer technology, is equivalent to the brain and nerve center of the whole IOT system.

Building information modeling (BIM) technology, a data tool used to engineering design, construction, and management, is applied in the virtual simulation system in this paper. It integrates the data-based and information-based model of the building. BIM technology is the digital representation of the actual building, which contains the space geometry information and attributes information. The virtual simulation system establishes a reliable channel for real-time monitoring of firefighting through the integration of data extraction and IOT sensors. Then, it combines the real-time location information provided by mobile phones and other devices to determine the exact location of the trapped people, so as to achieve real-time tracking.

3 Realization of Escape Strategy Decision

3.1 Assemble Sensor Nodes

The nodes are assembled according to the sensor type, microprocessor, communication module, alarm, and battery. The power design includes the configuration of input and output voltage, and the low-power design involves the determination of total power and sleep duty cycle.

A temperature alarm is designed in the intelligent firefighting IOT virtual simulation system. The specific parameters are shown in Table 1. In addition, humidity sensor nodes, smoke detectors, and harmful gas detectors with different parameters can also be designed.

3.2 Design Wireless Networking Mode

The appropriate network framework, network mode, and network protocol are determined according to different application environments. Message queuing telemetry transport (MQTT), which can support most platforms, is an instant messaging protocol based on TCP developed by IBM. Constrained application protocol (COAP) runs on user datagram protocol (UDP) and adopts datagram mode. Both of them are born to adapt to M2M. Compared with other communication protocols, they have greater advantages, which is the general trend of operating the Internet of Things.

Table 1. Basic information of the temperature alarm

Module	Type	Performance	Parameter
Sensor	DS18B20	Input voltage	5 V
Processor	STM8L152	Output voltage #1	3.6 V
Communication	BC35-G	Output voltage #2	3.3 V
Alarm	Active buzzer	Quantity of electricity	10.000 mAh
Battery	Lithium battery	Sleep duty cycle	50%

This virtual platform can simulate shops and wholesale markets, where consumers are scattered. Sensor nodes need to be deployed massively in these places; however, the transferred data quantity is small. It can also simulate specific intelligent firefighting system without network signal coverage, such as the basement of bank pump house. The various environment requirements give birth to different deployment patterns of network layer.

3.3 Simulate the Real Fire

BIM building information model technology is utilized to build virtual space in intelligent firefighting Internet of Things virtual simulation system, which integrates the IOT technology such as the sensors and network protocol architecture to obtain more comprehensive fire environment factors and dynamic parameters of trapped people. Through the simulation calculation of path planning algorithm, the effective escape strategy and escape route are determined. The mainstream planning algorithms include artificial potential field algorithm, ant colony planning algorithm, artificial bee colony planning algorithm, etc. On the premise that the parameters are in line with the actual situation, a mathematical model with high efficiency can be established. Then repeated simulation and optimization of machine learning and artificial intelligence can be carried out. This strategy has been widely used in international intelligent fire protection solutions.

The definition of general gravitational potential field function is shown in the following formula.

$$U_{att}(x) = \frac{1}{2} \cdot k_{att} \cdot l^2(x - x_g) \tag{1}$$

Thus, the gravitational function can be deduced by negative gradient operation of the gravitational potential field function.

$$F_{att}(x) = -\nabla U_{att}(x) = k_{att} \cdot l(x - x_g) \tag{2}$$

The general repulsion potential field function is likewise defined as follows.

$$U_{rep}(x) = \begin{cases} \frac{1}{2} \cdot k_{rep} \cdot \left(\frac{1}{l(x,x_0)} - \frac{1}{l_0}\right)^2, & l(x, x_0) \le l_0 \\ 0, & l(x, x_0) > l_0 \end{cases} \tag{3}$$

Then, the repulsion function can be derived by negative gradient operation of repulsion potential field function.

$$F_{rep}(x) = -\nabla U_{rep}(x) = \begin{cases} k_{rep} \cdot \left(\frac{1}{l(x,x_0)} - \frac{1}{l_0}\right) \cdot \frac{1}{l^2(x,x_0)} \cdot \frac{\partial l(x,x_0)}{\partial x}, & l(x, x_0) \leq l_0 \\ 0, & l(x, x_0) > l_0 \end{cases} \tag{4}$$

Ultimately, the resultant force is equal to F_{att} plus F_{rep}.

The building of teaching center for experimental electronic information in Nankai University is selected as the virtual reality and firefighting exercise scene in this system. Each laboratory is equipped with four smoke detection nodes, two harmful gas detection nodes, and one temperature detection node (shown in Fig. 2). Meanwhile, smoke sensors are deployed in the corridor outside the laboratories. All sensors are networked by means of NB-IOT, and the detected data is transmitted to the central processor for fire monitoring.

Fig. 2. Deployment of sensor nodes

In the fire drill of teaching building, the system randomly generates the location of user and fire, as well as the trend of fire spread. The system calculates and processes the data collected by the sensors, assists the user to make the decision quickly, accurately, and efficiently, and works out the measures to deal with the fire. The virtual simulation system guides the user to access the surrounding firefighting facilities to put out the fire in time, as illustrated in Fig. 3. Conversely, if the system judges that the fire cannot be extinguished, it will show the escape route (shown in Fig. 4) and demand the user to stay away from the danger. The system can also display the firefighting facilities around the fire source and send SMS to inform the person in charge for fire of different laboratories in this building. In addition, the virtual simulation platform is also equipped with an exhibition hall (shown in Fig. 5) to help users get familiar with the relevant fire laws and regulations, as well as the use of firefighting facilities.

Fig. 3. Extinguish the fire

Fig. 4. Escape route

4 Conclusion

An intelligent firefighting Internet of Things virtual simulation system is designed in this paper. The real fire is simulated by virtual reality technology. The escape strategy and escape route for trapped people are determined quickly and accurately by the system. The method of analyzing and processing massive data, integrating and managing the fire service work in a certain area or even the whole city, is a crucial issue in the future.

Fig. 5. Firefighting exhibition hall

Acknowledgements. This work is supported by the 2020 Undergraduate Education Reform Project Fund of Nankai University (NKJG2020004); the 2020 Experimental Course Teaching Reform Project Fund of Nankai University (20NKSYJG01); the 2021 Self-made Experimental Teaching Instrument and Equipment Project Fund of Nankai University and Teaching Center for Experimental Electronic Information.

References

1. Hong SG, Son KH, Lee H et al (2018) Augmented IoT service architecture assisting safe firefighting operation. Global internet of things summit (GIoTS)
2. Divan A, Kumar AS, Kumar AJ et al (2018) Fire detection using quadcopter. In: Second international conference on intelligent computing and control systems (ICICCS)
3. Guowei Z, Su Y, Guoqing Z et al (2020) Smart firefighting construction in China: status, problems, and reflections. Fire Mater (7)
4. Palmiere SE, Riascos CEM, Riascos LAM (2015) Integration of energy and fire prevention systems in greenbuildings. In: IEEE 24th international symposium on industrial electronics (ISIE)
5. Wang KM, Hui L (2017) Effectiveness evaluation of Internet of Things-aided firefighting by simulation. J Supercomput 3:1–15
6. Diwanji M, Hisvankar S, Khandelwal C (2019) Autonomous fire detecting and extinguishing robot. In: 2nd international conference on intelligent communication and computational techniques (ICCT)
7. Harikumar K, Senthilnath J, Sundaram S (2019) Multi-UAV oxyrrhis marina-inspired search and dynamic formation control for forest firefighting. IEEE Trans Autom Sci Eng 16(2):863–873
8. Zhong G, Zhang K, Wei H, et al (2019) Marginal deep architecture: stacking feature learning modules to build deep learning models. IEEE Access 30220–30233
9. As I, Pal S, Basu P (2018) Artificial intelligence in architecture: generating conceptual design via deep learning. Int J Architect Comput 16(4):306–327
10. Shi D, Ding J H, Errapotu, et al (2019) Deep Q-network-based route scheduling for TNC vehicles with passengers' location differential privacy. IEEE Internet Things J 6(5):7681–7692

11. Qiu C, Yao H, Yu R et al (2019) Deep Q-learning aided networking, caching, and computing resources allocation in software-defined satellite-terrestrial networks. IEEE Trans Veh Technol 68(6):5871–5883
12. Kebria PM, Khosravi A, Salaken SM et al (2020) Deep imitation learning for autonomous vehicles based on convolutional neural networks. IEEE/CAA J Automatica Sinica 7(1):82–95
13. Han Z, Lei T, Lu Z, et al (2019) Artificial intelligence based handoff management for dense WLANs: a deep reinforcement learning approach. IEEE Access 31688–31701
14. Montori F, Bedogni L, Bononi L (2018) A collaborative internet of things architecture for smart cities and environmental monitoring. IEEE Internet Things J 5(2):592–605
15. Samie F, Bauer L, Henkel J (2019) From cloud down to things: an overview of machine learning in internet of things. IEEE Internet Things J 6(3):4921–4934
16. Ansere JA, Han G, Wang H et al (2019) A reliable energy efficient dynamic spectrum sensing for cognitive radio IoT networks. IEEE Internet Things J 6(4):6748–6759

Weather Identification-Based Multi-level Visual Feature Combination

Ziheng Li[1], Anliang Zhou[2(✉)], and Yilong Geng[2]

[1] University of California, Irvine, Irvine, CA 92697, USA
[2] Beijing Institute of Graphic Communication, No. 1, Xinghua Street, Daxing District, Beijing, China
zhouanliang@bigc.edu.cn

Abstract. Different weather conditions often affect people's life in all ways. Nowadays, a large part of systems must make decision depending on current weather. Vision-based weather recognition is an important way to know weather conditions. In this paper, we bring in multi-level features: high-level features from Convolutional Neural Network (CNN) and classic low-level features, i.e., Scale-Invariant Feature Transform (SIFT) and Histogram of Gradient (HOG) for the weather recognition task. Furthermore, Support Vector Machine (SVM) is applied in this work. To evaluate the effectiveness of the combination of multi-level features and classifiers in the weather recognition, experimental studies are conducted on the public datasets. Experimental results demonstrate that the reasonable features combination in different conditions of data sizes contribute to efficiency performance.

Keywords: Weather classification · Convolutional neural network introduction · Scale-invariant feature transform

1 Introduction

At present, deep learning is sweeping across all fields. More and more people are engaged artificial intelligence and deep learning. We might ask, isn't just deep learning enough for nowadays? The answer is negative. In my opinion, although deep learning automatically extracts features, salient features may not be prominent in the process, resulting in a pan-suboptimal situation. Therefore, the performance of the final task may not be as good as traditional machine learning methods. Traditional machine learning methods, however, combined with human prior knowledge and intuitive feelings, design, and select several features that are strongly related to the task, and will greatly exceed the deep learning in terms of implementation efficiency. If feature extraction and selection are done properly, the results will also outperform deep learning. In addition, deep learning relies on big data, big models, and large computations. It faces many challenges, such as expensive labeling data, inconvenience to use on mobile devices, and expensive costs on material and time.

Q. Liang et al. (eds.), *Artificial Intelligence in China*, Lecture Notes in Electrical Engineering 653, https://doi.org/10.1007/978-981-15-8599-9_14

The work described in this article translates into contributions to the field of weather classification by exploring the use of a combination of traditional methods and deep learning as data augmentation techniques. When using general-purpose images to classify outdoor scenes in multi-class environments consider the difficulty on applying deep learning on limited equipment, enhancing learning results as much as possible with limited equipment. This article uses a dataset consisting of five types of weather. The images in datasets are selected and collected from online database, which called the Sun-Cloud-Rain-Fog-Snow (SCRFS) dataset.

This article is divided into five parts. In the first part, this topic was briefly introduced. The second section details the proposed dataset "SCRFS." The third section mentions the details of the evaluation using the corresponding background of CNN and SIFT and the chosen architecture. The fourth section evaluates the experimental results and the inferences that can be drawn. Finally, the last section sets out the conclusions that can guide future work on the topic.

2 SCRFS Dataset

Due to the limited equipment, totally 425 images are choosing from different open-source datasets, 85 images on each type of weather, to figure the performance on limited equipment. Since images in dataset are collected from Internet, the images are selected based on containing creative commons license [1]. The collected images become SCRFS weather dataset, named after the five categories that dataset includes. Figure 1 shows a group of sample images in SCRFS weather dataset.

Fig. 1 Sample images of dataset

3 Methodology

As introduced, the aim is to discuss that the best way to classify images by their weather condition, with high accuracy and as much efficiency as possible. Mostly focusing on CNN, SVM, extracting features by scale-invariant feature transform and histogram of oriented gradients and classified by SVM classifier. These algorithms offer advanced solution to the problem, and they can be efficient if used in correct using area. This paper introduced all the methods, general approach to classification, and conclusions at the end.

At the beginning, lots of related work are researched and examined on this task and other similar weather recognition tasks. Useful feature types are collected for scene recognition or specific weather classification tasks.

3.1 Resized and Cropped Images

Resizing and cropping images and arranging their files with related classes are the most essential part only for Convolutional Neural Network (CNN). The images are resized by keeping their aspect ratio same,

$$\text{Resize Percent} = \frac{\text{New basesize}}{\text{Oldsize}} \tag{1}$$

$$\text{Newsize} = \text{Oldsize} \times \text{ResizePercent} \tag{2}$$

Then, the cropped residual part from an image by detecting sky area and object area using "Canny Edge Detection" algorithm. After found the sky and object area in the images, the images' non-sky part is cropped as shown in Fig. 2.

Fig. 2 Image resized

3.2 Convolutional Neural Networks

Recently, CNNs are one of the most advanced and widely used classification algorithm in machine learning [2]. Many libraries that implement CNNs can be found. In this project,

Keras library which base uses Tensorflow back end and which is one of the most popular libraries in this field. Since CNNs is a complex algorithm, hyper-parameter changes and even little alterations affect their prediction and result much. Generally, increasing kernel size too much for convolution part can decrease the accuracy. ReLu activation function gives the best performance for the task in both convolution and fully connected layer. Pooling also provides efficiency and increases general accuracy. After 50 epochs, the network starts to overfit and this decreases overall accuracy as shown in Fig. 3.

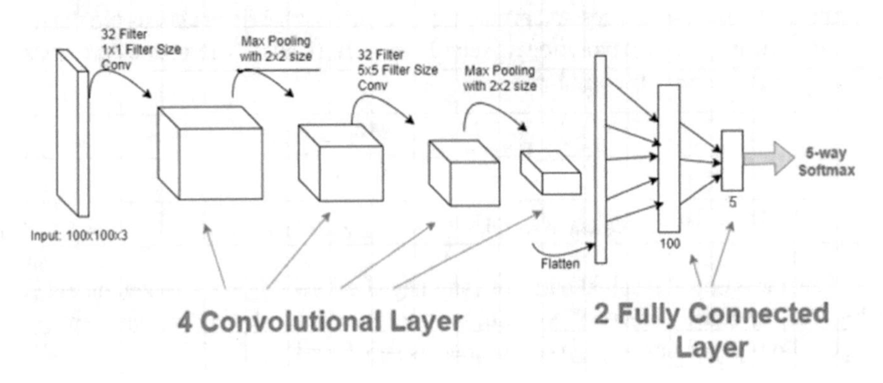

Fig. 3 CNN

3.3 Support Vector Machine

Support Vector Machine (SVMs) is a set of supervised learning methods used for classification, regression, and outliers detection. SVM [3] was well used before deep neural nets. Although nowadays neural nets outperform SVMs due to the complex and advanced structure for machine learning tasks, SVMs still in service and used in various machine learning projects. To run SVMs in the project, implementation in scikit-learn's "SVC" module are used.

3.4 Image Description—Feature Extraction

To apply SVM classifier, feature extraction from images is preprocessed.

3.4.1 Brightness Value

Brightness is the one of the most important pixel characteristics. It also explains weather images nicely and can take different values for different type of weather condition. After research, there is no conventional formula for brightness calculation. But there is a paper shows an effective and useful methods to calculate brightness substitutions. Luma brightness was used, which is widely used in image processing algorithm imitating performance of corresponding color TV adjusting knobs and brightness equivalent in MPEG and JPEG algorithms.

Luma brightness is calculated as:

$$Y\prime = 0.299r + 0.587g + 0.114b \tag{3}$$

where r, g, and b are stimulus sRGB coordinates [4].

3.4.2 Contrast Value

Like brightness, contrast value also explains weather images significantly. Contrast can be explained as the difference between the maximum and minimum pixel intensity in an image [5].

Contrast metric is calculated as:

$$a(x) = \min_{n \in r,g,b} I^n(x) \tag{4}$$

$$b(x) = \max_{n \in r,g,b} I^n(x) \tag{5}$$

$$n = \frac{\sum_{x \in X} a(x)}{S_x} \tag{6}$$

where x is each pixel in image, and S_x is total pixel count.

$$h = \frac{\sum_{x \in X} b(x)}{S_x} \tag{7}$$

where x is each pixel in image, and S_x is total pixel count.

Contrast value:

$$\mathbf{C} = n - h \tag{8}$$

3.4.3 HOG and SIFT Feature Extraction

HOG [6] and SIFT [7] feature extraction are used to combine with SVM classifier. The histogram of oriented gradients (HOG) is a feature descriptor used in computer vision and image processing for the purpose of object detection. By using HOG, the object appearance and shape in the image can be efficiently captured by edge directions.

The scale-invariant feature transform (SIFT) is a feature detection algorithm in computer vision to detect and describe local features in images. The SIFT method would first extract key points from sample dataset, then recognize the object in the new image by comparing features to key points.

For HOG and SIFT feature extraction, functions in libraries OpenCV [8] is used. The HOG feature extraction is as shown in Fig. 4.

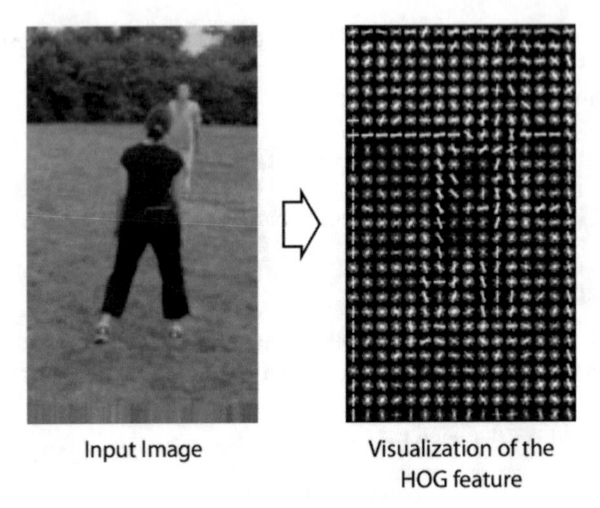

Fig. 4 HOG feature extraction

4 Results

We tried many models for each machine learning method and listed best results (accuracy and f1 score) for each method below. General accuracies are changing between 53.39 and 87.6%, depending on architecture used. The results are shown in Tables 1 and 2.

To evaluate the performance of each method, we used two different evaluation methods: accuracy (recall) and F1-score, the function as below:

Table 1 Accuracy compare

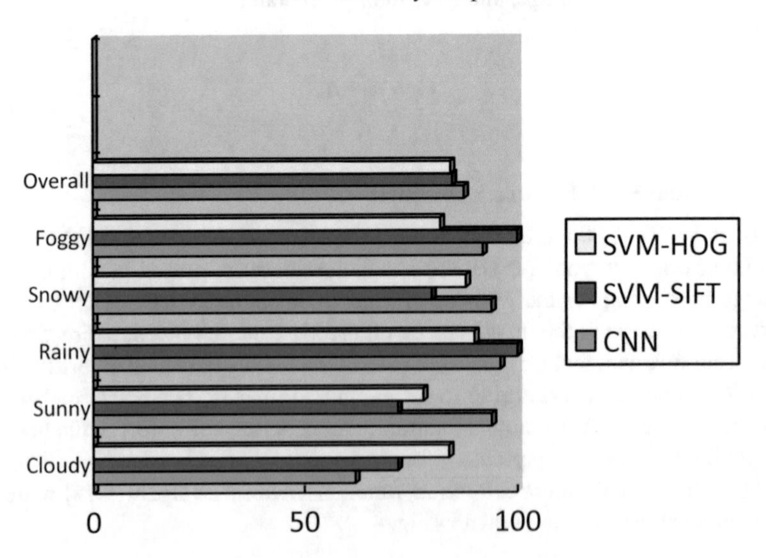

Table 2 F1-score compare

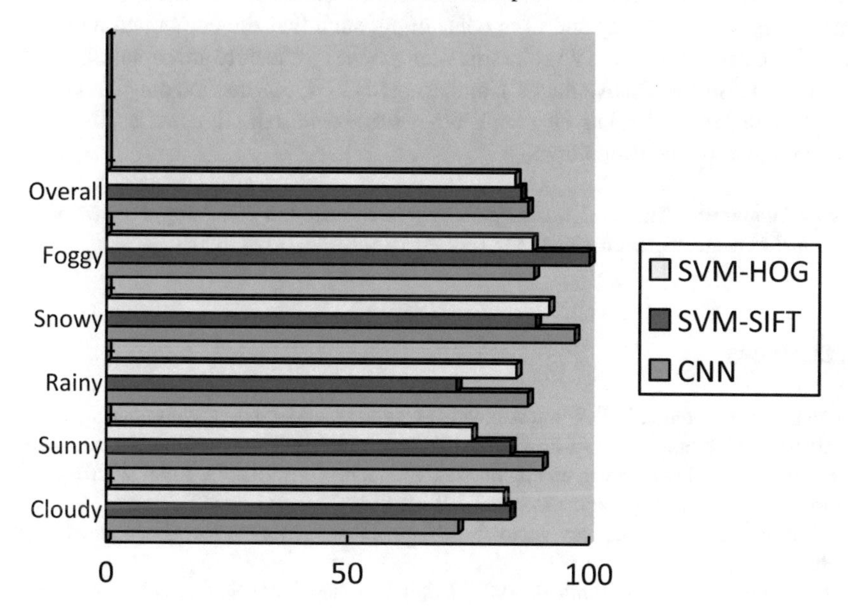

$$\text{precision} = \frac{\text{True - positives}}{\text{True - positives} + \text{False - positives}} \tag{9}$$

$$\text{recall} = \frac{\text{True - positives}}{\text{True - positives} + \text{False - negtives}} \tag{10}$$

$$\text{F-score} = 2 \times \frac{\text{precision} \times \text{recall}}{\text{precision} + \text{recall}} \tag{11}$$

CNN has overall the best accuracy. However, the time consuming is the most. SVM–SIFT have better accuracy on cloudy, rainy, and foggy than CNN SVM–HOG has better accuracy on cloudy than CNN. The result turns out that with the limited dataset, the performance of CNN is similar with the performance of the combination on SVM–HOG and SVM–SIFT. When considering both time efficiency and learning accuracy, we can conclude that the SVM–HOG and SVM–SIFT have better efficiency– with the same learning accuracy, the time cost on SVM–HOG and SVM–SIFT is much less than on CNN.

5 Conclusion

In this paper, we presented a comparison between different classifiers methods and the combination with traditional image feature extraction and ML method. In traditional mindset, deep learning always has overwhelming accuracy compared to traditional methodologies; however, experimental results indicate that the combination of traditional image process and machine learning can also have well result, and sometimes even surpass the overall result brought from one of the most ML method—CNN when considering both accuracy and time consuming, such that the combination of sift and hog feature extraction and SVM classification when right feature are captured. However, this method also has disadvantages. Comparing to CNN, this method only has better efficiency when dataset is relatively small. When the size of dataset increases, the accuracy of this method would drop down.

Acknowledgments. This work was supported by the Science and Technology Projects of Beijing Municipal Education Commission (03150120001/075, KM201911418003).

References

1. Creative Commons (2018) When we share, everyone wins—creative commons
2. Albawi S, Mohammed TA, Al-Zawi S (2017) Understanding of a convolutional neural network. In: International conference on engineering and technology (ICET), Antalya, 2017, pp 1–6. https://sci-hub.tw/10.1109/icengtechnol.2017.8308186
3. Scikit-learn.org. 1.4. Support Vector Machines. https://scikit-learn.org/stable/modules/svm.html
4. Bezryadin S, Bourov P, Ilinih D (2007) Brightness calculation in digital image processing. KWE Int.Inc., San Francisco, CA, USA. UniqueIC's, Saratov, Russia
5. Tutorialspoint. DIP—Brightness and Contrast. https://www.tutorialspoint.com/dip/brightness_and_contrast.htm
6. Dalal N, Triggs B (2005) Histograms of oriented gradients for human detection. In: IEEE computer society conference on computer vision and pattern recognition, San Diego, CA, USA
7. OpenCV. Introduction to SIFT (Scale-Invariant Feature Transform). https://opencv-python-tutroals.readthedocs.io/en/latest/py_tutorials/py_feature2d/py_sift_intro/py_sift_intro.html
8. OpenCV. ImageProcessing in OpenCV. https://docs.opencv.org/3.0-beta/doc/py_tutorials/py_imgproc/py_table_of_contents_imgproc/py_table_of_contents_imgproc.html

Railway Tracks Defects Detection Based on Deep Convolution Neural Networks

Zhong-Jun Wan[1(✉)] and Song-Qi Chen[2,3]

[1] Shanghai Maritime University, NO 1550 Haigang Av.Lingang New City, Pudong District, Shanghai 201306, China
wanzhongj@sina.com, 1627564825@qq.com

[2] Science and Technology on Underwater Vehicle Technology Laboratory, Harbin Engineering University, Harbin 150001, China
723488257@qq.com

[3] Information Department, Beijing City University, Beijing 101399, China

Abstract. With the burgeoning development of railway system throughout the world, accurate and efficient monitoring the state of rail tracks, a method serves one part of ensuring the safe operation of the railway system, are becoming increasingly important. Therefore, monitoring the health of railway track plays an indispensable role during the management of railway system. The appearance of convolutional neural network (CNN) greatly improves the problems of low accuracy and speed of traditional defect detection technology [1]. In this context, a method of rail health monitoring based on convolutional neural network is proposed in this paper. Series of You Only Look Once (YOLO) algorithms were applied to the detection of railway tracks defects, and a modified model (YOLOv3-M) based on YOLOv3 was proposed. In order to verify its effectiveness, experiments were conducted, and the results illustrated that the proposed method can effectively monitor the railway track state prior to its fail.

Keywords: Railway tracks · Defects detection · Deep learning · Convolutional neural networks · YOLOv3

1 Introduction

Rail systems are one of the most preferred transportation methods in today's world because they are fast, reliable, and cost-effective [2]. However, with the rapid development of high-speed railway in the world, the speed and loads of the trains rocket unexpectedly, which in turn increase the risk of producing railway tracks defects [3, 4]. Those railway tracks defects will endanger the safety of rail systems because abnormalities occurred on railway tracks can cause accidents like derailment. Therefore, the railway tracks must be periodically inspected with reliable and economical methods to avoid accidents and ensure the safety. Railway tracks defects detection is an important technical measure to accurately obtain the rail state, and the successful use of the technology will enable railway operators to conduct appropriate maintenance measures prior to the track fail [5].

Q. Liang et al. (eds.), *Artificial Intelligence in China*, Lecture Notes in Electrical Engineering 653, https://doi.org/10.1007/978-981-15-8599-9_15

The traditional railway track defects inspection is initially conducted manually. Railway inspectors conduct regular patrol inspection monthly or yearly based mainly on naked eyes, and their experience assisted by some simple instruments. Workers mainly inspect key parts of the high-speed railway track, recording the problems found during patrol inspection and then reporting to the maintenance department [6]. This kind of detection method is suitable for the railway track which is short or passed by a small number of trains. However, with the burgeoning development of high-speed railway throughout the world, manual detection is no longer applicable. Moreover, railway inspectors are easy to be affected by the surrounding environment and personal subjective impression in the process of manual railway track detection. Consequently, the detection efficiency is low, and it is easy to cause psychological and physiological problems of the railway inspectors.

At present, the detection of rail surface defects mainly uses electromagnetic wave, ultrasonic, or machine vision-based track detection technology. The electromagnetic wave transmits high-frequency electromagnetic wave in the detection of the interior of the medium. If the medium is uneven, the electromagnetic wave will carry structural information and be received by the antenna through reflection or scattering. The received electromagnetic signal can be processed in some ways, and then, the abnormal position, shape, and other information in the medium can be determined. But the information obtained by this method is not rich or accurate [7]. The ultrasonic flaw detection probe can detect the internal damage of high-speed rail by sending continuous ultrasonic pulse. However, the ultrasonic detect technology cannot detect the defects on the surface or defects near the surface. Moreover, the detection effects will be greatly affected by the cleanliness and roughness, and the detection speed will be correspondingly limited [8]. The above methods mainly detect the damage of the internal structure of high-speed rail, and the detection speed and accuracy cannot meet the requirements of rapid detection.

With the rapid development of laser technology and CCD technology, machine vision technology emerges and has been widely used in the detection field [9]. Machine vision technology is based on the image and can get the information such as position and texture after the algorithm analysis of the image. Machine vision detection boasts the advantages of fast detection speed, high-detection accuracy, and low cost and has been widely valued in the world. It is of great social and economic benefits to study the detection technology of the track surface defects based on machine vision. In recent years, deep convolutional neural network has been successfully applied to various computer vision and classification problems in different application fields [10]. The accuracy of these systems has exceeded the classical manual feature learning methods and they have achieved advanced performance in many cases. Some progress in these areas can be used to detect and identify faults in railway tracks, which is of great significance to the safety of railway systems.

To effectively extract image features of railway tracks defects and improve the detecting speed, we proposed a method based on YOLOv3 for railway tracks defects automatic localization. A railway tracks defects dataset was collected, and series of YOLO was used to construct models. Experiments proved that the proposed model was capable of detecting the defects of the track and achieved high accuracy. In the second part of this paper, we introduced the principle of YOLO series algorithm. Then, the method about how we trained models is illustrated. Finally, we analyzed the results and came to a conclusion that the proposed method is efficient.

2 Related Work

YOLO uses a single convolutional neural network to transform the problem of target detection into a regression problem [11]. Firstly, the input image is extracted by the feature extraction network to get a certain size feature map. Then, the input image is divided into $S \times S$ grid cells. If the central coordinate of an object in the ground truth is in which grid cell, the grid cell will predict the object. Each grid cell predicts B b-boxes (bounding box) and confidence score and C conditional class probabilities [12]. Each b-box contains five values(x, y, w, h, confidence). The x and y coordinates are the offsets of the corresponding elements. The w and h coordinates are the width and height of b-box [13]. The degree of confidence reflects the accuracy of the position when the object is included or not, and it is defined as:

$$\Pr(\text{Object}) \times IOU_{\text{pred}}^{\text{truth}}, \quad \Pr(\text{Object}) \in \{0, 1\} \tag{1}$$

The calculation of IOU is:

$$IOU = \frac{\text{Detection result} \cap \text{Ground truth}}{\text{Detection result} \cup \text{Ground truth}} \tag{2}$$

The structure of YOLOv1 network is shown in Fig. 1, which contains 24 convolution layers, four Maxpool layers and two full connection layers [11]. Convolution layer is used to obtain image features by convoluting the image processed by input layer. Its essence is to extract the feature information of input layer for subsequent classification and location processing. The Maxpool layer is used to reduce image pixels by sub-sampling the input data samples in the feature space. The full connection layer is used to predict the classification and location of the image. Its main function is to transform the two-dimensional matrix extracted from the feature into one-dimensional matrix.

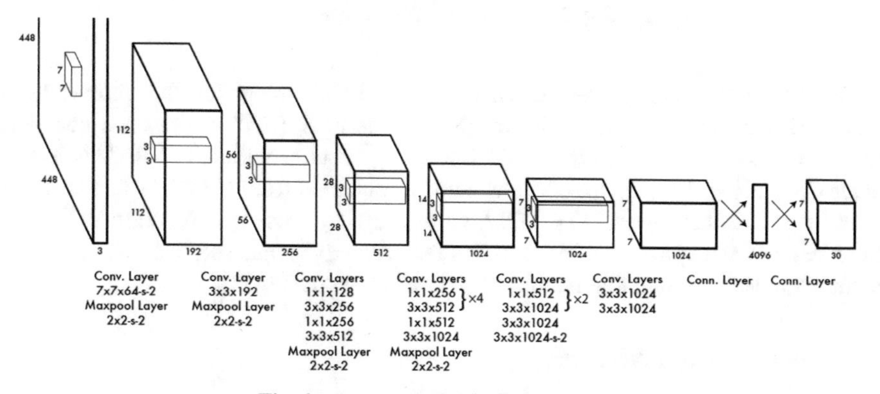

Fig. 1. Structure of YOLOv1 network

To improve the accuracy and the recall rate of object positioning, the author of YOLO put forward YOLOv2. A new structure, called Darknet-19, is used in YOLOv2, which is based on Google net. With 19 volume accumulation layers and 5 Max pooling layers, batch normalization is used in the network to accelerate convergence. Compared with YOLOv1, the resolution of training image is improved, the idea of anchor box in foster R-CNN is introduced to improve the design of network structure, and convolution layer is used to replace the full connection layer of YOLO in output layer [14, 15] (Fig. 2).

Type	Filters	Size/Stride	Output
Convolutional	32	3×3	224×224
Maxpool		$2 \times 2/2$	112×112
Convolutional	64	3×3	112×112
Maxpool		$2 \times 2/2$	56×56
Convolutional	128	3×3	56×56
Convolutional	64	1×1	56×56
Convolutional	128	3×3	56×56
Maxpool		$2 \times 2/2$	28×28
Convolutional	256	3×3	28×28
Convolutional	128	1×1	28×28
Convolutional	256	3×3	28×28
Maxpool		$2 \times 2/2$	14×14
Convolutional	512	3×3	14×14
Convolutional	256	1×1	14×14
Convolutional	512	3×3	14×14
Convolutional	256	1×1	14×14
Convolutional	512	3×3	14×14
Maxpool		$2 \times 2/2$	7×7
Convolutional	1024	3×3	7×7
Convolutional	512	1×1	7×7
Convolutional	1024	3×3	7×7
Convolutional	512	1×1	7×7
Convolutional	1024	3×3	7×7
Convolutional	1000	1×1	7×7
Avgpool		Global	1000
Softmax			

Fig. 2. Structure of YOLOv2 network

Based on the idea of residual network, YOLOv3 integrates the darknet-19 of YOLOv2, and the structure of feature pyramid network (FPN) proposes a new deep feature extraction network darknet-53 (including 53 convolutions) [16]. This is a 53 layer CNN using the skip connections network obtained from RESNET and uses 3×3 and 1×1 rollup layers [17]. YOLOv3 uses binary cross loss function as the loss function, and the upper two layers of the feature map sample the features and merge them with the corresponding feature map of the network (Fig. 3).

3 Our Proposed Method

In the aspect of basic image feature extraction, darknet-53 uses the method of residual network for reference and establishes a quick connection between layers. YOLO3 further

	Type	Filters	Size	Output
	Convolutional	32	3 × 3	256 × 256
	Convolutional	64	3 × 3 / 2	128 × 128
	Convolutional	32	1 × 1	
1×	Convolutional	64	3 × 3	
	Residual			128 × 128
	Convolutional	128	3 × 3 / 2	64 × 64
	Convolutional	64	1 × 1	
2×	Convolutional	128	3 × 3	
	Residual			64 × 64
	Convolutional	256	3 × 3 / 2	32 × 32
	Convolutional	128	1 × 1	
8×	Convolutional	256	3 × 3	
	Residual			32 × 32
	Convolutional	512	3 × 3 / 2	16 × 16
	Convolutional	256	1 × 1	
8×	Convolutional	512	3 × 3	
	Residual			16 × 16
	Convolutional	1024	3 × 3 / 2	8 × 8
	Convolutional	512	1 × 1	
4×	Convolutional	1024	3 × 3	
	Residual			8 × 8
	Avgpool		Global	
	Connected		1000	
	Softmax			

Fig. 3. Structure of YOLOv3 network

uses three different scale feature maps for object detection. With the change of the number and scale of the output feature map, the size of the previous box also needs to be adjusted accordingly [18]. YOLOv3 continues to use k-means clustering to get the size of the previous frame by setting three prior frames for each lower sampling scale and clustering a total of nine-size prior frames [15]. The output of logistic is used to predict object categories that support multiple label objects, not softmax. For the input image, YOLOv3 maps it to three-scale output tensors, indicating the probability that there are different objects in each position of the image [19]. In general, YOLOv3 uses the residual network structure for reference to form a deeper and multi-scale detection, which improves the detection effect of map and small target. The input and output forms of YOLOv3 are as follows: (1) Input 416 × 416 × 3 image and get three different scale prediction results through Darknet network. (2) Each scale corresponds to N channels, including prediction information. (3) Forecast results for each size of anchor for each grid.

Figure 5 shows the overall framework of railway tracks defects detection system. First, more than 2000 railway tracks defects images are collected and labeled. Then, the data augmentation method is used to increase the amount of training data and improve

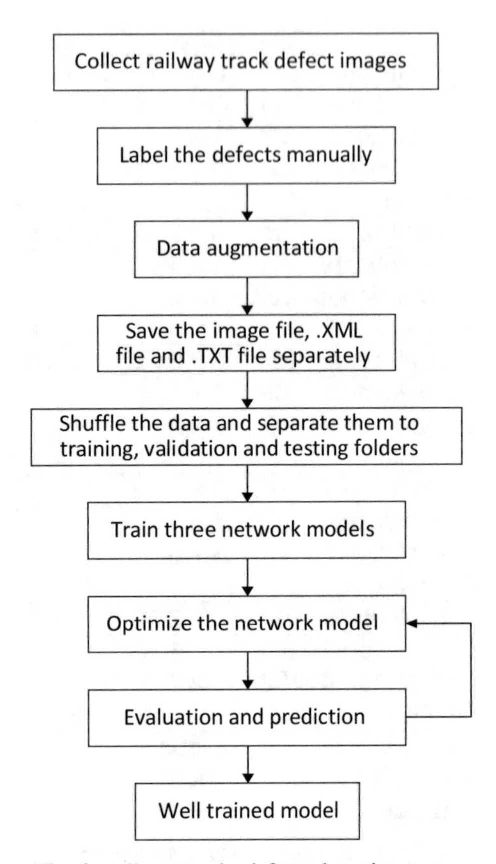

Fig. 4. railway tracks defects detection system

the generalization ability of the model [20]. Images are rotated, mirrored, and being changed brightness so that the original dataset is expanded by 20 times. After saving the image files, .XML files, and .TXT files, YOLOv1, YOLOv2, and YOLOv3 are used to train railway tracks defects detection models separately. By comparing their results, YOLOv3 is chose to optimize the network model. Finally, well-trained model is acquired after evaluation and prediction (Fig. 4).

$$\text{Recall} = \frac{TP}{TP+FN}$$

$$\text{Precision} = \frac{TP}{TP+FP}$$

Fig. 5. Recall and precision

4 Experiment

4.1 Experimental Setup

In this paper, the criteria for the test results are based on the values of average IOU, average recall, and average precision. Among them, TP refers to the positive class judged as positive class, FP refers to the negative class judged as positive class, and FN refers to the positive class judged as negative class [21]. The mean average precision (mAP) score is determined by calculating the mAP over all classes and all intersection over union (IOU) thresholds [22].

We conducted experiments mainly on a server equipped with a Ubuntu 16.04 system. The experimental platform Z390 UD has a CPU Intel i7-8700 K and this two NVIDIA GeForce 2080Ti 12G graphics cards and 64G memory.

Three different networks: 24 convolutional layers, Darknet-19, and Darknet-53 are applied for YOLOV1, YOLOv2, and YOLOV3 training. To ensure that the valuation results can be compared under the same conditions, we set the number of batches to 64 and the subdivisions to 8. The trained network models YOLOv1, YOLOv2, and YOLOv3 were all iterated 10000 times on the 2080Ti GPU.

4.2 Dataset Description

In this paper, 2533 railway defects images were collected and labeled. Part of the railway defects image is shown in Fig. 6. Then, data augmentation was applied here to increase the amount of training data and improve the generalization ability of the model to avoid over-fitting [23]. The dataset was expanded by 20 times, being rotated, mirrored, and changed brightness (Fig. 7).

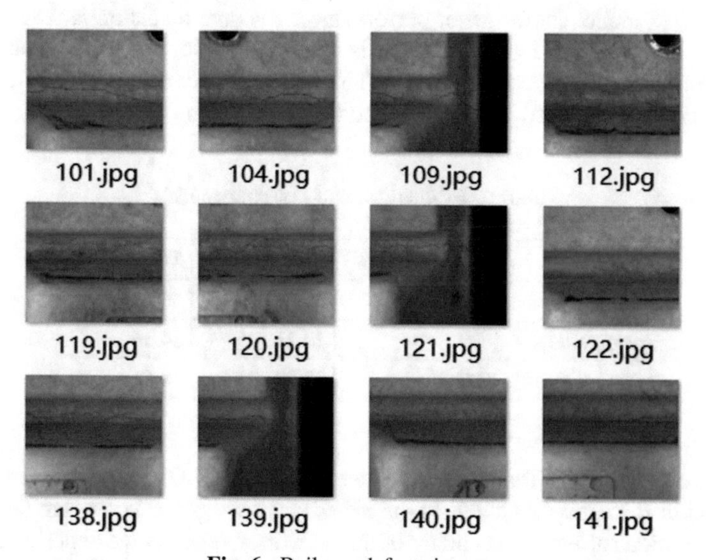

Fig. 6. Railway defects images

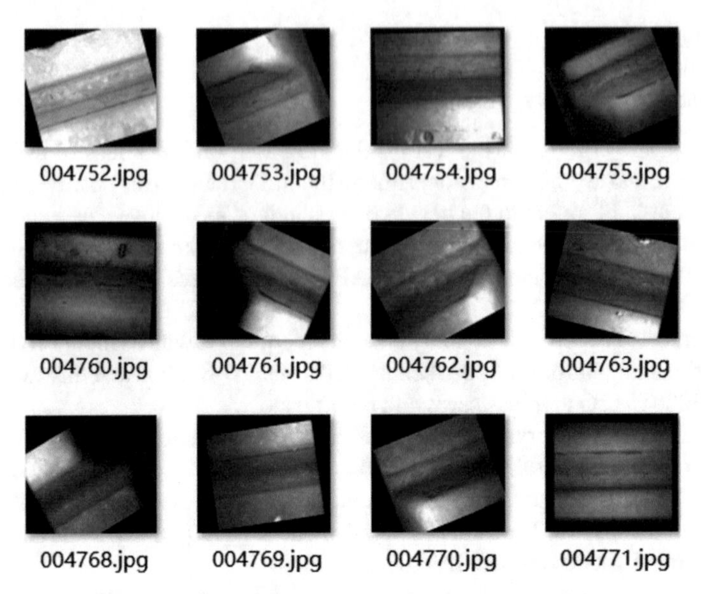

Fig. 7. Railway defects images after data augmentation

4.3 Results and Discussion

The valuation results of three versions of YOLO algorithm is showed in Table 1. YOLOv1's average precision is the lowest at 0.00075, and the IOU and recall values of YOLOv1 are 0. YOLOv2 has IOU at 0.5021 and recall values at 0.5050. The average precision of YOLOv2 is 0.20278, which is 0.20203 higher than that of YOLOv1; however, YOLOv3 model shows better performance: Average precision is 0.56222; value of IOU is 0.6456; and the value of recall is at about 0.8856. From the results, we can arrival at the conclusion that YOLOv1 and YOLOv2 are not suitable for detecting this dataset, while the proposed YOLOv3 is good at detecting the defects of it.

Table 1. Evaluation results of three models

Model	YOLOv1 (%)	YOLOv2 (%)	YOLOv3 (%)
Average precision	0.07532	20.27846	56.22197
Average IOU	0	50.21	64.56
Average recall	0	50.50	88.56

In order to get better training results, we selected YOLOv3 model to improve the parameters of the network model. In the experiment, we get to know that the learning rate and iterations affect the accuracy of the network model, so we experimented with changing the learning rate and iterations to optimize the model and find a better one by comparing the evaluation results.

The parameters of YOLOv3 were changed as follows: The number of iterations was set to 30,000, learning rate was set to 0.001, the policy of steps was used, and parameters are set to 20,000 and 25,000, and scales were set to 0.1 and 0.1. The experimental results were shown in Table 2. The results were shown in Fig. 8. It turns out that the average precision of the model is the best one when the number of batch and learning rate was set to 30,000 and 0.00001, separately.

Table 2. Evaluation results of three models

Model	YOLOv3 (%)	Change1 (%)	Change2 (%)
Average precision	59.42302	67.04187	68.84080
Average IOU	64.82	66.01	67.53
Average recall	89.15	87.07	86.75

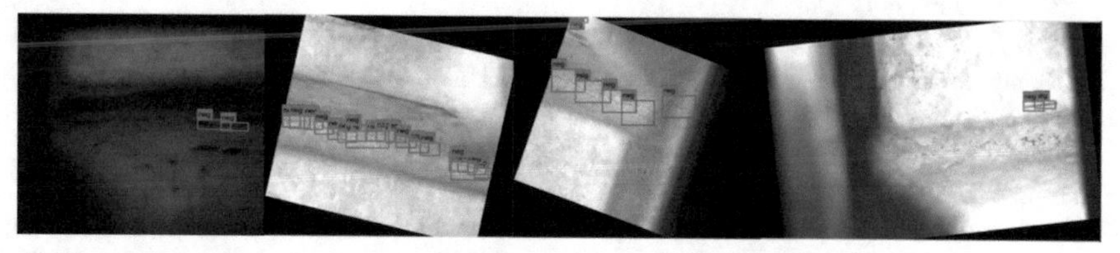

Fig. 8. Experimental test results

5 Conclusions

In this paper, a railway tracks defects detection method based on YOLOv3 was proposed. We used limited tracks defects images to conduct several experiments, and the results show that YOLOv3 model, the model achieved the best performance, is suitable for tracks defects detection. Continuing to optimize the loss function and improve the evaluation performance will be the future direction of our work.

Acknowledgements. This work was supported by the Research Fund from Science and Technology on Underwater Vehicle Technology Laboratory (No. 6142215190103).

References

1. Nie Y, Sommella P, O'Nils M, Liguori C, Lundgren J (2019) Automatic detection of melanoma with yolo deep convolutional neural networks. In: E-health and bioengineering conference (EHB), Iasi, Romania, pp 1–4. 10.1109/EHB47216.2019.8970033
2. Santur Y, Karaköse M, Akin E (2017) A new rail inspection method based on deep learning using laser cameras. In: International artificial intelligence and data processing symposium (IDAP), Malatya, pp 1–6. 10.1109/IDAP.2017.8090245

3. James A et al (2018) TrackNet—a deep learning based fault detection for railway track inspection. In: International conference on intelligent rail transportation (ICIRT), Singapore, pp 1–5. https://doi.org/10.1109/icirt.2018.8641608
4. Zhang X, Wang K, Wang Y, Shen Y, Hu H (2017) An improved method of rail health monitoring based on CNN and multiple acoustic emission events. In: IEEE international instrumentation and measurement technology conference (I2MTC), Turin, pp 1–6. 10.1109/I2MTC.2017.7969693
5. Sun Y, Liu Y, Yang C (2019) Railway joint detection using deep convolutional neural networks. In: IEEE 15th international conference on automation science and engineering (CASE), Vancouver, BC, Canada, pp 235–240. https://doi.org/10.1109/coase.2019.8843245
6. Yin X (2015) Research on machine vision detection algorithm of high speed rail track surface defects. Hunan University
7. Zhang B (2019) Comprehensive application of GPR method and ultrasonic array method in settlement detection of high-speed railway ballastless track. J Eng Geophys 16(05):700–705
8. Tian H (2016) Research on image detection system of track surface defects based on DSP. Shijiazhuang Railway University
9. Zhou Y (2014) High speed railway track detection based on image processing. Shijiazhuang Railway University
10. Liu J, Huang Y, Wang S, Zhao X, Zou Q, Zhang X (2019) Detection method of multi line rail fastener defect based on machine vision. China Railway Sci 40(04):27–35
11. Nie Y, Sommella P, O ' Nils M, Liguori C, Lundgren J (2019) Automatic detection of melanoma with yolo deep convolutional neural networks. In: E-health and bioengineering conference (EHB), Iasi, Romania, pp 1–4. 10.1109/EHB47216.2019.8970033.10
12. Redmon J, Divvala S, Girshick R, Farhadi A (2016) You only look once: unified, real-time object detection. In: IEEE conference on computer vision and pattern recognition (CVPR), Las Vegas, NV, pp 779–788. https://doi.org/10.1109/cvpr.2016.91
13. Engineering—fiber and fabric engineering (2020) New findings on fiber and fabric engineering from Xi'an Polytechnic University summarized (Fabric defect detection using the improved YOLOv3 model). J Eng
14. Redmon J, Farhadi A (2017) YOLO9000: better, faster, stronger. In: IEEE conference on computer vision and pattern recognition (CVPR), Honolulu, HI, pp 6517–6525. https://doi.org/10.1109/cvpr.2017.690
15. Joseph R, Farhadi A (2018) YOLOv3: an incremental improvement. arXiv preprint. arXiv: 1804.02767
16. Yang K, Sun Z, Wang A, Liu R, Wang Y, Sun L, Kang X (2020) Research on target detection method of material defect based on Yolo network system. J Syst Sci (03):70–75 (2020-05-15)
17. Le P, Guo S, Chen J, Lien JJ (2019) Ball-grid-array chip defects detection and classification using patch-based modified YOLOv3. In: International conference on technologies and applications of arti fifi cial intelligence (TAAI), Kaohsiung, Taiwan, pp 1–6. https://doi.org/10.1109/taai48200.2019.8959827
18. Yu S, Cheng Y, Xie L et al (2017) A novel recurrent hybrid network for feature fusion in action recognition. J Visual Commun Image Representation 49(nov.):192–203
19. Zhang X, Wang H, Zhou D, Li J, Liu H (2019) Abnormal detection of substation environment based on improved YOLOv3. In: IEEE 4th advanced information technology, electronic and automation control conference (IAEAC), Chengdu, China, pp 1138–1142. 10.1109/IAEAC47372.2019.8997957
20. Kashif J, Haroon B, Mehreen S (2012) Feature selection based on class-dependent densities for high-dimensional binary data. IEEE Trans Knowl Data Eng 24:465–477. https://doi.org/10.1109/TKDE.2010.263

21. Lu J, Ma C, Li L, Xing X, Zhang Y, Wang Z, Xu J (2018) A vehicle detection method for aerial image based on YOLO. J Comput Commun 6:98–107. https://doi.org/10.4236/jcc.2018.611009
22. Yang S, Xiong Y, Loy C Change, Tang X (2017) Face detection through scale-friendly deep convolutional networks
23. Zhang HW, Zhang LJ, Li PF et al (2018) Yarn-dyed fabric defect detection with YOLOV2 based on deep convolution neural networks. In: IEEE 7th data driven control and learning systems conference (DDCLS), IEEE

Efficiency Evaluation of Deep Model for Person Re-identification

Haijia Zhang[1,2], Sen Wang[1,2], Nuoran Wang[1,2], Shuang Liu[1,2], and Zhong Zhang[1,2](✉)

[1] Tianjin Key Laboratory of Wireless Mobile Communications and Power Transmission, Tianjin Normal University, Tianjin, China
haijia27zhang@gmail.com, wangsenzy@gmail.com, nuoran7607@163.com, shuangliu.tjnu@gmail.com, zhong.zhang8848@gmail.com

[2] College of Electronic and Communication Engineering, Tianjin Normal University, Tianjin, China

Abstract. In this paper, we evaluate the efficiency in training deep models for person re-identification (Re-ID) based on different experimental settings including the number of GPUs and the batch size. To this end, we employ the baseline and PCB to conduct amounts of experiments on Market-1501. The experimental results indicate that what experimental settings have important effects on the efficiency in training deep models.

Keywords: Convolutional neural network · Person re-identification · Efficiency evaluation

1 Introduction

In recent years, person re-identification (Re-ID) is widely applied in many works, such as multi-object tracking [1], cross-view action recognition [2,3], multi-camera body analysis [4] and so on. It focuses on finding specific pedestrians from the gallery set according to the given query images. Person Re-ID is easily affected by various factors, such as posture change,illumination intensity and so on, and therefore it is a challenging task.

Early in the person Re-ID, many researchers extract handcrafted features to represent the pedestrian images [5–9]. These methods usually exploit shallow clues including color and texture to describe pedestrian appearance, and therefore the performance is unsatisfactory. Recently, convolutional neural networks (CNNs) and deep learning are widely employed in VIoT [10], transboundary impact evaluation of Internet [11], social manufacturing [12], person Re-ID [13–16] and other fields, and many deep models based on CNNs and deep learning are proposed to improve the property of person Re-ID [17–22], such as the classification deep model (baseline) [17] and PCB [18]. These methods utilize

Q. Liang et al. (eds.), *Artificial Intelligence in China*, Lecture Notes in Electrical Engineering 653, https://doi.org/10.1007/978-981-15-8599-9_16

CNNs to learn deep features which are more robust to complex variations. Meanwhile, in order to train deep models for person Re-ID, many large-scale datasets are released, such as Market-1501 [23], CUHK03 [24], DukeMTMC-re-ID [25], MSMT17 [26] and so on. The dataset usually contains three subsets including training set, query set and gallery set, and each subset consists of thousands of pedestrian images. It is beneficial to use large-scale datasets for training deep models. However, as the number of pedestrian images grows, time consuming may increase from several hours to days. Hence, it is important to improve efficiency in training deep models.

In this paper, we conduct amounts of experiments to evaluate the efficiency in training deep models under different experimental settings including the number of GPUs and the batch size. Our contributions are summarized in two aspects:

- We elaborate the framework of the baseline and PCB so as to understand deep models.
- We validate that different experimental settings could influence the efficiency in training deep models.

The rest of the paper is divided into three sections. In Sect. 2, we briefly introduce the framework of the baseline and PCB. Then, we detail experimental settings and results in Sect. 3. We make summarization in Sect. 4.

2 Method

In order to more clearly understand the baseline and PCB, we detail the framework of them, respectively.

2.1 Baseline

The backbone of the baseline is ResNet-50 [27] which has 50 levels depth and compact structure so as to make the network possess very powerful capacity of representation. In the baseline, we modify the configuration of the fully connected (FC) layer by changing neurons number from 1000 to 512. Hence, when we feed pedestrian images into the baseline, we obtain features $g \in R^{512}$. Figure 1 shows the framework of the baseline.

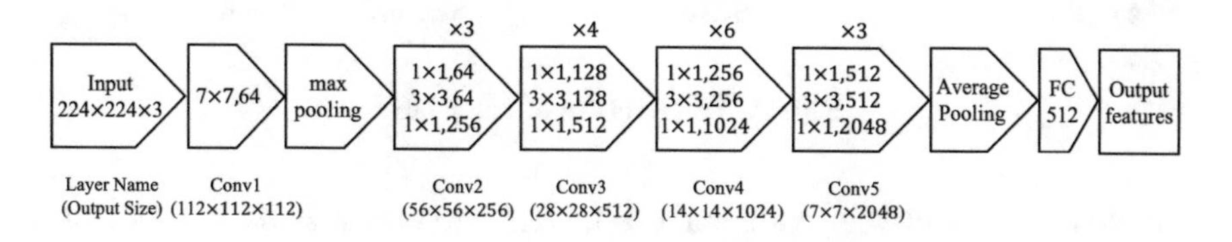

Fig. 1. Framework of the baseline

The cross-entropy loss function is utilized to compute the loss of baseline:

$$L_1 = -\sum_{n=1}^{N} p_n(g) \log q_n(g) \tag{1}$$

where N denotes identities (IDs) number, $p_n(g)$ is the true identity of features g, if g belongs to the n-th ID, then $p_n(g)$ is equal to 1, otherwise to zero, and $q_n(g)$ is prediction value.

During training, we preprocess each pedestrian image by resizing and cropped them into a fixed shape and then feed them into the baseline. In the test stage, we utilize features g extracted from the baseline to represent each pedestrian image and compute distance among them.

2.2 PCB

Similar to the baseline, ResNet-50 is also modified in the PCB to obtain feature maps from pedestrian images. Concretely, we retain the architecture before the average pooling layer, and we set the stride of Conv5_1 layer to 1 to extract larger size of feature maps. With these feature maps, PCB downsamples them in to P stripes along the channel axis to obtain features $h_p \in R^{2048}$ $(p = 1, 2, ..., P)$. Then 1×1 size independent convolutional layers are utilized to reduce the dimension of h_p to 256. Finally, we learn features $f_p \in R^{256}$ $(p = 1, 2, ..., P)$ and utilize them to compute the loss of the PCB:

$$L_2 = -\sum_{p=1}^{P} \sum_{n=1}^{N} p_n(f_p) \log q_n(f_p) \tag{2}$$

where $p_n(f_p)$ is the true identity of features f_p, if f_p belongs to the n-th identity, then $p_n(f_p)$ is equal to 1, otherwise to zero, and $q_n(f_p)$ is prediction value.

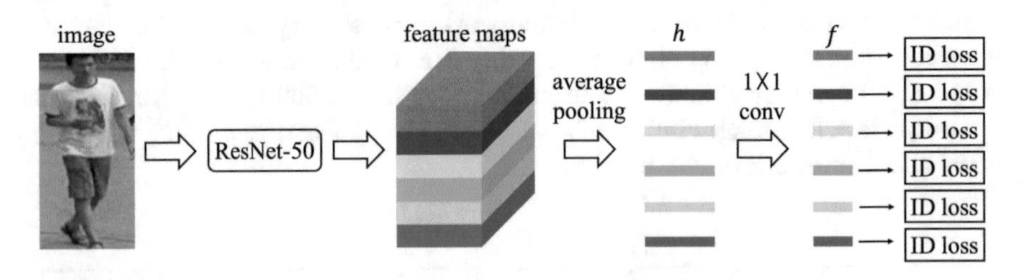

Fig. 2. Architecture of the PCB

In order to intuitively understand the structure of the PCB, we show it in Fig. 2. During training phase, we feed resized images into the PCB to obtain features f_p and then utilize these features to compute loss and optimize the PCB. As for the test phase, we concatenate f_p to produce the final representation of pedestrian images.

3 Experiments

In this section, we firstly introduce Market-1501 dataset, and then we present experiment settings of the baseline and PCB. Finally, we show the experimental results of the baseline and PCB.

3.1 Market-1501

There are totally 32,668 pedestrian images from 1501 IDs in the Market-1501 dataset. These images are divided into three subsets, i.e., training set, query set and gallery set. Concretely, the training set consists of 12,936 images from 751 IDs, the query set contains 3368 images from 750 IDs, and the gallery set is composed of rest 19,732 images. As shown in Fig. 3, some pedestrian images are sampled from Market-1501.

Fig. 3. Some sample images from Market-1501

3.2 Experiment Settings

In the training phase, the total epochs of baseline and PCB are both set to 60. The basic learning rate is initialized to 0.01 and 0.1 for the baseline and PCB, respectively, and they are decayed by 0.1× after 40 epochs.

3.3 Efficiency Evaluation

We evaluate the efficiency in training the baseline and PCB on Market-1501 under different experimental settings including the number of GPUs and the batch size. The time consuming is reported as evaluation metrics.

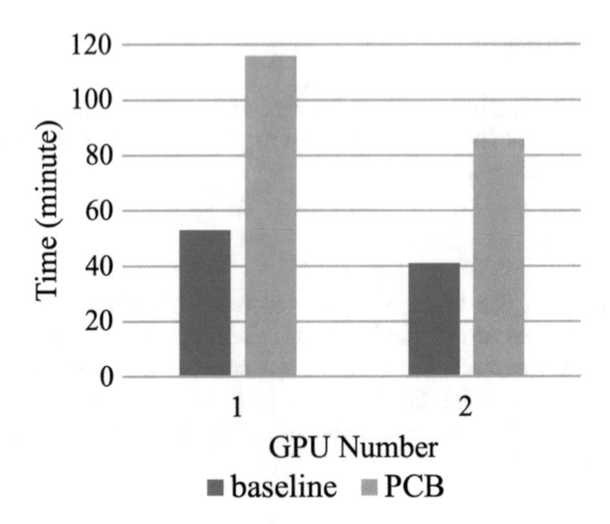

Fig. 4. Influence of GPU number

The influence of GPU number We evaluate the influence of GPU number by utilizing multi-GPUs to train the baseline and PCB. From Fig. 4, we can see that it is beneficial to utilize multi-GPUs in training the baseline and PCB.
The influence of batch size We set the batch size to 32 and 64 to evaluate the influence. As is shown in Table 1, where we can see that enlarging batch size can reduce the time consuming of the baseline and PCB.

Table 1. Influence of batch size

Method	Batch Size	Time (minute)
baseline	32	41.5
	64	34.4
PCB	32	86.3
	64	75.8

4 Conclusion

In this paper, we have evaluated the efficiency of deep models under different experimental settings including the number of GPUs and the batch size. We conduct lots of experiments on the baseline and PCB. We have proved that the two experimental settings have important effects on the efficiency of deep models. Our work is meaningful to help researchers training deep models in a more efficient way.

Acknowledgement. This work was supported by National Natural Science Foundation of China under Grant No. 61711530240, Natural Science Foundation of Tianjin under Grant No. 19JCZDJC31500 and No. 20JCZDJC00180, the Open Projects Program of National Laboratory of Pattern Recognition under Grant No. 202000002 and the Tianjin Higher Education Creative Team Funds Program.

References

1. Milan A, Leal-Taixé L, Reid I, Roth S, Schindler K (2016) MOT16: A benchmark for multi-object tracking. arXiv preprint arXiv:1603.00831
2. Zhang Z, Wang C, Xiao B, Zhou W, Liu S (2012) Action recognition using context-constrained linear coding. IEEE Signal Process Lett 19(7):439–442
3. Zhang Z, Wang C, Xiao B, Zhou W, Liu S, Shi C (2013) Cross-view action recognition via a continuous virtual path. In: IEEE conference on computer vision and pattern recognition, Portland, Oregon, USA, pp 2690–2697
4. Loy CC, Xiang T, Gong S (2009) Multi-camera activity correlation analysis. In: IEEE conference on computer vision and pattern recognition, Miami, Florida, USA, pp 1988–1995
5. Gheissari N, Sebastian TB, Hartley R (2006) Person reidentification using spatiotemporal appearance. In: IEEE conference on computer vision and pattern recognition, New York City, New York, USA, vol 2, pp 1528–1535
6. Farenzena M, Bazzani L, Perina A, Murino V, Cristani M (2010) Person re-identification by symmetry-driven accumulation of local features. In: IEEE conference on computer vision and pattern recognition, San Francisco, California, USA, pp 2360–2367
7. Cheng DS, Cristani M, Stoppa M, Bazzani L, Murino V (2011) Custom pictorial structures for re-identification. In: British machine vision conference, Dundee, Scotland, pp 1–11
8. Liao S, Hu Y, Zhu X, Li SZ (2015) Person re-identification by local maximal occurrence representation and metric learning. In: IEEE conference on computer vision and pattern recognition, Boston, Massachusetts, USA, pp 2197–2206
9. Tan S, Zheng F, Shao L (2015) Dense invariant feature based support vector ranking for person re-identification. In: IEEE global conference on signal and information processing, Orlando, Florida, pp 687–691
10. Zhang Z, Li D (2019) Hybrid cross deep network for domain adaptation and energy saving in visual internet of things. IEEE Int Things J 6(4):6026–6033
11. Xue X, Gao G, Wang S, Feng Z (2018) Service bridge: transboundary impact evaluation method of internet. IEEE Trans Comput Social Syst 5(3):758–772
12. Xue X, Wang S, Zhang L, Feng Z, Guo Y (2019) Social learning evolution (SLE): computational experiment-based modeling framework of social manufacturing. IEEE Trans Industr Inf 15(6):3343–3355
13. Ding S, Lin L, Wang G, Chao H (2015) Deep feature learning with relative distance comparison for person re-identification. Pattern Recogn 48(10):2993–3003
14. Cheng D, Gong Y, Zhou S, Wang J, Zheng N (2016) Person re-identification by multi-channel parts-based CNN with improved triplet loss function. In: IEEE conference on computer vision and pattern recognition, Las Vegas, Nevada, USA, pp 1335–1344
15. Su C, Li J, Zhang S, Xing J, Gao W, Tian Q (2017) Pose-driven deep convolutional model for person re-identification. In: IEEE international conference on computer vision, Venice, Italy, pp 3980–3989

16. Zhang Z, Huang M (2018) Discriminative structural metric learning for person reidentification in visual internet of things. IEEE Int Things J 5(5):3361–3368
17. Zheng Z, Zheng L, Yang Y (2018) A discriminatively learned CNN embedding for person re-identification. ACM Trans Multimedia Comput Commun Appl 14(1):13
18. Sun Y, Zheng L, Yang Y, Tian Q, Wang S (2018) Beyond part models: person retrieval with refined part pooling. In: European conference on computer vision, Munich, Germany, pp 480–496
19. Geng M, Wang Y, Xiang T, Tian Y (2018) Deep transfer learning for person re-identification. In: IEEE international conference on multimedia big data, China, Xi'an, pp 1–5
20. Yao H, Zhang S, Zhang Y, Li J, Tian Q (2019) Deep representation learning with part loss for person re-identification. IEEE Trans Image Process 28(6):2860–2871
21. Zhang Z, Zhang H, Liu S (2019) Coarse-fine convolutional neural network for person re-identification in camera sensor networks. IEEE Access 7:65186–65194
22. Zhang Z, Huang M, Liu S, Xiao B, Durrani T (2019) Fuzzy multilayer clustering and fuzzy label regularization for unsupervised person re-identification. IEEE Trans Fuzzy Syst. https://doi.org/10.1109/TFUZZ.2019.2914626
23. Zheng L, Shen L, Tian L, Wang S, Wang J, Tian Q (2015) Scalable person re-identification: A benchmark. In: IEEE international conference on computer vision, Santiago, Chile, pp 1116–1124
24. Li W, Zhao R, Xiao T, Wang X (2015) Deepreid: deep filter pairing neural network for person re-identification. In: IEEE conference on computer vision and pattern recognition, Boston, Massachusetts, USA, pp 152–159
25. Ristani E, Solera F, Zou R, Cucchiara R, Tomasi C (2016) Performance measures and a data set for multi-target, multi-camera tracking. In: European conference on computer vision, Amsterdam, Netherlands, pp 17–35
26. Wei L, Zhang S, Gao W, Tian Q (2018) Person transfer GAN to bridge domain gap for person re-identification. In: IEEE conference on computer vision and pattern recognition, Salt Lake City, Utah, USA, pp 79–88
27. He K, Zhang X, Ren S, Sun J (2016) Deep residual learning for image recognition. In: IEEE conference on computer vision and pattern recognition, Las Vegas, Nevada, USA, pp 770–778

Cloud Recognition Using Multimodal Information: A Review

Linlin Duan[1,2], Jingrui Zhang[1,2], Yaxiu Zhang[1,2], Zhong Zhang[1,2], Shuang Liu[1,2(✉)], and Xiaozhong Cao[3(✉)]

[1] Tianjin Key Laboratory of Wireless Mobile Communications and Power Transmission, Tianjin Normal University, Tianjin 300387, China
linlinduan07@gmail.com, 19zhangjr@gmail.com, 13453276296@163.com, zhong.zhang8848@gmail.com, shuangliu.tjnu@gmail.com

[2] College of Electronic and Communication Engineering, Tianjin Normal University, Tianjin 300387, China

[3] Meteorological Observation Centre, China Meteorological Administration, Beijing 100081, China
xzhongcao@163.com

Abstract. The cloud recognition technology is helpful in weather analysis, climate research, and communication, and various automatic cloud recognition methods are proposed. Among them, the cloud recognition methods using the multimodal information to show their superiority. In this paper, we give a review of these methods and divide them into two categories, i.e., separate fusion and end-to-end fusion. The separate fusion means directly fusing the multimodal information with features of cloud image, and the end-to-end fusion is to train the fusion of mutlimodal features and visual features of cloud images in an unified framework. Finally, we list experiment results on the cloud datasets to evaluate their performances.

Keywords: Cloud recognition · Multimodal information · Feature fusion

1 Introduction

Clouds are important natural phenomenon, which are closely related to weather, climate, atmospheric radiation, etc. Consequently, reliable cloud observation technology is of great significance to weather analysis, climate research, and communication. Cloud observation methods include two types, i.e., satellite observation and ground-based observation. The satellite observation method captures cloud pictures with wide observation range but low pixel resolution, which is not suitable for further researches of clouds. The ground-based observation method captures cloud images from the ground, and it can capture the local information of clouds which is valuable for related researches.

Q. Liang et al. (eds.), *Artificial Intelligence in China*, Lecture Notes in Electrical Engineering 653, https://doi.org/10.1007/978-981-15-8599-9_17

There are three observation factors of clouds including cloud cover, cloud height, and cloud type. Among the three observation factors, the recognition of cloud type is the most challenging because clouds are consistently changing. Originally, the recognition of cloud type mainly depends on manpower. However, the artificial judgement lacks of objectivity because the subjective judgment varies from person to person. In order to improve the cloud recognition technology, amounts of automatic cloud recognition methods [1–6] are proposed. For example, Liu et al. [1] proposed to extract structural features for cloud recognition, which contains information of cloud gray mean value, estimated cloud fraction, edge sharpness, and cloud mass and gaps information. In [2], the Salient Local Binary Pattern (SLBP) was proposed to obtain texture descriptors of cloud images, which is robust to noise. Zhuo et al. [3] utilized the Color Census Transform (CCT) and a novel automatic block assignment method to learn texture and structure features simultaneously. Ye et al. [4] proposed to extract multi-view features of superpixels in cloud images, which includes color, inside texture, neighbor texture, and global relation.

The Convolutional Neural Network (CNN) [7–11] has shown its superiority in many fields. Thus, many researches generalize CNN to the field of cloud recognition. Ye et al. [12] proposed DeepCloud which first utilizes the pre-trained CNN model to extract deep features, and then Fisher Vector (FV) is applied to encode deep features to represent cloud images. The CloudNet [13] which is composed of five convolutional layers and two Fully Connected (FC) layers was proposed to extract features for cloud recognition. Zhang et al. [14] utilized CNN to learn features maps, and then Local Binary Patterns (LBP) is utilized based on features maps to obtain final representations. 3D-CNN [15] proposed the CNN model to learn features of cloud images, which considers the visual information and temporal information.

The methods mentioned above mainly concentrate on learning features of cloud images, however, cloud types are relevant to many external factors which include temperature, humidity, pressure, wind speed, etc. For instance, the temperature and humidity are closely relevant to the formation of clouds, pressure can influence the movement of clouds, and the wind speed could influence the shape of clouds. We called these external factors as multimodal information. Several cloud recognition methods using the multimodal information has been proposed, and they have obtain better performance in cloud recognition.

In this paper, we give a review of cloud recognition methods using the multimodal information. These methods utilize multimodal cloud samples which are comprised of cloud images and corresponding multimodal information as the inputs, and the learned features contain visual information and multimodal information. From the perspective of vector level fusion, we group these methods into two categories, i.e., separate fusion and end-to-end fusion. Specifically, the separate fusion means directly fusing the multimodal information with features of cloud images [16]. On the other hand, the end-to-end fusion methods [17–19] train the fusion of mutlimodal features and visual features of cloud images in an unified framework.

2 Approach

2.1 Separate Fusion

In this section, we review the Deep Multimodal Fusion (DMF) [16] which belongs to the separate fusion method. DMF fine-tunes the pre-trained CNN model to extract deep visual features of cloud images. The multimodal information in DMF includes six multimodal factors.

Figure 1 is the flowchart of DMF obtaining the final representation. The CNN model extracts deep visual features, and then the multimodal information is concatenated to deep visual features. The fused feature is the final representation of the multimodal cloud sample.

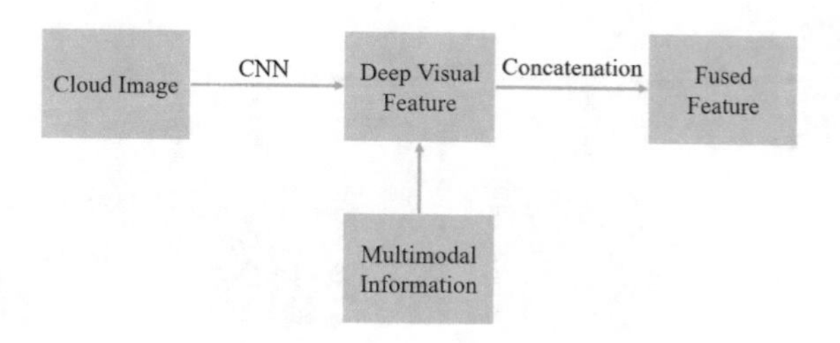

Fig. 1. The flowchart of DMF

DMF considers the multimodal information for cloud recognition, and their experiment results show that the multimodal information is useful in cloud recognition.

2.2 End-to-End Fusion

The end-to-end fusion methods learn deep features of cloud images and multimodal information simultaneously, and they fuse the deep visual features and deep multimodal features in the framework of CNN. As for the fusion of deep visual features and deep multimodal features, there are two fusion strategies, i.e., the single fusion and multiple fusion. We will review three end-to-end fusion methods which apply the two fusion strategies in this section.

Single Fusion Joint Fusion Convolutional Neural Network (JFCNN) [17] contains two subnetworks. As shown in Fig. 2, the vision subnetwork learns deep visual features, and the multimodal subnetwork is utilized to learn deep multimodal features. The proposed joint fusion layer is applied to fuse the two heterogeneous features.

JFCNN feeds the fused features to a FC layer with softmax function, and the cross-entropy loss is applied to train the overall framework. After training, the two heterogeneous features are concatenated to obtain final representations for recognition. The final representation of JFCNN contains the complementary

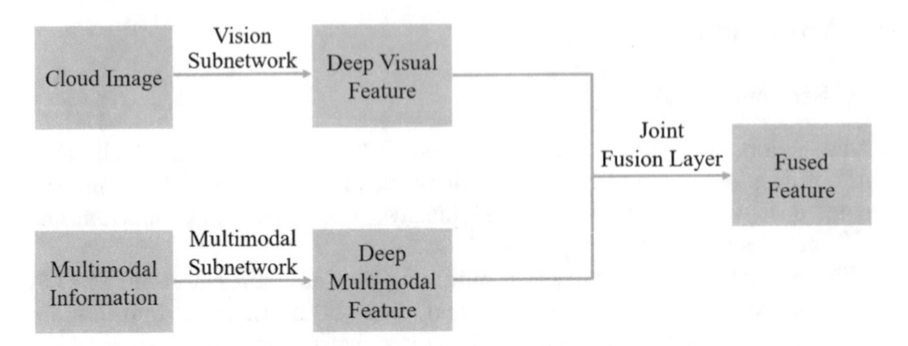

Fig. 2. The flowchart of feature fusion in JFCNN

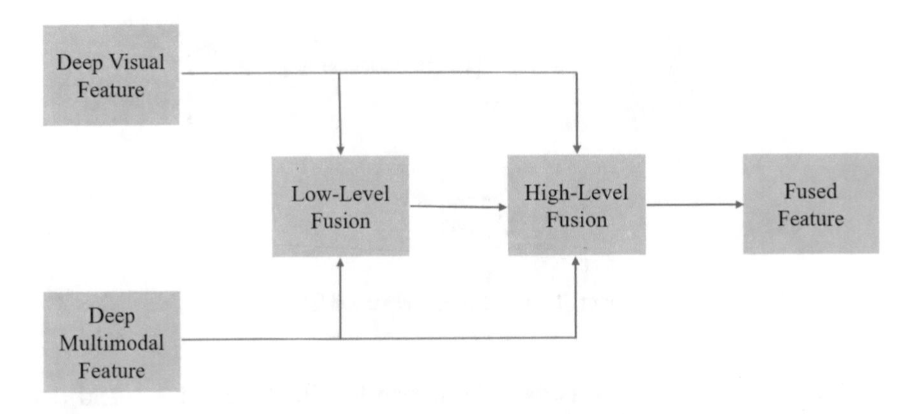

Fig. 3. The multiple fusion in HMF

information of the visual information and multimodal information, which makes the representations more discriminative.

Multiple Fusion In order to capture more correlations between deep visual features and deep multimodal features, Hierarchical Multimodal Fusion (HMF) [18] and Multi-Modal Fusion Network (MMFN) [19] propose a multiple fusion of the two heterogeneous features. HMF proposes the low-level fusion and high-level fusion to fuse the two heterogeneous features as shown in Fig. 3. The deep visual feature and deep multimodal feature are obtained by the visual subnetwork and multimodal subnetwork in HMF. HMF defines the output of high-level fusion as the final representation of multimodal cloud sample, which could mine rich semantic information from the two heterogeneous features, and describe the multimodal cloud sample completely.

MMFN proposes the multi-evidence and multi-modal fusion strategy, which is depicted in Fig. 4. MMFN consists of three subnetworks, i.e., multi-modal network, main network, and attentive network, which have the output of global visual features, local visual features and multi-modal features, respectively. concat1 is used to fuse global visual features and multi-modal features, and concat2 fuses local visual features and multi-modal features. Note that the output of concat1 and concat2 are fed into two FC layer and two cross-entropy losses to train framework of MMFN. After training, the output of concat1 and concat2 are concatenated to represent the multimodal cloud sample.

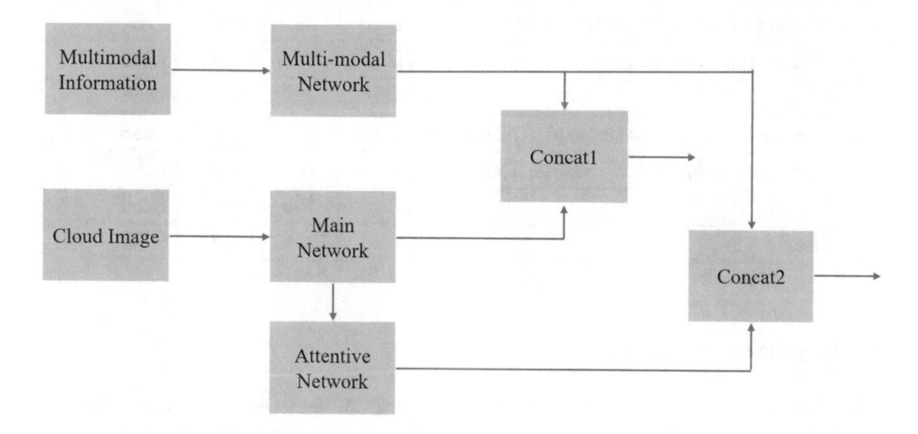

Fig. 4. The flowchart of multi-evidence and multi-modal fusion in MMFN

3 Experiments

3.1 Datasets

The above four cloud recognition methods using multimodal information are experimented on the dataset which includes amounts multimodal cloud samples. Each multimodal cloud sample is captured by a fisheye lens and small weather station from the ground. DMF utilizes the Multimodal Ground-based Cloud (MGC) dataset which has 1,720 multimodal cloud samples, and the cloud samples are divided into 7 types according to the World Meteorological Organization (WMO) [20]. Table 1 lists the sample number in each cloud type. The cloud images in MGC dataset are with the size of 1056×1056, and the multimodal information includes six factors which include the temperature, humidity, pressure, wind speed, average wind speed, and maximum wind speed in one minute. JFCNN performs experiments on the expanded MGC (MGC^l) which has 3177 multimodal cloud samples.

The HMF and MMFN utilize the Multimodal Ground-based Cloud Database (MGCD), in which the multimodal cloud samples are captures by a fisheye lens and small weather station. MGCD has 8000 multimodal cloud samples, and the size of cloud image is 1024×1024. The multimodal information in MGCD includes four factors, i.e., temperature, humidity, pressure, and wind speed. The sample number of each cloud type is listed in Table 1.

Table 1. The sample number of each cloud type on MGC, MGC^l, and MGCD

Cloud Type	Label	Sample Number		
		MGC	MGC^l	MGCD
Cumulus	1	160	330	1438
Altocumulus and Cirrocumulus	2	300	585	731
Cirrus and Cirrostratus	3	340	537	1323
Clear Sky	4	350	699	1338
Stratocumulus, Stratus and Altostratus	5	250	534	963
Cumulonimbus and Nimbostratus	6	140	543	1187
Mixed	7	180	483	1020
Total		1720	3711	8000

3.2 Experiment Results

In this part, we show the recognition accuracy of the four methods on their own datasets in Table 2. These methods obtain higher recognition accuracies than other methods which extract hand-crafted features of cloud samples or not consider the multimodal information. Consequently, considering the visual information and multimodal information contributes to the promotion of cloud recognition technology.

Table 2. Recognition accuracy of the reviewed methods on their own dataset

Method	Accuracy (%)
DMF	86.30
JFCNN	93.37
HMF	87.90
MFNN	88.63

4 Conclusion

In this paper, we have reviewed some cloud recognition methods using the multimodal information. We group these methods according to whether the multimodal information is fused separately or in an end-to-end manner. The experiment results tested on multimodal cloud samples demonstrate that the multimodal information is of great value to cloud recognition. It is worth making profound study on the cloud recognition using the multimodal information.

Acknowledgments. This work was supported by National Natural Science Foundation of China under Grant No. 61711530240, Natural Science Foundation of Tianjin under Grant No. 19JCZDJC31500 and 20JCZDJC00180, the Open Projects Program of National Laboratory of Pattern Recognition under Grant No. 202000002, and the Tianjin Higher Education Creative Team Funds Program.

References

1. Liu L, Sun X, Chen F, Zhao S, Gao T (2011) Cloud classification based on structure features of infrared images. J Atmospheric Oceanic Technol 28(3):410–417
2. Liu S, Wang C, Xiao B, Zhang Z, Shao Y (2013) Salient local binary pattern for ground-based cloud classification. Acta Meteorologica Sinica 27:211–220
3. Zhuo W, Cao Z, Xiao Y (2014) Cloud classification of ground-based images using texture-structure features. J Atmosphere Oceanic Technol 31(1):79–92
4. Ye L, Cao Z, Xiao Y, Yang Z (2019) Supervised fine-grained cloud detection and recognition in whole-sky images. IEEE Trans Geosci Remote Sens 57(10):7972–7985
5. Liu S, Duan L, Zhang Z, Cao X, Durrani TS (2020) Multimodal ground-based remote sensing cloud classification via learning heterogeneous deep features. IEEE Trans Geosci Remote Sens. https://doi.org/10.1109/TGRS.2020.2984265
6. Liu S, Li M, Zhong Z, Cao X, Durrani TS (2020) Ground-based cloud classification using task-based graph convolutional network. Geophys Res Lett 47(5):e2020GL087338
7. Kong S, Kim M, Hoang LM, Kim E (2018) Automatic LPI radar waveform recognition using CNN. IEEE Access 6:4207–4219
8. Yuan L, Wei X, Shen H, Zeng L, Hu D (2018) Multi-centerbrainimaging classification using a novel 3D CNN approach. IEEE Access 6:49925–49934
9. Yuan C, Xia Z, Jiang L, Cao Y, Jonathan Wu QM, Sun X (2019) Fingerprint liveness detection using an improved CNN with image scale equalization. IEEE Access 7:26953–26966
10. Xue X, Wang S, Zhang L, Feng Z, Guo Y (2019) Social learning evolution (SLE): Computational experiment-based modeling framework of social manufacturing. IEEE Trans Industr Inf 15(6):3343–3355
11. Xue X, Gao J, Wang S, Feng Z (2018) Service bridge: Transboundary impact evaluation method of internet. IEEE Trans Comput Social Syst 5(3):758–772
12. Ye L, Cao Z, Xiao Y, Li W (2017) DeepCloud: Ground-based cloud image categorization using deep convolutional features. IEEE Trans Geosci Remote Sens 55(10):5729–5740
13. Zhang J, Liu P, Zhang F, Song Q (2018) CloudNet: Ground-based cloud classification with deep convolutional neural network. Geophys Res Lett 45(16):8665–8672
14. Zhang Z, Li D, Liu S, Xiao B, Cao X (2018) Multi-view ground-based cloud recognition by transferring deep visual information. Appl Sci 8(5):748
15. Zhao X, Wei H, Wang H, Zhu T, Zhang K (2019) 3D-CNN-based feature extraction of ground-based cloud images for direct normal irradiance prediction. Sol Energy 181(15):510–518
16. Liu S, Li M (2018) Deep multimodal fusion for ground-based cloud classification in weather station networks. EURASIP J Wireless Commun Networking 2018(1):48
17. Liu S, Li M, Zhang Z, Xiao B, Cao X (2018) Multimodal ground-based cloud classification using joint fusion convolutional neural network. Remote Sens 10(6):822

18. Liu S, Duan L, Zhang Z, Cao X (2019) Hierarchical multimodal fusion for ground-Based cloud classification in weather station networks. IEEE Access 7:85688–85695
19. Liu S, Li M, Zhang Z, Xiao B, Durrani TS (2020) Multi-evidence and multi-modal fusion network for ground-based cloud recognition. Remote Sens 12(3):464
20. World Meteorological Organization (2017) International cloud atlas: Manual on the observation of clouds and other meteors. Accessed: Mar. 22, 2020. [Online]. Available:https://cloudatlas.wmo.int/home.html/

Graph Convolution Network for Person Re-identification

Wenmin Huang[1,2], Yilin Xu[1,2], Zhong Zhang[1,2], and Shuang Liu[1,2](✉)

[1] Tianjin Key Laboratory of Wireless Mobile Communications and Power Transmission, Tianjin Normal University, Tianjin, China
[2] College of Electronic and Communication Engineering, Tianjin Normal University, Tianjin, China
{huangwenmin2018,xuyilin416,zhong.zhang8848,shuangliu.tjnu}@gmail.com

Abstract. This paper introduces the application of Graph Convolutional Network (GCN) for person re-identification (re-ID). Specifically, GCN approaches for person re-ID are divided into two groups including global features and local features. We explain the motivation and implementation details for each approaches, and compare the GCN approaches with the Convolutional Neural Network (CNN) methods. Experimental results demonstrate the effectiveness of the GCN approaches.

Keywords: Graph convolutional network · Person re-identification · Graph structure data

1 Introduction

In recent years, a new deep learning model called Graph Convolutional Network (GCN) is proposed to process graph structure data [1–3]. The key idea is to implement the aggregation of features between nodes using convolution operation so as to update the features of nodes. According to the definition of convolutional operation, GCN can be divided into two groups, i.e., spectral methods and spatial methods. Spectral methods [4,5] realize convolution operation in frequency domain by means of convolution theorem. As for the spatial methods [6–8], they implement convolution operation directly on the nodes of graph.

GCN has been utilized for various tasks due to its effectiveness in processing graph structure data, and achieves great success. For example, Wang et al. [9] use GCN to complete link prediction between face samples, thus achieving better face clustering. Chen et al. [10] employ GCN to generate inter-dependent classifiers to settle the problem of image recognition. Shi et al. [11] propose two-stream Adaptive Graph Convolutional Network (2s-AGCN) to improve the accuracy of action recognition. These recent work shows that GCN has the property of universality and it is a valuable research direction to solve the existing problems [12–16].

Q. Liang et al. (eds.), *Artificial Intelligence in China*, Lecture Notes in Electrical Engineering 653, https://doi.org/10.1007/978-981-15-8599-9_18

This work aims to introduce the application of GCN for person re-identification (re-ID). Person re-ID, identifying a specified pedestrian at different time or scenes, is a hot research topic in computer vision [17–20]. It plays a significant role in intelligent surveillance system. In the recent literature, some work [21–23] has applied GCN to person re-ID with satisfactory results. Unfortunately, there is little literature to fully describe these latest studies and this is not conducive to the promotion and application of GCN for person re-ID community.

For this paper, we try to explain the latest research progress of GCN for person re-ID by introducing several representative methods. Specifically, for the sake of clarity, we first classify GCN methods for person re-ID into two categories: global features and local features. The global feature methods include Similarity-Guided Graph Neural Network (SGGNN) [21] and Context Graph Model (CGM) [22]. They take the global feature of pedestrian image as the node feature. The local feature methods include Part-based Hierarchical Graph Convolutional Network (PH-GCN) [23]. It takes the local feature of pedestrian as the node feature. Then, we explain the motivation and implementation details for each approach. Finally, we compare the experimental results of above methods with the Convolutional Neural Network (CNN) methods, which indicates that the application of GCN to person re-ID is a valuable research direction.

To summarize, the contributions of this paper can be summarized as the following two aspects:

(1) We introduce the motivation and implementation details of two kinds of GCN methods for person re-ID.

(2) We compare the GCN methods with the CNN methods, and the experimental results show the effectiveness of the GCN methods.

2 Approach

2.1 GCN for Global Feature

In this subsection, we introduce the motivation and implementation details for SGGNN and CGM.

2.1.1 SGGNN Motivation. To estimate the similarity of prose-gallery image pairs, the previous methods only consider information from the two images. However, for hard sample pairs, it is difficult to make a correct judgment only using this limited information. To overcome this limitations, we require to find useful internal connections between images.

Implementation details. Figure 1 gives the framework of SGGNN. Firstly, the image pairs consisting of probe and each gallery image are input into a Siamese-CNN based on ResNet-50 [24] and feature pairs are obtained. Then, element subtraction and square operation are performed for each feature pair to obtain the residual features. Finally, these residual features are processed by a Batch Normalization layer [35] and are fed into the GCN for graph learning.

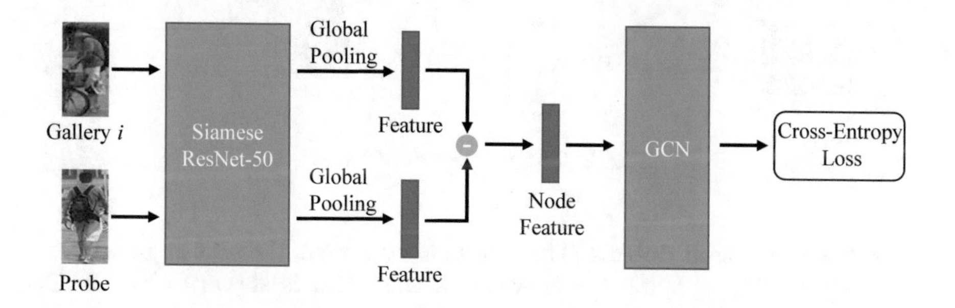

Fig. 1. The framework of SGGNN

In order to improve the contributions of hard sample pairs in the learning process, SGGNN creates graph structure data to model the relationship between image pairs. The residual feature of the probe-gallery pair is taken as the node feature, and the edge weight between node u and node v is defined as:

$$w_{u,v} = \begin{cases} 0,\ u = v \\ \dfrac{e^{S(g_u, g_v)}}{\sum_v e^{S(g_u, g_v)}},\ u \neq v \end{cases} \tag{1}$$

where g_u and g_v indicate u-th and v-th features of gallery image, respectively. $\mathrm{S}(g_u, g_v)$ represents the similarity between g_u and g_v.

Furthermore, the update strategy of node features is written as:

$$f_u^{out} = \eta f_u^{in} + (1 - \eta) \sum_v w_{u,v} Y(f_v^{in}), u \neq v \tag{2}$$

where f_u^{in} and f_u^{out} represent the input and output features of node u, respectively. f_v^{in} represents the input feature of node v, η is the hyperparameter, and Y is a mapping function composed of two fully connected layers.

To explain the graph learning mechanism, we assume that (p, g_u) and (g_u, g_v) are both simple positive pairs, while (p, g_v) is a hard positive pair. According to Eq. 1, (p, g_u) and (p, g_v) have large edge weight. As the message propagation on the graph, (p, g_u) can improve the similarity score of (p, g_v).

2.1.2 CGM Motivation.

Most existing approaches focus on learning pedestrian individual feature. However, pedestrian always appear in groups. Hence, considering the presence of other neighboring pedestrians with the target pedestrian can help us make more confident judgment.

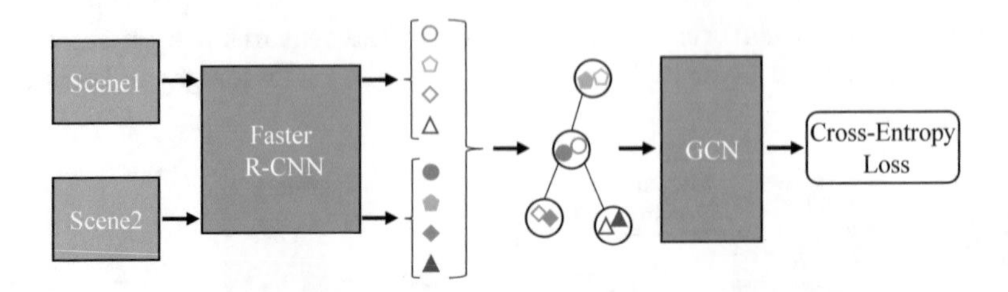

Fig. 2. The framework of CGM

Implementation details. This work only discusses the part of graph learning in CGM. The simplified framework of the CGM is shown in Fig. 2. The instance detection model based on Faster R-CNN [25] firstly detects the target pedestrian and three neighboring pedestrians in the background from each scene and obtains their global features. Then, the target pedestrian feature pair is taken as the target node feature, and the matching neighboring pedestrian feature pairs are taken as the context node feature to construct the graph structure data. Specifically, each context node is connected to the target node. If the target node is taken as the first node in the graph, the adjacent matrix is defined as:

$$A_{u,v} = \begin{cases} 1, \ u = 1 \ or \ v = 1 \ or \ u = v \\ 0, \ otherwise \end{cases} \tag{3}$$

Further, the i-th graph convolutional layer is formulated as:

$$X^{i+1} = \sigma(\widehat{A}X^i W^i) \tag{4}$$

where σ is the activation function, $\widehat{A}$ is the normalized A, W^i is the weight matrix that can be learned, and X^i is the node feature matrix.

Finally, all node features on the graph are combined into a 1024-dim feature and is fed into classifier to speculate whether the target pedestrian is the same pedestrian.

2.2 GCN for Local Feature

In this subsection, we introduce the motivation and implementation details for PH-GCN.

2.2.1 PH-GCN Motivation. The effectiveness of methods based on the part feature for person re-ID has been certified. However, most of them independently extract the local features of pedestrian, which ignores the relationship among local features. To overcome such shortcoming, PH-GCN establishes a hierarchy graph to model the relationship among local features.

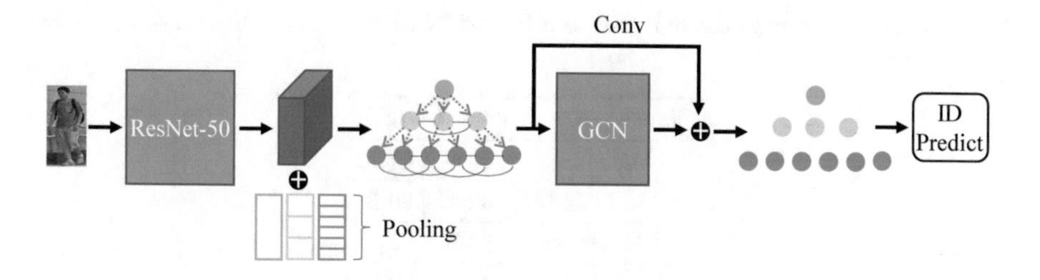

Fig. 3. The framework of PH-GCN

Implementation details. The framework of PH-GCN is given in Fig. 3. We first input image into ResNet-50 and obtain the convolutional activation map with 2048 channels. Then, PH-GCN performs three different forms pooling on the convolution activation map to obtain the hierarchical local features. Furthermore, PH-GCN applies local feature as node to establish the graph structure data. Specifically, the connection between nodes is pre-defined as shown in Fig. 3. If there is edge between node u and node v, the edge weight between them is defined as:

$$A_{u,v} = e^{(-\frac{\|x_u - x_v\|_2}{\beta})} \tag{5}$$

where x_u and x_v represent the features of node u and node v, respectively, and β is a hyperparameter.

The i-th graph convolutional layer is formulated as:

$$X^{(i+1)} = \sigma[((1-\eta)\widehat{A}X^i + \eta X^i)W^i] \tag{6}$$

where X^i is the node feature matrix, σ is the activation function, η is the hyperparameter, W^i is the learnable weight, and $\widehat{A}$ is normalized A.

Finally, the hierarchical local features are convolved and added to the node features output by GCN to acquire the final part feature representation.

3 Experiments

3.1 Dataset

Market-1501 [26] is a widely used person re-ID dataset. It contains 32,668 images of 1501 pedestrians. **CUHK-SYSU** [27] is a large-scale pedestrian retrieval dataset. It consists of 96,143 bounding boxes with 8432 pedestrians. We present the experimental results of SGGNN and PH-GCN on Market1501. As for the CGM, we report its experimental results on CUHK-SYSU.

3.2 Comparison with Other Approaches

The experimental results of SGGNN and PH-GCN on Market-1501 are listed in Table 1. SGGNN obtains 82.8% on mAP and 92.3% on top-1 accuracy. PH-GCN

Table 1. Comparison of the SGGNN and PH-GCN with the other methods on Market-1501 dataset

Methods	top-1	mAP
CADL [28]	73.8	47.1
OIM Loss [27]	82.1	60.9
SVDNet [29]	82.3	62.1
JLML [30]	85.1	65.5
HA-CNN [31]	91.2	75.7
SGGNN	**92.3**	**82.8**
PH-GCN	**93.5**	**79.0**

Table 2. Comparison of the CGM with the other methods on CUHK-SYSU dataset

Methods	top-1	mAP
CNN+DSIFT+Euclidean [32]	39.4	34.5
CNN+BoW+Cosine [26]	62.3	56.9
OIM Loss [27]	78.7	75.5
NPSM [33]	81.2	77.9
CNN+MGTS [34]	83.7	83.0
CGM	**86.5**	**84.1**

achieves 79.0% on mAP and 93.5% on top-1 accuracy. Their results are superior to other CNN methods, which shows that GCN can enhance the performance of person re-ID.

Table 2 lists the comparative results on CUHK-SYSU databset, where CGM gains the best performance. Compared with the best CNN methods, CGM achieves +1.1% on mAP and +2.8% on top-1 accuracy, which shows that GCN can effectively model the context information of pedestrians.

4 Conclusion

In this work, we have explained the application of GCN for person re-ID by introducing the motivation and implementation details of three representative methods including SGGNN, CGM and PH-GCN. Furthermore, we have compared GCN methods with CNN methods, and the experimental results show that GCN is beneficial to enhance the performance of person re-ID.

Acknowledgments. This work was supported by National Natural Science Foundation of China under Grant No. 61711530240, Natural Science Foundation of Tianjin under Grant No. 19JCZDJC31500 and 20JCZDJC00180, the Open Projects Program of National Laboratory of Pattern Recognition under Grant No. 202000002, and the Tianjin Higher Education Creative Team Funds Program.

References

1. Kipf TN, Welling M (2016) Semi-supervised classification with graph convolutional networks [Online]. Available: https://arxiv.xilesou.top/abs/1609.02907
2. Hamilton W, Ying Z, Leskovec J (2017) Inductive representation learning on large graphs. Neural Inf Process Syst, pp 1024–1034
3. Velickovic P, Cucurull G, Casanova A, Romero A, Lio P, Bengio Y (2017) Graph attention networks [Online]. Available: https://arxiv.xilesou.top/abs/1710.10903
4. Defferrard M, Bresson X, Vandergheynst P (2016) Convolutional neural networks on graphs with fast localized spectral filtering. Neural Inf Process Syst, pp 3844–3852
5. Bruna J, Zaremba W, Szlam A, LeCun Y (2013) Spectral networks and locally connected networks on graphs [Online]. Available: https://arxiv.xilesou.top/abs/1312.6203
6. Ma X, Zhang T, Xu C (2019) GCAN: Graph convolutional adversarial network for unsupervised domain adaptation. In: IEEE conference on computer vision and pattern recognition, pp 8266–8276
7. Monti F, Boscaini D, Masci J, Rodola E, Svoboda J, Bronstein MM (2017) Geometric deep learning on graphs and manifolds using mixture model CNNs. In: IEEE conference on computer vision and pattern recognition, pp 5115–5124
8. Niepert M, Ahmed M, Kutzkov K (2016) Learning convolutional neural networks for graphs. In: International conference on machine learning, pp 2014–2023
9. Wang Z, Zheng L, Li Y, Wang S (2019) Linkage based face clustering via graph convolution network. In: IEEE conference on computer vision and pattern recognition, pp 1117–1125
10. Chen ZM, Wei XS, Wang P, Guo Y (2019) Multi-label image recognition with graph convolutional networks. In: IEEE conference on computer vision and pattern recognition, pp 5177–5186
11. Shi L, Zhang Y, Cheng J, Lu H (2019) Two-stream adaptive graph convolutional networks for skeleton-based action recognition. In: IEEE conference on computer vision and pattern recognition, pp 12026–12035
12. Zhang Z, Li D (2019) Hybrid cross deep network for domain adaptation and energy saving in visual internet of things. IEEE Int Things J 6(4):6026–6033
13. Zhang Z, Wang C, Xiao B, Zhou W, Liu S, Shi C (2013) Cross-view action recognition via a continuous virtual path. In: IEEE conference on computer vision and pattern recognition, pp 2690–2697
14. Zhang Z, Wang C, Xiao B, Zhou W, Liu S (2012) Action recognition using context-constrained linear coding. IEEE Signal Process Lett 19(7):439–442
15. Xue X, Wang S, Zhang L, Feng Z, Guo Y (2019) Social learning evolution (SLE): computational experiment-based modeling framework of social manufacturing. IEEE Trans Industr Inf 15(6):3343–3355
16. Xue X, Gao J, Wang S, Feng Z (2018) Service bridge: transboundary impact evaluation method of internet. IEEE Trans Comput Social Syst 5(3):758–772
17. Chen D, Xu D, Li H, Sebe N, Wang X (2018) Group consistent similarity learning via deep CRF for person re-identification. In: IEEE conference on computer vision and pattern recognition, pp 8649–8658
18. Zheng L, Yang Y, Hauptmann AG (2016) Person re-identification: past, present and future [Online]. Available: https://arxiv.xilesou.top/abs/1610.02984
19. Zhang Z, Huang M, Liu S, Xiao B, Durrani T (2019) Fuzzy multilayer clustering and fuzzy label regularization for unsupervised person re-identification. IEEE Trans Fuzzy Syst. https://doi.org/10.1109/TFUZZ.2019.2914626

20. Zhang Z, Huang M (2018) Discriminative structural metric learning for person re-identification in visual internet of things. IEEE Int Things J 5(5):3361–3368
21. Shen Y, Li H, Yi S, Chen D, Wang X (2018) Person re-identification with deep similarity-guided graph neural network. In: European conference on computer vision, pp 486–504
22. Yan Y, Zhang Q, Ni B, Zhang W, Xu M, Yang X (2019) Learning context graph for person search. In: IEEE conference on computer vision and pattern recognition, pp 2158–2167
23. Jiang B, Wang X, Luo B (2019) PH-GCN: Person re-identification with part-based hierarchical graph convolutional network [Online]. Available: https://arxiv.xilesou.top/abs/1907.08822
24. He K, Zhang X, Ren S, Sun J (2016) Deep residual learning for image recognition. In: IEEE conference on computer vision and pattern recognition, pp 770–778
25. Ren S, He K, Girshick RB, Sun J (2015) aster R-CNN: Towards real-time object detection with region proposal networks. Neural Inf Process Syst, pp 91–99
26. Zheng L, Shen L, Tian L, Wang S, Wang J, Tian Q (2015) Scalable person re-identification: a benchmark. In: IEEE international conference on computer vision, pp 1116–1124
27. Xiao T, Li S, Wang B, Lin L, Wang X (2017) Joint detection and identification feature learning for person search. In: IEEE conference on computer vision and pattern recognition, pp 3376–3385
28. Lin J, Ren L, Lu J, Feng J, Zhou J (2017) Consistent-aware deep learning for person re-identification in a camera network. In: IEEE conference on computer vision and pattern recognition, pp 5771–5780
29. Sun Y, Zheng L, Deng W, Wang S (2017) Svdnet for pedestrian retrieval. In: IEEE international conference on computer vision, pp 3800–3808
30. Li W, Zhu X, Gong S (2017) Person re-identification by deep joint learning of multi-loss classification [Online]. Available: https://arxiv.xilesou.top/abs/1705.04724
31. Li W, Zhu X, Gong S (2018) Harmonious attention network for person re-identification. In: IEEE conference on computer vision and pattern recognition, pp 2285–2294
32. Zhao R, Ouyang W, Wang X (2013) Unsupervised salience learning for person re-identification. In: IEEE conference on computer vision and pattern recognition, pp 3586–3593
33. Liu H, Feng J, Jie Z, Karlekar J, Zhao B, Qi M, Jiang J, Yan S (2017) Neural person search machines. In: IEEE international conference on computer vision, pp 493–501
34. Chen D, Zhang S, Ouyang W, Yang J, Tai Y (2018) Person search via a mask-guided two-stream CNN model. In: European conference on computer vision, pp. 764–781
35. Ioffe S, Szegedy C (2015) Batch normalization: accelerating deep network training by reducing internal covariate shift. In: International conference on machine learning, pp 448–456

Cross-Domain Person Re-identification: A Review

Yanan Wang[1,2], Shuzhen Yang[1,2], Shuang Liu[1,2], and Zhong Zhang[1,2](✉)

[1] Tianjin Key Laboratory of Wireless Mobile Communications and Power Transmission, Tianjin Normal University, Tianjin, China
[2] College of Electronic and Communication Engineering, Tianjin Normal University, Tianjin, China
{yananwang585,ysz2020zhen,shuangliu.tjnu,zhong.zhang8848}@gmail.com

Abstract. Cross-domain person re-identification (Re-ID) is a challenging field which has a huge space to improve. With the progress of deep learning, many representative methods of cross-domain person Re-ID have emerged one after another. In this paper, we comprehensively describe and discuss the existing methods and make a simple classification for them. Meanwhile, we compare the performance of these methods on public datasets.

Keywords: Person re-identification · Domain adaptation

1 Introduction

Cross-domain person re-identification (Re-ID) [1–4] is a hot topic of person Re-ID, which aims to make the Re-ID model robust to cross domain scenarios. However, there are many challenges of cross-domain person Re-ID, such as variety of camera types, different styles between domains, and relatively small pedestrian dataset.

Compared with cross-domain person Re-ID, single-domain person Re-ID [5–9] mainly conducts the training of model using one labeled dataset, and the accuracy of this kind of methods is high. But when the well-trained model is used to identify the image from a strange dataset, the performance of the model is significantly reduced. Meanwhile, single-domain person Re-ID requires sufficient labeled samples, but the cost of labeling is high. Therefore, researchers are gradually committed to cross-domain person Re-ID which is more flexible applied in practice, such as intelligent security, image retrieval and so on [10–12].

In this paper, we review the existing cross-domain person Re-ID methods which roughly possess two categories: GAN-based and non GAN-based cross-domain person Re-ID methods. Afterwards, we introduce many representative methods of the two categories, and compare their performance on public datasets.

Q. Liang et al. (eds.), *Artificial Intelligence in China*, Lecture Notes in Electrical Engineering 653, https://doi.org/10.1007/978-981-15-8599-9_19

2 Cross-Domain Person Re-identification

Cross-domain person Re-ID usually employs a labeled source dataset and an unlabeled target dataset as the training samples to complete the domain adaptation. At present, researchers have made great progress in this field, including GAN-based methods and non GAN-based methods.

2.1 The GAN-Based Methods

The most difficult problem of person Re-ID is that the existing dataset is relatively small for deep learning, in which cross domain scenarios have a higher demand for sample diversity. Hence, it can be said that the emergence of GAN [13] opens a new door to cross-domain person Re-ID. Due to the good performance of generators, GAN and the improved version of GAN, such as cycleGAN [14], starGAN [15] and PTGAN [16], are frequently used for cross-domain person Re-ID.

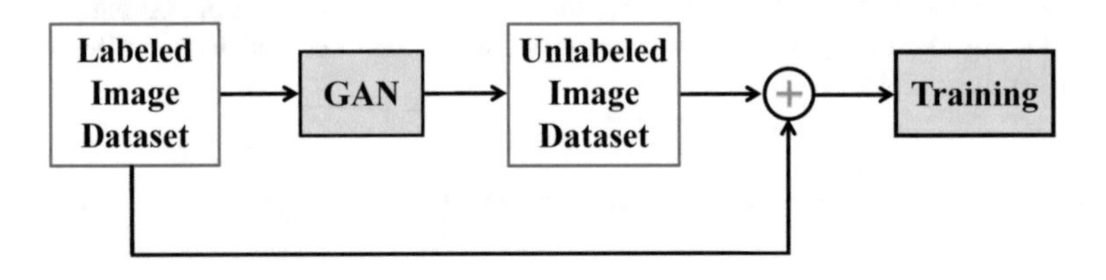

Fig. 1. The network framework of [17], where the unlabeled image dataset is obtained by randomly generating false samples for the labeled images

The GAN-based methods [17–22] can reduce the gaps between different domains through style translations. Specifically, Zheng et al. [17] first propose to apply GAN to person Re-ID, in which GAN is employed to generate unlabeled images, and then the generated and original images are mixed to train the Re-ID model, as shown in Fig. 1. Since training samples are added with unlabeled generated images, the authors propose a label smoothing regularization method to be suitable outliers, which makes the label distribution of unlabeled images in each class consistent and normalized. Meanwhile, the cross-entropy loss is applied for the labeled real images.

Inspired by [17], many representative methods employ GAN in the field of cross-domain person Re-ID. For example, Zhong et al. [18] aim to reduce the influence of camera style deviation for person Re-ID, in which they adopt cycleGAN to complete camera style expansion for the training dataset. Assuming that the training images are captured by N cameras, each camera style image set is regarded as a different domain. By generating $N-1$ camera style images for each pedestrian image, a model with robustness to camera style can be obtained by training with original and generated images. At the same time, because there may be errors and noises in the generated images, the authors choose LSR to train the generated images. Comparing with [17], the generated image inherits the label of the original image rather than randomly generating the unlabeled false image, which makes the generated image available for camera style adaptation.

In Deng et al. [19], describe the Similarity Preserving cycle-consistent Generative Adversarial Network (SPGAN) which is realized by adding a Siamese network on the basis of cycleGAN. CycleGAN can transform the style of the source dataset to target-style and keep the identity label, and the Siamese network can preserve more identity information for generated images. Using the style-transferred images, SPGAN can wisely avoid the influence of different domain styles on the model training, which makes the model more suitable for identifying target-style pedestrian images. Meanwhile, the local max pooling (LMP) is applied as the feature pooling method to reduce the interference of noises carried by the generated images.

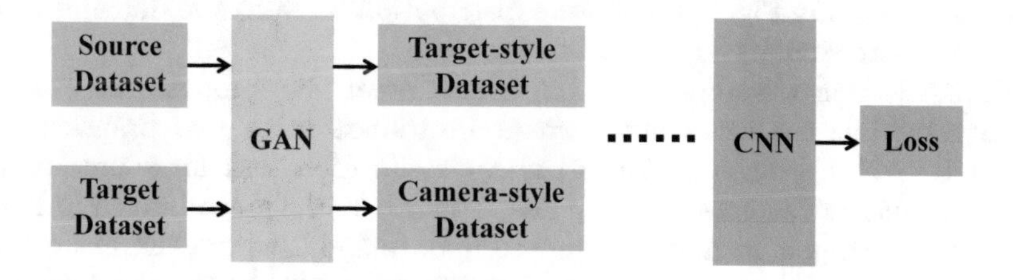

Fig. 2. The pipeline of the GAN-based cross-domain person Re-ID methods

The above two methods respectively focus on the camera-style adaptation and domain-style adaptation for the model. To train a model with the properties of domain invariance and camera invariance, Zhong et al. [20] present a method called Hetero-Homogeneous Learning (HHL), in which the model has two branches. For the one branch, cross-entropy loss is employed to optimize the model with the labeled source dataset, which can be regarded as a classification task. For another branch, the authors utilize the triplet loss to handle the similarity learning. Specifically, the second branch uses starGAN to generate camera-style images for the unlabeled target dataset, and then employs the source dataset, the target dataset and the camera-style dataset to train the domain invariance and the camera invariance of the model.

In a word, all the above methods employ GAN to complete data augmentation, which includes transforming images of the source dataset into target-domain style or expanding the target dataset with camera-style translation. As shown in Fig. 2, after the style translation, the generated datasets and the original datasets can be further applied in the training stage.

2.2 The Non GAN-Based Methods

Some non GAN-based methods for cross-domain person Re-ID achieve domain adaptation [23–25] by aligning the sample distribution, such as [26–29].

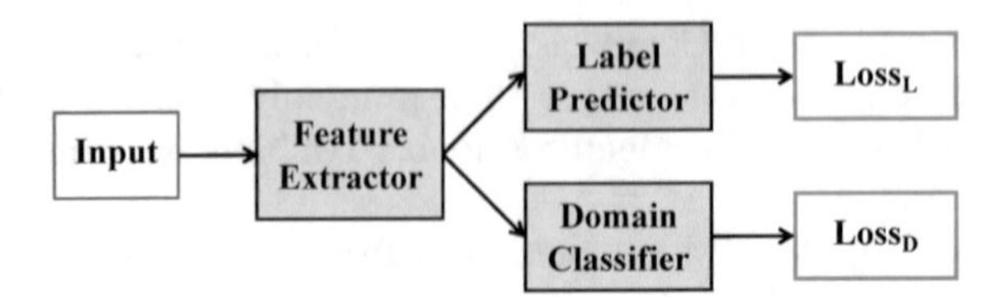

Fig. 3. The network framework of [26] is composed of a feature extractor, a label predictor and a domain classifier, where the $Loss_L$ and $Loss_D$ represent the corresponding optimization function

In detail, Ganin et al. [26] focus on the feature alignment between two domains. As shown in Fig. 3, the model consists of the feature extractor, the label predictor and the domain classifier, where the label predictor can predict identity label and the domain classifier can distinguish different domains. By minimizing $Loss_L$ and $loss_D$ in Fig. 3, the feature distribution between two domains can be more and more similar.

Different from the above method, Zhong et al. [30] propose the exemplar memory module to learn intra-domain invariance using two branches. One branch (the blue branch in Fig. 4) is a classification task for source dataset which calculating the cross-entropy loss. The other (the green branch in Fig. 4) adopts the exemplar memory module to update and save the features of unlabeled target samples, in which the updated features are obtained by calculating the similarity between the current features of the mini-batch and the stored features of the exemplar memory.

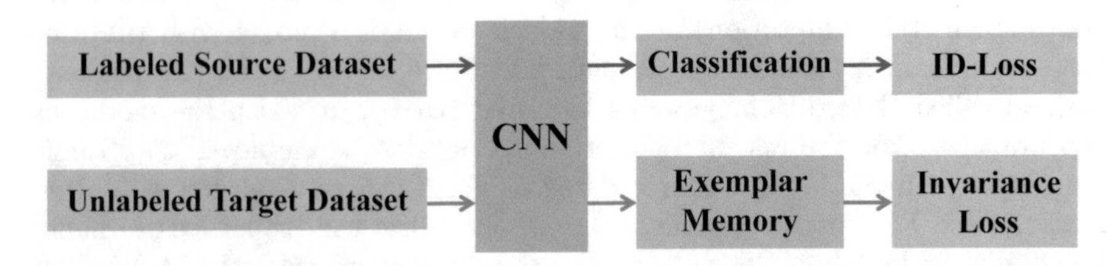

Fig. 4. The network framework of [30] is composed of a classification module and an exemplar memory module, which correspond to the ID-loss (i.e., the cross-entropy loss) and the invariance loss respectively

In addition, there are some methods to adapt the source dataset in the pre-training stage, and then focus on the target dataset to complete further training [31–33]. For example, in [31], the authors propose a method called self-similarity grouping. After pre-training with the source dataset, the clustering algorithm is applied to classify the unlabeled target dataset. Meanwhile, since the whole, upper and lower information of pedestrian characteristics are considered in this method, each pedestrian image can get three labels after clustering, which can be utilized to further optimize the pre-trained model in a supervised manner.

Table 1. The evaluation results of representative methods, where M, D and $M17$ denote Market-1501, DukeMTMC-reID and MSMT17 respectively

Methods	M-D		M-M17		D-M		D-M17	
	MAP	Rank-1	MAP	Rank-1	MAP	Rank-1	MAP	Rank-1
PTGAN [16]	–	27.4	–	10.2	–	38.6	–	11.8
SPGAN [19]	22.3	41.1	–	–	22.8	51.5	–	–
SPGAN+LMP [19]	26.4	46.9	–	–	26.9	58.1	–	–
HHL [20]	27.2	46.9	–	–	31.4	62.2	–	–
CR-GAN [21]	48.6	68.9	–	–	54.0	77.7	–	–
ECN [30]	40.4	63.3	8.5	25.3	43.0	75.1	10.2	30.2
PDA-Net [27]	45.1	63.2	–	–	47.6	75.2	–	–
SSG [31]	53.4	73.0	13.2	31.6	58.3	80.0	13.3	32.2

3 Experiments

Cross domain person Re-ID is usually trained by three large-scale pedestrian datasets, i.e., Market-1501 [34], DukeMTMC-reID [35] and MSMT17 [16]. Market has 12936 training images of 751 identities and its gallery set and query set respectively have 19,732 and 3368 images. Duke is composed of 16,522 training images for 702 identities, the gallery set has 17,661 images and the query set has 2228 images. MSMT17 includes 32,621 training images of 1041 identities and 93,820 test images of 3060 identities which randomly selects 11,659 images from the test set as the query set.

We list the evaluation results of representative methods in Table 1, where the GAN-based methods are recorded in the upper part, and the non GAN-based methods are recorded in the lower part. From Table 1, we find that non GAN-based methods are slightly better than GAN-based methods. Specifically, comparing CR-GAN and SSG, when Market and Duke are regarded as the source and target datasets respectively, the Rank-1 accuracy of CR-GAN is lower 4.1% than SSG. In other datasets, the similar accuracy differences occur. Meanwhile, since MSMT17 contains more identities, cameras, and more complex scenes and backgrounds, the evaluation results on MSMT17 relatively low. For example, when MSMT17 is applied as the target dataset, the rank-1 accuracy of SSG is about 31%.

4 Conclusion

In this paper, we have introduced some representative methods of cross-domain person Re-ID, including GAN-based methods and non GAN-based methods. Through specific introduction and discussion, we can find that these two kinds of methods have different advantages. The GAN-based methods apply GAN to generate images of different styles, which not only expands the dataset but also reduces the differences of domain styles. Meanwhile, the non GAN-based methods are more inclined to extract higher quality features and align them.

Perhaps, considering these two kinds of methods comprehensively can further improve cross-domain person Re-ID.

Acknowledgment. This work was supported by National Natural Science Foundation of China under Grant No. 61711530240, Natural Science Foundation of Tianjin under Grant No. 20JCZDJC00180 and No. 19JCZDJC31500, the Open Projects Program of National Laboratory of Pattern Recognition under Grant No. 202000002, and the Tianjin Higher Education Creative Team Funds Program.

References

1. Ma AJ, Li J, Yuen PC, Li P (2015) Cross-domain person reidentification using domain adaptation ranking SVMs. IEEE Trans Image Process 24(5):1599–1613
2. Peng P, Xiang T, Wang Y, Pontil M, Gong S, Huang T, Tian Y (2016) Unsupervised cross-dataset transfer learning for person re-identification. In: IEEE conference on computer vision and pattern recognition, pp 1306–1315
3. Lv J, Chen W, Li Q, Yang C (2018) Unsupervised cross-dataset person re-identification by transfer learning of spatial-temporal patterns. In: IEEE conference on computer vision and pattern recognition, pp 7948–7956
4. Liu J, Zha ZJ, Chen D, Hong R, Wang M (2019) Adaptive transfer network for cross-domain person re-identification. In: IEEE conference on computer vision and pattern recognition, pp 7202–7211
5. Prosser BJ, Zheng WS, Gong S, Xiang T, Mary Q (2010) Person re-identification by support vector ranking. In: British machine vision conference, vol 2, no 5, pp 1–11
6. Li W, Zhao R, Xiao T, Wang X (2014) Deepreid: Deep filter pairing neural network for person re-identification. In: IEEE conference on computer vision and pattern recognition, pp 152–159
7. Zhang Z, Si T, Liu S (2018) Integration convolutional neural network for person re-identification in camera networks. IEEE Access 6:36887–36896
8. Zhang Z, Huang M (2018) Discriminative structural metric learning for person reidentification in visual internet of things. IEEE Int Things J 5(5):3361–3368
9. Lin Y, Zheng L, Zheng Z, Wu Y, Hu Z, Yan C, Yang T (2019) Improving person re-identification by attribute and identity learning. Pattern Recogn 95:151–161
10. Zhang Z, Wang C, Xiao B, Zhou W, Liu S (2012) Action recognition using context-constrained linear coding. IEEE Signal Process Lett 19(7):439–442
11. Xue X, Gao J, Wang S, Feng Z (2018) Service bridge: Transboundary impact evaluation method of internet. IEEE Trans Comput Social Syst 5(3):758–772
12. Xue X, Wang S, Zhang L, Feng Z, Guo Y (2019) Social learning evolution (SLE): computational experiment-based modeling framework of social manufacturing. IEEE Trans Industrial Informatics 15(6):3343–3355
13. Goodfellow I, Pouget-Abadie J, Mirza M, Xu B, Warde-Farley D, Ozair S, Courville A, Bengio Y (2014) Generative adversial nets. Adv Neural Inf Process Syst, pp 2672–2680
14. Zhu JY, Park T, Isola P, Efros AA (2017) Unpaired image-to-image translation using cycle-consistent adversarial networks. In: IEEE international conference on computer vision, pp 2223–2232
15. Choi Y, Choi M, Kim M, Ha J, Kim S, Choo J (2018) StarGAN: Unified generative adversarial networks for multi-domain image-to-image translation. In: IEEE conference on computer vision and pattern recognition, pp 2223–2232

16. Wei L, Zhang S, Gao W, Tian Q (2018) Person transfer GAN to bridge domain gap for person re-identification. In: IEEE conference on computer vision and pattern recognition, pp 79–88
17. Zheng Z, Zheng L, Yang Y (2017) Unlabeled samples generated by GAN improve the person re-identification baseline in vitro. In: IEEE international conference on computer vision, pp 3754–3762
18. Zhong Z, Zheng L, Zheng Z, Li S, ang Y (2018) Camera style adaptation for person re-identification. In: IEEE conference on computer vision and pattern recognition, pp 5157–5166
19. Deng W, Zheng L, Ye Q, Yang G, Jiao J (2018) Image-image domain adaptation with preserved self-similarity and domain-dissimilarity for person re-identification. In: IEEE conference on computer vision and pattern recognition, pp 994–1003
20. Zhong Z, Zheng L, Li S, ang Y (2018) Generalizing a person retrieval model hetero- and homogeneously. In: European conference on computer vision, pp 176–192
21. Chen Y, Zhu X, Gong S (2019) Instance-guided context rendering for cross-domain person re-identification. In: IEEE international conference on computer vision, pp 232–242
22. Liu J, Li W, Pei H, Wang Y, Qu F, Qu Y, Chen Y (2019) Identity preserving generative adversarial network for cross-domain person re-identification. IEEE Access 7:114021–114032
23. Zhang Z, Wang C, Xiao B, Zhou W, Liu S, Shi C (2013) Cross-view action recognition via a continuous virtual path. In: IEEE conference on computer vision and pattern recognition, pp 2690–2697
24. Long M, Zhu H, Wang J, Jordan IM (2016) Unsupervised domain adaptation with residual transfer networks. Adv Neural Inf Process Syst, pp 136–144
25. Zhang Z, Li D (2019) Hybrid cross deep network for domain adaptation and energy saving in visual internet of things. IEEE Int Things J 6(4):6026–6033
26. Ganin Y, Lempitsky V, Unsupervised domain adaptation by backpropagation. https://arxiv.xilesou.top/abs/1409.7495
27. Li YJ, Lin CS, Lin YB, Wang YCF (2019) Cross-dataset person re-identification via unsupervised pose disentanglement and adaptation. In: IEEE international conference on computer vision, pp 7919–7929
28. Chen C, Xie W, Huang W, Rong Y, Ding X, Huang Y, Xu T, Huang J (2019) Progressive feature alignment for unsupervised domain adaptation. In: IEEE conference on computer vision and pattern recognition, pp 627–636
29. Chen C, Chen Z, Jiang B, Jin X (2019) Joint domain alignment and discriminative feature learning for unsupervised deep domain adaptation. In: AAAI conference on artificial intelligence, vol 33, pp 3296–3303
30. Zhong Z, Zheng L, Luo Z, Li S, Yang Y (2019) Invariance matters: exemplar memory for domain adaptive person re-identification. In: IEEE conference on computer vision and pattern recognition, pp 598–607
31. Fu Y, Wei Y, Wang G, Zhou Y, Shi H, Huang TS (2019) Self-similarity grouping: A simple unsupervised cross domain adaptation approach for person re-identification. In: IEEE international conference on computer vision, pp 6112–6121
32. Zhang X, Cao J, Shen C, You M (2019) Self-training with progressive augmentation for unsupervised cross-domain person re-identification. In: IEEE international conference on computer vision, pp 8222–8231
33. Zhang Z, Huang M, Liu S, Xiao B, Durrani TS (2019) Fuzzy multilayer clustering and fuzzy label regularization for unsupervised person re-identification. IEEE Trans Fuzzy Syst. https://doi.org/10.1109/TFUZZ.2019.2914626.8222-8231

34. Zheng L, Shen L, Tian L, Wang S, Wang J, Tian Q (2015) Scalable person re-identification: A benchmark. In: IEEE international conference on computer vision, pp 1116–1124
35. Ristani E, Solera F, Zou R, Cucchiara R, Tomasi C (2016) Performance measures and a data set for multi-target, multi-camera tracking. In: European conference on computer vision, pp 17–35

A Weighted Least Square Support Vector Regression Method with MPP-GGP Based Sequential Sampling for Efficient Reliability Analysis

Yang Guo[1](✉), Nan-nan Wang[2], and Gen-shen Kai[3]

[1] PLA Academy of Military Science, 100091 Beijing, China
guoyangnudt@gmail.com
[2] China Aerospace Academy of Systems Science and Engineering, 100042 Beijing, China
20583541@qq.com
[3] University of Electronic Science and Technology of China, 611731 Chengdu, China
genshen_kai@163.com

Abstract. Due to the huge computational cost of the limit state function, an approximate model is generally used to replace the limit state function in engineering reliability analysis. In this paper, weighted least squared support vector regression method, which is originally proposed to enhance the robustness of LSSVR, is introduced to reliability analysis. A weighting function that increases the importance of the training samples around the failure surface is proposed. In addition, sequential modeling strategy is combined in the method to further reduce the number of training samples required. The most probable point (MPP) is sampled to identify the region with high densities of probabilities, and the greatest gradient point (GGP) on the failure surface is sampled to identify the nonlinear region. A WLSSVR method based on MPP-GGP sequential sampling is established and tested on an uncertainty-based flutter analysis problem of composite wing. The results show that the proposed method which keeps a balance between the global and local approximate is very efficiency on solving the engineering reliability analysis problem.

Keywords: Reliability analysis · WLSSVR · Sequential modeling · Greatest gradient point · Most probable point

1 Introduction

The main task of reliability analysis is to calculate the probability of failure. Generally, the probability of failure is defined as

$$P_f = P\{g(x) < 0\} = \int_{g(X) \leq 0} f_x(x) dX \tag{1}$$

Q. Liang et al. (eds.), *Artificial Intelligence in China*, Lecture Notes
in Electrical Engineering 653, https://doi.org/10.1007/978-981-15-8599-9_20

where x is the variable in the reliability problem; $f_x(x)$ represents the probability density function in x-space; $g(x)$ is the limit state function. $g(x) < 0$ stands for the failure region. Likewise, $g(x) = 0$ stands for the limit state surface, and $g(x) > 0$ indicates the safe region. Several methods have been proposed to solve Eq. (1) in last decades, which can be divided into two categories: moment-based methods such as first-order reliability method and second-order reliability method and simulation methods such as Monte Carlo simulation (MCS) [1].

Because of the huge amount of calculation directly using the limit state function, researchers usually use the method of constructing approximate function to replace the limit state function. In this way, a balance can be achieved between the computational cost and the accuracy of the model, and a better solution can be obtained within the acceptable computational cost. In reliability analysis, commonly used approximation methods include RSM [2], PCE [3], Kriging [4], and NN [5], et al. Among them, the least square support vector machine (LSSVM) [6–11] has been widely investigated in recent years due to its high-approximation accuracy and robustness.

Although SVM/LSSVM has many advantages, its performance is highly dependent on training samples. Therefore, some high-efficiency sequential sampling strategies are still needed to be developed for the reliability analysis of complex limit state function. In this paper, WLSSVR method is introduced to the reliability analysis problem. In Sect. 2, the WLSSVR method based on most probable point (MPP)—GGP sequential sampling is proposed, which can be used for effective reliability analysis. In Sect. 3, the application of this method to an uncertainty-based flutter analysis problem of composite wing is presented.

2 Methods

2.1 Sequential Sampling Method

Since the failure probability calculation only cares about whether the limit state function is positive or negative, it is necessary to collect points with high probability of crossing the failure surface and put them into the training dataset. One of the important points on failure surface is the most probable point (MPP). MPP-based sequential sampling strategy can improve the local approximate precision around MPP. However, if only the MPP is sampled into the training dataset, it may reduce the global approximation quality of the failure surface, especially for the highly nonlinear limit state. As gradient information can help identifying the high nonlinear region, the point with the greatest gradient can be sampled into the training dataset to improve the quality of global approximation [12]. In this section, MPP- and GGP- based sequential sampling strategies are proposed.

(1) MPP-based sequential sampling strategy

The MPP-based sequential sampling strategy can be summarized as the following optimization problem,

$$\min_{x} \|u\|_2$$

$$\text{s.t. } \widetilde{g}(x) = 0$$
$$l_{\mathbf{x}} \geq \alpha \sqrt[d]{V/N}$$
$$x \in [a, b] \quad (2)$$

where u is the standard Gaussian variables transformed from x; a and b are the minimum and maximum values of x, respectively. The inequality constraint $l_x \geq \alpha \sqrt[d]{V/N}$ is used to avoid the oversampling, where l_x is the shortest distance of x from the existing training samples, V is the hypervolume of the d dimensional space, N is the number of training samples, and $0 \leq \alpha \leq 1$.

(2) GGP-based sequential sampling strategy

We identify the nonlinear part of the failure surface by calculating the relative derivative. The relative derivative can be calculated using the following Formula,

$$\frac{\partial x_n}{\partial x_i} = -\frac{\partial g(x)}{\partial x_i} / \frac{\partial g(x)}{\partial x_n} \quad (i = 1, \ldots n - 1) \quad (3)$$

where x_n is the last design variable.

The GGP is defined as the point with the largest norm of the relative derivative vector (marked as x_n'). The sequential sampling strategy based on GGP can be summarized as the following optimization problem,

$$\max_x \left| x_n' \right|$$
$$\text{s.t. } \widetilde{g}(x) = 0$$
$$l_x \geq \alpha \sqrt[d]{V/N}$$
$$x \in [x_l, x_u] \quad (4)$$

2.2 Weighting Function

For the reliability problem, the objective is to estimate the probabilities in a space separated by the failure surface. An accurate approximation of the limit state function is not necessary, as only its sign matters. The training samples which locate near the failure surface are therefore more important than other training samples. In order to quantify the importance of each training sample, a weighting function is established, which is

$$W_k = N \frac{\mathrm{e}^{-|g(x_k)|/2}}{\sum_{k=1}^{N} \mathrm{e}^{-|g(x_k)|/2}} \quad (5)$$

2.3 Procedure

The main steps of this method are:

Step 1: Generate the sample dataset for testing. Monte Carlo method is used to randomly generate uniformly distributed test sample points in the design space to form a test sample set. It should be noted that the limit state function values of these points are not calculated here, and the test sample set remains unchanged throughout the sequential modeling process.

Step 2: Generate the training samples (Design of experiments). According to the experimental design principles, the training sample set should be distributed as uniformly as possible, that is, with a large entropy value. Since the quasi-random number method can generate a set of sample points with larger entropy values, the initial training sample points are generated using the Sobol algorithm [13]. The training sample is calculated by the following Formula:

$$x_{ij} = a_{ij} + (b_{ij} - a_{ij})s_{ij}, \quad i = 1, 2, \ldots, E;\ j = 1, 2, \ldots, h \tag{6}$$

where E is the number of training samples; h is the dimension of x; s_{ij} is the Sobel number; a_{ij} and b_{ij} are the minimum and maximum values of the variable x in the jth dimension, respectively.

Step 3: Build the WLSSVR model. Evaluate $g(x)$ and W on all training samples and construct the WLSSVR model. RBF kernel is used as the kernel function.

Step 4: Calculate the estimation of the probability of failure. First, use the WLSSVR model to evaluate $\widetilde{g}(x)$ on the Monte Carlo population S. The probability of failure P_f at iteration k is then estimated by

$$P_f^k = \frac{1}{2}\left(N_s - \sum_{K=1}^{N_s} \text{sign}(\widetilde{g}(x_k))\right) \tag{7}$$

Step 5: Check the stopping condition. A stopping criterion which describes the relative change of the approximate model is proposed here. It can be computed as Eq. (8) at the iteration k.

$$\varepsilon_k = \frac{1}{2N_s}\sum_{i=1}^{N_s} |\text{sign}(\widetilde{g}_k(x_i)) - \text{sign}(\widetilde{g}_{k-1}(x_i))| \tag{8}$$

If $\epsilon_k \le \epsilon_0(k < 5)$ or $\frac{1}{5}\sum_{i=0}^{4} \epsilon_{k-i} \le \epsilon_0(k \ge 5)$, where ϵ_0 is a given value depending on the precision required, go to ***Step 8***, else go to ***Step 6***.

Step 6: Find the MPP or GGP. If the iteration number k is an odd number, find the MPP x_{MPP}, else find the GGP x_g. The MPP and GPP are sampled into the training sample set in a sequential way.

Step 7: Update the WLSSVR model. Add x_{MPP} or x_g to the training dataset and update the WLSSVR model according to Eq. (7).

Step 8: Output the final estimation of the probability of failure $\widetilde{P}_f$, End. If $k \ge 5$, $\widetilde{P}_f = \frac{1}{5}\sum_{i=0}^{4} P_f^{k-i}$, else $\widetilde{P}_f = P_f^k$.

3 Application Examples

In this section, one numerical reliability analysis examples and an uncertainty-based flutter analysis problem of composite wing are used to test the method we proposed. The WLSSVR parameters (γ, σ^2) are set to be $(10^4, 2)$. The estimated error of $\widetilde{P_f}$ is defined to quantify the performance of WLSSVR, which is

$$\zeta = \frac{1}{2N_s}\sum_{i=1}^{N_s}|\text{sign}(g(x_i)) - \text{sign}(\widetilde{g}(x_i))| \tag{9}$$

3.1 Complex Nonlinear Function

The limit state function is defined as

$$f(x) = \frac{1}{2}\sin(3.5 - 2x_1) + 0.2x_1 + \exp(\frac{x_2 - 1}{4}) - 3 \tag{10}$$

where x_1 and x_2 are independent standard Gaussian variables which belong to $N(5, 0.8)$. The box for generating the Sobel numbers is set as $a_{ij} = 0$ and $b_{ij} = 10$ $(i = 1, 2)$. The initial size of the training sample set is 40. The required precision ϵ_0 is set to be 3×10^{-4}. Monte Carlo population of 10^5 sampling points is generated to simulate the probability of failure, which is 0.1079. Table 1 presents the reliability analysis results of WLSSVR method with sequential sampling. The final sample number is about 116. If we directly use Sobol algorithm to generate 116 training samples, the estimated error of the probability of failure will be 0.1189. It is far bigger than ζ_{final} obtained from the sequential modeling. We can see that the sequential modeling has greatly improved the approximate quality. From Table 1, we can also get the conclusion that MPP-GGP-based sequential sampling is better than the MPP- or GGP-based sequential sampling.

Table 1 Reliability analysis results of WLSSVR method with sequential sampling

Approximate method	$\widetilde{P_f}$	ζ_{final}	Iteration	Sample number
WLSSVR (MPP-GGP)	0.1063	0.0068	76	116
WLSSVR(MPP)	0.0996	0.010	72	112
WLSSVR(GGP)	0.1035	0.0111	76	116

The final WLSSVR approximation failure surface and training samples are shown in Fig. 1, where β is the safety index (corresponding to the reliability $\Phi(\beta)$). It is easy to observe from Fig. 1a that WLSSVR model fits well with the real model at the region around the MPP, but the approximate error at the region far away from the MPP is big. In Fig. 1b, the global approximate quality of WLSSVR model is improved thanks to the

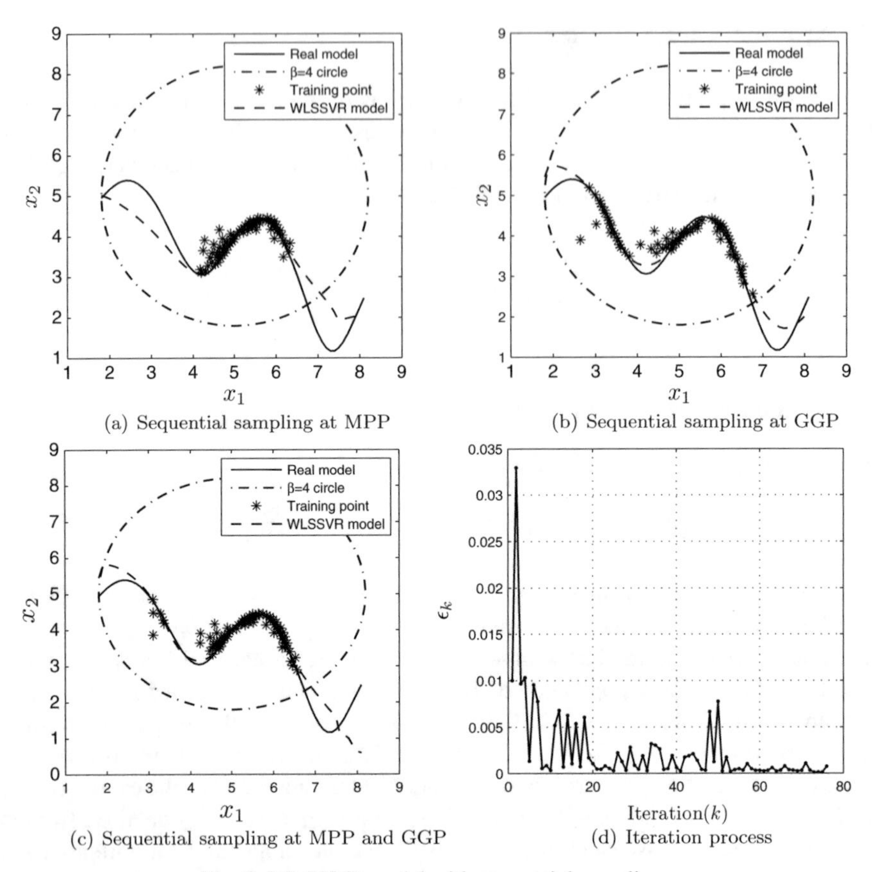

(a) Sequential sampling at MPP

(b) Sequential sampling at GGP

(c) Sequential sampling at MPP and GGP

(d) Iteration process

Fig. 1 WLSSVR model with sequential sampling

GPP which greatly effects on identifying the nonlinear region. However, the approximate precision of $\widetilde{P_f}$ is low because of the poor approximation around the region with high densities of probabilities.

Figure 1c shows the results under the MPP-GGP-based sequential sampling strategy. The MPP-GPP-based sequential sampling strategy keeps a balance between the global and local approximate and thus greatly improves the approximate quality and computational efficiency. Figure 1d presents the iteration process of the WLSSVR method with MPP-GGP-based sequential sampling. Convergence reaches at the 76th iteration.

3.2 Flutter Analysis of Composite Wing

In this paper, the composite wing model established in reference [14] is selected as the research object. The composite plate contains six layers with the fiber direction angle layup of $(\theta_1, \theta_2, \theta_3)_s$, as shown in Fig. 2.

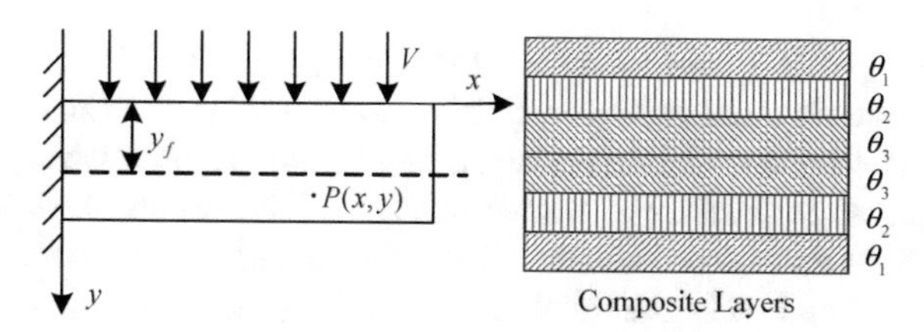

Fig. 2 Composite wing model

The deflection outside the plate is expressed as

$$\bar{\omega} = \sum_{i=1}^{n} m_i(x, y)k_i(t) \tag{11}$$

where $k_i(t)$ is the generalized displacement in different modes; $m_i(x, y)$ stands for different modes which are set as the following functions $x^2, x^3, x^4, x^2 \cdot (y - y_f), x^3 \cdot (y - y_f)$, $x^4 \cdot (y - y_f)$, $x^2 \cdot (y - y_f)^2$, $x^3 \cdot (y - y_f)^2$, and $x^4 \cdot (y - y_f)^2$. A modified strip theory approach with unsteady aerodynamics having a torsional velocity term [15] is used to build the aerodynamic model. The aeroelastic equation is as follows

$$\left[Fs^2 + (\rho VM + L)s + (\rho V^2 N + O)\right] \cdot j = k \tag{12}$$

where F, M, N, L, and O are the structural inertia, aerodynamic damping, aerodynamic stiffness, structural damping, and structural stiffness matrices, respectively, j represents generalized coordinate vector, and k represents generalized force. The Laplace variable is denoted as $s = g + i\omega$, where ω represents circular frequency, and g represents damping. The flutter speed is determined by increasing V until one of the damping values becomes negative. The nominal value of composite material properties are given in Table 2.

Table 2 Characteristics of composite materials for wing

Properties	Numerical value
longitudinal Young's modulus	98.0 GPa
transverse Young's modulus	7.9 GPa
Ply thickness(h)	0.134 mm
Poisson's ratio	0.28
Density(ρ_m)	1520 Kg/m^3
Shear modulus	5.6 GPa

Firstly, longitudinal Young's modulus, shear modulus, and ply thickness are considered as independent standard Gaussian variables with the coefficient of variation 0.04, 0.04, and 0.02, respectively. Two sets of data are selected to investigate the effect of the fiber laying angle in the composite material on the wing flutter. One is

$(-33.62°, 47.82°, 67.9°)_s$; the other is $(-45°, -45°, 0°)_s$. The Monte Carlo population size N_s is 10^4. The required precision ϵ_0 is set to be 1×10^{-4}. The number of initial training points is 40. The reliability analysis results are given in Table 3. The iteration process of sequential modeling is shown in Fig. 3. The results show that WLSSVR method with MPP-GGP-based sequential sampling can approximate the failure surface with high precision and make the reliability analysis quickly converge.

Table 3 Composite wing example with three stochastic variables: reliability analysis results

Test	$(\theta_1, \theta_2, \theta_3)_s$	Design flutter speed	P_f	$\widetilde{P_f}$	ζ_{final}	Iteration	Sample number
$T1$	$(-33.62°, 47.82°, 67.9°)_s$	32 m/s	0.5833	0.58372	5.4×10^{-4}	34	74
$T2$	$(-45°, -45°, 0°)_s$	22 m/s	0.1446	0.14484	4×10^{-4}	12	52

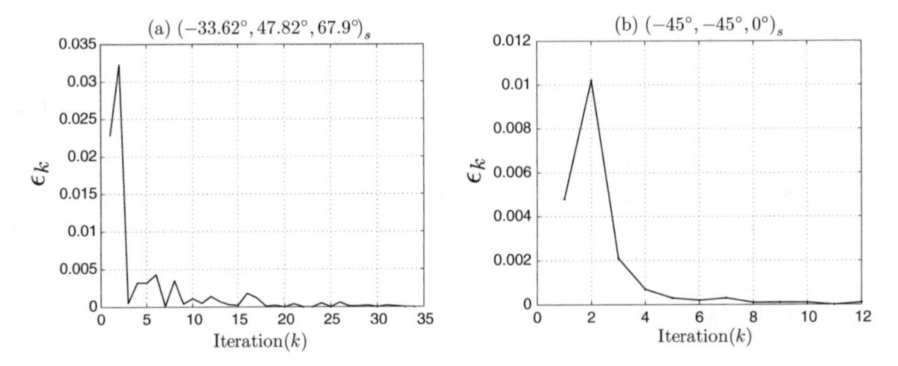

Fig. 3 Composite wing example with three stochastic variables: iteration process

Secondly, the fiber direction angles θ_1 and θ_2 are further treated as independent standard Gaussian variables based on the test $T2$. The coefficient of variation of θ_1 or θ_2 is 0.5. Then, there are five stochastic variables in this example. The probability of failure obtained from the MCS is 0.1498. The initial size of the training sample set is set as 60. The predicted failure probability is 0.14908, which is close to the result calculated by MCS method. Figure 4 shows the iteration process of this example. A total of 125 training points are needed to reach the convergence, and $\zeta_{\text{final}} = 1 \times 10^{-3}$.

4 Conclusions

A WLSSVR method based on MPP-GGP sequential sampling was developed for reliability analysis in this paper. On the one hand, a sequential sampling strategy based on GGP is proposed to solve the problem that the nonlinear region of the failure surface is difficult to approximate. On the other hand, MPP-based sequential sampling strategy is

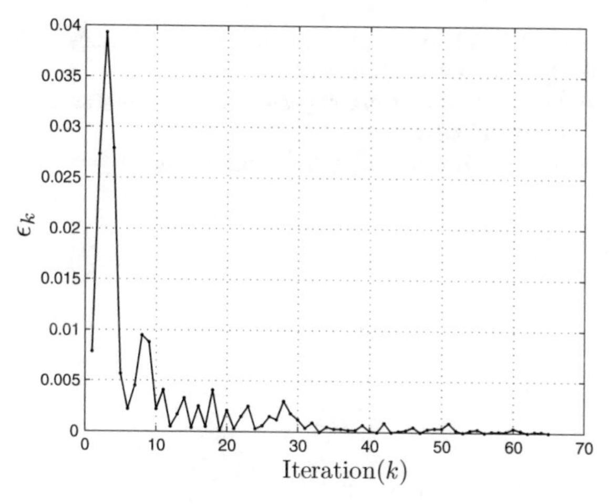

Fig. 4 Composite wing example with five stochastic variables: iteration process

used for the local approximate around the MPP. The test results on an uncertainty-based flutter analysis problem of composite wing shows that MPP-GGP-based sequential sampling strategy keeps a balance between the global and local approximate and provides an efficiency way to calculate the probability of failure as it only uses small amount of training samples.

References

1. Choi S, Grandhi RV, Canfield RA (2007) In: Reliability-based structural design, Springer
2. Allaix DL, Carbone VI (2011) An improvement of the response surface method. Struct Safety 33:165–172
3. Blatman G, Sudret B (2011) Adaptive sparse polynomial chaos expansion based on least angle regression. J Comput Phys 230:2345–2367
4. Echard B, Gayton N, Lemaire M (2011) AK-MCS: an active learning reliability method combining Kriging and Monte Carlo simulation. Struct Safety 33:145–154
5. Hurtado JE, Alvarez DA (2001) Neural-network-based reliability analysis: a comparative study. Comput Methods Appl Mech Eng 191:113–132
6. Papadrakakis M, Lagaros ND (2002) Reliability-based structural optimization using neural networks and Monte Carlo simulation. Comput Methods Appl Mech Eng 191:3491–3507
7. Zhu P, Zhang Y, Chen G (2011) Metamodeling development for reliability-based design optimization of automotive body structure. Comput Ind 62:729–741
8. Tan X, Bi W, Hou X, Wang W (2011) Reliability analysis using radial basis function networks and support vector machines. Comput Geotech 38:178–186
9. Moura MDC, Zio E, Lins ID, Droguett E (2011) Failure and reliability prediction by support vector machines regression of time series data. Reliab Eng Syst Safety 96:1527–1534
10. Dai H, Zhang H, Wang W (2012) A support vector density-based importance sampling for reliability assessment. Reliab Eng Syst Safety 106:86–93
11. Suykens JAK, Van Gestel T, De Brabanter J, De Moor B, Vandewalle J. (2002) In: Least squares support vector machines, World Scientific Publishing Co.Pre.Ltd, London
12. Yao W, Chen X, Luo W (2009) A gradient-based sequential radial basis function neural network modeling method. Neural Comput Appl 18:477–484

13. Hurtado JE, Alvarez DA (2010) An optimization method for learning statistical classifiers in structural reliability. Probab Eng Mech 25:26–34
14. Manan A, Cooper JE (2009) Design of composite wings including uncertainties: a probabilistic approach. J Aircraft 46:601–607
15. Wright JR, Cooper JE (2007) Introduction to aircraft aeroelasticity and load. Wiley, Chichester

Design of Quadrotor Automatic Tracking UAV Based on OpenMV

Hai Wang[1], Ying Zhang[1](✉), Guiling Sun[1], Zhihong Wang[1], Binghao Tian[2], and Penghui Li[2]

[1] Teaching Center for Experimental Electronic Information, College of Electronic Information and Optical Engineering, Nankai University, Tianjin 300350, China
caroline_zy@nankai.edu.cn
[2] Tianjin Tengyun Aviation Corporation of China Ltd., Tianjin 300300, China

Abstract. 32-bit microcontroller TM4C123G of TI is applied as the core control chip in this design. The data detected by the OpenMV machine vision module are calculated and processed by the cascade PID control algorithm combined with the dynamics model of the drone. In this paper, the attitude and control of the quadrotor UAV is studied. An intelligent vehicle with the characteristics of environment perception, automatic tracking, decision-making, and the autonomous flight is manufactured, which fulfills the design requirements.

Keywords: Quadrotor UAV · OpenMV · Cascade PID control algorithm · Dynamics model

1 Introduction

Unmanned aerial vehicle (UAV) was originally a military weapon to replace or assist human beings to participate in the battlefield [1–3]. Initially European and American countries utilized UAV to perform high-risk military tasks. With the change of the international situation and the vigorous development of sensor technology, communication technology, information processing technology, and intelligent control technology, UAV has weakened its military attribute [4]. Instead, the drone becomes a high-tech operation platform or industry tool in the information era. Due to the gradual maturity of the development technique and the sharp reduction of manufacturing cost, UAV has been widely applied in various fields, such as agricultural plant protection, electric power inspection, police enforcement, geological exploration, environmental monitoring, forest fire prevention, film, and television aerial photography and other civil fields [5–8].

The four arms of the quadrotor UAV have the same length, which ensures the same force of each motor distributed at the end of the arm. The flight control system and power supply are installed in the middle of the fuselage to maintain the center of gravity in the middle of the aircraft. The electronic speed controller (ESC) controls the rotation speed of each motor, so as to change the lifting force generated by driving the propeller.

Q. Liang et al. (eds.), *Artificial Intelligence in China*, Lecture Notes in Electrical Engineering 653, https://doi.org/10.1007/978-981-15-8599-9_21

Therefore, the drone can fly stably and achieve the pitch, roll, and yaw motions as required [9–11].

This design is based on the X-mode quadrotor UAV, which moves flexibly and responds quickly. Combined with microcontroller and various types of sensors, especially OpenMV machine vision module, the intelligent functions of environment perception and automatic tracking are realized by the cascade PID control algorithm [12].

2 Hardware System Design

TM4C123Gis utilized as the main control chip in the hardware system. In addition, there is OpenMV machine vision module in charge of collecting image data information, inertial sensor ICM-20602, barometer sensor SPL06, magnetic sensor AK8975, motor module, power supply, etc. in this design. The overall diagram is illustrated in Fig. 1. The microcontroller communicates with the sensors and deals with the detected data according to the different task requirements. TM4C123G also controls the operation situation of ESC through four PWM channels.

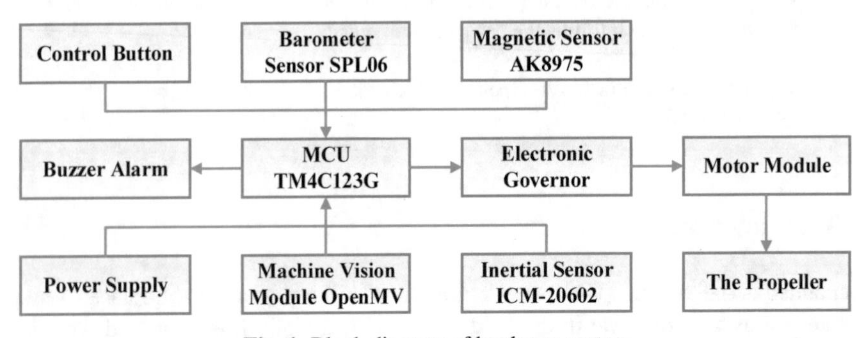

Fig. 1 Block diagram of hardware system

2.1 Tm4c123gmicrocontroller

The TM4C123G LaunchPad Evaluation Kit (shown in Fig. 2) is a low-cost evaluation platform for ARM Cortex-M4F based microcontrollers from Texas Instruments. The design highlights the TM4C123GH6PM microcontroller with a USB 2.0 device interface and hibernation module. The 80 MHz 32-bit ARM Cortex-M4-based microcontroller CPU contains 256 KB Flash, 32 KB SRAM, 2 KB EEPROM, and two Controller Area Network (CAN) modules. It has up to 24 PWM outputs and 12-bit high precision ADC, which is mainly responsible for sensor data acquisition and processing [13].

This development platform also features programmable user buttons and an RGB led for custom applications. The stackable headers of the TM4C123G make it easy and simple to expand the functionality when interfacing with other peripherals.

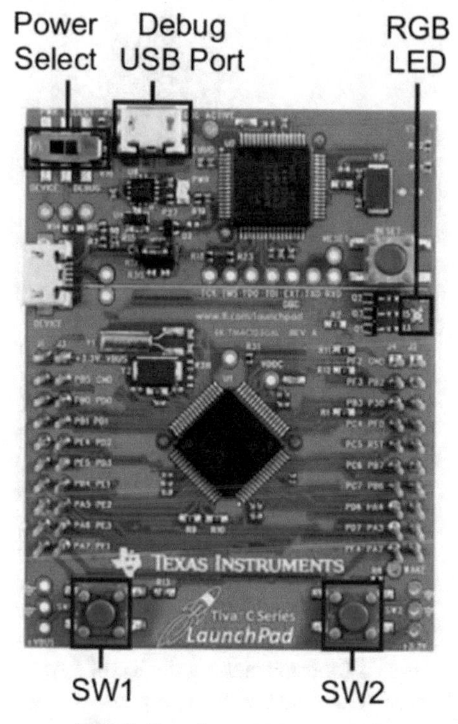

Fig. 2 Top view of TM4C123G

2.2 OpenMV Machine Vision Module

OpenMV is an open-source, low cost, multi-functional machine vision module, which is indicated in Fig. 3. It is an integrated OV7725 camera chip on the small hardware module. With STM32H743VI CPU as the core, the main machine vision algorithm is implemented efficiently with C language. Moreover, the python programming interface is also provided. The functions that the smart module can realize include frame differencing, color and marker tracking, face detection, QR code detection, and decoding, data matrix detection and decoding, linear barcode decoding, template matching, image capture, video recording, etc. [14–17].

The STM32H743VI ARM Cortex M7 processor running at 480 MHz with 1 MB SRAM and 2 MB of flash. All I/O pins output 3.3 V and are 5 V tolerant. The OpenMV Cam H7 comes with an OV7725 image sensor is capable of taking 640 × 480 8-bit Grayscale images or 640 × 480 16-bit RGB565 images at 60 FPS when the resolution is above 320 × 240 and 120 FPS when it is below. Most simple algorithms will run at above 60 FPS. Your image sensor comes with a 2.8 mm lens on a standard M12 lens mount. OpenMV has abundant hardware resources, and the UART, I2C, SPI, PWM, ADC, DAC as well as GPIO interfaces are introduced to facilitate the expansion of peripheral functions. USB interface is used to connect OpenMV IDE, which is an integrated development environment on the computer, to assist in programming, debugging, and firmware updates.

Fig. 3 OpenMV module

2.3 Other Modules

ICM-20602, one type of 6-axis motion tracking device of InvenSense, is selected as the inertial sensor in the system. The features are low power consumption, low noise, and extremely precise sensitivity error. The sensor incorporates 3-axis gyroscope and 3-axis accelerometer in a 3 × 3 × 0.75 mm (16-pin LGA) package. There are on-chip 16-bit ADC, programmable digital filter, embedded temperature sensor, and programmable interrupt in ICM-20602. The working voltage range of this device is as low as 1.71 V. Communication ports include I^2C and high-speed SPI (10 MHz).

Barometer module SPL0606 cooperates with optical flow module to intelligently make quadrotor UAV flying at a fixed altitude. The magnetic sensor AK8975 can obtain the data of magnetic field in each axis of x, y, and z, so as to provide a sense of direction for the drone. Compared with the brush motor, the BLDCM (Brushless Direct Current Motor) has better controllability and lower power consumption, so as to be utilized in this design. Meanwhile, a 100 nF capacitor is connected in parallel to the key switch circuit to realize the anti-shake function of the takeoff button.

3 The Realization of Tracking Function

The quadrotor automatic tracking UAV is designed to be able to realize the functions of one key start, one-meter height setting, linear tracking identification, automatic inspection, stable landing, etc.

3.1 Dynamic Model Establishing

As shown in Fig. 4, the coordinate system o-xyz of the drone is established ω_i is defined as the rotational speed of the brushless motor M_i, and F_i ($i = 1,2,3,4$) is the pull force provided by M_i to drive the propeller. Then, the calculation Formula of F_i can be expressed as belows.

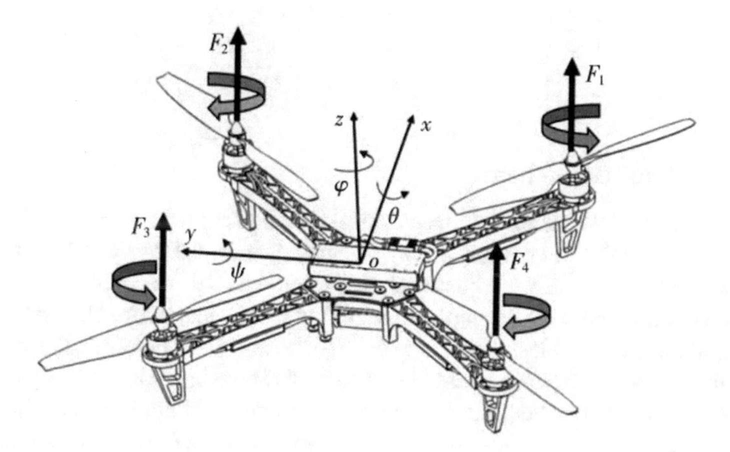

Fig. 4 Coordinate system of quadrotor UAV

$$F_i = \frac{1}{2}\rho \cdot C_t \cdot \omega_i^2 = k_t \cdot \omega_i^2 \tag{1}$$

ρ is the air density and C_t is the lift coefficient. So the tension is proportional to the square of the motor speed, and the ratio coefficient is k_t. In addition, the vertical lifting, the roll, the pitch, and the yaw control quantity are U_1, U_2, U_3, and U_4, respectively.

$$\begin{bmatrix} U_1 \\ U_2 \\ U_3 \\ U_4 \end{bmatrix} = \begin{bmatrix} F_1 + F_2 + F_3 + F_4 \\ 2(F_1 - F_2) \\ 2(F_2 - F_3) \\ F_1 - F_2 + F_3 - F_4 \end{bmatrix} \tag{2}$$

θ, ψ, φ are the roll angle, pitch angle, and yaw angle of quadrotor UAV. According to the Euler equation, the roll angular velocity $\dot{\theta}$, pitch angular velocity $\dot{\psi}$, and yaw angular velocity $\dot{\varphi}$, that is to say, the equations of each attitude angular velocity are obtained as belows.

$$\begin{bmatrix} \dot{\theta} \\ \dot{\psi} \\ \dot{\varphi} \end{bmatrix} = \begin{bmatrix} (U_2 - K_1 \cdot \theta) \cdot d/I_x \\ (U_3 - K_2 \cdot \psi) \cdot d/I_y \\ (U_4 - K_2 \cdot \varphi) \cdot d/I_z \end{bmatrix} \tag{3}$$

K_i is the air resistance coefficient. I_x, I_y, I_z are, respectively the moment of inertia of the drone around the *x*, *y*, and *z* axes in the coordinate system. d is the vertical distance from the UAV center to the x-axis of the coordinate system. If the air drag is ignored, the dynamic model equation of attitude control can be simplified as belows, combined with the above three Formulas.

$$\begin{bmatrix} \dot{\theta} \\ \dot{\psi} \\ \dot{\varphi} \end{bmatrix} = \begin{bmatrix} 2k_t \cdot d \cdot \left(\omega_1^2 - \omega_2^2\right)/I_x \\ 2k_t \cdot d \cdot \left(\omega_2^2 - \omega_3^2\right)/I_y \\ k_t \cdot d \cdot \left(\omega_1^2 - \omega_2^2 + \omega_3^2 - \omega_4^2\right)/I_z \end{bmatrix} \tag{4}$$

3.2 PID Attitude Regulation

The classical PID closed-loop feedback control method is utilized to control the attitude of the quadrotor UAV. The proportion (P), integral (I), and differential (D) of the deviation are combined to form the adjustment quantity to control the controlled object. The nonlinear coupling model is decoupled into four independent control channels by introducing four control variables, which are U_1, U_2, U_3, and U_4. Therefore, the system can be described as two subsystems: angular motion and translational motion. The former will affect the latter, however, the movement of the latter will not influence the former.

The data detected by the accelerometer and gyroscope will be distorted due to the disturbances, such as the wind and magnetic field. In this case, the Euler attitude angle cannot be accurately calculated. The cascade PID algorithm, shown in Fig. 5, is thus introduced in this design to solve the problem. The external loop is used for angle PID control and the interior loop for angular velocity PID control. The angular velocity loop can not only represent UAV attitude very accurately but also has a stronger ability to resist external interference. The system is more stable than the traditional single-loop angle PID controller.

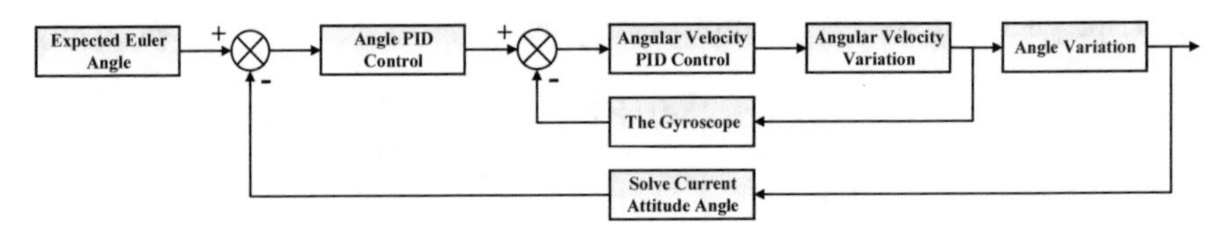

Fig. 5 The cascade PID algorithm

3.3 Function Realization

Connect the flight controller and OpenMV with the upper computer, debug the programs with Keil and python, respectively. Read the data and ensure the regular work operation of the drone's sensors, BLDCM, and motors. Observe the flight stability, record the results, and adjust the flight control parameters in time. Ultimately, this quadrotor UAV realizes the functions of one key start, one-meter height setting, linear tracking identification, automatic inspection, and stable landing.

4 Conclusion

The quadrotor UAV system is a high precision, multi-attitude, strong dynamic coupling flight system. In this design, TM4C123G and OpenMV machine vision module are selected as the main control chip and image acquisition device, respectively. The functions of height control, attitude control, and automatic tracking of the drone are successfully realized with the dynamic model and PID algorithm. The quadrotor model and theoretical basis of this design provide more development direction of multi-rotor UAV in the future.

Acknowledgements. This work is supported by the 2020 Undergraduate Education Reform Project Fund of Nankai University (NKJG2020004); the 2020 Experimental Course Teaching Reform Project Fund of Nankai University (20NKSYJG01); the 2021 Self-made Experimental Teaching Instrument and Equipment Project Fund of Nankai University; Tianjin Key Laboratory of Optoelectronic Sensor and Sensing Network Technology; Transformation and Extension Project of Agricultural Scientific and Technological Achievements in Tianjin (201901090); 2020 Undergraduate Education and Teaching Reform Project of Nankai University (NKJG2020326); First Class Undergraduate Education and Teaching Reform Project of Nankai University in 2019 (NKJG2019012); 2019 Experimental Teaching Curriculum Reform Project of Nankai University (2019NKSYJXKCGG02) and Teaching Center for Experimental Electronic Information.

References

1. Ye J, Zhang C, Lei HJ et al (2019) Secure UAV-to-UAV systems with spatially random UAVs[J]. IEEE Wireless Commun Lett 8(2):564–567
2. Zhang SH, Zhang HL, Di BY et al (2019) Cellular UAV-to-X communications: design and optimization for multi-UAV networks[J]. IEEE Trans Wireless Commun 18(2):1346–1359
3. Wang Z, Duan LJ, Zhang R (2019) Adaptive deployment for UAV-aided communication networks[J]. IEEE Trans Wireless Commun 18(9):4531–4543
4. Baek J, Han SI, Han Y (2020) Optimal UAV route in wireless charging sensor networks[J]. IEEE Internet of Things J 7(2):1327–1335
5. Ahmed S, Chowdhury MZ, Jang YM (2020) Energy-efficient UAV-to-user scheduling to maximize throughput in wireless networks[J]. IEEE Access 8:21215–21225
6. Motlagh NH, Bagaa M, Taleb T (2019) Energy and delay aware task assignment mechanism for UAV-based IoT platform[J]. IEEE Internet of Things J 6(4):6523–6536
7. Mukherjee A, Misra S, Chandra VSP et al (2019) Resource-optimized multi-armed bandit based offload path selection in edge UAV swarms[J]. IEEE Internet of Things J 6(3):4889–4896
8. Yang SZ, Deng YS, Tang XX et al (2019) Energy efficiency optimization for UAV-assisted backscatter communications[J]. IEEE Commun Lett 23(11):2041–2045
9. Cabecinhas D, Naldi R, Silvestre C et al (2016) Robust landing and sliding maneuver hybrid controller for a quadrotor vehicle[J]. IEEE Trans Control Syst Technol 24(2):400–412
10. Gageik N, Benz P, Montenegro S (2015) Obstacle detection and collision avoidance for a UAV with complementary low-cost sensors[J]. IEEE Access 3:599–609
11. Ryll M, Bulthoff HH, Giordano PR (2015) A novel overactuated quadrotor unmanned aerial vehicle: modeling, control, and experimental validation[J]. IEEE Trans Control Syst Technol 23(2):540–556
12. Wang H, Wang ZH, Sun GL et al (2019) Design of wind pendulum control system based on STM32F407[C]. Commun Signal Process Syst 571:1045–1054
13. Al-Haija QA, Samad MD (2020) Efficient LuxMeter design using TM4C123 microcontroller with motion detection application[C]. In: 2020 11th International conference on information and communication systems (ICICS)

14. Zhou SW, Gong JY, Zhou HY et al (2019) OpenMV based cradle head mount tracking system[C]. In: 2019 6th International conference on dependable systems and their applications (DSA)
15. Mu ZX, Li ZF (2018) Intelligent tracking car path planning based on hough transform and improved pid algorithm[C]. In: 2018 5th International conference on systems and informatics (ICSAI)
16. Shao YH, Tang XF, Chu HY et al (2019) Research on target tracking system of quadrotor UAV based on monocular vision[C]. In: Chinese automation congress (CAC)
17. Huang R, Wang H, Liu F, et al (2019) Design of cricket automatic control system based on computer vision technology[C]. In: International conference on robots and intelligent system (ICRIS)

A Reliability Evaluation Model of Intelligent Energy Meter in Typical Environment

You Gong, Huiying Liu(✉), Xin Yin, Heng Hu, and Guorui Wu

State Grid Heilongjiang Electric Power Co., Ltd. Electric Power Research Institute, Harbin 150000, Heilongjiang, China
sutfrank@126.com

Abstract. In order to solve the problem of reliability evaluation of intelligent watt-hour meter in a typical environment of our country, according to the reliability operation data of smart meters in typical areas, the index system which takes the basic error of intelligent energy meter under various typical environmental as the evaluation index is established in this paper. In this paper, an improved grey relation analysis method is used to put forward a comprehensive evaluation model of the reliability of the intelligent energy meter in a typical environment. Finally, the evaluation model is used to evaluate the reliability of a single-phase intelligent watt-hour meter from three manufacturers in a typical environment. The evaluation results show that the evaluation model can be used for reference in the field of reliability evaluation of smart meters.

Keywords: Reliability · Evaluation model · Grey relation analysis · Electric energy meter

1 Introduction

China is a vast country with a complex environment and climate. The typical regional environment for the application of smart meters mainly includes the high cold in Heilongjiang; the high altitude in Tibet; the high humidity, heat, salt fog in Fujian, and the high dry heat in Xinjiang. In all kinds of the harsh natural environment, whether the operation of the intelligent energy meter is reliable and whether the measurement is accurate is one of the key issues for power enterprises, industries, and users [1]. Therefore, it is very important to study the comprehensive evaluation method of the reliability of smart meters in a typical environment for discovering the hidden danger of the reliability of smart meters in time and improving the reliability test standard of smart meters.

At present, there are rich contents in the field of intelligent energy meter evaluation at home and abroad. Some authors evaluate the software quality of intelligent watt-hour meter [2]. In view of the reliability evaluation of intelligent energy meter, it mainly analyzes and evaluates the existing inspection items [3]. Some authors evaluate the life

This paper is supported by the science and technology project of the headquarters of the state grid corporation of China (project no.: 5230HQ19000F).

Q. Liang et al. (eds.), *Artificial Intelligence in China*, Lecture Notes in Electrical Engineering 653, https://doi.org/10.1007/978-981-15-8599-9_22

cycle quality of smart meters [4]. There are few researches on the reliability evaluation of smart meters in a typical environment. This paper analyzes the environmental factors that affect the reliability of smart meters. Through the analysis and research of the historical data of smart meters reliability in a typical environment, a comprehensive evaluation index system of smart meters reliability in a typical environment is established, and the gray correlation analysis method is used to evaluate the smart meters of various manufacturers. The reliability evaluation model established in this paper can evaluate the reliability of the whole intelligent electric energy meter truly and reasonably, and provide technical support for State Grid Corporation to carry out differentiated bidding and manufacturing process control of electric energy meter.

2 Establishment of Evaluation Model

2.1 The Construction Process of Evaluation Model

In the process of a comprehensive evaluation of the reliability of intelligent energy meter in a typical environment, there are three steps to build the evaluation model of the system: the first step is to build the index system of the evaluation system according to the environmental characteristics of typical regions by using the analytic hierarchy process; the second step is to analyze and process the reliability data of the energy meter by using MATLAB, and construct the initial matrix of the evaluation model; the third step is to evaluate the system synthetically by using grey relation method. The evaluation process flow is shown in Fig. 1.

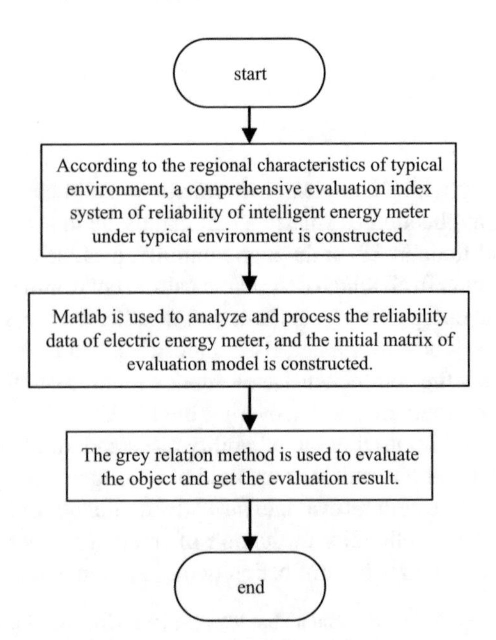

Fig. 1 Process chart of evaluation model establishment

2.2 Establishment of Index System

According to the four typical regional characteristics introduced in the introduction of this paper, a comprehensive reliability evaluation index system of intelligent energy meter under a typical environment is constructed. According to the principle of index system construction [5], a comprehensive evaluation index system for the reliability of intelligent energy meter under a typical environment is established by using the AHP theory, as shown in Fig. 2.

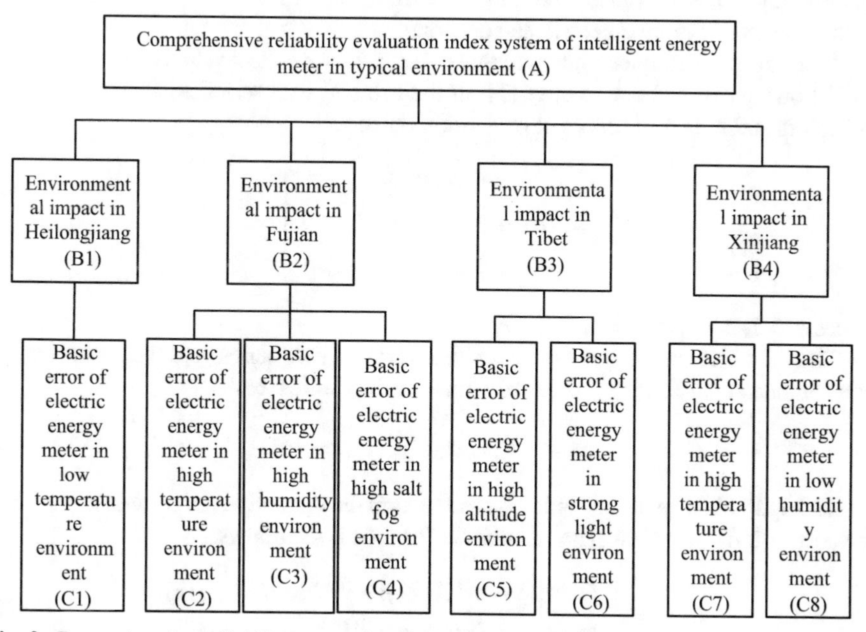

Fig. 2 Comprehensive reliability evaluation index system of intelligent energy meter in typical environment

As shown in Fig. 2, the target layer of the evaluation system is the comprehensive evaluation index system of the reliability of intelligent energy meter under a typical environment. The base layer is the influence degree of typical environmental parameters in Tibet, Heilongjiang, Xinjiang, and Fujian. Because the environmental parameters of each typical area have different influences on the reliability of the intelligent energy meter, the index layer is composed of the basic error parameters of the energy meter under the environment of temperature, humidity, air pressure, light, and salt fog in each typical area.

2.3 Improved Grey Relation Analysis Method

Grey relation analysis is an evaluation method of grey system theory, which is a quantitative description and comparison method of system development and change trend. Its basic idea is to judge the closeness of different data series and reference data series according to the geometric similarity between reference data series and data series. In the comprehensive evaluation of various fields, the positive reference data column is constructed, and the correlation between each evaluation object and the reference data column is calculated. The higher the correlation degree is, the better the evaluation result is [6].

Grey relation degree is a model for evaluating the unknown information system, which is a comprehensive evaluation model combining quantitative analysis and qualitative analysis. The model can solve the problem that evaluation indexes are difficult to quantify and count accurately, eliminate the influence of human factors, and make the evaluation results more objective and accurate.

According to the index system of this paper, data are collected and analyzed to form the following original data matrix [7]. In the original data series, M is the number of indicators and N is the number of evaluation objects.

$$X_a(b) = \begin{bmatrix} x_1(1) & \cdots & x_n(1) \\ \vdots & \ddots & \vdots \\ x_1(m) & \cdots & x_n(m) \end{bmatrix}$$

where $a = 1, 2, \ldots, n$; $b = 1, 2, \ldots, m$.

The reference data series can be an ideal forward data series, and the optimal value of each index constitutes the forward reference data series. Note:

$$x_0 = (x_0(1), x_0(2), \ldots, x_0(m)) \tag{1}$$

The optimal value of each index is used to form the reference data column to calculate the correlation coefficient. The calculation formula is as follows:

$$\zeta_i(k) = \frac{\min_i |x_0(k) - x_i(k)| + \rho * \max_i |x_0(k) - x_i(k)|}{|x_0(k) - x_i(k)| + \rho * \max_i |x_0(k) - x_i(k)|} \tag{2}$$

where $k = 1, 2, \ldots, m$; $i = 1, 2, \ldots, n$. ρ is the resolution coefficient, $0 < \rho < 1$. If ρ is smaller, the difference between correlation coefficients is larger, and the discrimination ability is stronger. Generally, ρ is 0.5.

For each evaluation object, the mean value of the correlation coefficient between each index and the corresponding element of the reference data column is calculated to reflect the correlation between each evaluation object and the reference data column, which is called the correlation degree. The calculation formula of correlation degree R is recorded as follows:

$$r_i = \frac{1}{m} \sum_{k=1}^{m} \zeta_i(k) \tag{3}$$

where $k = 1, 2, \ldots, m$.

In the process of analyzing the original data matrix, this paper improves the steps of the original algorithm, omits the dimensionless process, and improve the calculation efficiency of the algorithm.

3 Case Study

First, the reliability history data of intelligent energy meter in a typical environment is screened, analyzed, and processed. According to the typical environmental characteristics of each region, the range of environmental characteristic parameters is determined. According to the range of environmental characteristic parameters, the reliability history data of intelligent energy meter is screened, and then the curve fitting of the screened data is carried out with MATLAB, and the curve of the error of energy meter with the environmental stress is obtained [8]. According to the fitting curve, the basic error of the intelligent electric energy meter under environmental stress can be obtained.

Next, the reliability data of single-phase smart meters of three different manufacturers in four typical areas are analyzed and processed. Figure 3 is the fitting curve of reliability data of single-phase intelligent energy meter of manufacturer 1 in the high-temperature environment of Fujian. Figure 4 is the fitting curve of reliability data of single-phase intelligent energy meter of manufacturer 3 in the low humidity environment of Xinjiang.

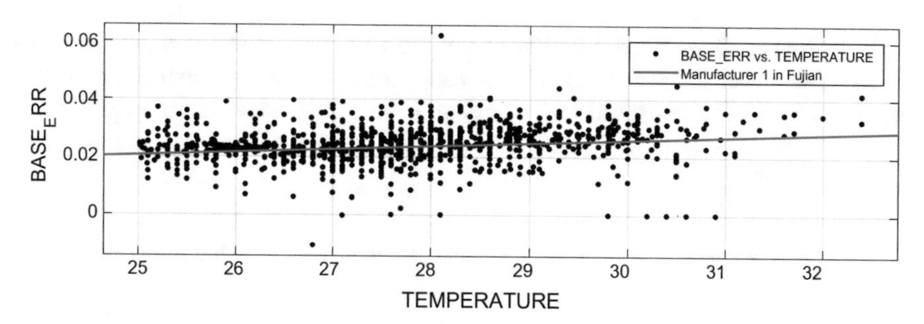

Fig. 3 Error-curve of electric energy meter of manufacturer 1 in the high temperature environment of Fujian

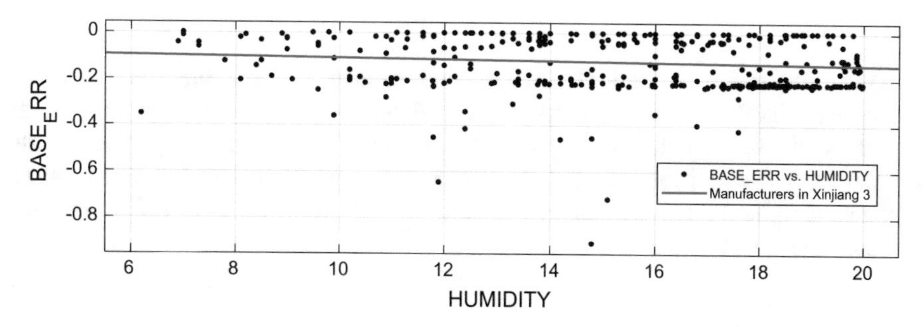

Fig. 4 Error-curve of electric energy meter of manufacturer 3 in the low humidity environment of Xinjiang

According to Fig. 3, the basic error of the intelligent electric energy meter changes little with the temperature under the high humidity and heat environment with the temperature above 25 °C, and the basic error is 0.02. According to Fig. 4, it can be analyzed that under the high dry and heat environment with humidity lower than 20, the basic error of the intelligent energy meter changes little with humidity, and the basic error is 0.1.

Through the curve fitting of the reliability data of three manufacturers' single-phase smart meters in each typical area, the basic errors of three manufacturers' single-phase smart meters under each typical environmental stress are obtained, and the specific parameters are shown in Table 1.

Table 1 basic error of electric energy meter of each manufacturer in each typical environment

	C1	C2	C3	C4	C5	C6	C7	C8
Manufacturer 1	0.23	0.07	0.05	0.02	0.02	0.03	0.08	0.08
Manufacturer 2	0.11	0.02	0.01	0.015	0.04	0.07	0.015	0.02
Manufacturer 3	0.13	0.02	0.03	0.02	0.02	0.02	0.1	0.1

According to Table 1, the initial matrix of the evaluation model is constructed. The parameter sequence is the optimal value of each index (the ideal optimal solution of the basic error is 0). The improved grey correlation analysis method in this paper is used for analysis and calculation. Finally, the correlation degree of the three manufacturers is shown in Table 2.

Table 2 Basic error correlation degree of electric energy meters of each manufacturer in each Table 2 typical environment

	Manufacturer 1	Manufacturer 2	Manufacturer 3
Correlation degree	0.670	0.899	0.786

According to the correlation degree of each manufacturer in Table 2, it can be seen that among the three manufacturers, the correlation degree of manufacturer 2 is the highest, so the single-phase intelligent energy meter of manufacturer 2 performs better in comprehensive performance; the correlation degree of manufacturer 1 is the lowest, so the single-phase intelligent energy meter of manufacturer 1 performs the worst in a comprehensive performance.

4 Conclusion

According to the environmental characteristics of typical regions, a comprehensive evaluation system for the reliability of intelligent energy meters in typical environments is established, which consists of four first-class indexes and eight second-class indexes. According to the reliability data of smart meters in four typical areas in 2018–2019, the reliability historical data of single-phase smart meters from three manufacturers are processed and analyzed, respectively, and the influence degree of each index is analyzed and evaluated by using the grey correlation algorithm. The results show that the comprehensive evaluation of manufacturer 2's single-phase intelligent energy meter is high; the comprehensive evaluation of the single-phase intelligent energy meter of manufacturer 1 is the lowest. In the same way, this evaluation model can be used to comprehensively evaluate the reliability of smart meters of various manufacturers, which has a certain application prospect.

References

1. Feng S (2012) The research on electrical performance evaluation method of single-phase smart meter [D]. North China Electric Power University, Bao Ding
2. Ji J, Hou X, Chen H et al (2015) Quality evaluation of the smart meter software based on AHP method [J]. Electri Meas Instrum 52(8):5–9
3. Xue Y, Zhang P, Wang Y et al (2016) Study and exploration on reliability assessment method for smart meters [J]. Electri Meas Instrum 53(13):90–94
4. Ju H, Yuan R, Ding H et al (2015) Research on smart meter full lifecycle quality evaluation method [J]. Electrical Meas Instrum 52(16A):55–58
5. Qin Y (2018) A study on the maturity evaluation of "the fusion of industrialization and informatization" based on hierarchy-gray relation [D]. Northeast Forestry University, Harbin
6. Jiang J, Zhou F, Zhong K et al (2019) Comprehensive evaluation method of power quality based on combination empowerment and improved grey correlation analysis [J]. Proceedings of the CSU-EPSA
7. Yanling W, Mengkai W, Ziqing Z et al (2017) Quantitative analysis model of power load influencing factors based on improved grey relational degree [J]. Power Syst Technol 41(6):1172–1178
8. Yin X, Yibiao L, Gong Y et al (2017) The error model of the smart meter under the influence of temperature [J]. Electri Meas Instrum 54(8):85–88

Analysis and Prediction of the Resettlement for Climate Refugees in the Maldives

Jiasong Mu(✉) and Hao Ma

College of Physical and Electronic Information, Tianjin Normal University, Tianjin, China
mujiasong@163.com

Abstract. With the increasing level of environmental pollution and rising global sea level, this situation has seriously threatened the survival of Maldives, Tuvalu, Kiribati, Marshall Islands and other island countries. This means that EDPs (Environmentally displaced persons) are at risk of losing their homes and culture. Recently, a ruling of the United Nations expressed the rational recognition of EDPs. If these island countries disappear due to sea-level rise, then the main body should take reasonable measures to settle these EDPs. These measures should at least include the placement of personnel and cultural protection. We built a model to predict the time it would take for the Maldives to complete the resettlement of climate refugees. We define an important indicator: Environmental saturation degree (θ). To solve the resettlement problem of EDPs, we consider many aspects (although not all aspects). To get a reasonable and fair solution, we need to establish a mathematical model, which is multi-level and multi-objective. Through the prediction results, the remaining time of the resettlement plan is directly reflected, which provides a good reference for the United Nations and the Maldives government.

Keywords: Gray-scale model · Equal dimension progressive model · Population forecast

1 Introduction

Climate change has forced some people to consider immigration. In this paper, we define an important indicator: Environmental saturation degree (θ). This model can be used to deal with many problems such as behaviour selection, comparison before and after stages and population migration. We predict the degree of land loss and population growth of island countries, and then we analyze the two together, taking environmental saturation as an important measure to get the population growth of EDPs. In the prediction of land loss of island countries, we regard each island as a cone with a height of 1.2 m and a bottom angle of 0.0387 °, which is calculated according to the annual rise of sea level of 3.3 mm. In the prediction of population growth, we use the equal dimension progressive model which can automatically update the uncertainty. Through the analysis, we can see that the land on the island will reach the maximum carrying capacity between 2030 and

Q. Liang et al. (eds.), *Artificial Intelligence in China*, Lecture Notes
in Electrical Engineering 653, https://doi.org/10.1007/978-981-15-8599-9_23

2035, and about 250,000 residents will move out in 2045, while the Maldives Islands are likely to disappear in 2060 or so. In our other work, we use principal component analysis to reduce the dimensions of various cultural quantitative indicators of EDPs, and take the first five indicators whose cumulative variance contribution rate reaches 95%. They are original reliability belief, original language inheritance, eating habits, fishing technology, navigation technology and dispersion of EDPs [2, 5].

In our future work, we will further analyze and predict the migration plan. More parameters are added to the Maldives population model to make it more reasonable.

2 Prediction of Sea Level Rise and EDPs Population

The idea of this part is to take Maldives as an example (the geographical location of the Maldives islands is shown in Fig. 1), predict the degree of sea-level rise and the total population of Maldives, and then calculate the degree of environmental saturation of Maldives θ (That is, the percentage of the population that has been accommodated per km^2 to the maximum population that can be accommodated), $\theta \in (0, 1)$, it is stipulated that once the environmental saturation degree exceeds 100%, this area will not be suitable for survival. [4].

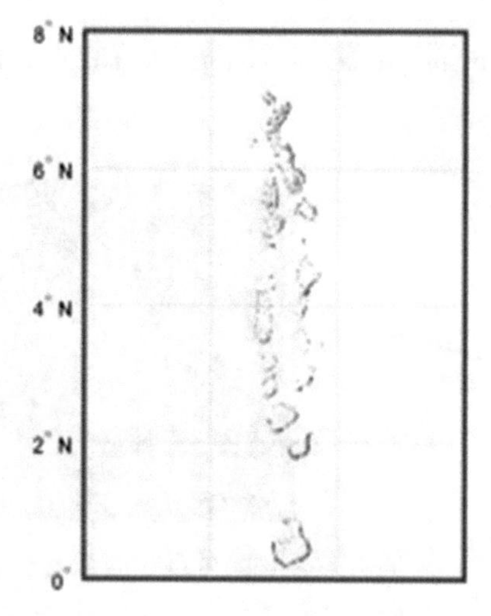

Fig. 1 The shape of Maldives data *Source* NASA elevation data drawn by MATLAB

2.1 Land Loss

For the prediction of global sea level, we use NASA Goddard Space flight center and think that the annual sea level rise rate is 3.3 mm/year, as shown in Fig. 2. Next, the land area loss caused by sea-level rise in Maldives is calculated.

At the same time of sea-level rise, the land area of Maldives Islands is shrinking. For the convenience of theoretical analysis, we make the following hypothesis:

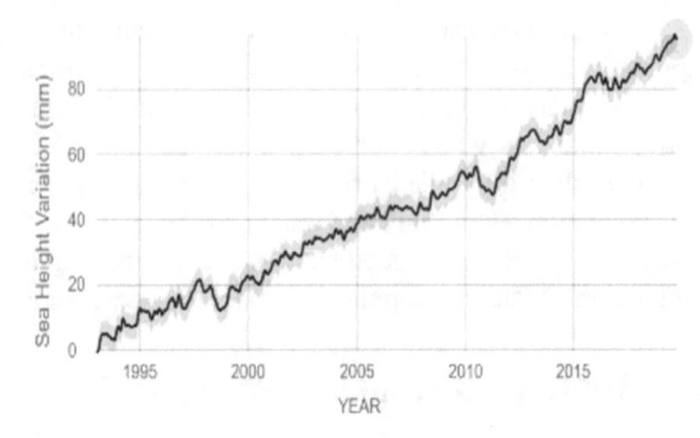

Fig. 2 Sea level rise time over time figure from NASA's climate database [1]

Hypothesis 1 to simplify the calculation, we assume that there are 200 inhabited islands in the Maldives, each of which has the same area and is 1 km^2.

Hypothesis 2 consider the shape of each island as a cone with a height of 1.2 m.

The land area considered in this paper refers to the area that can be inhabited by human beings.

Based on the above assumptions, we can get the shape of the island, as shown in Fig. 3.

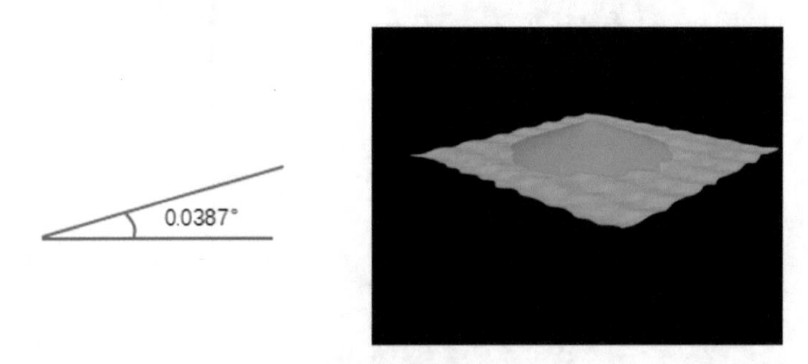

Fig. 3 The shape of the island

Figure 4 shows the process of sea-level rise. With the annual rate of sea-level rise of 3.3 mm, we can calculate the area lost by the island, as shown in Fig. 5.

2.2 EDPs Population

Our preliminary estimate is that the number of EDPs is constantly changing, what is to say, it will increase year by year. In order to predict the number of EDPs in Maldives, it is necessary to analyze the population growth of Maldives first, and then combine the sea level rise in front to get the results.

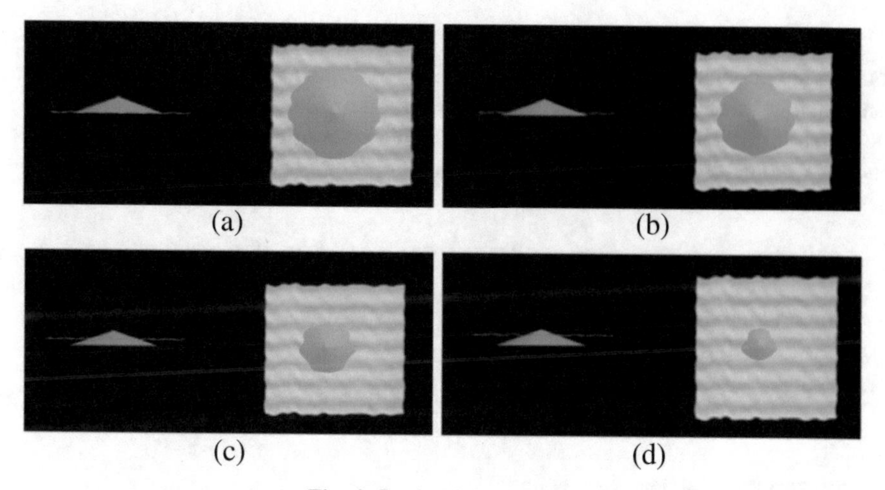

Fig. 4 Sea level rise process

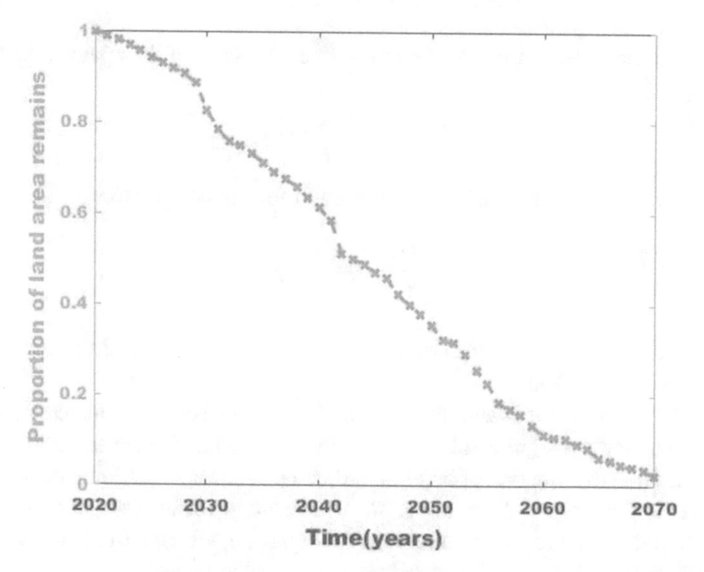

Fig. 5 Proportion of the land area remains

For the prediction of EDPs in Maldives, we adopt the gray model with high precision of equal dimension supplement. Compared with other models, the advantage of this model is that after the initial gray model is established, new known messages are introduced to fill the gap in the gray area in the process of repeated iterations. Each step is modified with the parameters of the previous step to make the prediction result more accurate. [3].

We use the initial data from the United Nations and the National Bureau of statistics, mainly the population data of Maldives in 1950–2019 and the national population data, on which we predict the future population situation of Maldives.

First, in order to ensure that the original data can use the equal dimension progressive model, we need to test the data, here we use the smoothness test.

Set the initial data row vector as

$$X_0 = [x_0(1), x_0(2), \ldots, x_0(n)] \tag{1}$$

The original data is accumulated item by item, that is

$$x_1(j) = \sum_{i=1}^{j} x_0(i)\,,\; j = 1, 2, \ldots, n \tag{2}$$

So, we can get a new row vector

$$X_1 = [x_1(1), x_1(2), \ldots, x_1(n)] \tag{3}$$

If (4) is a monotone decreasing function, then this kind of data passes the test.

$$\beta(j) = \frac{x_0(j)}{x_1(j)}\,,\; j = 2, 3, \ldots, n \tag{4}$$

Second, we establish the differential equation for the new vector

$$\frac{dX_1}{dt} + \eta X_1 = \gamma \tag{5}$$

$\eta \cdot \gamma$ are parameters, where η is the degree of population development and γ is the degree of population growth control.

It should be noted that Maldives is facing the risk of being inundated by seawater, which is undoubtedly a significant impact on the population growth of Maldives. In this regard, we assume that the risk of being inundated by seawater in Maldives is increasing year by year, and the suppression effect on population growth is also increasing year by year. We will add this inhibition effect into the equation, which will be in each iteration Increase 4% based on the calculated value.

$$w_1(j) = \frac{x_1(j+1) + x_1(j-1)}{2} \tag{6}$$

Artificially defined mean (6), by replacing the differential equation in (5) with the difference equation, it can be concluded that

$$[\eta, \lambda]^T = (W^T W)^{-1} W^T Y \tag{7}$$

Among them,

$$W = \begin{bmatrix} -w_1(2) & -w_1(2) & \ldots & -w_1(n) \\ 1 & 1 & \ldots & 1 \end{bmatrix} \tag{8}$$

$$Y = [x_0(2), x_0(3), \cdots, x_0(n)] \tag{9}$$

Take the obtained parameters into Eq. (5) and get the solution

$$\widehat{x}_1(j+1) = \frac{\gamma}{\eta} + (x_0(j) - \frac{\gamma}{\eta})e^{-\eta j} \quad , j = 1, 2, \ldots, n \tag{10}$$

We use MATLAB for programming according to the existing data and obtained the future population forecast results of Maldives and the world, as shown in Table 1. In order to make the results more intuitive, we drew a graph, as shown in Fig. 6.

Table 1 Maldives population forecast

Year	2020	2025	2030	2035	2040	2045	2050
Population (million)	0	0.12	0.2	0.29	0.35	0.43	0.47

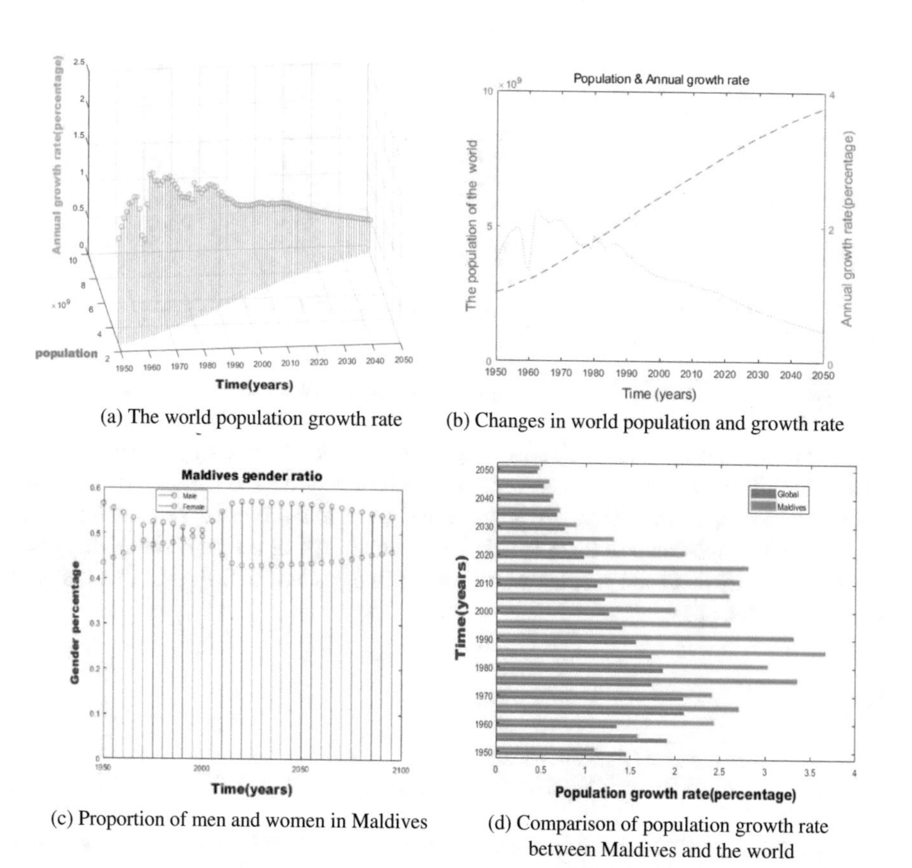

(a) The world population growth rate

(b) Changes in world population and growth rate

(c) Proportion of men and women in Maldives

(d) Comparison of population growth rate between Maldives and the world

Fig. 6 Maldives population and world population

According to the FAO, the potential carrying capacity of the earth is 26.4 billion people. Human habitation on earth should account for about 134,320,000 km^2. From this, we can calculate that the population per km^2 of the earth is 195. We use it to measure the population carrying situation of Maldives. Based on the prediction results of sea-level rise given above, we can get the relationship between future population and remaining land in Maldives, as shown in Fig. 7.

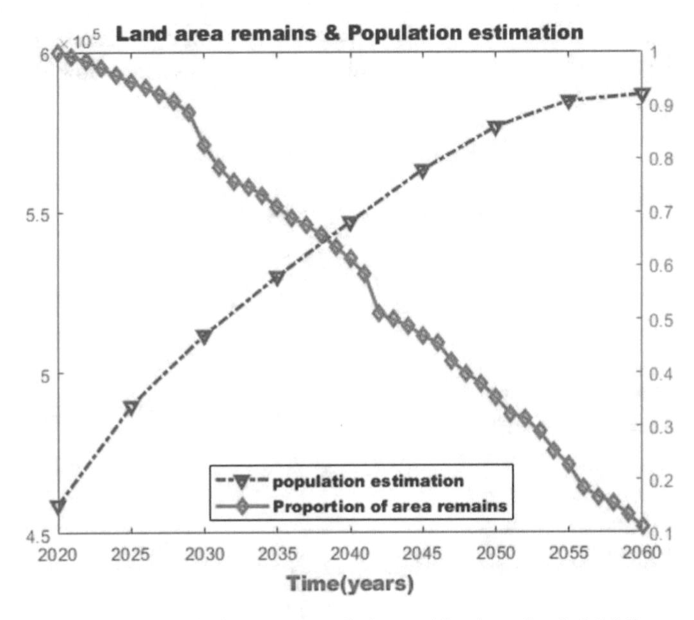

Fig. 7 Relationship between population and land surplus in Maldives

It can be seen from Fig. 7 that the population of Maldives is increasing and decreasing and the land area will be reduced by half in 2030–2035. About 2030–2035, the land on the island and the maximum carrying capacity must be relocated by some people to ensure the life of the remaining people on the island. With the increase of time, the number of people on the island will be more and more, and it is estimated that one-tenth of them will be relocated in 2025 In 2045, more than half of the population(about two hundred and fifty thousand people) will move out, and around 2060, the Maldives Islands are likely to disappear.

References

1. https://climate.nasa.gov/vital-signs/sea-level/
2. Burleson E (2010) Essay: climate change displacement to refuge. J Environ Law Litigation. 25(1):019–036
3. Ling Z, An X (2013) Research on internet users number prediction based on equal dimension and new information grey markov model. Comput Sci 40(4):119–121 China

4. Moerner NA, Tooley M, Possnert G (2004) New perspectives for the future of the maldives. Global Planet Change 40(1/2):177–182
5. Nicholls RJ, Cazenave R (2010) Sea-level rise and its impact on coastal zones. ENCE 328(5985):1517–1520

Smart Electricity Meters Test Data Management Service System

Liu Huiying[1], Yin Xin[1], Wang Xiaoyu[1], Wen Ruxin[1], Zhang Qiuyue[2](✉), and Du Bo[2]

[1] State Grid Heilongjiang Electric Power Co., Ltd. Electric Power Research Institute, Harbin 150000, Heilongjiang, China

[2] Heilongjiang Electrical Instrument Engineering Research Center Co., Ltd, Harbin 150000, Heilongjiang, China

gao3431128@sina.com

Abstract. According to the requirements of smart electricity meters test data management service system, on the basis of analyzing the insufficient of the traditional smart electricity meters quality inspection tests. A set of mobile phone APP software and smart electricity meters test data management service system based on the B/ S architecture is developed. A universal data management system platform is designed and implemented. The system architecture, system database and the design ideas of each component module in detail is introduced and introduced the technical tools and methods used in each part. Key technologies such as large data volume database management, data backup and recovery are researched. By partitioning smart electricity meters test data, load data and retrieval test on the system platform. Data management efficiency has been greatly improved compared to traditional methods, which meet the requirements for large data volume data management.

Keywords: Mobile APP · Smart electricity meters test · Oracle · B/S architecture · Spring MVC

1 Introduction

During smart electricity meters quality inspection tests, different types of tests are to be done. There must be many records during the test. The traditional smart electricity meters quality inspection test requires the quality inspector to manually record the test items and test results. It will also process the result data and report files. How to systematically and effectively manage these complex and diverse data is a huge and tedious task [1, 2].

This article develops a set of mobile APP software and test management service system for smart electricity meters based on B/S architecture. The system replaces the paper-based manual recording of the smart energy meter quality inspection test process and results. In this process, the quality inspection staff can check the test progress of the

This paper is supported by The Smart Electricity Test Management Service System Based on Energy Internet of state grid corporation of China.

Q. Liang et al. (eds.), *Artificial Intelligence in China*, Lecture Notes in Electrical Engineering 653, https://doi.org/10.1007/978-981-15-8599-9_24

smart electricity meters at any time, conduct the next test as required and upload it to the cloud platform. Shorten working time, improve work efficiency, and save and query data more conveniently. A general test data management platform was constructed to achieve flexible and unified management of a large number of test data.

2 System Design

The quality inspector completes the relevant tests of the smart electricity meters through the mobile APP and uploads the test results to the cloud platform. At the same time, the related test information can be queried through the mobile APP or the web-based test management service system. The system block diagram is shown in Fig. 1.

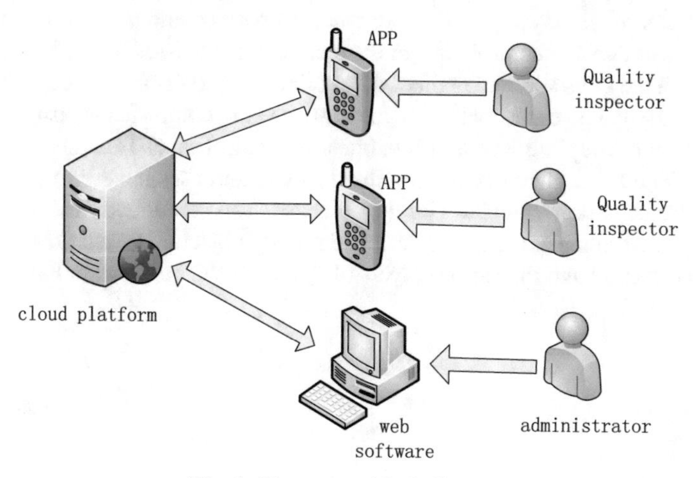

Fig. 1 The system block diagram

2.1 Android Phone APP Module

Android phone APP uses Android Studio as the development platform and uses open-source framework Android-async-http to communicate with server, JSON data as communication format. Android-async-http is an asynchronous network request processing library based on the Apache HttpClient library. Network processing is based on Android's non-UI threads, Handling request results through callback methods. Its main characteristics are as follows: Handle asynchronous HTTP requests and process callback results through anonymous inner classes. HTTP asynchronous requests are located in non-UI threads and do not block UI operations. Concurrent requests are processed through the thread pool. File uploads and downloads are handled. Response results are automatically packed the JSON format automatically handles reconnecting when a connection is broken.

2.2 Smart Electricity Meters Test management Service System Based on B/S Architecture

This system is based on B/S architecture and is implemented with Spring MVC as a framework [3]. Browser/ server mode (B/S mode) is to install and maintain a server, the client uses a web browser to get the relevant content. The B/S architecture develops into a three-layer architecture, which is the presentation layer, business logic layer, and data access layer [4].

This article applies the current popular MVC development model and uses Spring MVC as the framework. Its structure is divided into four layers [5]: View layer, Controller layer, Service layer and DAO layer. The layers are described as follows: The View layer is the browser page that sends different function requests to the server. After the server receives the request from the client, it sends the data requested by the client to the view layer; The Controller layer plays a core role, it operates entity classes and business layer objects to achieve control, design different control classes for different modules, interacts with the interface, and realizes data collection and display; The service layer is the business layer. It uses the method of the DAO layer to complete the business process, which plays a decoupling role and separates the control layer from specific database operations. The DAO layer operates on the database and uses the API provided by the Hibernate framework to interact with the database. You can directly call the functions provided by the framework to add, delete, and modify objects. The relationship between the B/S architecture and the various layers of Spring MVC is shown in Fig. 2 [6].

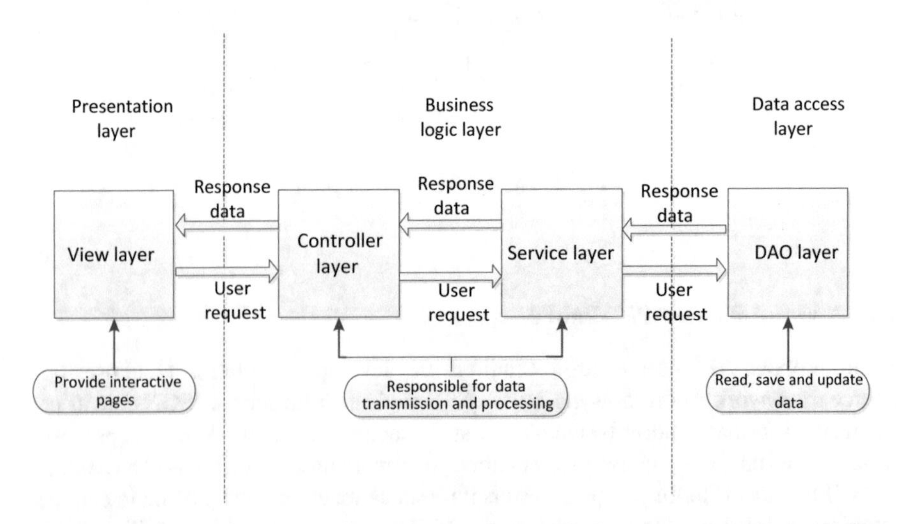

Fig. 2 The relationship between the B/S architecture and the various layers of Spring MVC

3 Database Design

This system selects Oracle database according to a large amount of test data and high-security requirements [7]. The system database is designed as follows: The system sets up a system database user to maintain fixed system information and project information data. These data include user table in system management; user role table, and corresponding user authority table; test item table in test management, Smart Electricity Meters basic information table, dictionary table (equipment type and test category).

In addition, because there are different test equipment, each test project is relatively independent, the data structure is relatively different, and data processing is less crossover. Therefore, a separate database user and a default tablespace are established for each test equipment in order to create a database table for the experimental data of various structures under the experiment.

In actual test projects, repeated tests are often conducted for the same project, resulting in massive data (GB-level or even TB-level) data. From the perspective of data management performance and operation convenience, partitioned tables provided by the Oracle database system are used to store these data.

A partitioned table is a partition of a large table or index into relatively small, independently manageable sections. Partitioning further subdivides a table, index, or index-organized table into segments. The segments of these database objects are called partitions. Each partition has its own name and can choose its own storage characteristics. Oracle's table partitioning feature improves the performance of certain queries and maintenance operations by improving manageability, performance, and availability [8].

There are many types of table partitions, including range partitions, list partitions, hash partitions, and combined partitions [9]. Range partitioning refers to partitioning according to a certain value in the data table and designating storage in different partitions according to different ranges of this value. It is a more commonly used partitioning mechanism. This article adopts such a partitioning method and saves the data by partition according to the time when the experimental data is stored in the database.

4 Back-End Technology

4.1 Cloud Server

Cloud server is also called cloud host. Compared with traditional servers, it has the advantages of low on-demand cost, stable performance, exclusive broadband, convenient management, multi-level backup, fast deployment, and strong scalability. This article uses Tencent cloud server as the hardware platform. The hardware configuration is 2 cores and 4 GB. All the code is stored on this cloud host and is operated by Xshell5 remote login.

4.2 Django Framework

Django is a full-stack web framework written in Python. The reasons for choosing Django as the back-end framework are as follows:

(1) The self-service management background is very perfect. No matter from the perspective of the convenience of the early testing or the later data maintenance, using the Django framework with a comprehensive background management system will greatly reduce the time for viewing and modifying data.

(2) With the built-in ORM, there is basically no need to write SQL statements. Each record is an object, and reading the data in the object is simple and fast.
(3) It is very convenient to extend the APP.
(4) Obvious error prompts, quickly and accurately locate bugs, and significantly improve efficiency.

5 System Implementation

5.1 Android Phone Module

The Android mobile phone uses Android Studio as the development platform and uses the open-source framework Android-async-http to communicate with the server to create an AsyncHttpClient; set the request parameters through the RequestParams object; after calling the get method of AsyncHttpClient, it actually encapsulates the request into a Runnable object and submit this task to the thread pool. Finally, the callback interface is called, and the result is returned to the main thread. Flow chart of android-async-http is shown in Fig. 3.

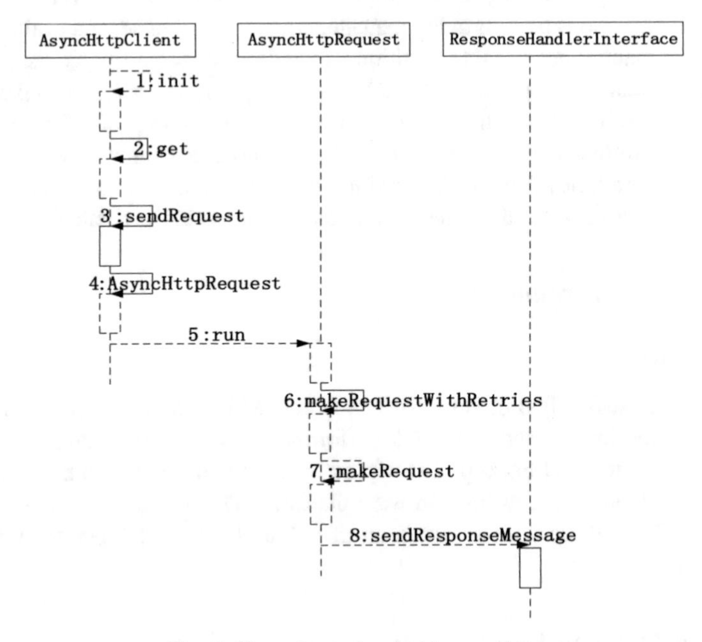

Fig. 3 Flow chart of android-async-http

5.2 B/S Architecture-Based Intelligent Energy Meter Test Management Service System

Eclipse is used as a compilation tool, developed in accordance with the J2EE specification, and the database is stored and managed by Oracle. Based on this, the experimental data management system is designed and implemented. The presentation layer page is implemented by HTML + CCS [10, 11]. In terms of page requests, jQuery technology is used together with Ajax asynchronous requests to give users a better experience. The request interception of the control layer is implemented through the annotations of Spring MVC. The business module process control is performed through different controllers, and the request processing and data preparation are performed. The data access layer is implemented through the hibernate framework + JDBC connection and interacts with the database to complete operations such as data acquisition and update. The database connection pool uses C3P0. Multiple database connections are established in advance. When the database needs to be accessed, the database connection in the idle state is taken from the connection pool. After the access is completed, the idle connection is automatically recovered.

Database backup is generally divided into physical backup and logical backup. Physical backup refers to the way of physically copying data files, control files, online log files, etc., and is performed at the file layer; logical backup exports the data to a file by means of the export tool provided by Oracle, and use the tool to restore the data into the database during recovery, which is performed at the data layer. This article uses logical backup, and uses expdp/ impdp tools to complete data backup and recovery.

6 System Test

Compare the query performance of the partitioned table with the query performance of the non-partitioned table. Create two database tables with the same table structure (including 24 fields), where test_tablePar is a table partitioned by the entry time, and test_tableNot is an ordinary table. A total of 13,602,460 pieces of data were written into this table at different times (10 writes, 1,306,246 each), and then a conditional query was performed according to the write time. Each time a field in the table was found, 1,306,246 were found. The data is executed 5 times in succession. The query time of the two tables is shown in Table 1.

Table 1 Data table field query performance comparison results

Serial number	Partition table query (time/s)	Unpartitioned table query (time/s)
1	2.215	3.232
2	1.511	3.098
3	1.789	2.758
4	1.712	2.746
5	2.068	3.350
Mean	1.859	3.037

In addition, the statistical efficiency of the data is compared, and the total data of each test is counted. The comparison results are shown in Table 2.

Table 2 Data statistics performance comparison results

Serial number	Partition table query (time/s)	Unpartitioned table query (time/s)
1	0.143	22.533
2	0.141	21.062
3	0.122	22.171
4	0.151	23.012
5	0.126	21.040
Mean	0.136	21.963

It can be seen from the test results in Table 2 that compared with the non-partitioned table, the query efficiency of the table fields in the partitioned table is improved by nearly two times, and the statistical effect of the data in Table 2 is increased by 160 times.

7 Conclusion

This article develops a set of mobile APP software and smart electricity meters test management service system based on the B/S architecture to complete the relevant quality inspection tests of smart electricity meters. The mobile APP uses the open-source framework Android- async-http, B/S architecture test management service system adopts Spring MVC framework, combined with the Oracle database. Finally, the relevant quality inspection tests of smart electricity meters are completed successfully. The overall structure, module design, and database design of the system fully consider the versatility of test data management and the future expansion of the system.

The system is convenient to use in practical test applications, to a certain extent, it solves the problem of inconvenience in the use and maintenance of traditional smart electricity meters quality inspection tests. The management system has better scalability and compatibility. It performs performance tests on data loading and data query functions to meet the functional requirements of a large amount of data management. The data query performance has been greatly improved compared to traditional methods.

References

1. Dong D, Zhu C, Hu Y (2014) Design of test data management platform[J]. J Rocket Propulsion 40(4):67–72
2. Zhao X (2016) The design and implementation of JingHai power fixed assets management system[D]. School of information and software engineering, ChengDu
3. Wang D (2017) Design and implementation of online shopping system based on spring and hibernate framework[D]. Wuhan Institute of Posts and Telecommunications, Wuhan
4. Shan D, Zhang X, Wei R (2014) Sharp JQuery[M]. Posts and Telecom Press, Beijign
5. Sun W (2013) Mastering hibernate: let Java objects hibernate in the relational database[M]. Publishing House of Electronics Industry, Beijing

6. Li G (2014) Crazy HTML5/ CCS3/ javascript handout[M]. Publishing House of Electronics Industry, Beijing
7. Wang S, Li Z (2011) Research and implementation of Oracle partition technology[J]. Technol Square 9:86–89
8. Xiaoya X, Yanhua X (2014) Backup and recovery analysis based on oracle database[J]. Inf Secur Technol 5(3):62–64
9. Wang J, Cheng Z, Zhang Z, He W (2015) Research on the database optimization method of large management information system[J]. Electric Power Inf Commun Technol 8
10. Zhiyuan D, Shuai L, Liao X (2019) The framework of data distribution management with region match method of middleware in joint test platform[]. In: The 31st Chinese control and decision conference (2019 CCDC)
11. Duan J, Pengcheng F, An G, Zhengfan Z (2015) Design of test data management system architecture based on cloud computing platform. In: Proceedings of 2015 international conference on circuits and system (CAS 2015)

Power Equipment Identification Based on Single Shot Detector

Hanwu Luo[1], Wenzhen Li[1], Qirui Wu[2,3](✉), Hailong Zhang[2,3], and Zhonghan Peng[2,3]

[1] State Grid East Inner Mongolia Electric Power Supply Co., Ltd, Hohhot, People's Republic of China

[2] NARI Group Corporation Ltd, Nanjing, People's Republic of China
8278799@qq.com

[3] Wuhan NARI Limited Liability Company, State Grid Electric Power Research Institute, Wuhan, People's Republic of China

Abstract. In recent years, more and more deep learning technologies have been extensively used in various fields, such as intelligent medical, intelligent manufacturing, to realize the intelligence of various systems. To solve this traditional manual inspection of power equipment, we propose a method of intelligent identification of electrical equipment using a Single Shot Detector based on the light network—MobileNet V2. Through data enhancement and other technologies to make up for the lack of training data often encountered in deep learning. The average precision is 96.3%, and the average recall rate is 85.9% in this paper. It also has a good recognition effect for complex scenes. We provide a solution to equip unmanned aerial vehicles, remote cameras, inspection robots, and other resource-constrained devices with the power equipment identification system based on deep convolution network.

Keywords: MobileNet single shot detector · Power equipment · Data enhancement

1 Introduction

In recent years, artificial intelligence has rapidly developed. More and more deep learning technologies are being applied in various fields, such as intelligent medical, intelligent manufacturing. The power system is also beginning to move in the direction of intelligence.

Traditional power equipment supervision mainly relies on manual patrols. With the increasing number of power equipment, it is obvious that this method will consume a lot of manpower and resources. Nowadays, many power companies are equipped with equipment such as unmanned aerial vehicle and remote cameras to improve the management efficiency of power equipment. However, these devices only have image acquisition function. After the image information is obtained by the staff, manual identification is still needed. Therefore, how to intelligently identify the collected images is an important research aspect of the intelligent electric grid.

Q. Liang et al. (eds.), *Artificial Intelligence in China*, Lecture Notes
in Electrical Engineering 653, https://doi.org/10.1007/978-981-15-8599-9_25

Object detection is one of the classic assignment of computer vision. Its purpose is to locate and classify the target object. The images collected in the power system often include the surroundings (such as trees, buildings, etc.) and various devices (such as transformers, insulators, etc.), the target device needs to be located and identified, so the target detection technology can be applied to the supervision of power equipment to realize the intelligence of the power system.

Traditional target detection work is mostly based on well-designed manual features, while hand-designed features are not very robust to diversity changes. The form of power equipment is diverse, and the image data collected at different moments is also affected by environmental factors. Therefore, the traditional target detection technology is hard to satisfy the demands of intelligent identification of power equipment. With the development of deep learning technology, object detection has been rapidly developed. Object detection has shifted from the original traditional manual extraction feature method to the feature extraction based on CNN. The feature based on convolutional neural network solves the above problem because it is learned from the data.

When deep learning came along, the object detection has two most popular directions [1]. One is the deep learning target detection algorithm on account of proposed region represented by Faster R-CNN [2], and the other is deep learning on account of regression method represented by Single Shot Detector (SSD) [3]. At present, Faster R-CNN has been applied to vehicle identification (Ruan and Sun [4]), remote sensing images identification (Wang et al. [5]), and power equipment identification (such as Wang [6]). SSD is faster than Faster R-CNN. However, there are few industrial applications for SSD, so we will study the identification of power equipment based on SSD.

2 Related Work

The traditional power equipment identification uses the template matching method. In [7], the SIFT algorithm is used to fetch the feature points of the template image and the test image, and coarse matching is performed for the feature points. Then, use the RANSAC algorithm to remove the mismatched points. Hou et al. [8] improves the matching efficiency by reducing the dimension of feature point descriptors, and then improves the accuracy by two-phase matching. Zhang et al. [9] proposed a quadratic template matching algorithm for quickly identifying target images. The algorithm performs rough matching on the data of the interlaced columns interlace row/ column and performs exact matching on the basis of this, which greatly reduces the amount of data calculated. In [10], the improved SIFT algorithm is used for image matching, and the visual attention model is introduced into the Bow model for equipment identification. In addition, traditional power device identification also has a method of image segmentation and recognition. In [11], the infrared image is segmented by the method of region growing, and then the invariant moment features of the power equipment in the segmented image are extracted to classify by support vector machine.

In order to address the shortcomings of traditional manual features-based power equipment identification methods, scholars began to turn their research goals into deep learning. Li et al. [12] extracted the depth features of the image based on the AlexNet and used the trained random forest for classification during the test. In [13], the visual saliency model based on deep neural network is further established to detect the equipment targets in the image. On this basis, the equipment targets segmentation based on the maximum entropy method is carried out and identified by the method in [12]. Chen et al. [14] use a similar idea to the [11] to extract image features. Finally, the backpropagation neural network is used to design the classifier, and the extracted invariant moment feature vectors are used to classify and identify. Wang [6] proposed a power equipment identification network named CWGCNet which improved the accuracy of recognition.

3 SSD-Based Power Equipment Identification

In this paper, the collected power equipment images are firstly denoised, and then use the image enhancement method to enhance the contrast between the target and the background. Then the generalization ability of the model was improved by data enhancement technology. Finally, the SSD model based on the light deep convolution network MobileNets V2 [15] is used to perform the identification of power equipment.

3.1 Image Preprocessing

Image Noise Reduction

Influenced by environmental factors such as light and wind, the collected pictures of power equipment will inevitably be disturbed by noise and the characteristic information of the target will be affected, which will reduce the accuracy of recognition. We will use the following methods for noise reduction.

(1) *Mean filtering method.*

The mean filtering method is a spatial domain noise reduction method. The algorithm assigns the average of all pixels of a pixel and its neighborhood to the output image's corresponding pixel. After the averaging process, the noise is weakened to the surrounding pixel points, and the noise amplitude is reduced, but the noise area becomes larger. To solve this problem, a threshold can be set. Comparing noise and neighborhood pixel grayscale, it is considered to be noise only when the difference is higher than some threshold. This method is simple in algorithm and fast in operation.

(2) *Median filtering method.*

The median filtering method is also a spatial domain noise reduction method. The principle of the algorithm is to first determine an odd pixel window, sort the pixels in the window from large to small, and take the median value to replace the pixel of the center point. The edge information of the image can be protected in this way effectively, and the algorithm is simple and shows real-time performance.

(3) *Wavelet denoising method.*

Wavelet denoising is a transform domain denoising method. On account of the situation that larger wavelet coefficients are generally actual signals, The principle of the algorithm could be achieved while smaller coefficients are likely to be noise. Therefore, the image signal can be firstly decomposed by wavelet, the appropriate threshold is set, the wavelet coefficient smaller than the threshold is set to zero, and the wavelet coefficient larger than the threshold remains unchanged; then the estimated coefficient is obtained through the threshold function mapping; At last, the estimated coefficient is inversely transformed to achieve image denoising and reconstruction [16]. This algorithm can better protect the details of pictures, and it is a widely used image denoising method.

Image Enhancement

Histogram equalization can be used to adjust the image histogram contrast. It transforms the image of a known gray probability distribution into an image with uniform distribution of gray probability distribution through a transformation based on cumulative distribution function, to achieve the purpose of enhancing contrast, enhance the difference between the foreground and background gradations.

Data Enhancement

Because the data collected is very limited, and the in-depth learning network is data-driven and requires a large amount of data for training, this paper will expand the sample size through data enhancement technology. In addition, this method can also enhance the generalization ability of learning network. The way SSD achieves data enhancement by randomly sampling training data is as follows:

Use input image but not modification.

Random sampling.

If the center of the real bounding box is within the sampling block, then the overlap of the real bounding box and the sampling block is preserved. After using the above sampling steps, illumination transformation, and geometric transformation are also employed. The illumination transformation includes random adjustment of brightness, contrast, hue, saturation, and color channel switching. Geometric transformations include random expansion, cropping, and mirroring. Each transformation is randomly used with a certain probability (the general probability is 0.5).

3.2 Feature Extraction

Feature extraction is an important part of power equipment identification. On account of the in-depth learning network, the characteristics extracted solve the problem of manual features. We will use MobileNets V2 for feature extraction.

The MobileNet series of networks is a new neural network architecture developed by Google for mobile and embedded devices. It can run efficiently with limited resources and maintain the highest possible accuracy. Therefore, we choose this network as the model for feature extraction. As Fig. 1.

To decompose the original standard convolution layer into two layers is the key to the MobileNet model. The previous layer is a deep convolutional layer, which is optically filtered by applying a convolution filter to each convolutional channel. The latter layer

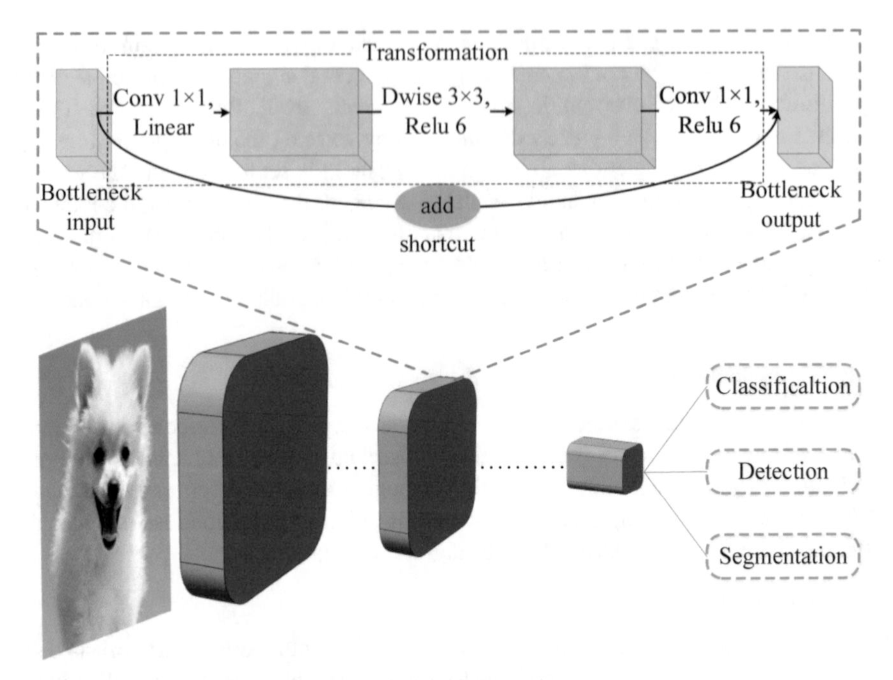

Fig. 1 MobileNets V2

is a pointwise convolutional layer, a convolution of 1 × 1 kernels, which constructs new features by calculating the linear combination.

MobileNet V2 has made two major improvements on this basis, one is to solve the linear bottleneck, and the other is to invert the residuals. The former is based on the theory that the activation function can destroy the feature in low-dimensional space and effectively increase the nonlinearity in high-dimensional space. Since the main function of pointwise convolution after depthwise convolution is dimension reduction, the activation function ReLU6 is removed. While the latter adds a pointwise convolution before the depthwise convolution. The main function of this pointwise convolution is to upgrade the dimension so that the depthwise convolution can work in the high-dimensional space.

3.3 SSD Model Based on Deep Learning Network Framework

SSD is based on the regression method, which is a classic object detection algorithm. It is different from others based on proposed regions. The target detection is divided into three steps: hypothesize bounding boxes, resample pixels or features for each box, and apply a high-quality classifier. SSD saves the first two steps and gets predictions directly from the image, which greatly improves detection speed. As Fig. 2.

SSD is a truncated deep convolutional network (we use MobileNet V2), and then adds the auxiliary structure for detection:

(1) **Multi-scale feature maps.** To make it possible to predict at multiple scales, adding convolutional layers with decreasing scales.

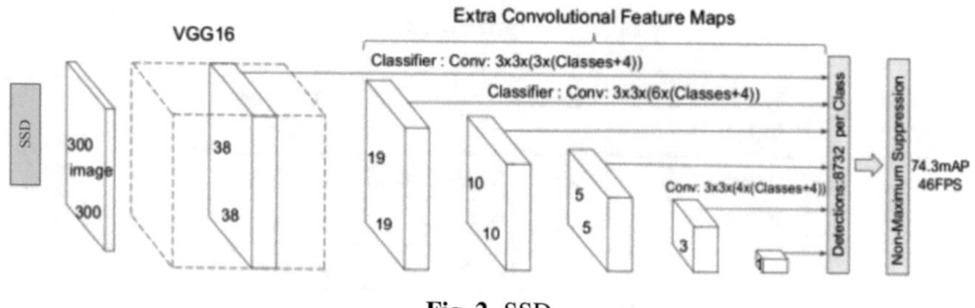

Fig. 2 SSD

(2) **Convolution predictors.** Each convolution feature layer generates a fixed series of predictions by use convolution kernels.
(3) **Aspect ratios and default boxes.** In each feature map cell, predict the offset from the default bounding box shape in the feature cell and the score of each category that appears in these boxes. For multiple feature maps at the top of the network, a set of default bounding boxes is associated with each feature map cell.

4 Experiment and Analysis

4.1 Introduction to the Dataset

The dataset of this paper contains 3241 images, each of which contains many types of equipment. There are seven types of equipment in our dataset. They are transformer cooler, transformer bushing, transformer conservator, transformer main body, transformer wall bushing, insulator, and isolating switch. In this paper, samples of different positions, different angles, different exposure levels (weather and equipment factors) were collected for each equipment and they are randomly divided into training set and test set in 9:1 ratio.

4.2 Introduction to the Experimental Platform

The operating system of this experiment is Ubuntu1.18.0.04, the GPU is NVIDIA GeForce GTX 1080Ti, the deep learning framework is tensorflow1.13.1, and the programming language is python.

4.3 Experimental Results and Analysis

The average precision and average recall rate were used in this paper to analyze the experimental results. The precision is measured by the prediction result and the actual overlap degree (0.5:0.05:0.95 refers to 10 thresholds with 0.05 steps), the recall rate is the result of the maximum recall rate in a given fixed number of test results. We can see the experimental results in the table below. It's obvious that our scheme has an improvement over the average accuracy of 92.37% of the model selected in [6] (Table 1).

In this paper, two complex scenes (including multiple equipment in one picture) are randomly selected for performance display. The meanings of the labels in the figure

Table 1 Experimental results

Evaluation index	Metrics		Our scheme
Average precision	IoU	0.5:0.05:0.95	0.819
		0.5	0.963
		0.75	0.928
Average recall rate	#Dets	1	0.663
		10	0.859

are: transformer cooler(vt01_cooler), transformer bushing (vt01_bushing), transformer conservator (vt01_conservator), transformer mainbody (vt01_mainbody), transformer wall bushing (vt01_wbushing), insulator (dc01_insulator). It can be seen from Figs. 3 and 4 that our scheme can detect and identify the power equipment well. Figure 5 selects a scene with insufficient illumination. It can be seen that our scheme can maintain a good recognition effect under the condition of poor illumination.

Fig. 3 Complex scene recognition

5 Conclusion

The method of using the Single Shot Detector model based on the light network MobileNet V2 to realize the intelligent identification of power equipment has high accuracy and recall rate and can achieve good results even in complex scenes. In the future, it will be considered to be mounted on a remote camera or unmanned aerial vehicle or a patrol robot to realize the intelligent inspection of power equipment.

Fig. 4 Complex scene recognition

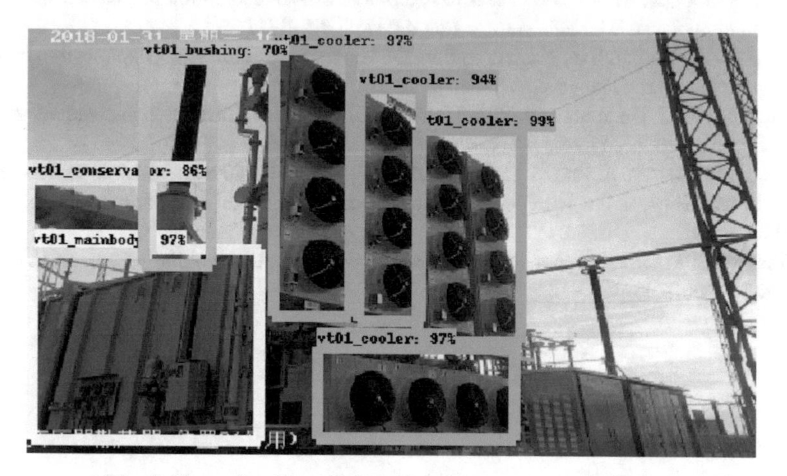

Fig. 5 Example of Insufficient illumination scene recognition

Acknowledgements. This work was Funded by the State Grid Science and Technology Project (Research on Key Technologies of Intelligent Image Preprocessing and Visual Perception of Transmission and Transformation Equipment).

References

1. Huang J, Rathod V, Sun C et al (2017) Speed/accuracy trade-offs for modern convolutional object detectors[C]. IEEE CVPR 4
2. Ren S, He K, Girshick R et al (2015) Faster r-cnn: towards real-time object detection with region proposal networks[C]. In: Advances in neural information processing systems. pp 91–99
3. Liu W, Anguelov D, Erhan D, et al. (2016) Ssd: single shot multibox detector[C]. In: European conference on computer vision. Springer, Cham, pp 21–37

4. Ruan H, Sun H (2018) Vehicle multi-attribute recognition based on Faster r-cnn [J]. Comput Technol Develop 28(10):136–141
5. Wang JC, Tan XC, Wang ZH et al (2018) Faster R-CNN deep learning network based object recognition of remote sensing image[J]. J Geo-Inf Sci 20(10):1500–1508
6. Wang YR (2018) Substation equipment recognition based on deep convolution neural network [D]. Shenyang Agricultural University
7. Chen WH (2018) Research on the application of image recognition technology in the online monitoring of power equipment[D]. North China Electric Power University
8. Hou YM, Chen YH, Liu YQ (2014) Research on the application of bi-directional fast SIFT matching for power equipment identification [J]. Manuf Autom 36(01):62–65
9. Zhang H, Wang W et al (2010) Application of image recognition technology in electric power equipment on-line monitoring [J]. Power Syst Prot Control 38(06):88–91
10. Wang CY (2017) Research on power equipment identification based on infrared and visible images[D]. North China Electric Power University
11. Shi JY (2017) Substation equipment recognition and thermal fault diagnosis based on infrared image processing[D]. Shanghai DianJi University
12. Li JF, Wang QR, Li M (2017) Electric equipment image recognition based on deep learning and random forest [J]. High Voltage Eng 43(11):3705–3711
13. Li JF (2018) Research on electric equipment image recognition and its application based on deep learning[D]. Guangdong University of Technology
14. Chen JY, Jin LJ, Duan SH et al (2013) Power equipment identification in infrared image based on Hu invariant moments. J Mech Electri Eng 30(01):5–8
15. Sandler M, Howard A, Zhu M, et al. Mobilenetv2: inverted residuals and linear bottlenecks[C]. In: Proceedings of the IEEE conference on computer vision and pattern recognition. pp 4510–4520
16. Wang X (2011) Image denoising based on adaptive fuzzy thresholding approach[J]. Comput Simul 28(06):296–298 + 31

RETRACTED CHAPTER: Power Equipment Defect Detection Algorithm Based on Deep Learning

Hanwu Luo[1], Qirui Wu[2,3](✉), Kai Chen[2,3], Zhonghan Peng[2,3], Peng Fan[2,3], and Jingliang Hu[2,3]

[1] State Grid East Inner Mongolia Electric Power Supply Co. Ltd, Hohhot, People's Republic of China

[2] NARI Group Corporation Ltd, Nanjing, People's Republic of China
8278799@qq.com

[3] Wuhan NARI Limited Liability Company, State Grid Electric Power Research Institute, Wuhan, People's Republic of China

Abstract. Ensuring the safety of power equipment has an important meaning to the development of the country and people's lives. The safety of power equipment cannot be separated from the timely detection of power equipment defects. The current mainstream power equipment defect detection method is to take a picture of the power equipment in a complex environment through a drone and manually select from the video. This method requires a great deal of manpower and resources. This paper intends to use a target detection algorithm, which can replace the manual detection of power equipment defects based on UAV pictures. Most of the traditional power equipment defect detection algorithms can only target specific scenarios and do not have strong generalization. This paper compares the current target detection algorithm commonly used in industry, investigates its structure, performance, advantages and disadvantages, and its performance in public data, and summarizes the basic target detection network architecture which is most suitable for power equipment defects. In this paper, a new target detection algorithm is used to self-learn through deep neural networks to perform defect inspection of power equipment in UAV pictures with different backgrounds, illuminations, and scales. Through the reasonable selection of the target detection algorithm framework and the improvement of multiple steps, the proposed algorithm for detecting power equipment defects is improved by nearly 60% compared with the conventional deep learning target detection algorithm.

Keywords: Target detection · Power equipment defects · Convolutional neural network

---

The original version of this chapter was retracted: The retraction note to this chapter is available at https://doi.org/10.1007/978-981-15-8599-9_71

Q. Liang et al. (eds.), *Artificial Intelligence in China*, Lecture Notes in Electrical Engineering 653, https://doi.org/10.1007/978-981-15-8599-9_26

1 Introduction

1.1 Research Background

With the continuous development of the national economy, the demand for electricity in various industries has grown rapidly, and the pace of China's power infrastructure construction has continued to accelerate. The normal operation of the guaranteed power grid and the good condition of the power equipment have become a very critical issue.

At present, the operation and maintenance of the power grid, safety supervision, and disaster prevention rely heavily on image and video surveillance. However, a large amount of image and video information mainly depends on manual processing. There are many inherent problems in manual processing. Image intelligent recognition technology is used in more and more scenes. This project intends to establish the application of target detection algorithms based on deep learning in power equipment defect detection, build high-quality image algorithms, support power transmission, and transformation equipment defect recognition, and improve the application level of power equipment image intelligent detection.

Using deep learning-based target detection technology to detect defects in power equipment has difficulties: First, the types of defects are different, and the algorithm needs to be able to accurately detect and locate different types of defects in the drone image; Second, the outdoor environment where the power equipment is located is different, and the algorithm needs to overcome the problem of changing backgrounds, and it has good performance in complex backgrounds. Third, UAV images often shoot objects from different heights and angles. The recognition algorithm needs to be able to detect targets at multiple scales, and the algorithm needs to be highly robust (Fig. 1).

Fig. 1 Common power equipment defects

1.2 Research Status at Home and Abroad

Traditional power equipment defect detection algorithm

In 1997, Kazuo Yamamoto and others in Japan analyzed the infrared image to detect power line defects [1]. Australia's Yuee Liu and others also studied the method of separating power lines in real-time from video images taken by drones in 2012 [2]. Japan also uses thermal imaging scanning to detect the operating temperature of power lines and provides an early warning when overheating occurs [3]. In 2010, Yang Yong hui of the Chinese Academy of Sciences and others mainly studied how to monitor transmission lines through video and images [4] In 2012, Tongji University researched the automatic detection method of transmission line faults based on visible light images and ultraviolet [5], which can efficiently detect the transmission line broken strands, correct the image after detecting the end of the transmission line according to the characteristics of the endpoint. In 2010, Yu min Ge of North China Electric Power University proposed an insulator surface state detection method based on aerial visible image [6]. In 2013, Yang Hao and others of Chongqing University used computer binocular vision technology to collect pictures of the same insulator through two cameras with different positions [7]. In 2009, Li zuo sheng of Hunan University used infrared images to study the pollution level of insulator [8]. In general, there are some preliminary studies on image analysis and status detection of transmission lines, but most of the relevant studies are still at the preliminary exploration stage. There is no large number of application images for actual application scenarios to test and verify, and their practicability and robustness are relatively poor.

2 Research on Improvement of Power Equipment Defect Detection Algorithm

Faster-RCNN structure algorithm is widely used in industry because of its high accuracy. This paper adjusts the basic network structure and adds a multi-scale feature extraction module to detect defects of different sizes in pictures of different scales and adapt to different UAV resolutions and camera distance.

2.1 Extraction of Multi-Scale Features

In the pictures input to the detection model, the target to be detected often exists at various scales due to the different shooting distance and the size of the target itself. This also presents a challenge to the target detection algorithm-the target needs to be detected at various scales. As deep feature extraction networks tend to extract more detailed features at shallow levels, these features will contain rich spatial information and have a greater response to small scale targets. However, due to the lack of receptive fields, there will be a lack of context-sensitive semantic information. As the feature extraction network deepens, the network will continue to downsample, thereby increasing the receptive field corresponding to each area in the feature map. At this time, the extracted features will contain rich semantic information, which will be more effective for large-scale targets. The large response, but often lose a lot of details in the picture. Therefore, we need to comprehensively apply multi-dimensional feature information to achieve multi-scale feature extraction. This paper extracts multi-scale features by top-down extraction of multi-scale features and top-down fusion of pyramid structures in different stages.

Bottom-up feature extraction

The operation of the deep feature extraction network is bottom-up feature extraction. The hierarchical structure consists of down-sampling with a scale of 2 from bottom to top, thus containing feature maps at multiple scales. Between downsampling, there will be multiple convolutional layers to perform convolution operations, thereby extracting many feature maps of the same size. As down sampling progresses, the size of the feature map will continue to decrease, and the spatial information contained will be lost, but the receptive field corresponding to each area will increase, thereby obtaining rich contextual information. We take the feature map output from the last convolutional layer of each stage as the feature output of this stage, because the deepest layer of each stage will have the richest features.

(1) Top-down feature fusion

The top-down process is actually obtained by up-sampling and fusing bottom-up feature maps. First, the feature maps of each stage originally obtained from low to top are convolved with 1 × 1, and the number of channels is fixed to 256. By upsampling the convolutional upper layer feature map twice, scaling to the same size as the lower layer feature map, and then fusing the bottom layer feature map with the upsampled upper layer feature, that is, adding and convolving, so that The high-level semantic information is fused with the low-level detail information, and finally the 3 × 3 convolutions is used to convolve the added feature map to eliminate the aliasing effect. Through the top-down feature fusion, the low-level feature map will have a more accurate response to the detection target because of the high-level semantic information obtained.

Candidate region generation based on feature pyramid

In this paper, the multi-scale feature extraction network of the pyramid structure is used to generate features of different scales and merged as a candidate area generation network, which effectively improves the accuracy of target detection at different scales.

2.2 For Detecting Defects of Different Shapes

Convolutional neural networks have always had a relatively inherent limitation-the shape of the convolution kernel is fixed. Such characteristics will result in that for a given convolution kernel, the area covered by its receptive field is fixed. Targets in different positions often have different shapes and scales. The fixed shape of the convolution kernel will limit the area covered by the receptive field, thus limiting the detection of targets of different shapes.

Among the defects of power equipment, a very important defect category is the corrosion of conductive wires. The convolution of the traditional square structure still has a certain ability to deal with such defects. Difficulty, so this paper uses deformable convolution to accept different scales and deformations. Learning to adjust the scale adaptively and the range of receptive field coverage will be of great help to the detection.

Variable convolution

The conventional feature map volume integration is divided into two steps: (l) sampling the input feature map x with a conventional grid R; (2) the feature map part sampled through the grid and the convolution kernel (weight w) Perform inner product summation. Let us use a 3 × 3 convolution kernel for convolution, R can be expressed as (Fig. 2):

$$R = \{(-1, -1), (-1, 0), \ldots, (0, 1), (1, 1)\}$$

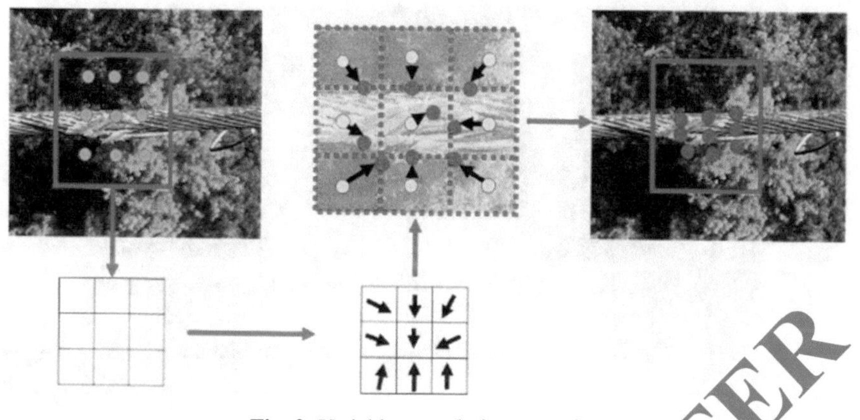

Fig. 2 Variable convolution operation

where (0,0) represents the center of the grid R, and the remaining values represent the horizontal and vertical offsets from the center. At each position p_0 of the output feature map y, we can get

$$y(p_0) = \sum_{pn \in R} w(p_n) \cdot x(p_n + p_0)$$

where p_n is an enumeration of every position of R.

In variable convolution, the conventional grid R is expanded by the offset $\{\Delta p_n | n = 1, \ldots, N\}$ where $N = |R|$, the above Formula is further transformed into:

$$y(p_0) = \sum_{pn \in R} w(p_n) \cdot x(p_n + p_0 + \Delta p_n)$$

After such changes, sampling is no longer limited to the area corresponding to the conventional grid R, but an irregular area is formed by the offset. What offset should be used will be independently learned by the network. When the corresponding value after the position shift is a decimal, the value corresponding to the shifted position is obtained by performing bilinear interpolation on the input picture.

Figure 3 shows the general flow of the variable convolution operation. Before the convolution operation, the network predicts the offset of each convolution position through the learned information. After the offset, the convolution position changes. Let the network pick the most suitable place to extract features for convolution, and make the convolution area focus on the defects as much as possible. In the experiment, in the last three stages of the deep feature extraction network Resnet101, deformable convolution is adopted.

2.3 Optimized Learning Strategies

Online difficult sample mining

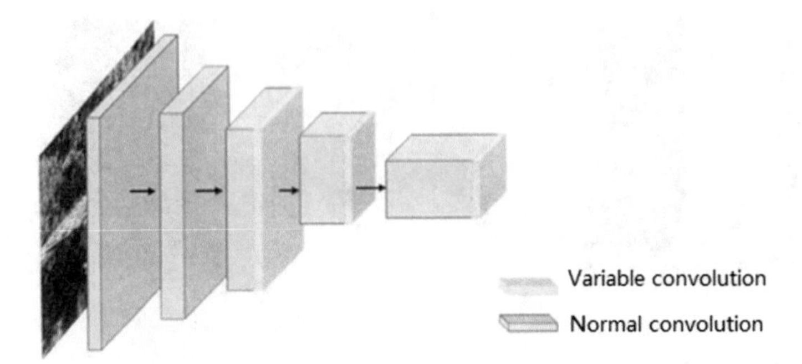

Fig. 3 Feature extraction using variable convolution

(1.1) Category imbalance problem

The so-called category imbalance means that much data in the sample belongs to the same category, while other categories of data only occupy a small part. When the classifier is faced with such data, if the training strategy is not carefully formulated, the classifier will aim to improve the accuracy and tend to classify all samples into the category of a large number of data.

(1.2) Uneven balance of categories in target detection

The candidate area generation network will generate a large number of area candidate frames based on the feature map, but in the picture of power equipment defects, a picture often has only one or a few defects to be detected or even no defects to be detected. This will cause "category imbalance problems".

The Faster-RCNN detector alleviates this problem. First of all, in the candidate boxes generated by the area generation network, we will remove a large number of candidate boxes belonging to the background through the filtering rules, and select the background boxes before and after proportionally. This process can reduce the gap between the number of foreground and background candidate boxes to a certain extent. However, the number of candidate boxes belonging to the background is much larger than the candidate boxes belonging to the foreground. Most of the receptive fields of these boxes on the feature map are backgrounds, so the candidate boxes in the background area that are screened out are highly likely to belong to "easy training samples". What the target detection algorithm needs are to be able to correctly distinguish the background before and after and to correctly classify the positive samples.

(1.3) Online difficult sample mining

In the Faster R-CNN framework, the candidate region generation network generates many candidate regions, many of which are negative sample frames that do not intersect or rarely intersect with the target frame to be detected. These samples are "easy to train" samples. The "difficult to train" negative sample refers to the negative sample frame that has a larger intersecting value with the foreground target frame but does not exceed the set positive sample threshold. Therefore, we need to focus on training those negative sample frames that have a high intersection with the target frame to be detected.

The online hard sample mining method can find out the hard-to-train samples and increase the weight of these samples to the loss function. The online difficult sample mining method first calculates the loss of all candidate frame refinement and classification areas, sorts the loss from high to low, and selects a total of 512 positive samples according to the ratio of 1:3, using these samples as training samples to train the network.

Data augmentation

Data flipping: Considering that the drone will take pictures of power equipment from different directions, and the defects of power equipment have the invariance of flipping. Therefore, during the training process, we will randomly flip the pictures and the equipment defect frames marked by the pictures at the same time, so that the amount of data can be increased.

Data zooming: UAVs often do not maintain the same altitude during the shooting of power equipment. The resolutions of drones of different brands are also quite different. In the training process, this article simulates the real environment by randomly zooming the picture, so as to improve the model's ability to respond to defects of power equipment of different scales.

3 Experimental Results

In this experiment, 5847 images of power equipment taken by drones with different resolutions were collected. There are foreign objects in the tower, damaged insulators, large-scale metal corrosion, and abnormal ground conductors. The data is divided into training, verification, and test sets according to the number of 8:1:1.

3.1 Experimental Results

In this paper, a total of four stages of experiments were conducted to study whether the improvement of the traditional Faster-RCNN network is effective. In this paper, Faster-RCNN based on Resnet101 as the feature extraction network is selected as the benchmark algorithm, corresponding to the algorithm number① in the table. In this paper, this algorithm is used as a benchmark algorithm to optimize and improve. The number② corresponds to the result of adding multi-scale feature extraction on the benchmark algorithm Faster-RCNN network. You can see that the result will be significantly improved. The number③ corresponds to the addition of deformable convolution on the basis of experiment②. It can be seen that the detection accuracy is further improved. Finally, experiment④ added difficult sample mining, the algorithm model became easier to converge during the training process, and by learning from more difficult samples to obtain stronger classification detection capabilities.

Because the wire anomaly itself is small, this category can hardly be detected on the original benchmark algorithm. The foreign body of the tower itself has a large target and is easy to detect. The benchmarked algorithm can already achieve high detection accuracy, but due to the fact that the shooting distance of the drone is far away in the data, there are still some small for size targets. Variable convolution can better deal with slender wires, broken insulators, and foreign objects with different shapes and towers.

In Table 1, the benchmark algorithm has greatly improved the problem that small defects in the high-resolution power equipment pictures can hardly be detected, and the accuracy of the three major types of defects is improved for foreign objects in the tower, broken insulators, and large-scale rust. About 30, 15, and 70%. In the end, the average accuracy of each category is improved by about 60%.

Table 1 Comparison between the final algorithm and the benchmark algorithm

	Abnormal wire	Foreign body	Damaged insulator	Large size rust	MAP
Benchmark algorithm	3.8	63.7	64.7	39.9	43.2
Optimal algorithm	50.2	85.9	73.6	68.2	69.5

Figure 4 shows some of the results of the algorithm for the detection of four major types of defects, where tower corresponds to the foreign body in the tower, cable corresponds to the abnormal wire, insulator corresponds to the damage of the insulator, and rust corresponds to the corrosion of the metal.

Fig. 4 Visual display of defect detection

4 Summary and Outlook

In this paper, the target detection algorithm based on deep learning is mainly used to replace the manual detection of high-resolution pictures obtained by UAV acquisition, and it is optimized for some special problems in the task. Based on the evaluation of existing target detection algorithms, a multi-scale extraction of image features was carried out on how to detect targets of different scales on images of different resolutions. At the same time, in view of the variability of the shape of power equipment, a study on optimizing the convolution operation was carried out. It also studies the attention setting of the object to be detected in the labeled data.

Acknowledgements. This work was funded by the State Grid Science and Technology Project (Research on Key Technologies of Intelligent Image Preprocessing and Visual Perception of Transmission and Transformation Equipment).

References

1. Yamamoto K, Yamada K (1997) Analysis of the infrared images to detect power lines[A]. In: IEEE Tencon 97 IEEE region 10 conference speech and image technologies for computing and telecommunications[C], vol 1. Brisbane, Queensland, Australia, IEEE, pp 343–346
2. Li Z, Bruggemann TS, Ford JJ, Mejias L, Liu Y (2012) Toward automated power line corridor monitoring using advanced aircraft control and multisource feature fusion[J]. J Field Robot 29(1):4–24
3. Bologna F, Mahatho N, Hoch DA (2003) Infra-red and ultra-violet imaging techniques applied to the inspection of outdoor transmission voltage insulators[A]. In: Africon conference in Africa[C]. IEEE
4. Yang Y, Liu C, Huang L (2010) Application of image and video analysis in power equipment monitoring system [J]. Comput Appl 30(1):282–301
5. Liu K, Wang B, Chen X (2012) [J] Electromech Eng 29(2):211–214
6. Yumin G, Baoshu L, Shutao Z (2010) Insulator surface condition detection based on aerial images [J]. High Voltage Electri Appl 46(4):65–68
7. Yang H, Wu W (2013) Image monitoring of insulator ice coating based on 3D reconstruction [J]. Electric Power Autom Equip 33(2):92–98
8. Li Z (2009) Research on the key technology of infrared thermal image detection of insulator pollution level [D]. Hunan University

Research on Active Learning Method Based on Domain Adaptation and Collaborative Training

Wenzhen Li[1], Qirui Wu[2,3](✉), Hanwu Luo[1], Guoli Zhang[1], Zhonghan Peng[2,3], and Kai Chen[2,3]

[1] State Grid East Inner Mongolia Electric Power Supply Co. Ltd, Hohhot, People's Republic of China

[2] NARI Group Corporation Ltd, Nanjing, People's Republic of China
8278799@qq.com

[3] Wuhan NARI Limited Liability Company, State Grid Electric Power Research Institute, Wuhan, People's Republic of China

Abstract. In recent years, more and more deep learning technologies have been widely applied in various fields, such as intelligent medical, intelligent manufacturing, to realize the intelligence of various systems. In order to solve the problems of traditional manual inspection of power equipment, we propose a Convolutional Neural Network (CNN) has brought a revolutionary change to computer vision, but the ability of CNN to study relies heavily on the amount of labeled data. Currently, most of the labeled datasets are made up of natural images, which makes it a difficult problem to acquire labeled data in specific fields such as biomedical because of the high cost of human labeling. Active learning is an important method to solve this problem. This thesis is oriented to the problem of image classification and studies the current challenges and solutions of active learning. How to select the samples to be labeled so that the best neural network model can be learned at the minimum labeling cost is the core of the active learning algorithm. This thesis focuses on the selection strategy of the samples to be labeled, analyzes, and summarizes the advantages and disadvantages of current active learning methods, and designs a learning model.

Keywords: Active learning · Co-training · Fine-tuning · Sample diversity

1 Introduction

1.1 Research Background and Significance

In the process of training the network model, if we only use a small number of labeled samples for training, it will make the trained model generalization is very poor, cannot be applied to a variety of practical tasks; at the same time if we Not using these unlabeled samples that can be easily obtained is also a waste of data. Therefore, how to use label-free

Q. Liang et al. (eds.), *Artificial Intelligence in China*, Lecture Notes
in Electrical Engineering 653, https://doi.org/10.1007/978-981-15-8599-9_27

samples to reduce the labeling cost of samples required for model training has become a more concerning issue in machine learning. And active learning is an important means to solve this problem.

Active learning is a research hotspot in the field of machine learning in recent years. Its algorithm is mainly composed of two parts: a classifier network module and a sample selection module. In the classifier network module, we use limited labeled samples to train the deep convolutional neural network repeatedly, and save the network model as the resulting output after it reaches a certain accuracy; In the sample selection module, an "oracle" is needed to label the sample. We can pick out a batch of important data from the unlabeled data in an iterative form for labeling, so as to obtain more labeled data, and add it to the training set for training by the classifier network. The goal of the sample selection module is to provide the best-performing sample training set and maximize the performance of the classifier with the minimum label cost.

Active learning has been widely used in data mining, risk control, and other fields, and has been proven to save up to 44% in manual labeling costs [1]. Therefore, it is of great research value and practical significance to study the sample selection strategy of active learning to further reduce the cost of manual annotation and improve the actual application effect.

This topic has studied active learning algorithms based on pools in recent years and improved on them. The idea of collaborative training and fine-tuning techniques in domain adaptation are combined into active learning [2]. The active learning model with better accuracy under the manual labeling data. For the trainer part of the active learning model, we use the fine-tuning skills and domain adaptation ideas to use the deep network model trained on ImageNet as our training starting point and use the sample data on the target data set to train, making our model It is easier to train and increases the generalization ability of the model; for the sample selection module part, we introduced a high-confidence sample from another view based on the idea of collaborative training based on the traditional uncertainty-based sample selection strategy. And these high-confidence samples are labeled with pseudo labels output by another view classifier [3]. At the same time, we designed a sample batch selection algorithm to reduce sam3ple feature overlap, so that the deep model can learn more sample features. During each training, the classifier receives the samples selected by the sample selection module, where the high-confidence samples are labeled by the selection module and the low-confidence samples are manually labeled by the user and trained to train the network weights Update, and then use the updated network to predict the remaining samples, and enter the results into the sample selection module for the next round of screening training. Through this iterative active learning method, the accuracy of the model is continuously increased to achieve the purpose of training a high-precision model with a small number of manually labeled samples.

2 Active Learning Model

2.1 The Idea of Collaborative Training

Cooperative training is a method based on divergence. It believes that the same sample can extract features and classifications from different perspectives, and different classifiers can be trained from different perspectives so that the classifiers trained from different perspectives complement each other. Most of the current collaborative training algorithms are based on the following premises: first, the collaborative training algorithm requires two views of the data with sufficient, redundant, and conditionally independent; second, during the entire iteration, the data tagged with pseudo-labels by the classifier has high confidence; Third, the classifier will not label the data with false pseudo-labels; Fourth, the data with pseudo-labels will not be removed from the training set in the subsequent training process. In the early stages of collaborative training, the trained classifier does not have high accuracy. If we directly use these low-confidence labels to train the network, it will inevitably reduce the accuracy of the network and interfere with the performance of collaborative training; If we continue to use these wrong labels in subsequent training without modifying their labels or removing them from the training set, it will also reduce the accuracy of the classifier.

In response to the above problems, Ma et al. [4] proposed the SPaCo (Self-Paced Co-training) method in 2017. SPaCo proposed the following solutions.

For the problem of sample confidence interference accuracy in the early training process, a threshold λ is set in SPaCo. Only when the loss between the predicted value of the classifier and the true label is less than the threshold, the classifier will label the sample with a false label and Join the training set.

Replace traditional parallel training strategies with alternative training strategies. The main highlight of the idea of alternating training is: in a round of training, we will train the two networks *A* and *B* successively. When training the *B* network, we will first use the sample screening vector v^B of the *B* network in this state to update the screening vector v^A of the previously trained *A* network. Finally, add the samples selected by the two updated filter vectors to the training set.

A draw with replacing strategy is proposed to replace the traditional draw without replacing strategy. During the training process, as the classifier continues to learn and update, the classifier's predicted value for samples that have been pseudo-labeled will also change. This was mentioned in the previous alternating training, that is, we will update the screening vectors of the two networks during each training, and remove the samples that do not reach the threshold from the training set. The update strategy of the screening vector also empirically reflects the idea that the high-confidence samples selected by the two networks should be more consistent.

Since the data requirements of the collaborative training algorithm are too stringent, most of the data of real-world tasks currently have only one attribute set, which is single-view data, and the conditional independence of views is an excessively strong assumption. The training work basically revolves around the independent assumption of relaxing or avoiding conditions.

2.2 The Active Learning Model Proposed in This Topic

The framework of the active learning model proposed in this topic is shown in Fig. 1, which shows the overall process of our active learning method, including the following three stages: deep convolutional network model initialization phase, high-confidence sample and to-be-marked sample screening phase, and network model update phase based on collaborative training ideas. The following will introduce the above learning stages separately.

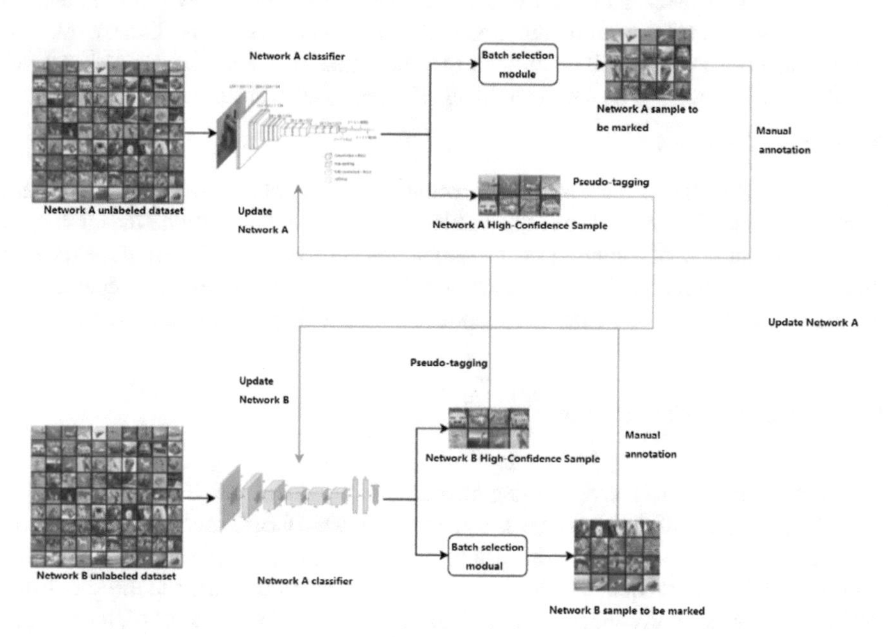

Fig. 1 Model framework

CNN model initialization

In the active learning method proposed in this topic, there are two views of the main view and the sub-view, which requires two deep convolutional neural networks with different structures. We selected VGG16_bn [5] and ResNet50 [6] as the classifier network for the main view and the sub-view, respectively. After building the network model, if we directly send the first batch of image data into the model to classify the images, then we will get a completely random distribution of classification results, and this result is completely invalid for the subsequent sample screening module. At this time, the network does not have any prior knowledge, so any input images are also randomly classified.

We chose to randomly extract 10% of the total number of images in the training set as the initial data set. But at this time, if we use the initial training set to train the network directly, we will find that the network convergence speed is very slow or does not converge completely within a certain number of training iterations. Or there may be overfitting of the network.

In order to solve the above problems, we introduced the "fine-tuning" technique in domain adaptation, replacing the network model with randomly initialized weights with the network model pre-trained on ImageNet as the starting point for active learning tasks. In the subsequent active learning process, the network model update stage is fine-tuned based on the pre-trained model. Because our target dataset is a smaller homogeneous dataset, we can use the model pre-trained on the larger dataset to obtain rich and generalized underlying features. With the foundation of these low-level features, when we transition to a smaller data set for fine-tuning, we can use these low-level features to combine higher-level discriminative features. On the contrary, if we directly train on these smaller data sets, it will be difficult to obtain these rich and generalized low-level features, and we will not be able to train general upper-level abstractions.

Screening of samples

A batch of image data will get the classification probability of each image for each class through the classifier network, and form a classification result matrix with the number of rows and the number of columns into the sample selection module. The sample screening module selects high-confidence samples and the most uncertain samples according to the result matrix. The following will introduce the screening process of these two types of samples.

(1) Screening of high-confidence samples

Based on the traditional active learning method, we introduced the CEAL method [4] to consider the high-confidence samples. By reproducing the CEAL method and testing on the target data set, we found that the disadvantage of this high-confidence sample selection strategy is that it requires a lot of experiments to determine different thresholds for different data sets; With continuous iterations, the classification accuracy of the network is continuously improved, and the threshold needs to be continuously reduced. For different data sets, the attenuation factor that controls the threshold attenuation is also very difficult to set. The above problems will eventually lead to the uncertainty of the number of high-confidence samples selected by the CEAL method in each round of training, and whether the confidence of these samples is high enough is uncertain, Therefore, the CEAL method needs to sample more samples under the premise of achieving the same network accuracy compared to other active learning methods. The classification accuracy of the classifier network trained under the same number of samples is lower than other active learning methods. Because the pseudo labels given to high-confidence samples have too many wrong labels, the label contains too much noise.

In response to the above problems, we have improved the screening strategy for high-confidence samples. Let the number of samples in the current training set be n. For the classification result matrix output by the classifier network, each row represents the probability that the network classifies the image into each class. We choose the category index with the largest posterior probability as the final classification result of the image. That is, for an input image sample x_i, the corresponding predicted label is

$$\hat{y}_i = \arg\max p(y_i|x_i; W, b)$$

where y_i is the classification probability of the sample x_i output by the network for each sample, W is the network weight matrix, and b is the network bias parameter.

We sort the entire classification result matrix according to the prediction probability of the final classification result of each image from large to small, and set a hyperparameter α, select the first $n * \alpha$ samples as high-confidence samples and submit them to manual annotation. The benefits of this strategy have the following two points: (1) The parameter settings for different data sets are simple and do not need to spend a lot of time to adjust the parameters; (2) At the beginning of the experiment, the classification accuracy of the classifier network output is low, at this time the number of training set samples is small, the value of $n * \alpha$ is also small, at this time, too many high-confidence samples will not be selected and the number of false pseudo-labels will be increased, affecting the experimental results; As the experiment progresses, the active learning framework continues to iterate, more and more samples are added to the training set, the value of $n * \alpha$ is constantly increasing, and the accuracy of the classifier network is also constantly improving, so more and more high-confidence samples can be added to the training with more accurate pseudo-labels.

(2) Screening of samples to be marked

In the CEAL method and the traditional active learning method, a certain number of samples with the lowest classification probability output by the classifier network are often used as the most uncertain samples and hand-marked. In this way of sample selection, there will be the problem of redundant samples to be marked. Take the five balls in the sample pool in Fig. 1 as an example. Suppose the classifier screens these five samples as uncertain samples At this time, the classifier cannot determine whether the five samples are football or volleyball. According to our previous selection method, the five samples will be handed over to the expert label next. But among the five samples, the third and fifth samples have high similarity. If both of these samples are submitted to expert labeling, it will cause redundancy in labeling costs. There are a lot of samples, so experts will continue to label very similar samples with the same label.

In order to solve the above problem of sample redundancy, we introduced the thinking of collaborative training and designed a batch sample selection strategy. Assuming that the number of samples to be annotated by each batch of screening and user annotations is K, the sample screening process we designed is as follows:

1. After a round of training, let two different networks predict the remaining unlabeled samples at the same time to obtain the predicted probability of the two networks and the predicted label for each sample;
2. Select the samples predicted as different labels in the two networks, and join the candidate unlabeled sample set;
3. According to the predicted labels of the main network, cluster the unlabeled samples, and the predicted probability of each sample output by the network for each class, calculate the average value of the variance of the predicted probability of each type of sample $v_1, v_2, \ldots, v_n$:

$$v_s = \frac{1}{m * (m-1)} \sum_{i=1}^{m} \sum_{j=1}^{m} D(y_j - y_i)$$

4. Set a super-parameter β, with β_{v1}, β_{v2},..., β_{vn}, as the threshold of each class;
5. For all remaining unlabeled samples, calculate the information entropy from the predicted probability of the network output in the main view, and sort the samples according to the information entropy from high to low;
6. Traverse the sorted unlabeled samples, if there is no sample in the selected unlabeled sample that is predicted to be of the same category, it will be added directly; if there is a sample of the same category, it will be calculated one by one with the selected sample Variance and take the mean as the sample similarity, and then compare with the threshold. The definition of sample similarity is as follows:

$$u = \frac{1}{q}\sum_{i=1}^{q} D(y_i - y_p)$$

where q is the number of similar samples to which the candidate unlabeled sample set has been added, and y_p is the predicted probability vector of the currently screened samples. If the degree of similarity is greater than the threshold, it means that it is slightly different from the selected sample, which needs to be manually marked and added to the selected sample set; If it is less than the threshold, it means that the sample has a high similarity with the selected sample, and no need to mark it, it will be ignored.

Through the above screening strategy, more features of the sample can be considered, and the distribution of the selected samples to be marked in the sample space will be more scattered.

3 Experiment

In this experiment, CEAL method and VAAL method [7] are used as comparison methods. The VAAL method is a method based on sample performance. It uses an encoder to learn the potential distribution of sample data and obtain a low-dimensional feature space, then uses the decoder to reconstruct the encoded data, and then use the adversarial network to cluster (classify) the labeled and unlabeled data in the space, and then select the most difficult sample against the network as the most informative sample, label and join the training.

In order to compare the fairness of the experiment to effectively verify the effectiveness of our method, we set the classifier network in the CEAL method and the VAAL method to pre-trained VGG16_bn and set the learning rate in the same way. We tested the three methods on the above three data sets, and obtained the following results (As Table 1):

4 Summary

This paper proposes an active learning model based on the idea of collaborative training and combined with fine-tuning skills in domain adaptation. This model selects samples

Table 1 CIFAR10 experimental results

CIFAR10	10%	15%	20%	25%	30%	35%	40%
VAAL	72.26	76.28	78.46	80.42	81.84	82.91	83.91
CEAL	72.12	76.33	78.44	80.83	82.16	83.38	84.24
Ours	72.18	76.83	79.83	81.66	83.04	84.15	85.31

with high uncertainty and small feature overlap through collaborative training and a batch sample screening module we designed. It is hand-labeled, combined with the consideration of high-confidence samples and labeled with pseudo labels, and added to the training to increase the amount of sample information in the training set and enable the network to learn more features of the samples. Then we conducted a horizontal comparison experiment on the CIFAR and CALTECH data sets of the CEAL method with the highest accuracy in the model and the same type of methods in recent years and the latest method of different classes. At the same time, we conducted an ablation analysis experiment on our model itself. Experiments show that the model trained by our method has a certain degree of improvement in accuracy, and verifies the effectiveness of the sample selection core module in the algorithm.

Acknowledgements. This work was funded by the State Grid Science and Technology Project (Research on Key Technologies of Intelligent Image Preprocessing and Visual Perception of Transmission and Transformation Equipment).

References

1. Thompson CA, Califf ME, Mooney RJ (1999) Active learning for natural language parsing and information extraction[A]. ICML[C] 406–414
2. Xu Y, Zhang H, Miller K et al (2017) Noise-tolerant interactive learning using pairwise comparisons[A]. Adv Neural Inf Process Syst[C] 2431–2440
3. Wang Z, Du B, Zhang L et al (2017) On gleaning knowledge from multiple domains for active learning [A]. IJCAI [C] 3013–3019
4. Ma F, Meng D, Xie Q et al (2017) Self-paced co-training[A]. In: Proceedings of the 34th international conference on machine learning, vol 70[C]. pp 2275–2284
5. Simonyan K, Zisserman A (2014) Very deep convolutional networks for large-scale image recognition[J]. arXiv preprint arXiv:1409.1556,2014
6. He K, Zhang X, Ren S et al (2016) Deep residual learning for image recognition[A]. In: Proceedings of the IEEE conference on computer vision and pattern recognition[C] 770–778
7. Sinha S, Ebrahimi S, Darrell T (2019) Variational adversarial active learning[A]. In: Proceedings of the IEEE international conference on computer vision[C] 5972–5981

Evolution Analysis of Research Hotspots in the Field of Machine Learning Based on Complex Network

Tala, Cui Yimin(✉), Li Junmei, and Su Xiaoyan

Academy of Military Sciences, 100101 Beijing, China
tsui-min@163.com

Abstract. In this paper, Web of Science Core Collection is used as the data source and the complex network analysis method is adopted to study the evolution process of machine learning research hotspots in different historical stages, The characteristics of the focuses are summarized, which has a good reference for more accurate understanding of the development of the field.

Keywords: Machine learning · Co-occurrence network · Word cloud · Hotspots evolution

1 Introduction

Machine learning is one of the core technologies of artificial intelligence, which involves multi-disciplinary knowledge such as probability theory, statistics and approximation theory. It mainly studies how to enable machines to acquire human learning ability by simulating human beings. It helps understanding rules from a large amount of historical data and acquiring new knowledge and skills [1].

The field of machine learning is getting widely used [2]. Studying the development history of machine learning will be of great significance for accurately understanding the development characteristics of machine learning and predicting the future development trends.

In this paper, the papers of Wos (Web of Science) database in machine learning field is used as data source to demonstrate and evaluate the evolution process of hotspots in machine learning field by complex network analysis method.

2 Data and Method

2.1 Data Source

Web of Science Core Collection was taken as the source database, and the retrieval Formula is: TS = "machine learning". The retrieval time span was set as 1958–2019, and 142,325 articles were finally retrieved.

Q. Liang et al. (eds.), *Artificial Intelligence in China*, Lecture Notes
in Electrical Engineering 653, https://doi.org/10.1007/978-981-15-8599-9_28

2.2 Method

This article is carried out from micro and meso levels. First of all, build a keyword co-occurrence network under different time periods, and secondly, calculate the measurement indicators of keywords in each stage from the micro level, and analyze the importance and change process of keywords from both horizontal and vertical perspectives. Then, through clustering networks in each stage, at the meso level, finds the combination of keywords in each stage from the resulting community, analyzes the hot issues at this stage, and finally combines the phenomena at the micro layer to evaluate the evolution of research hotspots in the field of machine learning.

3 Keywords Co-Occurrence Network Analysis

3.1 Build Keyword Co-Occurrence Networks

Keywords are the core generalization of the literature. It is generally believed that if two keywords appear frequently in the same paper, it indicates that there is a close relationship between the two [3]. By building keyword co-occurrence networks for the literature in different periods, the spatial structure of the research content in the field of machine learning in a specific period can be found. The specific steps are as follows:

Cleaning data Based on the factors of citation and time, all literatures are divided into four parts according to publication time: before 2016, 2017, 2018 and 2019, and the literatures with citation less than 10 (excluding 10, the same later), 9, 8 and 3 are deleted respectively. There were 33,719 articles in the remaining literatures, accounting for 23.7% of the total. The remaining citations were 1,427,813, accounting for 84.6% of the total.

Time slicing and network building Considering the small number of literatures before 1999, the first stage was set as 1958–1999, and the time span of the last four stages was 5 years, respectively 2000–2004, 2005–2009, 2010–2014 and 2015–2019. Five keyword co-occurrence networks were constructed according to the five stages. For the fair participation of all nodes, "machine learning" was not used as a node in the construction of the network.

Figure 1 shows the network structure of keyword co-occurrence in each stage. In the figure, each node represents a keyword, the edge represents the co-occurrence relationship between the keywords, and different node colors represent different communities.

3.2 Identify Research Hotspots

Network index analysis of keywords

Tables 1, 2 and 3 are the top 20 keywords of frequency, degree centrality and structural hole index score respectively. Word frequency is the total number of occurrences of keywords in a historical stage. Degree centrality measures the degree of connection between a node and other nodes in the network [4]. If two nodes in the network are not

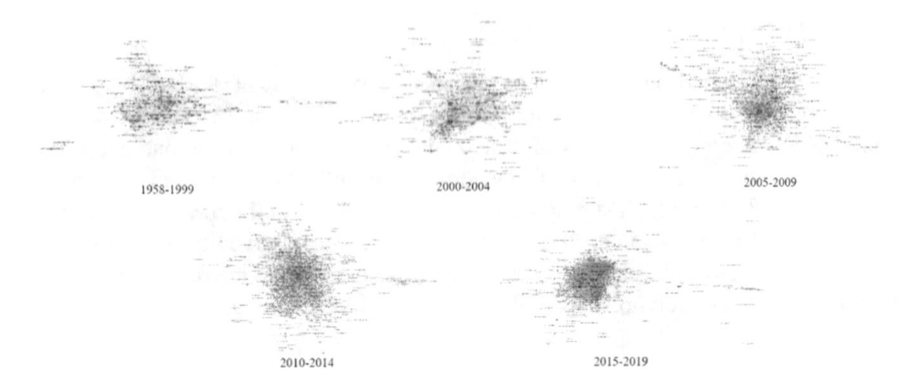

Fig. 1 Keyword co-occurrence network in each stage

directly connected, there is a structural hole between them. According to Burt's structural constraint algorithm [5], Table 3 lists the 20 keywords with the lowest structural hole index at each stage.

From the micro level, it can be seen from tables that "Classification" has always been the core problem of machine learning. Combining the three indicators, it can be found that the keyword "SVM" not only occupies the central position in the whole field of research for a long time, but also has penetrated into various fields of machine learning since its inception. In contrast, the high ranking of the frequency of the keyword "Deep learning" and the bottom ranking of the structural hole index in 2015–2019 show that although its research has a high attention, it involves relatively narrow fields. "CNN" and "RNN" share the same characteristics.

Community division and hotspots evolution analysis

Research hotspots refers to the field with internal connection and high attention in a certain period of time. In the keyword co-occurrence network, it is expressed in the form of keywords with a certain spatial structure. Therefore, the observation from the micro level is the emergence and change of a single keyword, which has a certain reference significance. However, it cannot well show the hot research situation in the field of science and technology.

This section uses VOSviewer [6] to analyze the structure of keyword network in the way of complex network, finds out the different correlation forms of keywords in different periods, and evaluates the research hotspots situation in machine learning field in each stage.

The "Formalization method" was selected as "Fractionalization", and the min cluster size parameter was set as 20, 30, 30, 50 and 80 according to the number of keyword nodes in different stages. The default clustering method was adopted and the five stages were obtained 6, 12, 19, 21 and 18 communities. By visualizing the clustering results through word cloud map, the combination of keywords in a certain stage is compared horizontally, and the combination, separation and recombination of keywords at different periods are compared vertically, so as to obtain the hot evolution of machine learning field in the past few decades.

Table 1 Word frequency

Stage	1958–1999	2000–2004	2005–2009	2010–2014	2015–2019
	NEURAL NETWORK	SVM	SVM	SVM	DEEP LEARNING
	GENETIC ALGORITHM	NEURAL NETWORK	CLASSIFICATION	CLASSIFICATION	SVM
	AI	DATA MINING	NEURAL NETWORK	ELM	ELM
	ANN	CLASSIFICATION	DATA MINING	FEATURE SELECTION	CLASSIFICATION
	CLASSIFICATION	GENETIC ALGORITHM	FEATURE SELECTION	NEURAL NETWORK	CNN
	PATTERN RECOGNITION	PATTERN RECOGNITION	ANN	DATA MINING	RANDOM FOREST
	DECISION TREES	FEATURE SELECTION	GENETIC ALGORITHM	ANN	NEURAL NETWORK
	CASE-BASED REASONING	AI	PATTERN RECOGNITION	RANDOM FOREST	FEATURE SELECTION
Keyword	KNOWLEDGE ACQUISITION	ANN	KERNEL FUNCTIONS	SCHEDULING	ANN
	EXPERT SYSTEMS	KERNEL FUNCTIONS	SCHEDULING	PATTERN RECOGNITION	AI
	DATA MINING	DECISION TREES	LEARNING EFFECT	LEARNING EFFECT	DATA MINING
	FUZZY LOGIC	KNOWLEDGE DISCOVERY	CLUSTERING	SEMI-SUPERVISED LEARNING	BIG DATA

(*continued*)

Table 1 (*continued*)

Stage	1958–1999	2000–2004	2005–2009	2010–2014	2015–2019
	KNOWLEDGE DISCOVERY	REINFORCEMENT LEARNING	PREDICTION	CLUSTERING	PREDICTION
	GENERALIZATION	ROUGH SETS	TEXT CLASSIFICATION	SVR	FEATURE EXTRACTION
	INDUCTIVE LEARNING	CLUSTERING	DECISION TREES	GENETIC ALGORITHM	ENSEMBLE LEARNING
	RULE INDUCTION	SUPERVISED LEARNING	REGRESSION	ACTIVE LEARNING	MAGNETIC RESONANCE IMAGING
	INDUCTION	REGRESSION	RANDOM FOREST	FEATURE EXTRACTION	FAULT DIAGNOSIS
	REINFORCEMENT LEARNING	FEATURE EXTRACTION	REINFORCEMENT LEARNING	KERNEL FUNCTIONS	REMOTE SENSING
	INDUCTIVE LOGIC PROGRAMMING	GENETIC PROGRAMMING	SVR	SUPERVISED LEARNING	NATURAL LANGUAGE PROCESSING

Table 2 Degree centrality

Stage	1958–1999	2000–2004	2005–2009	2010–2014	2015–2019
	NEURAL NETWORK	SVM	SVM	SVM	SVM
	GENETIC ALGORITHM	NEURAL NETWORK	CLASSIFICATION	CLASSIFICATION	DEEP LEARNING
	CLASSIFICATION	CLASSIFICATION	NEURAL NETWORK	FEATURE SELECTION	ELM
	PATTERN RECOGNITION	DATA MINING	DATA MINING	DATA MINING	CLASSIFICATION
	AI	PATTERN RECOGNITION	FEATURE SELECTION	NEURAL NETWORK	RANDOM FOREST
	CASE-BASED REASONING	GENETIC ALGORITHM	GENETIC ALGORITHM	ELM	NEURAL NETWORK
	KNOWLEDGE ACQUISITION	AI	ANN	ANN	CNN
	KNOWLEDGE DISCOVERY	KNOWLEDGE DISCOVERY	PATTERN RECOGNITION	PATTERN RECOGNITION	ANN
	ANN	FEATURE SELECTION	KERNEL FUNCTIONS	RANDOM FOREST	FEATURE SELECTION
Keyword	EXPERT SYSTEMS	DECISION TREES	CLUSTERING	FEATURE EXTRACTION	AI
	DECISION TREES	ANN	RANDOM FOREST	CLUSTERING	DATA MINING

(continued)

Table 2 (*continued*)

Stage	1958–1999	2000–2004	2005–2009	2010–2014	2015–2019
	RULE INDUCTION	KERNEL FUNCTIONS	BOOSTING	SEMI-SUPERVISED LEARNING	BIG DATA
	DATA MINING	CLUSTERING	REGRESSION	GENETIC ALGORITHM	PREDICTION
	FUZZY LOGIC	SUPERVISED LEARNING	PREDICTION	REGRESSION	FEATURE EXTRACTION
	GENERALIZATIO N	REGRESSION	MODEL SELECTION	PREDICTION	ENSEMBLE LEARNING
	REINFORCEMENT LEARNING	FEATURE EXTRACTION	SVR	ALGORITHMS	CLUSTERING
	INDUCTIVE LEARNING	REINFORCEMENT LEARNING	SEMI-SUPERVISED LEARNING	SVR	REMOTE SENSING
	ROUGH SETS	PATTERN CLASSIFICATION	DECISION TREES	UNSUPERVISED LEARNING	PATTERN RECOGNITION
	INDUCTION	BIOINFORMATICS	TEXT CLASSIFICATION	ACTIVE LEARNING	MAGNETIC RESONANCE IMAGING
	NEAREST NEIGHBOR	BOOSTING	SUPERVISED LEARNING	SUPERVISED LEARNING	PCA

Table 3 Structural hole

Stage	1958–1999	2000–2004	2005–2009	2010–2014	2015–2019
	NEURAL NETWORK	SVM	SVM	SVM	ELM
	GENETIC ALGORITHM	NEURAL NETWORK	DATA MINING	DATA MINING	SVM
	CASE-BASED REASONING	DATA MINING	NEURAL NETWORK	NEURAL NETWORK	DATA MINING
	CLASSIFICATION	GENETIC ALGORITHM	CLASSIFICATION	UNSUPERVISED LEARNING	MACHINE LEARNING ALGORITHM
	PATTERN RECOGNITION	CLASSIFICATION	REINFORCEMENT LEARNING	OPTIMIZATION	PREDICTION
	DECISION TREES	AI	MODEL SELECTION	ELM	RANDOM FOREST
	AI	REINFORCEMENT LEARNING	FEATURE SELECTION	CLASSIFICATION	BIG DATA
	KNOWLEDGE DISCOVERY	DECISION TREES	CLUSTERING	COMPUTER VISION	AI
	ANN	OPTIMIZATION	RANDOM FOREST	SVR	PATTERN RECOGNITION
Keyword	DATA MINING	INDUCTIVE LOGIC PROGRAMMING	PATTERN RECOGNITION	AI	CLASSIFICATION
	INDUCTIVE LEARNING	SUPERVISED LEARNING	COMPUTER VISION	REINFORCEMENT LEARNING	CLUSTERING
	FUZZY LOGIC	BOOSTING	BOOSTING	GENETIC ALGORITHM	DEEP LEARNING

(continued)

Table 3 (*continued*)

Stage	1958–1999	2000–2004	2005–2009	2010–2014	2015–2019
	KNOWLEDGE ACQUISITION	INFORMATION RETRIEVAL	GENETIC PROGRAMMING	CLUSTERING	DIAGNOSIS
	INDUCTION	ANN	SEMI-SUPERVISED LEARNING	BREAST CANCER	SUPERVISED LEARNING
	RULE INDUCTION	PATTERN RECOGNITION	GENETIC ALGORITHM	FEATURE EXTRACTION	CLOUD COMPUTING
	EXPERT SYSTEM	CLUSTERING	SIMULATION	PATTERN RECOGNITION	OPTIMIZATION
	ROUGH SETS	BIOINFORMATICS	AI	PREDICTION	TIME SERIES
	NEAREST NEIGHBOR	RULE INDUCTION	PCA	SUPERVISED LEARNING	BIOMARKERS
	FEATURE SELECTION	GENETIC PROGRAMMING	OPTIMIZATION	NATURAL LANGUAGE PROCESSING	SUPERVISED MACHINE LEARNING

Due to the limitation of space, this paper only shows the word cloud map of all communities in the first stage, and gives a brief description of each community. Based on this, the paper analyzes the changes of communities at different stages to show the changes of machine learning research focuses.

1958–1999. Six communities are found through clustering, and the cloud diagram of each community word is shown in Fig. 2 (the community is named in order of the number of keywords it contains, and the size of keywords is proportional to its degree, the same as below).

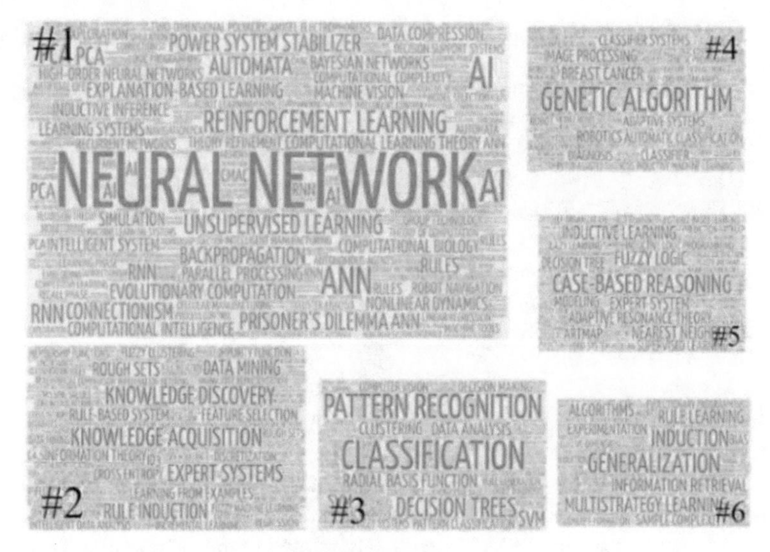

Fig. 2 1958–1999

The 1st community focused on the keyword “Natural network”, which was the core issue of machine learning in the field of “AI” (Artificial Intelligence) during that period. People want to directly simulate human neural network to build intelligent learning models of machines. It also includes the main keywords “Rules”, “RNN”, “Reinforcement learning” and so on. The Second community, mainly formed by the keywords “Knowledge acquisition”, “Knowledge discovery” and “Expert system”, shows the hot issue of the early application research of machine learning. Based on the core problem solved by machine learning, “Classification”, the Third community was formed, including popular keywords such as “Decision tree”, “SVM” and “Pattern recognition”. The Fourth community surrounding the “Genetic algorithm” is a focus for early algorithm research. The Fifth and Sixth communities are relatively small in scale, which are about logical reasoning, induction and learning strategies.

2000–2004. At this stage, a total of 12 communities were obtained. Three key research directions have been separated from Third community in the previous stage, namely the First, the Third and the Nineth communities. The core keywords of the research are “SVM”, “pattern recognition” and “decision tree”, as shown in Fig. 3a. It can be seen from the word cloud diagram that, in this stage, the study of “SVM” is closely related to "Kernel function" and "Feature selection". A new research focus, "Data mining" problem, has evolved from the Second community in the last stage, as shown in Fig. 3b. The main body of the First, Fourth and Fifth communities in the previous stage is integrated into the unified class centering on the "Neural network", as shown in Fig c. In addition, the main hotspots are the Sixth community related to natural language processing and the Seventh, Eighth, 12th communities surrounding the keywords "Hidden Markov Models", "Supervised learning" and "Reinforcement learning", as shown in Fig. 3d.

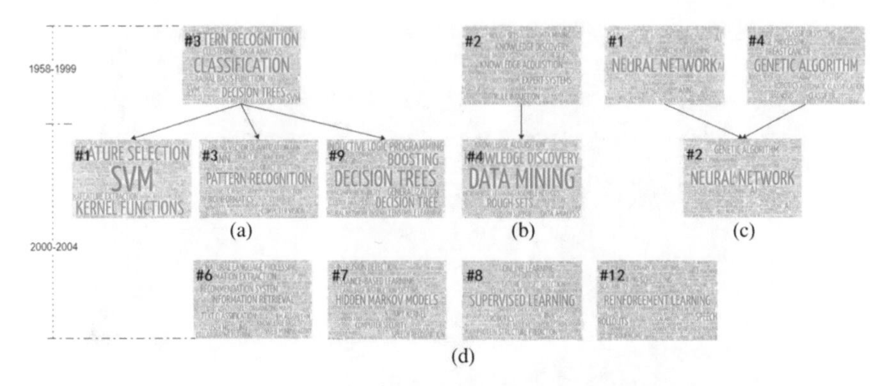

Fig. 3 2000–2004

2005–2009. This phase was divided into 19 communities. "SVM" is still the center of 1st community. The key point of the change is that "ELM"(Extreme Learning Machine) appears, and "Adaboost" and "SVR" are closely related to the 1st community, as shown in Fig. 4a. Natural language processing, formed by the integration of the Fourth and Sixth communities, has become one of the most important research directions, as shown in Fig. 4b. The Fourth community centered on image processing as the keyword for "pattern recognition" is still a hot topic, but the content of the research has changed. At this time, the main focus is "Feature extraction", "Face recognition" and "Unsupervised learning", as shown in Fig. 4c. Figure 4d shows that "Random Forest" has become the new center of the community about the "Decision tree". In addition, major research focuses are the Eighth, Nineth, Tenth and 14th communities around the keywords "Neural network", "Semi-supervised learning", "Computer vision" and "PCA", as well as Seventh community related to the application of life science and medicine, as shown Fin Fig. 4e.

2010–2014. This phase includes 21 communities. The research on "SVM" is still the most popular. "ELM" is independent from the community of "SVM", which makes the neural network wake up from the downturn and become a research focus to compete with SVM, as shown in Fig. 5a and b. Compared with the previous stage, there are many life science and medicine related words around the community of "random forest", such as "Cancer", "Protein-protein interaction", "QSAR" (quantitative structure—activity relationship), which show the vitality of this algorithm in this field, as shown in Fig. 5c. In the Fifth community on "Natural language processing", the frequent occurrence of "Social

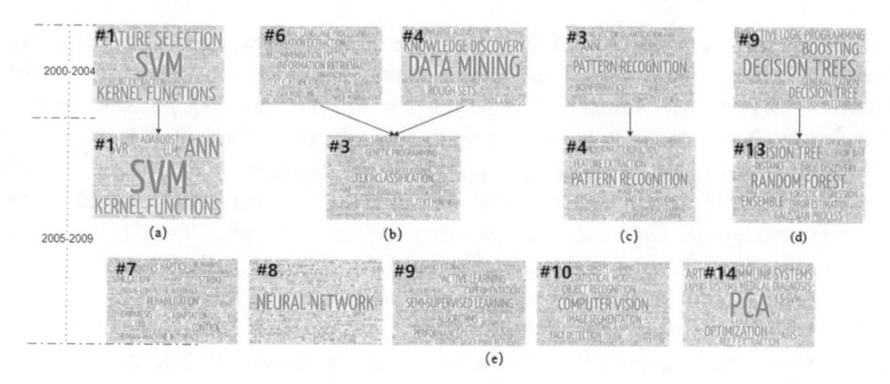

Fig. 4 2005–2009

network" has become the new focus of attention in this field, as shown in Fig. 5d. "Data mining" has become another hotspot at this stage, forming the Seventh community, but the focus has changed from the early "Text processing" to the problem closely related to the words "Big data", "Cloud computing", "Security" and "Python", as shown in Fig. 5d. In addition, the Eighth, Nineth and 13th communities are formed respectively around "Computer vision" and "Remote sense", "Deep learning" and "Unsupervised learning", "Schedule" and "Effect". There is also a research hotspot related to the application of life medicine, forming the Sixth community, as shown in Fig. 5e.

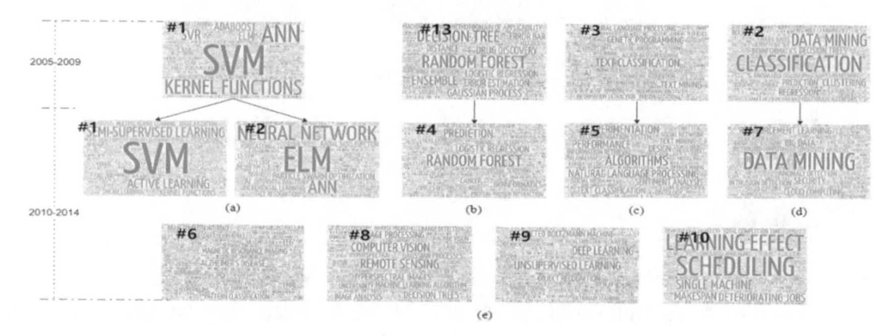

Fig. 5 2010–2014

2015–2019. This phase was divided into 18 communities. At this stage, significant changes took place in the network structure. "Deep Learning", as the most popular research direction, became the 1st community. In addition, the community of "ELM" also developed further, and the neural network entered another outbreak period, as shown in Fig. 6a and b. Third community shows that new phrases such as "Remote sense", "UAV" and "Landset" appear in the community on "Random forest", and the application in this field is further expanded, as shown in Fig. 6c. In the Fifth community, "Big data" has become the center of the problem of "Data mining" and a important research hotspot. In this community, in addition to the existing keywords in the previous stage (2010–2014), "Internet of thing" has become a key issue of this community, as shown in Fig. 6d.

The research hotspots on life medicine are divided into two directions. One is the "Prediction" problem, forming the Second community. The main keywords are "Cancer", "MRI", "Radiomics", "Lasso", etc.; the other is the problem about brain-computer, forming the Tenth community, the keywords include "EEG" (Electroencephalogram) "Brain-computer interface" "Eye tracking" and so on, as shown in Fig. 6e. In addition, compared with the previous stage, "SVM", "PCA", "K-NN", "Adaboost" and "LDA" are all integrated into the same class and become the Sixth community. There is a differentiation situation of "Neural network" method and non-"Neural network" method. It can be seen from Fig. 6f that natural language processing has more connections with "Twitter" and "Social network", which has become a new research trend. 11th community is composed of new application and theoretical directions such as "Recommendation system", "5G", "Sparse representation" and "Dictionary learning", as shown in Fig. 6f.

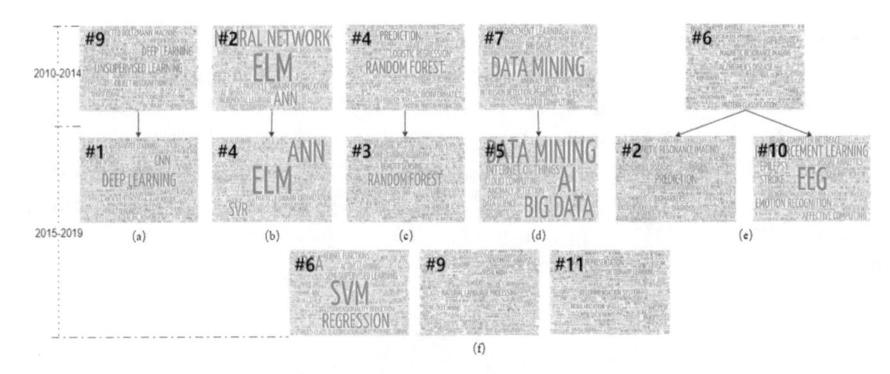

Fig. 6 2015–2019

4 Conclusion

In this paper, the complex network method is used to analyze the evolution path and characteristics of research hotspots in the field of machine learning from the horizontal and vertical perspectives at the micro and meso levels. It comprehensively shows the evolution of research hotspots in the field of machine learning at different stages. The specific conclusions are as follows:

(1) From the perspective of the whole development process, the overall skeleton of machine learning research has been formed as early as last century, and the extension, expansion and correlation of various branches on the skeleton in different periods are not the same, leading to different research hotspots. The formation and change of hotspots often result from a major "BREAKTHROUGH" on a branch.
(2) "Support vector machine", "Random tree" and other classical algorithms keep hot for a long time, which shows that people are inclined to deepen and expand the existing achievements.

(3) "Breakthrough" plays a decisive role in the situation structure and research hotspots. The birth of SVM has changed the previous research situation with neural network as the core. The "breakthrough" of ELM and deep learning rapidly changed the research trend of "SVM" as the core for more than ten years.

References

1. Xuegong Z (2000) Introduction to statistical learning theory and support vector machines [J]. Acta Automatica Sinica 01:36–46
2. Tejpal A, Kaur K (2019) Machine learning survey[J]. Int J Comput Sci Eng 7(2):453–457
3. Li H, An H, Wang Y et al (2016) Evolutionary features of academic articles co-keyword network and keywords co-occurrence network: Based on two-mode affiliation network[J]. Phys A Stat Mech Its Appl 450:657–669
4. Freeman LC (1979) Centrality in social networks conceptual clarification[J]. Social Netw 1(3):215–239
5. Lazega E, Burt RS (1995) Structural holes: the social structure of competition[J]. Revue Franaise de Sociologie 36(4):779
6. Eck NJV, Waltman L (2010) Software survey: VOSviewer, a computer program for bibliometric mapping[J]. Entometrics 84(2):523–538
7. Lee PC, Su HN (2010) Investigating the structure of regional innovation system research through keyword co-occurrence and social network analysis[J]. Innov Manage Policy Practice 12(1):26–40

Research on Index Network Construction and Effectiveness Evaluation Method of Air Traffic Control System Based on CDM Data

Zheng Li[1,2], Yinfeng Li[1(✉)], Xiaowen Wang[1], Meng Xu[1], and Shenghao Fu[1]

[1] College of Civil and Architectural Engineering, North China University of Science and Technology, Tangshan 063210, China
15030576997@126.com

[2] State Key Laboratory of Air Traffic Management System and Technology, Nanjing 210007, China

Abstract. The construction of the air traffic control system effectiveness evaluation system is one of the methods to evaluate the efficiency of air traffic control departments. From the perspective of the air traffic control business departments, this paper builds a networked air traffic control system performance evaluation model based on CDM data and solves the indicators through the factor analysis method Interdependence problem, and the feasibility and effectiveness of the method are verified by examples.

Keywords: Air traffic control effectiveness · Index network · Factor analysis

1 Introduction

The effectiveness of the air traffic control (ATC) system characterizes the ability of various departments of the ATC system to complete air traffic control tasks. The air traffic control command task is mainly completed by the cooperation of the three functional departments of tower control, approach control and regional control. The air traffic controller effectively monitors and commands the aircraft to effectively ensure the aircraft to complete the transportation task safely and efficiently. This paper starts from the air traffic control data, based on the key multi-dimensional air traffic control system performance index network of quantitative and measurable data, and applies factor analysis to reduce the dimension and calculate the evaluation results.

2 Indicator Set Construction

In order to comprehensively analyze the effectiveness of the ATC operating system, it is necessary to analyze the unit indicators of different levels in each component system, and on this basis, a component efficacy index set and a component performance index set should be established to provide a data basis for the construction of the air traffic control system effectiveness evaluation index network [1].

Q. Liang et al. (eds.), *Artificial Intelligence in China*, Lecture Notes in Electrical Engineering 653, https://doi.org/10.1007/978-981-15-8599-9_29

2.1 Basis for Selecting Indicators

The factors influencing the effectiveness of the air traffic control system and their manifestations are different.

The ICAO Performance Evaluation Specification Doc 9882 summarizes the key performance areas of the ATC system, including safety, security, cost-effectiveness, accessibility and fairness, capacity, environmental impact, and eleven aspects such as predictability, participation and collaboration, flexibility, flight efficiency, and global interoperability. FAA takes usability, cost efficiency, flexibility, predictability, and delay as performance indicators to measure the operational capacity and efficiency of air traffic management on the premise of ensuring safety. PRC defines 10 categories of safety, delay, cost efficiency, predictability, accessibility, flexibility, flight efficiency, effectiveness, environment and fairness to evaluate air traffic control effectiveness.

Based on Doc 9882 document, FAA and PRC's ATC effectiveness evaluation direction, this paper combines the design principles of the assessment indicators and the existing data to build an ATC operation system effectiveness indicator set [2].

2.2 Construction of Multi-dimensional Indicator System

Under normal conditions, the tower control commands the aircraft to taxi from the parking position to the runway and takes off and climb to a certain height or designated point. The regional control is responsible for the safety and efficiency of the aircraft during the cruise phase, and the approach control is responsible for the tower control and regional control of the approach and departure aircraft the connection part. The multi-dimensional index system considers the three dimensions of tower, approach, and region, and establishes an air traffic control effectiveness evaluation index system based on available CDM data.

1. Tower dimension

The construction of the index set of the tower dimension is designed from the three aspects of efficiency, delay and capacity, and serves as the initial emergent efficacy layer index.

Efficiency indicators include runway utilization rate, average taxi-in time for arrival, average taxi-out time for departure, arrival interval for continuous departure flights, release interval for continuous departure flights, release efficiency for continuous departure flights, CTOT execution compliance rate, runway waiting time, etc. The delay indicators include delayed flights of outbound, the delay time of outbound flights, and delay time of inbound flights. The capacity indicators include airport peak arrival capacity, airport peak departure capacity, visibility, etc. These indicators are used as component system efficacy indicators. The specific calculation parameters involved in the component system performance indicators, such as actual landing time (ALDT), actual take-off time (ATOT), runway capacity, the time of the last block, last COBT, take-off run-way, last CTOT, last LIN QUE For the first time, the standard time for ground taxiing, planned take-off time, planned landing time, etc. are used as component system performance indicators.

2. Approach dimension

The index set construction of approach dimension is designed from two aspects of efficiency and delay, as the initial emergent efficacy layer index.

Among the component efficiency indicators, the efficiency indicators include the average flight time of arrival flight and the average flight time of the departure flight. Delay indicators include delay time of departure flights, delayed flights of departure, delay time of arrival flights, and delayed flights of arrival. The component system performance indicators are ALDT, ATOT, and waypoint time.

3. Regional dimension

The construction of the index set in the regional dimension is designed from the two aspects of efficiency and delay, as the initial emergent efficacy layer index.

The efficiency indicators in the component performance indicators include the flight time of the segment, and the delay indicators include the delay time of the departing flights, the delay time of the incoming flights, and the delay time of the overflying flights. The performance index of the component system is the waypoint time.

The performance index items of each dimension component are shown in Table 1.

Table 1 Index items of ATC system component effectiveness index network

Index	Component efficacy index	Index	Component efficacy index
X_1	runway utilization rate	X_{13}	airport peak departure capacity
X_2	average taxi-in time for arrival	X_{14}	visibility
X_3	average taxi-out time for departure	X_{15}	Average flight time of arrival flight
X_4	arrival interval for continuous departure flights	X_{16}	Average flight time of departure flights
X_5	release interval for continuous departure flights	X_{17}	delay time of departure flights
X_6	release efficiency for continuous departure flights	X_{18}	delayed flights of departure
X_7	CTOT execution compliance rate	X_{19}	delay time of arrival flights
X_8	runway waiting time	X_{20}	delayed flights of arrival
X_9	delayed flights of outbound	X_{21}	flight time of the segment
X_{10}	delay time of outbound flights	X_{22}	regional delay time of departure flights
X_{11}	delay time of inbound flights	X_{23}	regional delay time of arrival flights
X_{12}	airport peak arrival capacity	X_{24}	delay time of overflying flights

2.3 Index Network Construction

The index of each component efficacy layer in Table 1 is not independent of each other. The calculation of the numerical results depends on the data of each component performance layer. The data of the component efficacy layer and the component performance layer are not one-to-one correspondence. Therefore, the index system used for evaluation in this paper is a mesh structure, and the effectiveness index evaluation system of the mesh air traffic control system constructed is shown in Fig. 1.

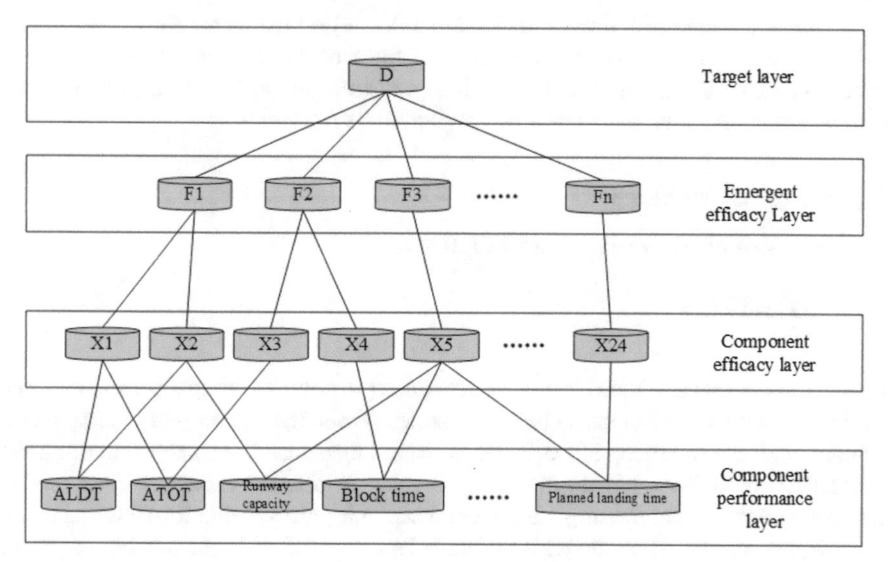

Fig. 1 Index network for the effectiveness evaluation of the air traffic control system

3 Factor Analysis

3.1 Factor Analysis

The effectiveness evaluation index system of the ATC operation system established in this paper is a mesh structure. Each index has different dimensions and is not independent of each other [3]. It reflects the performance of the ATC operation system in different aspects and cannot fully reflect the actual situation of ATC operation [4]

Factor analysis is a dimensionality reduction method. Its essence is to summarize the indicators with intricate relationships into a few new unrelated comprehensive factors, thereby simplifying a statistical analysis method of variables. The basic principle is to group variables according to the size of the correlation so that the correlation between the variables in the same group is high. Each group of variables represents a basic structure, called a common factor. There is no correlation between the public factors and the main information of the original indicators is retained.

Assuming that there are p evaluation indicators, n sample units are expressed as linearly weighted sums of m ($m < p$) common factors and a unique factor as shown in Eq. (1).

$$\begin{aligned} X_1 &= \omega_{11}F_1 + \omega_{21}F_2 + \cdots + \omega_{1m}F_m + \varepsilon_1, \\ X_2 &= \omega_{21}F_1 + \omega_{22}F_2 + \cdots + \omega_{2m}F_m + \varepsilon_2, \\ &\cdots \\ X_p &= \omega_{p1}F_1 + \omega_{p2}F_2 + \cdots + \omega_{pm}F_m + \varepsilon_p. \end{aligned} \tag{1}$$

where ω_{ij} is the load of the variable X_i (i = 1,2,...,p) on the factor F_j (j = 1,2,...,m), indicating the relative importance of the variable i on the common factor j. All the matrices formed by ω_{ij} are called factor load matrices, and ε_i (i = 1,2,...,p) are called special factors. Assume that $\varepsilon_i \sim N\left(0, \sigma^2\right)$ and σ^2 is the variance.

3.2 Factor Analysis Steps

The main steps of factor analysis are as follows.

1. KMO inspection

KMO (Kaiser–Meyer–Olkin) test is an index used to compare simple correlations and partial correlation coefficients between variables. When the square sum of simple correlation coefficients among all variables is much larger than the square sum of partial correlation coefficients, the closer the KMO value is to 1, the stronger the correlation between variables, and the original indicator variables are suitable for factor analysis. Generally speaking, when the KMO value is less than 0.6, it is not suitable for factor analysis.

2. Determining factors

The commonly used method of constructing factor variables is to solve the characteristic equation $|\lambda I - R| = 0$, find the eigenvalue and eigenvector of the correlation coefficient matrix, and then calculate the variance contribution rate and cumulative variance contribution rate.

According to the first m factors of the characteristic value size or the cumulative contribution rate reaching 85–95% to reflect the original evaluation index.

3. Calculate factor variable score

Construct the factor loading matrix as shown in Eq. (2).

$$A = \left(\sqrt{\lambda_i}\beta_{ij}\right)_{m \times p} \quad (i, j = 1, 2, \ldots, p) \tag{2}$$

where β_{ij} is the coefficient between the factor i and the original variable X_j. The maximum orthogonal rotation of variance is used to minimize the number of variables with the largest load on each factor and simplify the interpretation of the factors. Factor variables are represented as linear combinations of the original variables as shown in Eq. (3).

$$F_i = \beta_{i1}X_1 + \beta_{i2}X_2 + \ldots + \beta_{ip}X_p \quad i = 1, 2, \cdots, m \tag{3}$$

The linear combination of the original indicators can get the score value of each sample data on different factors.

4 Synthetic Data Examples

This paper extracts the data of one month's ATC CDM system of an airport, including ALDT, ATOT, runway capacity, the time of the last block, the last COBT, the takeoff runway, the last CTOT, the last LIN, the first QUE, and the standard time for ground taxiing, Planned takeoff time, planned landing time and waypoint passing time, as component performance indicators, used to calculate the 24 component performance index values in Table 1. The data is grouped and counted with the hour as the time granularity, and a total of 744 sets of sample values are obtained. The KMO test is performed by SPSS20 software, and the value is 0.722, which is greater than 0.7. Therefore, there is a strong correlation between the 24 index variables and factor analysis can be done.

Substituting 744 sets of data, you can get the characteristic value, contribution rate and cumulative contribution rate value of each variable, as shown in Table 2.

Table 2 Total variance explained

	Total	% of Variance	Cumulative %		Total	% of Variance	Cumulative %
X_1	4.603	19.180	19.180	X_{13}	0.728	3.033	83.050
X_2	2.446	10.192	29.371	X_{14}	0.588	2.451	85.501
X_3	1.765	7.355	36.726	X_{15}	0.577	2.402	87.903
X_4	1.602	6.674	43.400	X_{16}	0.522	2.173	90.076
X_5	1.553	6.471	49.870	X_{17}	0.451	1.880	91.956
X_6	1.313	5.470	55.340	X_{18}	0.417	1.735	93.691
X_7	1.103	4.595	59.935	X_{19}	0.375	1.561	95.253
X_8	1.044	4.352	64.287	X_{20}	0.346	1.441	96.694
X_9	1.010	4.207	68.494	X_{21}	0.299	1.246	97.939
X_{10}	0.976	4.067	72.561	X_{22}	0.245	1.020	98.959
X_{11}	0.909	3.786	76.347	X_{23}	0.228	0.948	99.907
X_{12}	0.881	3.670	80.016	X_{24}	0.022	0.093	100.000

According to the total variance explained in Table 2, it can be concluded that there are 9 variables with eigenvalues greater than 1, so the index system composed of 24 indexes is extracted with 9 factors, and the total variance of the original variables described by 9 factors 68.494%, more than 60%. It can be concluded that these 9 common factors can reflect most of the information contained in the original variables. The rotated component matrix can be obtained by factor rotation, as shown in Table 3. These 9 factors are the emergent effectiveness generated during the operation of the ATC.

Table 3 Rotated component matrix

	Component								
	1	2	3	4	5	6	7	8	9
X_9	0.816	0.119	−0.159	−0.285	0.075	−0.022	0.055	0.085	-0.058
X_3	0.784	0.147	0.031	0.237	−0.112	0.091	0.106	−0.03	0.104
X_8	0.708	0.232	−0.048	−0.222	−0.043	0.027	0.084	0.141	-0.092
X_{18}	0.666	0.072	−0.085	−0.103	0.419	−0.087	0.057	0.069	-0.121
X_{20}	−0.01	0.874	0.029	0.007	0.133	0.033	−0.013	−0.062	0.019
X_{19}	0.262	0.815	−0.049	−0.079	0.02	−0.03	0.073	−0.018	-0.044
X_{15}	0.345	0.755	−0.075	−0.194	−0.014	0.075	0.151	0.151	-0.033
X_5	−0.096	−0.044	0.937	0.211	−0.006	−0.05	0.002	−0.04	0.046
X_6	−0.083	−0.017	0.937	0.194	−0.022	−0.051	0.004	−0.034	0.047
X_{23}	−0.204	−0.166	0.12	0.717	0.215	−0.101	−0.025	0.064	-0.021
X_{21}	0.062	−0.019	0.268	0.695	−0.264	0.083	−0.176	−0.184	0.073
X_{24}	−0.236	0.043	0.126	0.525	0.269	0.084	0.284	0.163	-0.213
X_4	−0.041	−0.348	0.337	0.506	−0.055	−0.064	−0.087	−0.257	0.085
X_{17}	0.108	0.017	0.017	−0.069	0.766	−0.036	−0.064	0.034	0.086
X_{16}	−0.053	0.155	−0.057	0.242	0.622	0.11	−0.116	−0.11	-0.052
X_{12}	−0.018	0.035	−0.033	−0.051	−0.079	0.794	−0.063	−0.02	-0.186
X_{14}	0.033	0.045	−0.097	0.084	0.116	0.786	0.075	0.049	0.098
X_2	0.024	0.12	−0.023	−0.154	−0.128	0.026	0.842	−0.009	0.021
X_1	0.447	0.045	0.014	0.119	−0.031	0.005	0.731	−0.14	0.048
X_7	0.116	0.026	0.053	−0.033	−0.15	0.075	−0.178	0.691	-0.127
X_2	0.083	−0.043	−0.356	−0.09	0.344	0.008	0.197	0.6	0.079
X_{10}	−0.091	−0.033	0.031	0.022	0.043	−0.107	0.031	−0.149	0.743
X_{13}	0.025	−0.047	0.151	−0.123	0.002	0.44	0.054	0.175	0.501
X_{11}	−0.008	0.118	−0.279	0.147	−0.282	−0.273	−0.108	0.311	0.317

According to the output component score matrix of 9 influence factors, the following factor score function can be obtained by substituting into Formula (3). Considering the contribution rate of variance, the calculation Formula of the effectiveness evaluation value of the sample per unit time as shown Eq. (4).

$$D = 0.169F_1 + 0.141F_2 + 0.138F_3 + 0.122F_4 + 0.01F_5 + 0.098F_6 + 0.093F_7 + 0.074F_8 + 0.066F_9 \quad (4)$$

Substituting 744 sets of sample data, the evaluation score of each sample is shown in Fig. 2. The lowest three values appear at 2 am to 3 am on the 2nd, 1 am to 2 am on the 1st, and 3 am to 4 am on the 2nd, indicating that the ATC system performance is smooth; the highest three They appeared at 4 am to 5 am on the 16th, 17 am to 18:00 on the 21st, and 21 am to 22:00 on the 3rd, indicating that the efficiency of the air traffic control system was relatively slow.

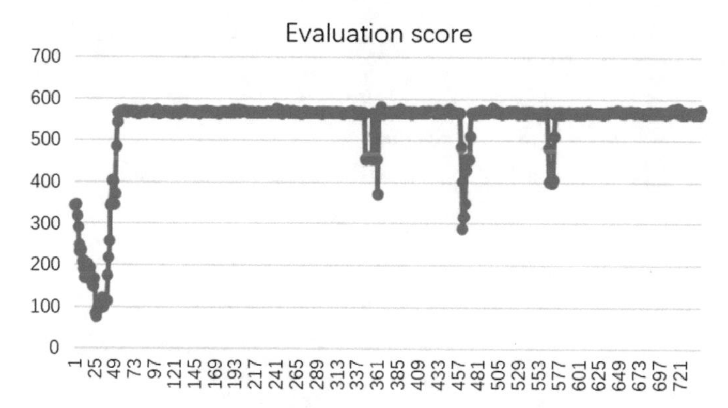

Fig. 2 ATC system effectiveness evaluation index score

5 Conclusion

In this paper, we extract 24 evaluation indexes from efficiency, capacity and delay, which are based on CDM data, to construct the ATC system performance evaluation index network. In order to solve the correlation between indicators, the factor analysis method is used to reduce the dimensionality of variable indicators and extract common factors to obtain evaluation scores. The feasibility and effectiveness of this method were verified by 744 sets of data from a certain instance.

Acknowledgements. This research is supported by State Key Laboratory of Air Traffic Management System and Technology (NO: SKLATM201802), CAAC North China Regional Administration Science and Technology Project (NO: 201803), and 2020 civil aviation safety ability building funding project:Research and application on air traffic operation management analysis technologies based on multi-source big data.

References

1. Chen P (2011) Equipment evaluation and index systems for ATC systems. Command Inf Syst Technol 2(3):10–13
2. Chen Z (2010) Air traffic control system effectiveness evaluation and equipment support planning. Nanjing University of Aeronautics and Astronautics, Nanjing
3. Liu X (2019) Quality evaluation of internal control report based on factor analysis method. Xi'an University of Technology, Xi'an
4. Yan Y, Ding H (2019) Big data value embodiment of air traffic management in business operation. Command Inf Syst Technol 10(1):7–12

Rerouting Path Planning Based on MAKLINK Diagram and MS-Genetic Algorithm

Tong Wei[1,3], Manzhen Duan[1,3], Bin Dong[2,3], Yinfeng Li[1,2,3](✉), and Shenghao Fu[2,3]

[1] College of Civil and Architectural Engineering, North China University of Science and Technology, Tangshan 063210, China
15030576997@126.com

[2] State Key Laboratory of Air Traffic Management System and Technology, Nanjing 210007, China

[3] China Civil Aviation Air Traffic Management Bureau in North China, Beijing 100621, China

Abstract. Aiming at the problem of global route planning for diversion in a static two-dimensional environment, a route planning method based on MS-genetic algorithm and MAKLINK graph is proposed. The MS algorithm combined with the genetic algorithm is optimized step by step, and take the shortest path optimized by genetic algorithm as the global optimal path. It solves the problem that the traditional algorithm is easily trapped in the local optimal solution in the two-dimensional path planning and can only find the approximate global optimal path. The MAKLINK graph theory is used to establish a two-dimensional space model, and MATLAB is used as the coding software tool to compare and verify the MS-genetic algorithm and the ant colony algorithm in path planning. The experimental results prove the feasibility and effectiveness of the algorithm scheme.

Keywords: Global path planning · MAKLINK diagram · Genetic algorithm · MS algorithm

1 Introduction

Severe weather is an important factor affecting flight safety, often leading to a large range of flight delays or cancellations. The key problem to be solved is to implement the strategy of diversion to avoid dangerous areas, establish safe and efficient temporary routes, and improve the utilization rate of spare space. In 2017, Taylor used Dijkstra shortest path algorithm to quickly generate different candidate paths, so that the generated reroute path must pass through the specified intermediate node [1]. In 2018, Ayo et al. proposed an intelligent platform for flight diversion optimization and simulation for adverse weather, Flight Parout. A genetic algorithm-based diversion method based on improved mutation technology was applied to the system to solve the problem of falling into a local optimal solution [2]. In China, Wang Fei applied MAKLINK map and genetic algorithm [3] in 2014 and used a three-stage method to carry out route planning research. In 2016, Wang Guanglei took the shortest route as the optimization goal, based on the directed

Q. Liang et al. (eds.), *Artificial Intelligence in China*, Lecture Notes in Electrical Engineering 653, https://doi.org/10.1007/978-981-15-8599-9_30

graph, through the improved Dijkstra algorithm to find the route that can avoid the no-fly areas such as dangerous weather areas and control areas [4]. In 2018, Luo Qinhan used a convolution neural network to extract its information features in two-dimensional grid image and then improved A* algorithm to find the heuristic function of path using a machine learning algorithm, realizing the design of an intelligent path planning algorithm [5].

In this paper, the previous algorithm can only find the approximate global optimal path in the MAKLINK graph, converging to the local suboptimal solution. The MS algorithm is combined with the genetic algorithm. The MS algorithm first calculates the first K shortest paths and then uses genetic algorithm to quickly optimize these paths. In the MAKLINK diagram, the global optimization capability of this method is significantly stronger than the traditional method.

2 Construction of Airspace Meteorological Mixed Environment Model

First, ignore the convective weather altitude information and expand the flight restricted area in all directions, so that when the aircraft is simplified into a particle and walks along the expanded restricted area edge, it will not interfere with the aircraft's safe flight conditions. Convex polygons are used to represent the restricted flight area, and the dangerous areas and obstacles predicted in the route are converted into convex hull problems. The improved Graham algorithm [6] is used to generate convex polygons and $h_{\min}$ is extended on the basis of the convex polygons. Figure 1 is a schematic diagram of the extended danger zone. The dashed line indicates the original danger zone, and the solid line indicates the expanded danger zone is recorded as a restricted flight zone. This ensures that even if the new route follows the boundary of the restricted flight zone in the path planning, it can ensure that the moving aircraft can cross the danger zone. Then construct the MAKLINK line to divide the space into obstacles and free areas. Finally, construct a link matrix, connect the midpoint, start point, and endpoint of the MAKLINK line to form an undirected network graph, generate a movable free space, and complete the environment modeling. Each free link line cannot pass through the restricted area. One endpoint of the link line is the vertex of the restricted area polygon, and the other endpoint is the vertex of the restricted area polygon or the point on the boundary of the free space, and each convex edge has at least two free link lines as selectable paths.

3 Establishment of a Mathematical Model for Rerouting Path Planning

The purpose of route planning is to avoid flight restricted areas and to minimize the navigation path. Therefore, in consideration of actual operation needs, factors such as the course change of the aircraft and the number of turning points should also be considered. Here, the angle of changing course and the number of turning points are restricted, and the corresponding mathematical model is established [7]. Assuming that the turning

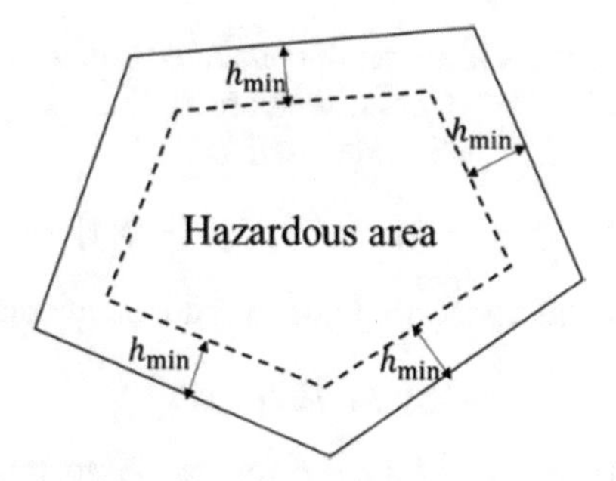

Fig. 1. Schematic diagram of extended danger zone

point sequence of the route to be planned is $(p_1, p_2, p_3, \ldots p_n)$, establish a mathematical mode.

$$\min \sum_{i=0}^{n} l(p_i, p_{i+1}) \tag{1}$$

$$s.t.(p_i, p_{i+1}) \cap S_j = \emptyset, i = 1, 2, \ldots, n, j = 1, 2, \ldots, m \tag{2}$$

$$\theta_{HC} \le 90^\circ \tag{3}$$

$$l(p_i, p_{i+1}) \ge 7.4km, i = 1, 2, \ldots, n \tag{4}$$

$$m \le m_{\min} \tag{5}$$

$$h \ge h_{\min} \tag{6}$$

Among them, Formula (1) means that the shortest path is the goal, (p_i, p_{i+1}) is the flight segment composed of points p_i and p_{i+1}, $l(p_i, p_{i+1})$ refers to the length of the segment; Formula (2) refers to the restriction of the restricted flight zone, that is, (p_i, p_{i+1}) the segment cannot pass through the restricted zone S_j; Eq. (3) indicates that the turning angle θ_{HC} cannot exceed 90 degrees; Eq. (4) indicates that the length of each flight segment is not less than 7.4 km [7], which can satisfy the completion of two turns within this distance; m is less than the set threshold $m_{\min}$, Formula (6) indicates that the distance h of the planned path from the danger zone is not less than the set threshold $h_{\min}$.

4 Rerouting Path Planning Based on MS-Genetic Algorithm

4.1 Initial Path Planning

The MS algorithm was proposed by Professor Martins and Professor Santos in 2000 [8], aiming to efficiently calculate the first k shortest paths in digraphs. Define directed graph $G(N, A)$, N is node set, A is arc set of connected nodes, $A = (i, j)$, i, j, respectively represent the first and last nodes of connected arc A, then the path of connected nodes s, t in directed graph $G(N, A)$ can be expressed as

$$p = \{s, (s, i), i, \ldots, j, (j, t), t\} \tag{7}$$

The first K shortest paths connecting node s, t are expressed as

$$P = \{p_1, p_2, \ldots, p_k\} \tag{8}$$

where $d(p_i) \leq d(p_{i+1})(i = 1, \ldots, k)$, $d(p_i)$ represents the total length of path p_i, and p_i, p_i is determined before p_{i+1}.

The principle of the MS algorithm is to first use the Dijkstra algorithm in the directed graph $G_1(N_1, A_1)$ to calculate the shortest path p_1 between the nodes S and T, and then "delete" the path p_1 (the deletion here is not really deleted, but through the directed graph Add additional nodes and corresponding arcs to achieve $G_2(N_2, A_2)$, then Dijkstra algorithm calculates the shortest path p_2 of S, T in $G_2(N_2, A_2)$, where p_2 is $G_1(N_1, A_1)$ The second shortest path in the middle; then "delete" the path p_2 at $G_2(N_2, A_2)$ to get $G_3(N_3, A_3)$, and so on, until the path p_k is calculated, and the $\{p_1, p_2, \ldots, p_k\}$ finally obtained is It is the first k shortest path of the nodes S, T in the directed graph $G_1(N_1, A_1)$.

According to the normal idea, for the established MAKLINK graph, use the Dijkstra algorithm to calculate the "suboptimal path", and then use the optimization algorithm to perform the second optimization to obtain the "optimal path", but the "optimal path" calculated using this method "Not necessarily the true optimal path. To this end, this paper proposes to use the MS algorithm to calculate the first K shortest paths, and then use genetic algorithms to perform secondary optimization on these K paths, and select the shortest path as the globally optimal path strategy, which greatly improves the algorithm's global optimization ability.

Before applying the MS algorithm to the MAKLINK graph, a corresponding link matrix needs to be established for the directed graph $G(N, A)$. The mathematical description of the link matrix is:

$$\text{adjm} = \begin{cases} \text{length}(N_i, N_j) & (N_i, N_j) \in A \\ \infty & (N_i, N_j) \notin A \end{cases} \tag{9}$$

$$\text{length}(N_i, N_j) = \sqrt{(N_{i,x}, N_{j,x})^2 + (N_{i,j}, N_{i,j})^2} \tag{10}$$

4.2 Path Optimization

The genetic algorithm (GA) is used to optimize the path. According to the construction of the actual problem, the optimal objective function is solved. Through the fitness function, multiple feasible solutions are brought into the function for calculation, and their sizes are compared. The optimal solution of the substituted function is selected as the optimal solution. Then, the appropriate individuals are retained to disturb the solution through crossover and mutation, so as to further improve the optimization degree of the solution and search for the most suitable solution for the environment. The basic calculation process of the genetic algorithm is as follows:

(1) Initialization: Set the maximum evolutionary generation T and the initial evolutionary generation; randomly generate M chromosome books as the initial population. To study the rerouting flight path of aircraft in strong convective weather, setting different flight paths is called chromosomes, and multiple initial solutions constitute a population. Assuming that the shortest path calculated by the MS algorithm is $S, V_1, V_2, \ldots, V_n, T$, where S and T are the starting point and ending point, respectively, and the second optimization is to adjust $V_1, V_2, \ldots, V_n$ The position of each point, search for the solution with the smallest overall objective function value, and then obtain the optimal path. By letting V_i slide freely on the link line according to the generated random number between 0 and 1, then the path point V_i at any point of the optimized MAKLINK line is:

$$V_i = V_{i1} + (V_{i2} - V_{i1}) \times t_i \tag{11}$$

Among them:V_{i1}, V_{i2} are the two ends of the free link line where V_i is located, t_i (the starting and ending positions do not slide), each group $(t_1, t_2, \ldots, t_n)$ corresponds to a path. The points on different MAKLINK lines can become the turning point of the course of the rerouting route. The chromosome is made up of elements that determine different points, where t_i is a genetic variable, and each t_i determines a point.

(2) Evaluation function-fitness: calculate the fitness of each chromosome according to the determined fitness function. According to the research of the paper on the diversion of aircraft under strong convective weather, in order to achieve the purpose of shorter flying distance, the evaluation function is set as shown in Eq. (12).

$$f(x) = \min \sum\nolimits_{i=1}^{N} V_i V_{i+1} \tag{12}$$

Among them:$V_i V_{i+1}$ the distance between nodes V_i, V_{i+1}, S is the starting point and T is the ending point.

(3) Selection: On the basis of calculating the fitness of individuals in the population, based on this, the optimized individuals will be inherited to the next generation, or new individuals will be generated through crossover and mutation, and then added to the population.
(4) Crossover operation: select a typical single-point crossover operator and randomly set a crossover point. The DNA is cut off at the same position on two chromosomes, and the two strings before and after are combined to form two new chromosomes, also known as gene recombination or hybridization.
(5) Mutation operation: Set a mutation operator to perform mutation operation on individuals in the next generation population according to mutation probability. A single point dual neighborhood mutation operator is used, as shown in Eq. (13).

$$t_i' = 1 - t_i \tag{13}$$

Among them t_i' a new gene after t_i mutated by a mutation operator.

4.3 Path Smoothing

After applying genetic algorithm to optimize the initial path, the globally shortest path based on MAKLINK graph is obtained, which satisfies the constraints of Formulas (1), (2), and (6) in the diversion model, but the number of turning points and the angle does not necessarily meet actual needs. Therefore, the method of "cutting and straightening" is adopted for the road section [7], so that the road section can reduce the turning points without intersecting the danger zone, and control the turning angle should not be too large.

5 Case Verification Analysis

The premise of the route modification plan is to carry out environment modeling, extract the information of the danger zone through sensors such as radar, and establish a restricted flight zone. In order to verify the practicability of the algorithm in the global path planning, this paper refers to "30 Case Analysis of MATLAB Intelligent Algorithms" to construct a 200 * 200 environment model [9]. There are 4 convex edges in the space as restricted areas. The starting point of the diversion is set to S (20, 180) and the target point T (160, 80). Figure 2 shows the undirected network diagram. The black polygon in the figure is a convex polygon generated by expanding the danger zone. The dotted line corresponds to the MAKLINK line, and the solid black line is the undirected network diagram obtained by connecting the midpoints of the MAKLINK line. Figure 3 shows the 22 × 22 order link matrix corresponding to the MAKLINK diagram.

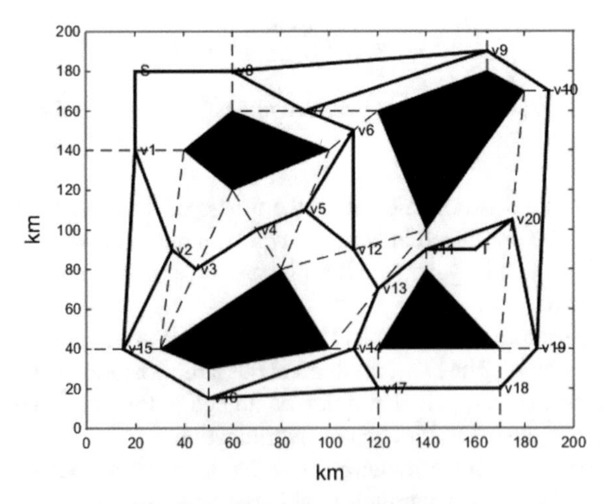

Fig. 2. undirected network diagram

$$\begin{bmatrix} \infty & 40.00 & \infty & \infty & \cdots & \infty \\ 40.00 & \infty & 52.20 & \infty & \cdots & \infty \\ \infty & 52.20 & \infty & 14.14 & \cdots & \infty \\ \infty & \infty & 14.14 & \infty & \cdots & \infty \\ \vdots & \vdots & \vdots & \vdots & \ddots & \vdots \\ \infty & \infty & \infty & \infty & \cdots & \infty \end{bmatrix}$$

Fig. 3. MAKLINK diagram link matrix

In the undirected network diagram, the MS algorithm is first used to plan the sub-optimal path. Figure 4 shows $K = 3$, and the starting and ending points are S and T, respectively. The first three shortest paths calculated by the MS algorithm. According to the length of the path, it is sorted as path 1 < path 2 < path 3, and the lengths are 229.061, 242.067, and 259.649, respectively.

Then the genetic algorithm is used to optimize the sub-optimal path. The parameters of the genetic algorithm are set as follows: the population number is 50, the genetic generation number is 200, the coding length is 10, the crossover probability is 0.7, and the mutation probability is 0.1.

The MS-genetic algorithm is compared with the ant colony algorithm to verify the effectiveness of the method. In Fig. 5, the MS-genetic algorithm corresponds to the dashed blue line, and the red solid line corresponds to the ant colony algorithm. The $\{t_1, t_2, \ldots, t_n\}$ corresponding to the two paths are {0.062561, 0.925709, 0, 0.921799, 0.062561, 0.250244}, {0.1, 0.9, 0.1, 0.9, 0.1, 0.3}.

In this paper, the first three shortest paths are optimized, and Choose the shortest path as the optimal solution. The experimental results of two paths calculated by MS-genetic algorithm and ant colony algorithm based on Dijkstra are shown in Fig. 2. The optimal path length calculated by MS-genetic algorithm is 168.998, which is significantly shorter

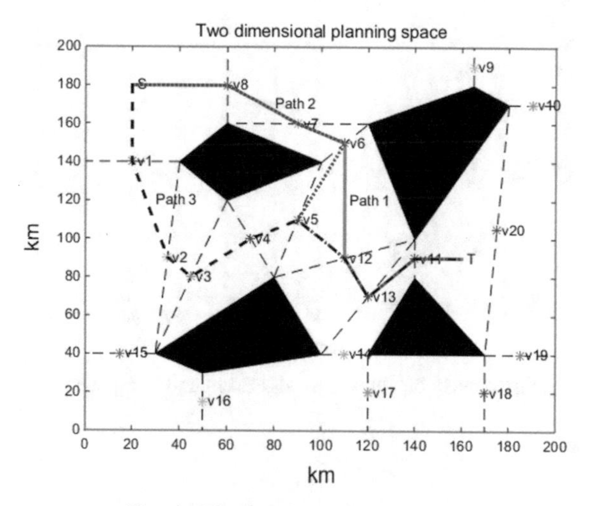

Fig. 4. The first three shortest paths

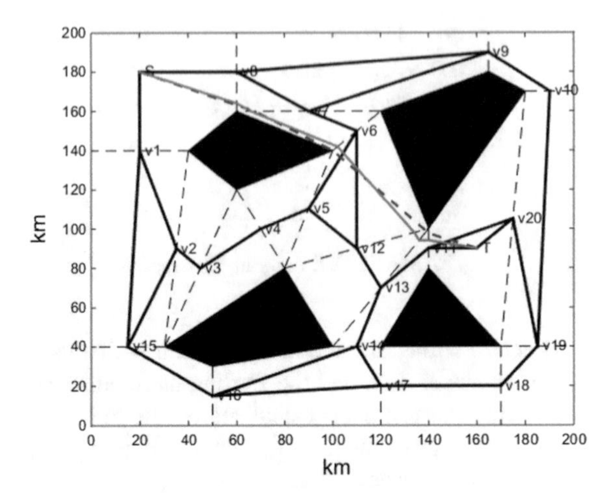

Fig. 5. Path planning comparison

than the path length optimized by ant colony algorithm 174.432. It can be seen from Fig. 6. that the convergence speed and calculation accuracy of the MS-genetic algorithm is also much stronger than the ant colony algorithm. Table 1 is the average optimal path length calculated by Dijkstra-ant colony algorithm and MS-genetic algorithm through multiple experiments, the average number of iterations and the average operating efficiency, the turning angle, and the number of turning points when reaching convergence and stability.

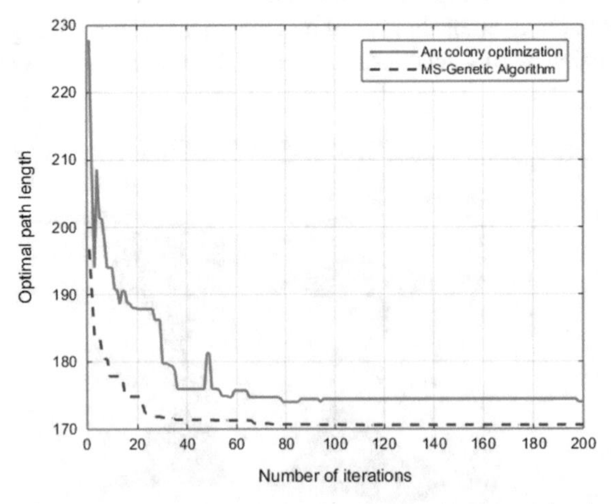

Fig. 6. Contrast between ant colony algorithm and MS-genetic algorithm

Table 1. Comparison of test results of ant colony algorithm and MS genetic algorithm

Algorithm	Optimal path length	Number of iterations	Number of turning points	Maximum turning angle (degrees)	Average operating efficiency
Ant colony optimization	174.432	85	3	31.5	<2 min
MS-Genetic algorithm	168.998	55	3	28.9	<2 min

6 Conclusion

In this paper, in the two-dimensional global static environment, the MAKLINK graph method is used to model the flight workspace, and the MS algorithm and the genetic algorithm are stepwise optimized for the shortcomings of the existing algorithm in the aircraft path planning. The MS algorithm calculates first The first K shortest path, and then the genetic algorithm for the second optimization, the optimal path after the second optimization is taken as the global optimal path, which solves the problem that the algorithm such as ant colony algorithm can only find the approximate optimal path in the MAKLINK graph. The data shows that the convergence speed of MS-genetic algorithm is slightly faster than ant colony algorithm, and the searched path length is shorter than ant colony algorithm. The success rate of MS-genetic algorithm is obviously superior to ant colony algorithm.

Acknowledgements. This research is supported by State Key Laboratory of Air Traffic Management System and Technology (NO: SKLATM201802);CAAC North China Regional Administration Science and Technology Project, (NO: 201904); 2020 civil aviation safety ability building funding project: Research and application on air traffic operation management analysis technologies based on multi-source big data.

References

1. Taylor CP, Liu S, Larsen D et al (2017) Designing flight-specific reroutes using network optimization. In: 17th AIAA Aviation technology, integration, and operations conference
2. Ayo BS, Hu YF et al (2018) Flight parout: a simulation platform for intelligent flight path reroutes for adverse weather. In: 37th AIAA/IEEE Digital avionics systems conference (DASC), IEEE
3. Wang F et al (2014) Research on route planning method based on MAKLINK chart and genetic algorithm. Transp Syst Eng Inf 000(005):154–160
4. Wang G, Wei W et al (2016) Route planning algorithm for transport aircraft. Command Inf Syst Technol 7(38) (02):47–50
5. Qinhan L, Li R, Shao W et al (2018) Intelligent path planning and design for tactical environment. Command Inf Syst Technol 9(06):53–58
6. Chen Kejia et al (2017) Planning and prediction of restricted flight areas in dangerous weather. J Nanjing Univ Aeronaut Astronaut (Social Science Edition), 019 (004), 59–63, 67

7. Martins EQV, Santos JLE (1999) A new shortest paths ranking algorithm[J]. InvestigacaoOperacional 20(1):47–62
8. Li X (2009) Research on route planning of flight diversion under dangerous weather conditions. Nanjing University of Aeronautics and Astronautics
9. Yu Lei (2011) Matlab intelligent algorithm 30 case analysis. Beijing University of Aeronautics and Astronautics Press, pp 217–227

Data-Driven Fault-Aware Multi-objective Optimization for Flexible Job-Shop Scheduling Problem

Zhibo Sui, Xiaoxia Li(✉), Jie Yang, and Jianxing Liu

College of Informatics, Huazhong Agricultural University, Wuhan 430070, PR China
lixiaoxia@mail.hzau.edu.cn

Abstract. Flexible job-shop scheduling problem (FJSP) is one of the most important optimization problems in smart manufacturing. Its optimal solution can significantly improve the performances of manufacturing processes. However, the robustness of the optimal solution for FJSP is still needed to be improved to decrease the losses caused by the unnecessary rescheduling. In this paper, a fault-aware multi-objective optimization method has been developed to obtain a robust solution for FJSP. In the method, the fault probability model is first built using big data analytics to evaluate the robustness of the FJSP solution. And then, the fault probability, machining energy consumption and makespan are simultaneously considered as the optimization objectives to obtain the Pareto optimal set for FJSP using NSGA-II. Finally, the Pareto solution with the minimal fault probability is chosen for FJSP. The experiments on the benchmark case show that the optimization results of this method are promising for FJSP.

Keywords: Flexible job-shop scheduling · Multi-objective optimization · Energy consumption · Fault probability · Big data analysis

1 Introduction

The increasing economic globalization drives companies to adopt optimization strategies to improve their manufacturing performances to remain competitive. One of the most commonly used optimization strategies is the optimization of FJSP. There are two manufacturing scenarios related to FJSP. One is the decision making of flexible shop scheduling where an initial optimum scheduling solution is obtained before machining. The other is the rescheduling of the remaining tasks where a new optimal scheduling solution is searched to respond to the faults occurred during machining. Obviously, if the initial scheduling solution offered by the former is robust enough, some rescheduling in the latter can be avoided. That is, the manufacturing performances can be further improved by a robust solution for FJSP due to less task delay or other unnecessary losses.

For FJSP, a large number of evolution algorithms have been used to search its optimum solutions because FJSP is a complex NP-hard issue [1]. During the search, multiple performances, including machining time, cost, quality, energy consumption and so on,

Q. Liang et al. (eds.), *Artificial Intelligence in China*, Lecture Notes
in Electrical Engineering 653, https://doi.org/10.1007/978-981-15-8599-9_31

are considered as the optimization objectives. However, facing with the dynamic manufacturing processes, most of the researches focus on rescheduling without considering a more robust initial solution for FJSP.

Nowadays, big data in manufacturing offers good potentials for manufacturing system to obtain robust solutions before decision making of shop floor scheduling. One of the most important potentials is high-fault predictability where manufacturing big data is analyzed to predict the possible faults occurring in manufacturing processes. Based on the fault prediction, the corresponding indicator can be considered in FJSP to obtain more robust optimization solutions.

In order to obtain more robust scheduling solutions for FJSP, the prediction model is built in this paper to evaluate the fault probability of the scheduling solution using the big data analysis. Based on the predicted fault probability, machining time, energy consumption and the solution's robustness are simultaneously considered to search for robust flexible workshop scheduling schemes using NSGA-II.

The remainder of this paper is organized as follows. Section 2 reviews the related work. In Sect. 3, the model of the fault probability is built, and then, the model of the fault-aware multi-objective FJSP is presented. In Sect. 4, the fault probability-based NSGA-II for FJSP is described. In Sect. 5, case studies and comparison with other methods are given. Finally, a conclusion is drawn in Sect. 6.

2 Related Work

In recent years, the research on FJSP has been extensively carried out, and the comprehensive surveys can be found from references [1, 2]. This paper focuses on big data-driven FJSP, and the related state-of-the-art research is summarized below.

Based on the big data collected in the manufacturing process, most of the research works were dedicated to modelling the energy consumed in manufacturing processes to obtain energy efficient scheduling solution by further using heuristic algorithms. For instance, the specific energy consumption (SEC) of machining processes was modelled as the function of the material removal rate by analyzing the energy consumption data collected from machine tools [3]. Based on the SEC model, Dai et al. [4] built the total energy consumption model for the flexible flow shop and applied the model to obtain satisfactory solutions using genetic-simulated algorithm. Li et al. [5] established an energy consumption model by analyzing the energy consumed in each stage of the machining processes and then developed a hybrid algorithm to search for the optimal scheduling solution. Zhang and Chiong [6] modelled the machining energy consumption and used a multi-objective genetic algorithm with enhanced local search for JSP. Sheng et al. [7] adopted artificial neural networks to build the complex nonlinear relationships between key machining parameters and machining energy consumption.

Different from the research works on modelling energy consumption, some other research works focused on building the models for predicting the faults occurred in manufacturing processes. For instance, Ji and Wang [4] modelled the fault probabilities of the scheduled tasks using decision trees built by analyzing the historical data of workshop machining. Liang et al. [8] used convolutional neural networks to establish potential dynamic failures during machining and scheduling optimization. Harris and Sychra [9] developed a prediction model based on the application of one or more statistical tools and pattern recognition technologies. Wang and ye [10] presented a digital twin reference model for rotating machinery fault diagnosis. The framework proposed by Yoo [11] uses a real-time quality assessment tool based on support vector machine (SVM) algorithms, which enable users to classify product quality patterns based on production data.

Though a large amount of research has been reported as above, the manufacturing process of a workpiece is in a dynamic environment where machine faults may occur. Thus, the robustness of the solution for FJSP should be emphasized to respond to machine faults occurred in machining processes.

3 Fault-Aware Multi-objective Optimization Modelling for FJSP

3.1 Fault Probability Prediction Modelling for FJSP

The fault probability is an indicator to be used to evaluate the machine failure occurrence during executing the scheduling solution. The evaluation is carried out before the implementation of the scheduling solution. Therefore, the factors that may give rise to machine faults in the manufacturing processes are explored from both the scheduling solution and the process plans.

As a major factor, the machining time of the machine resources assigned in the scheduling solution can be obtained using Eqs. (1 and 2).

$$T_{M_i} = T_{M_i\_\text{historical}} + \sum_{j=1}^{n} f_{O_j} \times T_{M_i}(O_j) \tag{1}$$

$$f_{O_j} = \begin{cases} 1 \text{ if the } i\text{th machine is assigned to } j\text{th operation} \\ 0 \text{ if not} \end{cases} \tag{2}$$

where T_{Mi} is the ith machine's machining time during the execution of the scheduling solution; $T_{Mi\_\text{historical}}$ is it's machining time that has been spent before the scheduling solution's execution;f_{oi} represents whether the jth operation executed on the ith machine; $T_{Mi}(O_j)$ is the machining time spent on executing the jth operation O_j. Additionally, some other factors such as the workpiece's material hardness, surface roughness requirements and so on may also influence the machining processes. Thus, all these above factors are considered as the features to predict the scheduling solution's fault probability.

Following the above features, the corresponding data and the machining state can be collected from the historical machining processes to obtain the historical feature data sets. Additionally, for a scheduling solution to be evaluated, its feature data subset can also be retrieved from the current scheduling solution and process plans. All these feature data sets are then combined into an integrated data set.

Based on the integrated data set, a k-means-based approach is developed to obtain the fault probability. The main steps of the approach are described as follows:

- Normalization: Eq. (3) is first used to normalize the integrated data set to avoid the effect of the different units.

$$F_{\text{norm}}(i,j) = \frac{F(i,j) - \min(F(j))}{\max(F(j)) - \min(F(j))} \tag{3}$$

where $F_{\text{norm}}(i, j)$ and $F(i, j)$ are the normal and unnormal value of the ith feature data set's jth feature, respectively, and $F(j)$ is the jth feature.

- Clustering: A k-means algorithm [10] is adopted to group the integrated data set.
- Evaluation: The cluster including the scheduling solution to be evaluated is selected to be used to evaluate the solution's fault probability using Eq. (4).

$$\text{FP} = \left(\text{FN}/N\right) \times 100\% \tag{4}$$

where FP is the fault probability; FN and N are the number of the feature data sets with failure occurrence and the total number of the feature data sets in the integrated data set, respectively.

3.2 Multi-objective Optimization Modelling for FJSP

The fault probability defined above is used to evaluate the robustness of a schedule. Additionally, the maximum interval time spent to machine all the parts (makespan) is chosen as a criterion to reflect the productivity. Energy consumption (EC) of manufacturing processes is used as another performance criterion to improve the energy efficiency of the manufacturing process. Let n and m be the number of the operations to be executed and the number of the machines, respectively. These two criteria can be defined in the following:

$$\text{Makespan} = \max(T(O_n)) \tag{5}$$

where n is the number of the operations to be executed; O_n is the nth operation of all the workpieces to be machined; $T(O_n)$ is its completion time.

$$\text{EC} = \sum_{j=1}^{m} \sum_{i=1}^{n_i} \int_{T_{i\_\text{start}}}^{T_{i\_\text{end}}} P_{\text{working}}(M_j)\mathrm{d}t \tag{6}$$

where m is the number of the machines; M_j is the jth machine; n_i is the number of the operations executed on the M_j; T_{i_start} and T_{i_end} are the start and end time for the ith operation; $P_{\text{working}}(M_j)$ is the jth machine's working power.

Based on the above objective functions, the multi-objectives optimization model for ÷ as follows:

$$\begin{cases} \min \text{Makespan} \\ \min \text{EC} \\ \min\text{FP} \end{cases} \tag{7}$$

Subject to: precedence constraints on the operations' machining sequence.

4 Fault Probability-Based NSGA-II for FJSP

4.1 Initialization

A non-dominated sorting genetic algorithm (NSGA-II) is employed in this paper to solve the FJSP. There are two aspects in the step of initialization. One is the initialization of the parameters used in NSGA-II, which includes the number of the iteration, the cross-probability, the mutation probability and so on. The other is the creation of the initial population where the encoding scheme for the chromosome and the creation method should be considered.

Each chromosome in NSGA-II indicates the solution of the FJSP. A chromosome is divided into two segments: the representation of the sequenced operations and the representation of the resources. In the former, each number represents the job ID which appears m times to demonstrate the processing sequence of the corresponding operations. The latter is the resource assigned to the corresponding operation.

Based on the encoding scheme for the chromosome, all of the chromosomes are created randomly to obtain the initial population of the NSGA-II. Firstly, the part of operation representation is filled by the job ID selected randomly from the job set to represent the corresponding operation sequence. Then, for each of the decided operation in the first part, the number of its machine is assigned randomly in the part of resource representation.

4.2 Basic Genetic Operations

Based on the parameters configuration and the initial population, a new population is produced using the basic genetic operations which are listed as follows:

- The selection operation: Based on the individuals' fitness, tournament selection mechanism [12] is employed to choose the individuals to be further used.
- The crossover operation: For the selected individuals, two-point crossover [13] in accordance with the encoding rule is adopted to achieve their crossover and produces a new individual.
- The mutation operation: In order to yield individual with new information, two genes in the segment of resource representation are chosen randomly and mutated according to the mutation probability.

4.3 Evaluation

In NSGA-II, individual solutions can be evaluated through non-dominated level and crowding distance during the evolution which is described as follows:

- The non-dominated sorting: Based on the Eqs. (5 and 6), the fitness of each individual is evaluated and compared to categorize the individuals into different non-dominated fronts. And then, a new generation can be obtained by choosing the individuals with better fitness values.

- The crowding distance: In order to preserve the population's diversity, the crowding distance is calculated using Eq. (8). Based on the crowding distance, the individuals in the new generation obtained on the non-dominated level are compared to choose those with lower rank or the same rank but a lesser crowded region as the more suitable for inheritance.

$$CD_n = \sum_{j=1}^{m} \left(\left| f_j^{i+1} - f_j^{i-1} \right| \right) \tag{8}$$

where CD_n represents the crowding degree of the nth individual; f_j^{i+1} m is the number of optimization goals; m represents the jth objective function value of the $i + 1$ point; f_j^{i-1} represents the jth objective function value of the $i - 1$ point.

4.4 Selection and Output

Based on the new population obtained by the above iteration, the new iteration of the genetic operations and the evaluation will be repeated until the iteration reaches the largest number. Finally, a Pareto set of the optimum scheduling solutions is obtained. However, only one solution can be selected and used to be executed in the manufacturing.

In order to select a proper robust solution to respond to the dynamic characteristic of the manufacturing, the fault probability of the solutions in the Pareto set is evaluated using Eq. (4). The solution with the lowest fault probability is selected as the final solution to be output and implemented in the practical manufacturing.

5 Case Studies and Discussion

To verify the effectiveness and feasibility of the approach, the proposed method has been used to solve a benchmark case of FJSP [14]. The experiment was performed on Windows 10 operating system with corei7-8550U cpu at 1.80 GHz and 8 GB of main memory. The simulation was carried out using python programming language. The selected parameters of the k-means algorithm are: k is 8; the iteration is 300; the tolerance is 0.0004. For NSGA-II, the selected parameters are: The population size and the iteration are 50 and 200, respectively, the cross-probability is 0.8, and the mutation probability is 0.02.

The feature data set for this case is shown in Table 1. Each column is the corresponding feature. Each line is a historical feature data set. Based on the historical feature data sets, each scheduling solution generated by the proposed NSGA-II was integrated into the historical feature data sets and clustered to evaluate its fault probability. The Pareto frontier obtained by NSGA-II is shown in Fig. 1. It can be observed that there exists conflicting relationships among the makespan, energy consumption and fault probability. The Pareto optimal solution with the least fault probability was chosen as the scheduling solution.

The comparisons between the proposed approach and the NSGA-II are shown in Table 2. It can be seen that the proposed approach can obtain a more robust scheduling solution than other approaches without considering the fault probability.

Table 1 Historical feature data sets

No.	Machining time						Hardness	Ra
	M_1	M_2	M_3	M_4	M_5	M_6		
1	460	137	142	426	479	352	2.00	0.15
2	486	183	83	140	9	193	1.57	0.02
3	328	178	420	149	394	423	1.96	0.01
4	23	135	349	6	255	235	0.08	0.09
5	284	288	178	405	343	44	2.27	0.26
6	379	65	252	305	28	131	0.28	0.25
7	65	471	222	309	474	9	1.12	0.27
8	476	433	217	332	389	349	1.96	0.22
…	…	…	…	…	…	…	…	…

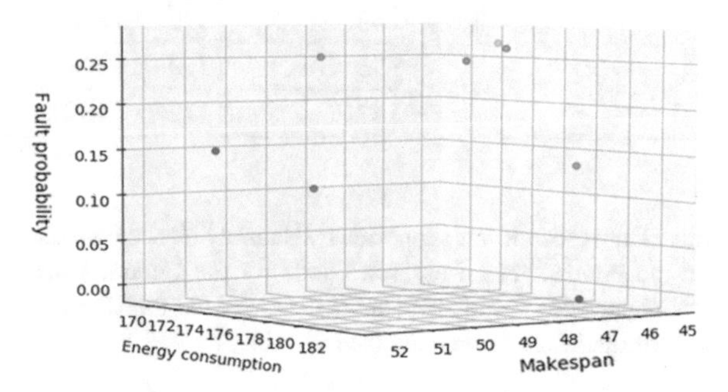

Fig. 1 Pareto frontier obtained by NSGA-II

6 Conclusions

It is critical for manufacturing companies to develop robust flexible scheduling solutions to become more suitable for the dynamic inherent in manufacturing processes. This research presents a data-driven fault-aware multi-objective optimization approach for FJSP. The experimental results demonstrate that the developed approach is promising. The contributions of the presented approach are summarized as following:

- A preliminary prediction model for the fault probability of a scheduling solution is established to support the fault-aware scheduling decision making.
- The fault probability of the scheduling solution is considered to improve the robustness of the optimal solutions identified by the evolutionary algorithm.

Since big data also offers good potentials to evolutionary algorithms, the data driven evolutionary algorithms for FJSP and other optimization problems will be investigated in the future research.

Table 2 Comparison studies

Method	Pareto optimal set		
	Makespan	EC	Fault probability (%)
Fault probability based NSGA-II	50	174	11.11
	47	183	0.00
	48	169	25.00
	52	173	15.38
	47	176	25.00
	45	173	26.67
	45	178	13.33
NSGA-II neglecting faulting probability	46	176	26.32
	48	169	25.00
	48	168	20.93
	46	176	18.18
	53	170	20.00
	47	172	18.18

Acknowledgements. This research was supported by Natural Science Foundation of China (grant No. 61803169) and the Fundamental Research Funds for the Central Universities (grant No. 2662018JC029). The paper reflects only the authors' views, and the Union is not liable for any use that may be made of the information contained therein.

References

1. Chaudhry IA, Khan AA (2016) A research survey: review of flexible job shop scheduling techniques. Int Trans Oper Res 23(3):551–591
2. Xie J, Gao L, Peng KK et al (2019) Review on flexible job shop scheduling. IET Collaborative Intell Manuf 1(3):67–77
3. Kara S, Li W (2011) Unit process energy consumption models for material removal processes. CIRP Annals—Manuf Technol 60(1): 37–40
4. Dai M, Tang D, Giret A, Salido MA et al (2013) Energy-efficient scheduling for a flexible flow shop using an improved genetic-simulated annealing algorithm. Robot Comput-Integrated Manuf 29(5):418–429
5. Li XX, Li WD, Cai XT et al (2015) A hybrid optimization approach for sustainable process planning and scheduling. Integrated Comput-Aided Eng 22(4):311–326
6. Zhang R, Chiong R (2016) Solving the energy-efficient job shop scheduling problem: A multi-objective genetic algorithm with enhanced local search for minimizing the total weighted tardiness and total energy consumption. J Clean Prod 112(4):3361–3375
7. Wang S, Lu X, Li XX et al (2015) A systematic approach of process planning and scheduling optimization for sustainable machining. J Clean Prod 87:914–929

8. Liang YC, Li WD et al (2019) Fog computing and convolutional neural network enabled prognosis for machining process optimization. J Manuf Syst 52:32–42
9. Harris D, Sychra J, Schmit L et al (2002) Method for predicting machine or process faults and automated system for implementing same: U.S. Patent Appl 9(755): 208
10. Huang X, Yin C, Dadras S et al (2017) Adaptive rapid defect identification in ECPT based on K-means and automatic segmentation algorithm. J Ambient Intell Humaniz Comput
11. Yoo A, Oh Y G, Park H, et al. (2013) A product quality monitoring framework using SVM-based production data analysis in online shop floor controls. In: IIE annual conference, proceedings. Institute of Industrial and Systems Engineers, 3255
12. Chu TH, Nguyen, Neill M O'(2018) Semantic tournament selection for genetic programming based on statistical analysis of error vectors. Inf Sci 436: 352–366
13. Spears WM, Anand V (1991) A study of crossover operators in genetic programming. In: International symposium on methodologies for intelligent systems, pp 409–418
14. http://people.idsia.ch/~monaldo/fjsp.html
15. Wang JJ, Ye LK, Gao RX et al (2019) Digital Twin for rotating machinery fault diagnosis in smart manufacturing. Int J Prod Res 57(12):3920–3934

Research on Airspace Conflict Resolution Algorithm Based on Dempster–Shafer Theory

Zelin Li[1](✉), Tianhao Tan[2], and Shiming Zhu[1]

[1] State Key Laboratory of Air Traffic Management System and Technology, Nanjing 221116, China
512954223@qq.com

[2] Air Force of the CPLA Unit 93735, Tianjin 301700, China

Abstract. The future trend of integrated joint operations makes the task of battlefield airspace management more important. However, in the face of users' demands for using airspace, airspace resources are more precious and tense. Especially the conflict resolution between different types of tactical airspace use needs is an important issue in wartime airspace management. The resolution of airspace conflict should be combined with the user attribute decision of airspace use demand—the airspace use demand with low priority attribute such as weapon platform and task type should avoid the airspace use demand with high priority. However, for a certain airspace use requirement, when the priority ranking of different attributes conflicts with the impact of avoidance decision, the decision dilemma will be caused. In this paper, priority information is fused by evidence theory to solve the problem of spatial conflict resolution.

Keywords: Airspace management · Conflict resolution · Dempster–Shafer theory

1 Introduction

The requirement of peacetime and wartime integration of national defense [1] provides a wide range of business needs and important subjects in the research direction of airspace management [2, 3]. Airspace use demand is also increasing. One of the important tasks of wartime airspace management is to rationally coordinate the airspace use needs of various users and to relieve the conflicts among the airspace use needs. Its principle is that airspace conflict resolution (ACR) should be combined with the airspace use requirements of user attributes such as weapon platform and mission types. The airspace use requirements with low priority attributes should avoid the airspace use requirements with high priority on the elevation level or the time range of use. However, the result of prioritizing based on weapon platform may not be the same as that of prioritizing based on task type—airspace use requirements with high priority of weapon platform may not have high priority of task type. It is difficult to comprehensively consider the attributes of airspace use requirements depending on a single priority order, while it is

Q. Liang et al. (eds.), *Artificial Intelligence in China*, Lecture Notes in Electrical Engineering 653, https://doi.org/10.1007/978-981-15-8599-9_32

easy to cause the decision-making dilemma in conflict resolution by comprehensively considering the conflicting priority order. Dempster put forward the Dempster–Shafer Theory (DST) in 1967 [4], and then, his student Shafer further developed and improved it in 1976 [5], making imprecise inference on uncertain information from multiple sources and obtaining the solution of the problem. The method of evidence theory can fuse the information of two priority scoring methods and solve the decision-making problem of airspace conflict resolution.

2 DST Definition of ACR

The evidence theory firstly determines the identification framework (or hypothesis space) Θ, which is defined as a complete set of incompatible events and regarded as a platform database (PDB) in data fusion [6]. Airspace conflict resolution can be regarded as the comprehensive priority setting of airspace use plan (AUP). Event A is "the highest comprehensive priority of airspace use plan A," so there should be N incompatible events for N airspace use plans, forming one recognition framework Θ. For each of the events, the basic probability allocation, namely the mass function, is set as a function that maps from domain of definition 2^{Θ} to codomain [0, 1], and satisfies:

$$\begin{cases} m(\phi) = 0 \\ \sum_{A \subseteq 2^{\Theta}} m(A) = 1 \end{cases}$$

where $m(A)$ is called the basic reliability assignment of event A, indicating the degree of support for event A [7]. For the problem of ACR, the basic reliability assignment of event A is defined as:

$$m(A) = \frac{p(A)}{\sum_{B \subset 2^{\Theta}} p(B)}$$

where $p(A)$ represents the priority score of event A. In other words, the higher the decision-maker's priority score for (combination of) airspace use plan A, the higher the support degree for it. The basic reliability assignment is in the same proportion as the priority score, and the sum of the basic reliability assignment satisfying the identification framework is 1. $m(\varphi) = 0$ indicated that no airspace use plan had the highest overall priority. At the same time, according to DST theory, the definition is:

2.1 Belief Function (Bel)

$$\begin{cases} \mathrm{Bel}(\phi) = 0 \\ \mathrm{Bel}(\Theta) = 1 \\ \mathrm{Bel}(A) = \sum\limits_{B \subseteq A} m(B) \end{cases}$$

The Belief function $\mathrm{Bel}(A)$ represents that the integrated priority of airspace use plan combination is equal to the sum of the integrated priority of each airspace use plan in the airspace conflict resolution problem. A which can satisfy $m(A) > 0$ is called the focal element of Belief function Bel. Since the priority scores are all positive, all events are focal elements. The union of all focal elements is called the kernel.

2.2 Plausibility Function (Pl)

$$\begin{cases} \text{Pl}(A) = 1 - \text{Bel}(\bar{A}) \\ \text{Pl}(A) = \sum_{\text{A} \cap \text{B} \neq \phi} M(B)\ \forall \text{A} \subseteq \Theta \end{cases}$$

The Plausibility function Pl(A) represents the wide support degree of airspace use plan combination A in airspace conflict resolution problem. And the Plausibility function of AUP A is the sum of the basic reliability assignment values of airspace use plan combination whose intersection with AUP A is non-empty.

2.3 Confidence Interval

According to the above definition, the following intervals are divided: the interval of supporting evidence [0, Bel(A)], the interval of Plausibility [0, Pl(A)] and the interval of rejecting evidence [Pl(A), 1]. The trust interval [Bel(A), Pl(A)] is defined to indicate the degree of determination (or uncertainty) of "the highest comprehensive priority of airspace use plan combination A."

3 Dempster Combination Rules

Assume that m_1 and m_2 are basic probability assignment under the same identification framework: m_1 represents the basic probability assignment of airspace use plan (combination) determined by the priority score of task type, while m_2 represents the basic probability assignment of the airspace use plan (combination) determined by the priority score of the weapon platform. If the focal elements of Bel_1 are $B_1, \ldots, B_K$, the focal elements of Bel_2 are $C_1, \ldots, C_K$, and applying rule of the orthogonal sum $m = m_1 \oplus m_2$, we can get

$$m(A) = \frac{1}{1-K} \sum_{i,j,B_i \cap C_j = A} m_1(B_i) m_2(C_j)$$

where $\frac{1}{1-K}$ is the normalization constant, if $\frac{1}{1-K} \neq 0$, $m(A)$ is also a basic probability assignment which is called the comprehensive probability assignment of m_1 and m_2 and satisfies with $\sum_{A \subseteq \Theta} m(A) = 1$ at the same time. If $\frac{1}{1-K} = 0$, $m_1 \oplus m_2$ is undefined and $m(A)$ does not exist. In this case, we say m_1 is conflict with m_2 [8], so

$$K = \sum_{i,j,B_i \cap C_j = \phi} m_1(B) m_2(C)$$

is called the collision term (or collision coefficient). In other words, the product of the basic probability assigned by the two priority scoring methods to the two completely mutually exclusive airspace use plans is not empty, indicating that the two priority scoring methods have different opinions on it. It is an opinion conflict, and the sum of all opinion conflicts becomes the conflict coefficient of the two priority scoring methods.

3.1 The Deformation of the Combination Rule

On the basis of this framework, later generations have different understanding and allocation of conflict evidence, resulting in different Dempster's rule of combination:

3.1.1 Smets Combination Rule

Smets believes that all evidence is reliable, and that the conflict comes from a flawed and imperfect identification framework [9]. In the rules of Smets design, the conflicting evidence after combination is retained without normalization, but it is assigned to the empty focusing element after evidence combination. This method satisfies both the commutative law and the associative law [10]:

$$\begin{cases} m(A) = \sum\limits_{B_i \cap C_j = A} m_1(B_i) m_2(C_j) \\ m(\phi) = \sum\limits_{B_i \cap C_j = \phi} m_1(B_i) m_2(C_j) \end{cases}$$

3.1.2 Yager Combination Rule

Yager believes that the identification framework is complete, and the conflicting evidence part represents unreliable information, so the conflicting part is treated as "unknown" and added to the assignment to the complete set [11]. The Yager rule is a fairly conservative approach that satisfies the commutative law rather than the associative law:

$$\begin{cases} m(\phi) = 0 \\ m(A) = \sum\limits_{B_i \cap C_j = A \neq \phi} m_1(B_i) m_2(C_j) \\ m(\Theta) = m_1(\Theta) m_2(\Theta) + \sum_{B_i \cap C_j = \phi} m_1(B_i) m_2(C_j) \end{cases}$$

3.1.3 Dubois and Prade Combination Rule

In the Dubois & Prade combination rule designed by Dubois et al., the conflicts generated by focus element B_i and C_j should be attributed to focus element $B_i \cup C_j$, rather than to the universal set of identification framework [12]. Dubois & Prade method is more specific, satisfying the commutative law and not the associative law:paying attention to the

$$\begin{cases} m(\phi) = 0 \\ m(A) = \sum_{B_i \cap C_j = A \neq \phi} m_1(B_i) m_2(C_j) + \sum_{B_i \cap C_j = \phi, B_i \cup C_j = A} m_1(B_i) m_2(C_j) \end{cases}$$

3.2 Problem of Element Explosion

We have to consider the problem of focal element explosion [13]; that is, the number of focal elements increases exponentially with the number of identification frame elements, and the calculation amount increases accordingly [14]. Especially for the airspace conflict resolution problem, this method uses evidence theory to solve the priority fusion problem of different sources. The establishment of identification framework depends on the airspace use plans, and the number of identification framework elements increases with the increase of the number of airspace use plans. However, from the perspective of airspace management matters, the demand for airspace use is large and complex, so the focal element explosion problem will become the focus of the evidence theory method of airspace conflict resolution in the future.

The solution is to design fast combination rules. Kennes proposed an optimization algorithm for Dempster's rule [15]. Barnett and Shafer's hierarchical computing method [14] and Moral's method based on the Monte Carlo both realize the combination of evidence. Wickramarathne proposed a simplified calculation of evidence combination based on Monte Carlo and statistical sampling [16]. Li Yuefeng proposed an approximate calculation method in the evidence theory by approximating the basic reliability and simplifying it [17].

3.3 DST Algorithm for Airspace Conflict Release

3.3.1 Algorithm Process

To sum up, according to the priority score of the mission or weapon platform, two basic probability assignments are obtained, which are then fused according to different combination rules to obtain the combined output, and the priority score considered as comprehensive is obtained for further conflict resolution operation. The process is as follows (Fig. 1).

3.3.2 Algorithm Simulation

Assuming that there are three airspace use plans $\{A, B, C\}$ in the context of tactical airspace management, the basic probability assigned according to the task priority is $m_1(A) = 0.99, m_1(B) = 0.01, m_1(C) = 0$, which means that the airspace use plan A is extremely important from the task priority. The basic probability assigned according to the priority of weapon platform is $m_2(A) = 0, m_2(B) = 0.01, m_2(C) = 0.99$, which means that the airspace use plan C is more important from the priority of weapon platform. Table 1 shows the fusion results of the combined formulas obtained by DS, Yager, Smets and Average method.

The simulation results are derived from the Zadeh paradox. It can be seen that there is a high conflict between the priorities of airspace use plans. And the priorities of missions and weapon platforms are not the same. The ordinary average method will ignore the effect of extremely low priority ranking, while the DS algorithm, in order to take both into account, results in the highest fusion result of airspace use plan B, which violates the original intention of design. Yager and Smets algorithm are better at handling fusion conflicts. Especially for the airspace management application background, the integration results of the three airspace use plans are basically flat and in line with the expectation without paying attention to the assignment of the total set Θ and ϕ.

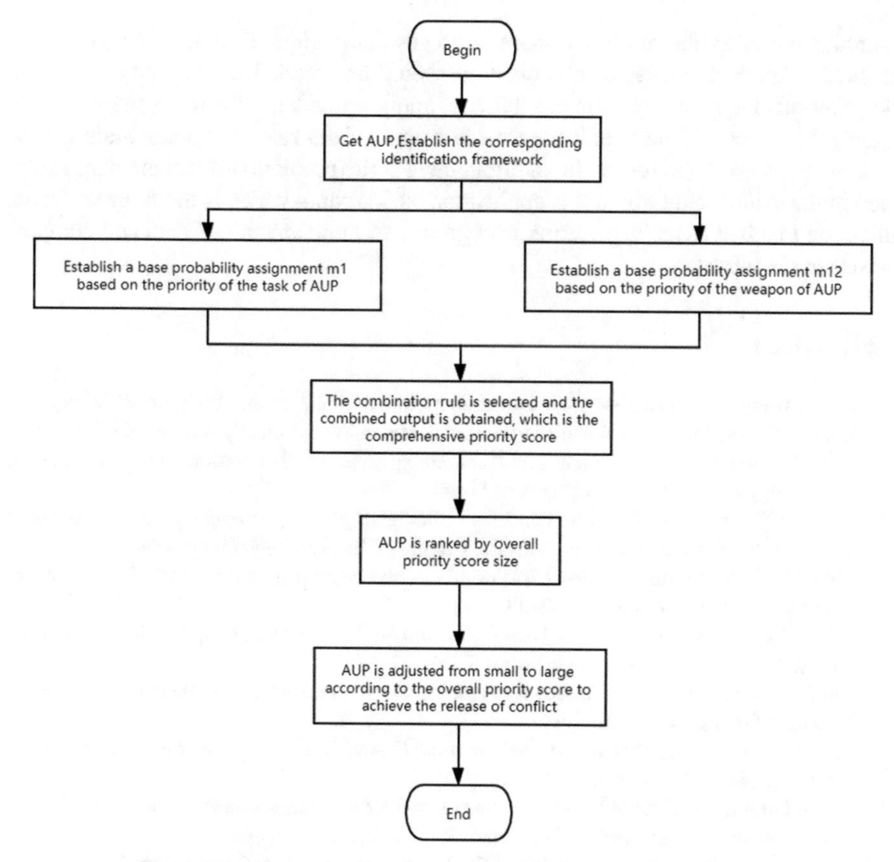

Fig. 1 Flow chart of evidence theory of airspace conflict resolution

Table 1 Comparison of algorithm fusion results

Results	Proposition				
	A	*B*	*C*	Θ	ϕ
DS	0.00	1	0.00	0	0
Yager	0	0.0001	0	0.9999	0
Smets	0	0.0001	0	0	0.9999
Average	0.495	0.01	0.495	0	0

4 Conclusion

Based on the background of tactical airspace management, this paper discusses the application value of evidence theory in the conflict resolution of airspace use plans and achieves the information fusion of multiple priority orders by establishing the basic probability assignment derived from the priority score. Due to the diversity of rules of combination of evidence theory, there is no absolute unique formula for this kind of methods. Applicable rules of combination should be selected in the face of different decision-making scenarios. The feasibility of applying evidence theory in airspace management is discussed here, and the pros and cons of the rules of combination can be further extended in the future. In addition, there is the problem of focal element explosion in the evidence theory in this application background, which is the future research direction to adapt to the application background of large amount of data and complex scenes in the future.

References

1. He W, Wang J (2011) Military-civilian cooperation, military-civilian integration, jointly promoting the construction of the military-civilian integration security system. In: The 123rd session of China engineering science and technology forum—2011 National Defense Science, Technology and Industry Development Forum
2. Xin M (2017) Precise full time control of intelligent global planning—battlefield airspace control in the form of intelligent warfare. Command Inf Syst Technol 8(5):38–42
3. Mengke X (2015) Study on the basic problems and countermeasures of the flexible use of airspace. Sci Technol Innov 16:32–33
4. Yuan l, Zeng G, Wang W (2006) Trust evaluation model based on Dempster-Shafer evidence theory. J Wuhan Univ (Science Edition) 52(5):627–630
5. Liu J (2014) Research on Pattern Recognition Based on Fractal Dimension of Reliability Function. Southwest University
6. Zhang Y (2007) Application of Information Fusion Technology in Ship Main Engine Inspection. Shanghai Maritime University
7. Tai W, Ding J, Liu Y et al (2018) An improved evidence synthesis method and its application in fuzzy evaluation. Firepower Command Control 43(278)(5):69–73
8. Zhou Z (2009) Application Research of Evidence Theory in WSN Data Fusion. Central South University
9. Han D, Yang Y, Han C (2014) Research progress of DS evidence theory and related issues. Control Decision-Making (1)
10. Yang H (2006) General investigation information fusion based on DSMT. Journal 6 46(10):129–132
11. Yang Y (2006) Classification, evaluation criteria and application of evidential reasoning combination method. Northwestern Polytechnical University
12. Wen Q, Xiaodan W, Jindeng Z et al (2012) A modified combination rule of Dubois-Prade evidential reasoning. Control Decision-Making 27(1):139–142
13. Sun R (2012)Research on information fusion based on D-S evidence theory and its application in reliability data processing. University of Electronic Science and Technology
14. Zhai S (2014) Application of information fusion technology in fault diagnosis of generator. East China University of Science and Technology
15. Kennes R, Smets P (1990) Fast algorithms for Dempster-Shafer theory. In: International conference on information processing & management of uncertainty in knowledge-based systems. IEEE, New York
16. Wichramarathne TL, Premaratne K, Murthi MN (2011) Monte-Carlo approximations for Dempster-Shafer belief theoretic algorithms. In: International conference on information fusion
17. Yuefeng L, Dayou L (1995) Approximate calculation method in evidence theory. J Natural Sci Jilin Univ 1:28–32

A Learning Based Automated Algorithm Selection for Flexible Job-Shop Scheduling

Xinyu Wang, Xiaoxia Li(✉), Rongyin Zhu, and Zetao Lv

College of Informatics, Huazhong Agricultural University, Wuhan, China
lixiaoxia@mail.hzau.edu.cn

Abstract. Flexible job-shop scheduling is an important stage in manufacturing process. A good scheduling solution can significantly improve multiple machining performances. In this paper, a learning-based method is presented to achieve the automated algorithm selection for flexible job-shop scheduling. In the method, the historical scheduling data generated by the evolutionary algorithms are used to build two BP neural network models for each algorithm to predict its optimization result and running time under different parameter configurations. Based on the models, a multi-objective genetic algorithm is used to obtain the optimal parameters for each evolutionary algorithms, and then, the results of the comparison among the algorithms are used to achieve the final algorithm selection. Experiments on the benchmarks of flexible job-shop scheduling problem were performed to verify the proposed method, and promising results were achieved.

Keywords: Flexible Job-shop scheduling problem · Automatic algorithm selection · Parameter configuration · Multi-objective genetic algorithm

1 Introduction

Flexible Job-shop Scheduling Problem (FJSP) is an extension of classic Job-shop Scheduling Problem (JSP). It has been proven that multiple machining performances such as productivity, machining energy consumption and so on can be improved by a good solution for FJSP. It is known that FJSP is a complex NP-hard problem. A large number of evolutionary algorithms have been developed to optimize FJSP. Thus, it is important for the machining companies to select a suitable algorithm to obtain the optimal solution for FJSP.

In recent years, with the development and continuous improvement of evolutionary algorithms, its advantages in solving combinatorial optimization problems have become increasingly apparent. It has been proven that good optimization results for the combinatorial optimization problems can be obtained using the evolutionary algorithms. As thus, evolutionary algorithms have been employed to handle FJSP. The most commonly used evolutionary algorithms include Simulated Annealing Algorithm (SAA) [1, 2], Particle Swarm Optimization (PSO) [3, 4], Artificial Bee Colony Algorithm (ABC) [5, 6], Ant Colony Optimization (ACO) [7, 8], Genetic Algorithm (GA), [9, 10] etc.

Q. Liang et al. (eds.), *Artificial Intelligence in China*, Lecture Notes in Electrical Engineering 653, https://doi.org/10.1007/978-981-15-8599-9_33

However, each evolutionary algorithm has its own limitations. It has been indicated that different algorithms perform best on different types of problem instances—a phenomenon also known as performance complementarity in convergence, diversity and so on. Therefore, it is needed to choose the appropriate evolutionary algorithm for different FJSP instances.

Also, it has been observed that the parameters of the evolutionary algorithm influence the optimization results for FJSP. That is, the evolutionary algorithm with the same parameter configuration cannot perform the same algorithms with different parameter configuration on all problem instances. Hence, suitable algorithm parameters' selection is also needed for different FJSP instances to keep performances.

In order to handle FJSP with a suitable evolutionary algorithm, a learning based automated algorithm selection approach for FJSP is proposed in this paper. In the proposed approach, a BP neural network model is built to characterize the relationship between the algorithms' parameters, the FJSP's instances and optimization results. Based on the model, the parameters fitting the FJSP's instance can be obtained for each algorithm. The algorithm with the best result is recommended to be used to handle FJSP.

The remainder of this article is structured as follows. In Sect. 2, the related work of this article is formally introduced. Next, in Sect. 3, it is explained in detail how to select algorithms and parameters for FJSP instances, and test cases are given in Sect. 4. Finally, Sect. 5 summarizes this article and looks forward to future work.

2 Related Work

In order to implement the automatic algorithm selection system for FJSP, a lot of related work has been done. First, the advantages and disadvantages of different intelligent optimization algorithms on FJSP have been acquired from lots of papers, and specific implementations have been given. Then, the research about automatic algorithm selection in common mathematical problems and job-shop scheduling problems has been done.

Today, many intelligent optimization algorithms are applied to solve the flexible job-shop scheduling problem, each of which has its own excellence and shortage. Genetic algorithms have been known as one of the best algorithms for dealing with global problems due to their excellent global optimization capabilities and strong robustness. However, ordinary GA also has problems such as slow convergence, easy precocity and insufficient diversity of population. In order to overcome the above-mentioned shortcomings, many scholars have modified the GA to solve FJSP algorithmically. For example, Liu et al. [11] generated multiple initial populations and used the mutual exchange between populations to overcome the premature convergence of GA; Sun [12] proposed a strategy of mixing genetic algorithm with Tabu search to speed up the speed of convergence and improve the ability of local search. Ant colony optimization is a random heuristic method characterized by universal and reckless that has penetrated into the field of workshop scheduling in recent years. In spite of the fact that ACO started late, some achievements are made when it is applied to FJSP. For example, Silva [13] found through research that ACO is more suitable for dynamic scheduling problems, and Yu et al. [14] proposed the use of ACO with mutation processing, so that it is not easy to fall into local optimum. Although particle swarm optimization is featured with fast convergence speed and simple calculation, it is easy to fall into local optimum. Jia [15] did a lot of work on the application of PSO. He used chaos to adaptively optimize the parameters of the algorithm and proposed a hybrid PSO to improve the search ability. The simulated annealing algorithm has also been widely used in FJSP due to its high stability, and in recent years, improved versions have been continuously proposed. Jun et al. [16] introduced the particle coding methods in PSO to simulated annealing and used three different local search methods to construct the domain structure of individuals.

In the past ten years, several surveys on automated algorithm selection have been published. For example, Kotthoff's survey in 2014 presents an extensive, valuable guide to the automated algorithm selection and provides answers to several important questions, such as (i) what are the differences between static and dynamic portfolios, (ii) what should be selected (single solver, schedule, different candidate portfolios), (iii) what are the differences between online and offline selection, (iv) how should the costs for using algorithm portfolios to be considered, (v) which prediction type (classification, regression, etc.) is most promising when training an algorithm selector, and (vi) what are differences between static and dynamic, as well as low-level and high-level features. The survey of Munoz Acosta et al. in 2013 mainly focused on the algorithm selection of continuous optimization problems and introduced the characteristics of the instances in detail. In addition, in their follow-up survey, Munoz Acosta et al. provided further insights into the existing components of algorithm selection in the domain of continuous optimization. Over the years, remarkable achievements about algorithm selection techniques have been made in several research areas—especially for discrete combinatorial problems. However, due to the significant differences between various problems, not only the respective instance features, but also solvers and algorithm selectors vary considerably. Therefore, it is impossible to cover all work in this area. Taking travelling salesman problem as an example, several studies have shown that at least three TSP solvers, EAX, LKH and MAOS, define state-of-the-art performance on different kinds of TSP instances. Kerschke et al. extended the original. They considered additional types of TSP instances, feature sets and solvers and furthermore adopted more sophisticated machine learning techniques, including various feature selection strategies for constructing algorithm selectors. Of course, per-instance algorithm selection has shown to be effective on several other discrete problems—such as the Travelling Thief Problem (TTP) where Wagner et al. recently presented the first study of algorithm selection, along with a comprehensive collection of performance and feature data, Propositional Satisfiability Problem(SAT) where Xu et al. developed SATZILLA2012 showing outstanding performance in multiple propositional SAT competitions, the problem of Quantified Boolean Formulae (QBF) to which Pulina and TAcchella demonstrated the successful application of algorithm selection, Quadratic Assignment Problem(QAP) for which Smith-Miles presented her algorithm selector and so on. However, it is a pity that we haven't found any survey about algorithm selectors of FJSP. The application field of FJSP model is very extensive, for which we also need to develop the algorithm selector of FJSP.

If there is no algorithm selector, you can only obtain the algorithm with the best optimization effect of a certain FJSP instance through past experience and multiple attempts, which is very inefficient and totally unreliable. Providing suitable algorithms for job-shop scheduling can not only reduce production costs, but also improve production efficiency, which has great research significance.

3 The Methodology

3.1 The Whole Framework

For different JSP instances, our algorithm selector not only selects the most suitable algorithm from the algorithm library, but also configures the parameters of the algorithm. So far, GA, ABC and SAA have been implemented. First, each algorithm is let run with variable values of every parameter, and the running time and optimization effect are recorded. Since the running time and the optimization effect are reference standards for algorithm selection and parameters' configuration, two models will be trained for each algorithm which predicts the running time and optimization effect, respectively. Then using those two models as fitness functions, the Pareto optimality of each algorithm in the algorithm library is obtained through the multi-objective genetic algorithm, and the most suitable one is selected as the unique solution. Finally, comparing unique solutions of different algorithms, the most perfect recommendation is selected for the user.

3.2 The Learning Based Modelling for Algorithm Evaluation

The TensorFlow machine learning library is used to build the models mentioned above, which is a BP neural network model with an input layer, a hidden layer and an output layer. The inputs of each model include the number of jobs, the number of machines and the values of several algorithm parameters; the output is the running time or the optimization effect.

Convolutional neural network had already been tried but it turned out that the fitting effect is not ideal. It is worth mentioning that while training the model, the maximum-minimum normalization and exponential decay learning rate are used. Different inputs often have different dimensions and units, which will affect the results of data analysis. In order to eliminate the dimensional impact between indicators, some inputs needs to be normalized, such as the number of machines, the number of jobs and population size. It is also important to set the size of the learning rate, because the learning rate is too small, which will greatly reduce the convergence speed and increase the training time; too large a learning rate may cause the parameters to oscillate back and forth on both sides of the optimal solution.

3.3 The Detailed Steps of the Method

A selection range for each parameter of each algorithm is provided by us, and the optional values in the range are called empirical values. The empirical values are obtained by consulting other articles. When an empirical value is selected as a value of a parameter, the algorithm will have better performance. The algorithm is run with all parameters assigned to empirical values. In order to prevent accidents, each case is run ten times, and the running time as well as optimization effect is averaged.

The random initial value generated each time the algorithm runs is different. If the processing time is used to represent the performance of the algorithm, we cannot objectively and correctly reflect the degree of optimization of the algorithm, which is why the optimization effect is used to represent the performance of the algorithm. The initial random individual is denoted as *A*, and the optimal solution individual found by the algorithm is denoted as *B*. A_t and B_t refer to the respective processing times. The optimization effect is defined as follow.

$$|A_t - B_t|/A_t$$

Each algorithm must find the Pareto optimality through multi-objective genetic algorithm, and the solutions of Pareto optimal are not the only, for which the most suitable unique solution needs to be selected. Taking the processing time as the *y*-axis and the optimization effects the *x*-axis, draw all the points of the Pareto optimality and connect them to get a rising curve. Figure out the growth rate of each point compared to the previous one and find out the ten points with the smallest growth rate. Of these ten points, the point with the greatest optimization effect is selected as the unique solution.

4 Case Studies and Discussion

BP neural network models and multi-objective genetic algorithm are both completed in the Anaconda. Here, this algorithm selector is tested with an example where the number of jobs is 20 and the number of machines is 10. As shown in Figs. 1, 2 and 3, each algorithm can obtain a series of Pareto optimality through multi-objective genetic algorithm, among which a unique solution can be selected for each algorithm according to the method in the third paragraph of Sect. 3.3. The method of selecting the unique solution can be adjusted according to the actual needs of the user. The unique solution and the values of the three parameters for each algorithm have been shown in Table 1. In spite of the fact that the meanings of the parameters of the three algorithms are different, they are not the focus here so they will not be shown. According to Table 1, it is obvious that not only does GA has the best optimization effect, but also runs fastest, for which we should choose GA, and assigns the three parameters 0.050, 0.154 and 100, respectively.

Without this selector, users could only attempt every algorithm and try to set parameters to each group of empirical values, which is time-consuming and laborious.

5 Conclusions

Although the FJSP algorithm selector have been implemented, the research on FJSP automatic algorithm selection has just begun, which means that our algorithm selector still has a lot of room for improvement. Our algorithm selector can be improved in the following ways:

(1) Increase the number of algorithms. Add more intelligent optimization algorithms and build models in the same way. Too few algorithms make selectors limited. The selector integrates more algorithms to provide users with more choices.

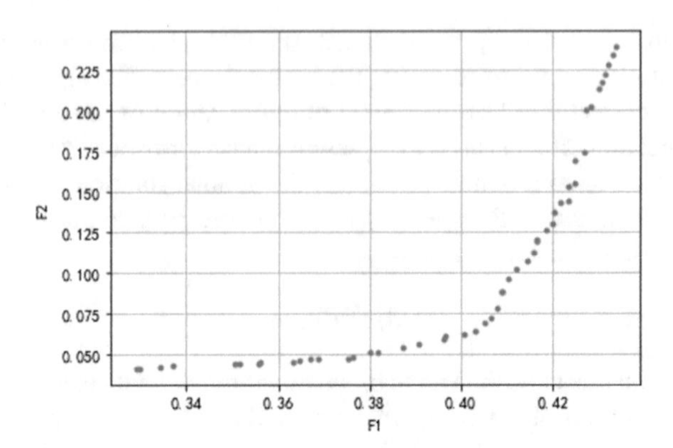

Fig. 1 Pareto optimality of GA

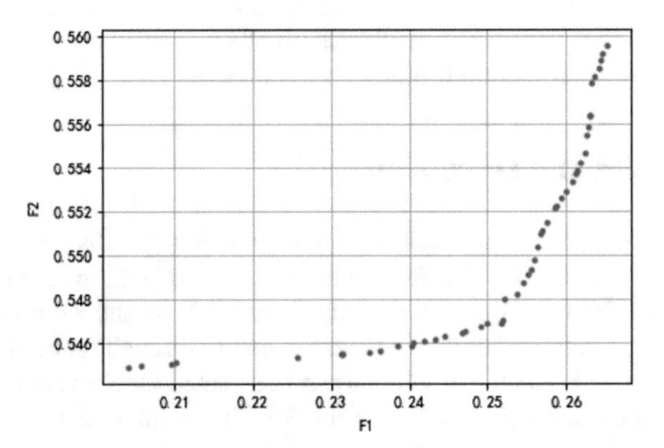

Fig. 2 Pareto optimality of SAA

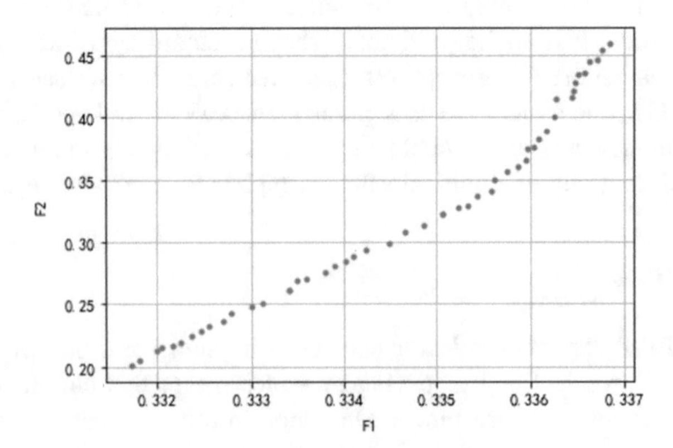

Fig. 3 Pareto optimality of ABC

(2) Find a model with better prediction effect. Try more updated neural network models to find models with better fitting results.

Table 1 Comparison studies of three algorithms

Algorithm	Optimization effect	Running time	$p1$	$p2$	$p3$
GA	0.414	0.116	0.050	0.154	100
SAA	0.255	0.549	927	338	0.980
SA	0.333	0.251	0.301	374	25

(3) Select the instance features with better generalization. At present, only the number of jobs and the number of machines are used as the characteristics of different examples. However, even if the number of jobs and the number of machines are the same, the number of processes of different jobs may be different and the processing time of different processes on different machines may be different. New features need to be found to summarize all of these factors, which can make the model fit better and make the selector's prediction accuracy better.

Acknowledgements. This research was supported by Natural Science Foundation of China (grant No. 61803169) and the Fundamental Research Funds for the Central Universities (grant No. 2662018JC029). The paper reflects only the authors' views, and the Union is not liable for any use that may be made of the information contained therein.

References

1. Seifollahi S, Bagirov A, Borzeshi EZ et al (2019) A simulated annealing-based maximum-margin clustering algorithm[J]. Comput Intell 35(1):23–41
2. Elmi A, Solimanpur M, Topaloglu S et al (2011) A simulated annealing algorithm for the job shop cell scheduling problem with intercellular moves and reentrant parts[J]. Comput Ind Eng 61(1):171–178
3. Kennedy J (2011) Particle swarm optimization[J]. Proc of 1995 IEEE Int Conf Neural Netw (Perth, Australia), Nov. 27–Dec. 4(8):1942–1948
4. Chrysopoulos A, Mitkas PA (2017) Customized time-of-use pricing for small-scale consumers using multi-objective particle swarm optimization[J]. Adv Building Energy Res 1–23
5. Lv L, Han L, Fan T et al (2016) Artificial bee colony algorithm with accelerating convergence[J]. Int J Wireless Mobile Comput 10(1):76
6. Gao WF, Huang LL, Liu SY et al (2017) Artificial bee colony algorithm based on information learning[J]. IEEE Trans Cybern 45(12):2827–2839
7. Randall M, Lewis A (2015) A parallel implementation of ant colony optimization[J]. J Parallel Distrib Comput 62(9):1421–1432
8. Dorigo M, Blum C (2005) Ant colony optimization theory: a survey[J]. Theor Comp Sci 344(2):243–278
9. Morris GM, Goodsell DS, Halliday RS et al (2015) Automated docking using a Lamarckian genetic algorithm and an empirical binding free energy function[J]. J Comput Chem 19(14):1639–1662

10. Akkan C, Gulcu A (2018) A bi-criteria hybrid genetic algorithm with robustness objective for the course timetabling problem[J]. Comput Oper Res 9:22–32
11. 刘爱军,杨育,邢青松 (2011) 多目标模糊柔性车间调度中的 多种群遗传算法[J]. 计算机集成制造系统 17(9):1954–1961
12. 孙艳丰 (2006) 基于遗传算法和禁忌搜索算法的混合策略及应用[J]. 北京工业大学学报 32(3):258–262
13. Silva CA, Sousa JMC, Runkler TA (2008) Rescheduling and optimization of logistic processes using GA and ACO[J]. Eng Appl Artif Intell 21(3):343–352
14. Yan-Hai H, Jun-Qi Y, Fei-Fan Y et al (2005) Flow shop rescheduling problem under rush orders[J]. J Zhejiang Univ 6(10):1040–1046
15. 贾兆红, 陈华平, 孙耀晖 (2008) 多目标粒子群优化算法在柔性车间调度中的应用[J]. 小型微型计算机系统 5(5):885–889
16. 李俊, 刘志雄, 张煜, 贺晶晶 (2015) 柔性作业车间调度优化的改进模拟退火算法[J]. 武汉科技大学学报 2:111–116

Controller's Workload and Sector Capacity Assessment Based on 4D Track

Changcheng Li[1(✉)] and Yuxin Hu[2(✉)]

[1] Nanjing University of Aeronautics and Astronautics, Nanjing, China
lichangcheng_1993@163.com
[2] The 28th Research Institute of China Electronics TEchnology Group Corporation, Nanjing, China
helenhu_yuxin@163.com

Abstract. Firstly, this paper analyzes the shortcomings of the current controller's workload calculation and sector capacity assessment methods. On this basis, attempts to establish a set of 4D track-based workload calculation and capacity assessment method. Secondly, the overview and pre-processing logic of 4D track data are discussed, so that the workload calculation can be performed later. Then, it divides into three parts: controller monitor workload, coordination workload, conflict detection and resolution workload and discusses their calculation methods and processing logic in detail. Finally, taking five area sectors in the East China Information Region as an example, the results of workload calculation and capacity assessment based on 4D track are given. As the published capacity values of the sectors, the effectiveness of the model and algorithm is verified.

Keywords: 4D track · Controller's workload · Sector capacity assessment

1 Introduction

Air traffic management mainly includes capacity and flow management, and sector capacity is the most important component of capacity management. Sector capacity is closely related to the workload of controllers. Common methods for controller workload calculation and sector capacity assessment include computer simulation, controller simulator, and machine learning [1–5]. The disadvantage of the method based on computer simulation is that it is difficult to simulate the real situation. The method based on the controller simulator can simulate the real situation, but the cost is extremely high (requires multiple controllers to spend a lot of time on the simulator to test). The generalization ability of the evaluation model based on machine learning is limited to the same type of sector, and the generalization ability between different types of sectors is weak.

Based on 4D trajectory data, this paper proposes a set of controller workload calculation and sector assessment methods based on 4D track.

Q. Liang et al. (eds.), *Artificial Intelligence in China*, Lecture Notes
in Electrical Engineering 653, https://doi.org/10.1007/978-981-15-8599-9_34

2 4D Track

2.1 4D Track Data Overview

4D track data includes: flight ID, callsign, aircraft type, longitude, latitude, altitude, speed, flight direction, airborne equipment, whether located on the ground, registration number, monitor time, insert time, departure airport, arrival airport, and VFID.

The flight ID is basically null. The callsign is the callsign of the aircraft corresponding to the 4D track. The longitude, latitude, altitude, speed, and course are the flight status of the aircraft corresponding to the 4D track at the instant of monitor time, including geographical position (longitude and latitude), altitude, speed, and course. Airborne equipment is basically null. Whether it is on the ground is basically false. The registration number is the registration number of the aircraft corresponding to the 4D track. The monitor time is the time point of the aircraft flight status corresponding to the 4D track. The insertion time is the time point when the 4D track data is inserted into the data system. The departure airport and arrival airport are the departure airport and arrival airport of the aircraft corresponding to the 4D track. The VFID is a special code for the flight between a certain departure and arrival of the aircraft corresponding to the 4D trajectory, which is unique to a continuous 4D trajectory.

Among them, the content related to controller workload calculation includes: longitude, latitude, altitude, monitor time, and VFID. Longitude and latitude are used to determine whether the 4D track is within the polygonal plane structure of a sector. Height is used to determine whether the 4D track is within the height of a sector. VFID is used to determine a continuous series of 4D trajectory data is selected from the 4D tracks. The monitor time is used to filter out a continuous series of 4D trajectory and then sorted according to time from before to after to form a series of 4D chronological order track data.

2.2 4D Track Data Pre-processing

The data input is the original 4D track data, and the data output is the 4D track data with sector marks.

Step 1: Read sector structure data and original 4D track data.
Step 2: Take a piece of original 4D track data and determine whether the track is within the height of the sector structure. If yes, go to step 3; if not, go to step 2.
Step 3: Determine whether the track is within the sector structure plane and use the "Ray Method" to determine the method. If yes, go to step 4; if not, go to step 2.
Step 4: Mark the track data with the sector name and group the processed track data according to VFID, and sort them according to the monitor time.

Because the number of 4D tracks is very large, it is very important to increase the processing speed when programming. The two judgments in step 2 and step 3 are height range judgment and plane range judgment, respectively, and their sequence must be to judge the height first and then judge the plane structure range. Only then can the program process the fastest speed. The reason is: To judge the altitude, you only need to compare the relationship between the height of a 4D track and the minimum height, and the highest height of the sector, which is much faster than the "Ray method" to determine the relationship between the 4D track and a polygonal area. By prioritizing the altitude, tens of millions or even hundreds of millions of 4D track data will filter out a large proportion of the data, only the 4D track whose altitude is between the minimum and maximum altitude of the sector enters the next screen for the scope of planar structure.

After filtering according to the height range and the plane polygon structure range, if the piece of 4D track data is within a certain sector range in the sector structure data, the name of the sector is added to the piece of 4D track as a mark. The average daily global track data is about 30 million pieces. After traversing all the track data, all the tracks within the sector range involved in the sector structure data are marked with the name of the sector and saved. Other track data will be omitted.

3 Monitor Workload

The monitor workload of the controller is the workload generated by the controller monitor the aircraft operating in the sector he controls. In a unit time period, the greater the number of aircraft operating in the sector, the greater the controller's monitor workload. Assuming that there are two identical sectors, the number of aircraft operating in the sector is equal in a unit time period, then the longer the average passing time of the aircraft, the greater the monitor workload of the controller of the sector. This paper uses the scale that the aircraft generates 2 s of monitor workload for the controller of the sector every 1 min of flight in the sector to measure the monitor workload of the controller. Because this paper selects the real 4D track for 24 h as the data source of the monitor workload, the data volume of the 4D trajectory data is large, and it is not easy to process. Therefore, the calculation process and processing logic of the controller's monitor workload based on the 4D track are mainly explained.

For the calculation of the controller's monitor workload, the data input is the track data with sector marks and the data output is the controller's monitor workload.

Step 1: Take a series of 4D track data from the track data with sector marks.
Step 2: Use interpolation to complete the string of track data to 1 s track data.
Step 3: Loop to find the first track with the target sector label and the last track in the target sector with the target sector label.
Step 4: Obtain the flight time of the track in the target sector according to the monitor time difference between the two pieces of track data in step 3.
Step 5: Obtain the corresponding target sector monitor workload according to the flight time of step 4.
Step 6: Repeat the process from step 1 to step 5 until all the track data with sector marks are traversed.

During the completion process in step 2, the flight ID, callsign, aircraft type, airborne equipment, whether it is on the ground, registration number, insert time, departure airport, arrival airport, and VFID remain unchanged. Longitude, latitude, altitude, speed, heading according to all kinds of information of the two pieces of original track data before and after using interpolation method to complete, the monitor time is the time point of the track data to be completed; the specific calculation process is shown in Eq. 1.1 to Eq. 1.5

$$T_{\text{Now}}^{\text{Longitude}} = \frac{T_{\text{After}}^{\text{Longitude}} - T_{\text{Before}}^{\text{Longitude}}}{T_{\text{After}}^{\text{MonitorTime}} - T_{\text{Before}}^{\text{MonitorTime}}} * \left(T_{\text{Now}}^{\text{MonitorTime}} - T_{\text{Before}}^{\text{MonitorTime}}\right) \tag{1.1}$$

$$T_{\text{Now}}^{\text{Latitude}} = \frac{T_{\text{After}}^{\text{Latitude}} - T_{\text{Before}}^{\text{Longitude}}}{T_{\text{After}}^{\text{MonitorTime}} - T_{\text{Before}}^{\text{MonitorTime}}} * \left(T_{\text{Now}}^{\text{MonitorTime}} - T_{\text{Before}}^{\text{MonitorTime}}\right) \tag{1.2}$$

$$T_{\text{Now}}^{\text{Height}} = \frac{T_{\text{After}}^{\text{Height}} - T_{\text{Before}}^{\text{Height}}}{T_{\text{After}}^{\text{MonitorTime}} - T_{\text{Before}}^{\text{MonitorTime}}} * \left(T_{\text{Now}}^{\text{MonitorTime}} - T_{\text{Before}}^{\text{MonitorTime}}\right) \tag{1.3}$$

$$T_{\text{Now}}^{\text{Speed}} = \frac{T_{\text{After}}^{\text{Speed}} - T_{\text{Before}}^{\text{Speed}}}{T_{\text{After}}^{\text{MonitorTime}} - T_{\text{Before}}^{\text{MonitorTime}}} * \left(T_{\text{Now}}^{\text{MonitorTime}} - T_{\text{Before}}^{\text{MonitorTime}}\right) \tag{1.4}$$

$$T_{\text{Now}}^{\text{Direction}} = \frac{T_{\text{After}}^{\text{Direction}} - T_{\text{Before}}^{\text{Direction}}}{T_{\text{After}}^{\text{MonitorTime}} - T_{\text{Before}}^{\text{MonitorTime}}} * \left(T_{\text{Now}}^{\text{MonitorTime}} - T_{\text{Before}}^{\text{MonitorTime}}\right) \tag{1.5}$$

4 Coordination Workload

The coordination workload of the controller is the workload that the controller is responsible for when the aircraft enters the sector or aircraft command he controls, and the aircraft leaves the sector he controls, or the aircraft command is handed over to other sectors. The greater the number of aircraft entering/leaving the sector in a unit time period, the greater the coordination workload of the controller. Since the aircraft entering the sector changes from the uncontrolled state of the sector controller to the controlled state of the sector controller, the controller needs not only pure radio call operations, but also observation and other coordination operations. Since the aircraft leaving the sector changes from the control status of the sector controller to the non-control status of the sector controller, the controller only needs to perform radio call operation with the pilot of the aircraft to complete all operations. In this paper, each time an aircraft enters the sector; it generates 5 s of observation workload, 8 s of radio call workload, and 2 s of other coordination workload on the controller of the sector. A total of 15 s of aircraft moves into the coordination workload; the sector generates 8 s of radio call workload for the sector controller; and a total of 8 s of aircraft moves out of the coordination workload. As shown in Table 1, the coordination workload of the controller is measured on this scale. Since this paper selects the real 4D track for 24 consecutive hours as the data source for coordination load, the data volume of 4D track data is large and difficult to process. Some calculation processes and processing logic similar to the controller monitor workload are already elaborated in Sect. 3, so this section focuses on the calculation process and processing logic specific to controller coordination workload.

The controller's coordination workload extraction is based on the track data with sector marks as the data source. The calculation process and logic using interpolation and the like are similar to the extraction process of the controller's monitor workload described in Sect. 3. The difference is that the output result is the controller's coordination

Table 1 Aircraft moved into/moved out coordination workload

Coordination workload types	The workload of a single aircraft moved into (s)	The workload of a single aircraft moved out (s)
Observation workload	5	0
Radio call workload	8	8
Other workload	2	0
Total	15	8

workload, and the relevant part of the output process is also different. Find out the first track with the target sector label and the last track with the target sector label in a series of track data with sector mark accuracy of 1 s (between the two pieces of track). Other track data is also in the target sector. The monitor time of the first track with the target sector label is the time point of the coordination workload moving into the sector, and the monitor time of the last track with the target sector label is the time point of the coordination workload moving out of the sector.

5 Controller Conflict Detection and Resolution

The controller's conflict detection and resolution workload are the controller's identification of potential conflicts between aircraft operating in the sector he controls and instructing the relevant aircraft to change the flight status to resolve the potential conflicts and the resulting workload. Within a unit time period, the more potential conflicts occur between the aircrafts in the sector, the greater the conflict detection and resolution workload of the controller.

The scale of conflict detection and resolution workload adopted in this paper is shown in Table 2. According to the different horizontal and vertical states of each two aircrafts, set the workload duration for single conflict detection and the workload duration for single conflict resolution. Since this paper selects the real 4D track for 24 consecutive hours as the data source for conflict detection and resolution workload, the data volume of 4D track data is large and difficult to process. Some calculation processes and processing logic similar to the controller monitor workload and coordination workload are already elaborated in Sects. 3 and 4, so this section focuses on the calculation process and processing logic specific to controller's conflict detection and resolution workload.

This paper is based on China's airspace management regulations and focuses on workload calculation and capacity assessment for sectors above mid and low altitude. The minimum horizontal safety interval for Chinese area sectors is 10 km, and the minimum vertical safety interval is 300 m.

Various types of conflict detection require the controller to provide a common conflict detection distance; that is, if the vertical height difference between the two aircraft is less than the minimum vertical safety distance of 300 m, and the horizontal distance between the two aircraft is less than the minimum conflict detection distance; the controller will conduct a conflict detection behavior, which will produce corresponding conflict

Table 2 Conflict detection and resolution workload

Horizontal state of aircraft	Vertical state of aircraft	Single conflict detection workload (s)	Single conflict resolution workload (s)
Same course	Both aircrafts flying horizontally	10	60
	One aircraft flying horizontally, the other aircraft climbing/descending	15	
	Both aircrafts climbing/descending	20	
Cross course	Both aircrafts flying horizontally	20	
	One aircraft flying horizontally, the other aircraft climbing/descending	25	
	Both aircrafts climbing/descending	30	
Opposite course	Both aircrafts flying horizontally	20	
	One aircraft flying horizontally, the other aircraft climbing/descending	25	
	Both aircrafts climbing/descending	30	

detection workload. Generally speaking, the conflict detection distance of the same course is smaller than that of the cross-course, and the conflict detection distance of the cross-course is smaller than that of the opposite course. The reason is that when the performance of the aircraft is not much different from the altitude, and the speed of the aircraft is mostly within a certain range. The relative convergence speed of two aircraft with the same course is generally less than the relative convergence speed of the two aircraft with cross-course. The relative convergence speed of two aircraft in cross-course is generally less than the relative convergence speed of two aircraft in opposite course. The conflict detection distance used in this paper is shown in Table 3.

All kinds of conflict resolution also need to communicate with the controller to establish a conflict resolution rule base. Because the 4D track data are all real, within a certain airspace, it may not happen for many days that the vertical separation of the two aircraft is less than the minimum vertical safety separation, and their horizontal

Table 3 Conflict detection distance

Aircraft horizontal state	Conflict detection distance (km)
Same course	30
Cross-course	40
Opposite course	50

separation is also less than the minimum horizontal safety separation. The conflict resolution rule base includes three major categories: changing speed, changing altitude, and changing course. To change the speed, the controller needs to provide a speed change threshold and the corresponding time period; that is, if the speed change of an aircraft involved in conflict detection exceeds the speed change threshold within the time period, the controller is presumed to use the speed change method to resolve conflicts. The logic for changing altitude and course is the same as the logic for changing speed described above.

5.1 Conflict Detection Database

The data input is the track data with sector marks, and the data output is the controller's conflict detection database.

Step 1: Take two series of 4D track data from the track data with sector marks. The index of the second series of track data in the track data with sector marks must be greater than the index of the first series of track data, because the conflict detection involves two aircraft and one conflict should not be recorded twice.
Step 2: Determine whether there is overlap between the start time and end time of the two series of 4D track. If there is no overlap, repeat step 1; if there is overlap, go to step 3.
Step 3: Use interpolation to supplement the two series of 4D track data with two series of sector-marked trajectory data with an accuracy of 1 s.
Step 4: Take two pieces of track data of the same monitor time in chronological order. The purpose is to traverse the overlapping flight time of each aircraft every 1 s as a data source for subsequent potential conflict judgment.
Step 5: Determine whether the two pieces of track data are less than the minimum vertical safety interval. If the two pieces of track data are less than the minimum vertical safety interval, proceed to step 6; if the two pieces of track data are not less than the minimum vertical safety interval, repeat step 4.
Step 6: Determine whether the two pieces of track data are less than the minimum horizontal safety interval. If the two pieces of track data are less than the minimum horizontal safety interval, proceed to step 7; if the two pieces of track data are not less than the minimum horizontal safety interval, repeat step 4.
Step 7: Mark the two pieces of track as potential conflicts detected by the conflict, and aggregate to form a track database with conflict detection marks. At this time, the track database with conflict detection marks has not been fully processed, and the operation in step 8 is performed.

Step 8: Starting from the track database with conflict detection marks, grouped according to the VFID of the two pieces of track, grouped according to the continuity of the monitor time, and sorted according to the monitor time of the two pieces of track.
Step 9: Traverse all track data with sector marks to form a conflict detection database.

5.2 Conflict Detection Workload

The data input is the conflict detection database, and the data output is the controller's conflict detection workload.

Step 1: Take a conflict detection record from the conflict detection database. Each conflict detection record includes two series of 4D track information with accuracy of 1 s, and their start time and end time are consistent.
Step 2: Take two pieces of track data with the same monitor time in the order of monitor time from before to after, and take two pieces of track data with the same monitor time according to the monitor time + 1 s. If the monitor time + 1 s is greater than or equal to the conflict detection start time and less than the conflict detection end time, proceed to step 3; if the monitor time + 1 s is equal to the conflict detection end time, repeat step 1.
Step 3: Compare the first two pieces of track taken in step 2 to find the absolute value of their heading angle difference. If the absolute value of the angle difference is greater than or equal to 0° and less than 45°, the horizontal state corresponding to the two aircrafts is the same course. If the absolute value of the angle difference is greater than or equal to 45 and less than 135°, the horizontal state corresponding to the two aircrafts is the cross-course. If the absolute value of the angle difference is greater than or equal to 135° and less than or equal to 180°, the horizontal state of the corresponding two aircraft at this time is the opposite course.
Step 4: Compare the four pieces of track taken in step 2 and calculate the expected climb/descent rates for the two aircrafts. The altitude difference between the two pieces of tracks divided by the monitor time difference between the two pieces of track is the expected climb/descent rate of the corresponding aircraft. If the expected climb/descent rate is greater than 0, the aircraft is in the climb state within the time range. If the expected climb/descent rate is less than 0, the aircraft is in the descent state during the time range. If the expected climb/descent rate is equal to 0, the aircraft is in the level flight state in the time range.
Step 5: According to the horizontal state and vertical state, combined with the conflict detection classification in Table 3, to determine the type of conflict detection. It should be noted that a conflict detection event usually includes a lot of track data, and the conflict type of each set of track data may not be exactly the same. For example, two aircraft may initially be a conflict detection type with opposite course, but because the controller directs the aircraft to change the heading to resolve the conflict, and the conflict detection type of the opposite course may change to a detection type of cross-course before the conflict is completely released. In this paper, the type of conflict detection of the first group of track is used to determine the type of conflict detection event.
Step 6: According to the type of conflict detection obtained in step 5, combined with the workload generated by various types of single conflict detection in Table 2, the corresponding controller conflict detection workload can be obtained. As shown

in Formula 1.6. $W^{\mathrm{ConflictDetection}}$ indicates the total conflict detection workload, $w_i^{\mathrm{ConflictDetection}}$ indicates the workload generated by a single type of conflict detection, and $\mathrm{Count}_i^{\mathrm{ConflictDetection}}$ indicates the number of times the type of conflict detection occurs.

$$W^{\mathrm{ConflictDetection}} = \sum w_i^{\mathrm{ConflictDetection}} \times \mathrm{Count}_i^{\mathrm{ConflictDetection}} \tag{1.6}$$

5.3 Conflict Resolution Workload

The data input is the conflict detection database, and the data output is the controller's conflict resolution workload. The processing logic is basically similar to the conflict detection workload; the difference is that the conflict resolution workload of the controller is calculated according to various types of single conflict resolution in Table 2, as shown in Eq. 1.7. $W^{\mathrm{ConflictResolution}}$ indicates the total conflict resolution workload, $w_i^{\mathrm{ConflictResolution}}$ indicates the workload generated by a single type conflict resolution, and $\mathrm{Count}_i^{\mathrm{ConflictResolution}}$ indicates the number of times the type conflict resolution occurs.

$$W^{\mathrm{ConflictResolution}} = \sum w_i^{\mathrm{ConflictResolution}} \times \mathrm{Count}_i^{\mathrm{ConflictResolution}} \tag{1.7}$$

The controller's conflict detection and resolution workload are the sum of the controller's conflict detection workload and the controller's conflict resolution load, as shown in Eq. 1.8. $W^{\mathrm{ConflictAll}}$ indicates the total conflict detection and resolution workload of the controller.

$$W^{\mathrm{ConflictAll}} = W^{\mathrm{ConflictDetection}} + W^{\mathrm{ConflictResolution}} \tag{1.8}$$

6 Example Verification of Controller Workload and Sector Capacity Assessment

According to the controller workload evaluation system established above, the track data of a certain week in 2019 is used to calculate the controller workload of the five Hefei area sectors in the East China Information Region. The fitting relationship between the hourly number of flights of the five area sectors and the hourly workload threshold is shown in Fig. 1.

When the hourly workload threshold reaches 70 (that is, 2520 s per hour), the hourly capacity of each sector can be obtained according to the fitting curve of each sector in Fig. 1 as shown in Table 4. The ZSOFAR01 sector has a capacity of 53 flights/hour, the ZSOFAR02 sector has a capacity of 33 flights/hour, the ZSOFAR03 sector has a capacity of 29 flights/hour, the ZSOFAR04 sector has a capacity of 43 flights/hour, and the ZSOFAR05 sector has a capacity of 43 sorties/hour.

After communicating with the Hefei area sector control facility, the capacity evaluation result is in line with the actual operating experience of each sector, and the results have been adopted as a reference value for the evaluation capacity of the Hefei regional sectors.

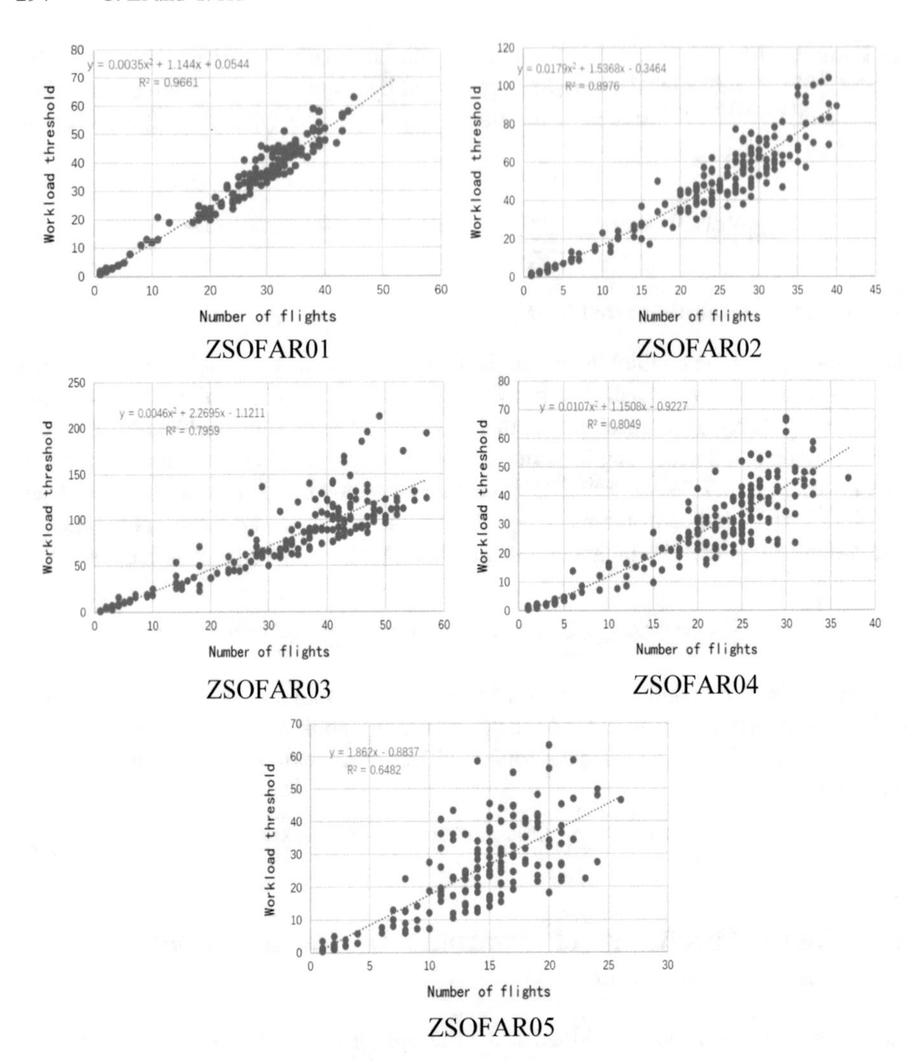

Fig. 1 Fitting figure of sector capacity assessment

Table 4 Capacity assessment results of five area sectors in Hefei, East China Information Region

Sector number	Sector name	Hourly capacity (orders)
ZSOFAR01	Area sector01	53
ZSOFAR02	Area sector02	33
ZSOFAR03	Area sector03	29
ZSOFAR04	Area sector04	43
ZSOFAR05	Area sector05	43

References

1. d'Engelbronner JG, Borst C, Ellerbroek J, Van Paassen MM, Mulder M (2015) Solution-space–based analysis of dynamic air traffic controller workload. J Aircraft 52(4):1146–1160
2. Gianazza D (2017, June). Learning air traffic controller workload from past sector operations
3. Corver SC, Unger D, Grote G (2016) Predicting air traffic controller workload: trajectory uncertainty as the moderator of the indirect effect of traffic density on controller workload through traffic conflict. Hum Factors 58(4):560–573
4. Rodríguez S, Sánchez L, López P, Cañas JJ (2015, September) Pupillometry to assess air traffic controller workload through the mental workload model. In: Proceedings of the 5th international conference on application and theory of automation in command and control systems, pp 95–104
5. Wang H, Gong D, Wen R (2015, May) Air traffic controllers workload forecasting method based on neural network. In: The 27th Chinese control and decision conference (2015 CCDC), pp 2460–2463. IEEE, New York

Classification of Tea Pests Based on Automatic Machine Learning

Heng Zhou[1,2], Fuchuan Ni[1,2(✉)], Ziyan Wang[1,2], Fang Zheng[1,2], and Na Yao[1,2,3]

[1] College of Informatics, Huazhong Agricultural University, Wuhan, Hubei 430070, China
zhouheng0918@qq.com, {fcni_cn,zhengfang}@mail.hzau.edu.cn, wzy_unique@163.com, ykn1103@163.com

[2] Hubei Engineering Technology Research Center of Agricultural Big Data, Wuhan, Hubei 430070, China

[3] College of Information Engineering, Tarim University, Alaer, Xinjiang 843300, China

Abstract. Tea pests, as one of the main threats in the tea gardens, cause considerable losses to the tea industry every year. In order to ensure the quality of the tea production process, it is of great importance to detect tea pests timely and use correct treatment to control tea pests. Recent developments in deep learning have dramatically improved the accuracy of classification. Especially the network learned by automatic machine learning (AutoML) technology, even surpasses the existing human-designed network in performance. In this paper, we present a deep-learning-based approach for tea pests classification. Our aim is to find out which deep neural network is more suitable for our task. Therefore, we consider four deep-learning architectures to build our classification system: PNASNet-5, ResNet-50, ResNeXt-101, and Inception-ResNet-V2, where the first one is obtained by AutoML and the latter three are human-designed mainstream neural networks. We train and test these deep neural networks end-to-end on our Tea Pests Dataset. Experimental results show that our proposed system can effectively classify 22 common types of tea pests and reach a 5.0609% top-1 error on the test set.

Keywords: Deep learning · Automatic machine learning · Tea pests detection

1 Introduction

Tea pests, as one of the main threats in the tea gardens, cause considerable losses to the tea industry every year. In order to ensure the quality of the tea production process, it is of great importance to detect tea pests timely and use correct treatment to control tea pests. In the actual management of tea gardens, due to the lack of understanding of the morphological characteristics and active period of tea pests, it is difficult to use pesticides in the best control period, resulting in unsatisfactory pest control effects and pesticide residues on tea leaves. In addition, the lack of expert experience in the recognition of tea pests and related control knowledge makes it difficult for tea garden managers to have proper use of pesticides, which in turn leads to pesticide residues exceeding national

Q. Liang et al. (eds.), *Artificial Intelligence in China*, Lecture Notes
in Electrical Engineering 653, https://doi.org/10.1007/978-981-15-8599-9_35

standards, directly reducing the sanitary quality of tea. Therefore, the rapid and accurate recognition of tea pests and the timely adoption of effective control measures has become a challenge in the tea industry.

We are currently witnessing the rapid development of convolutional neural networks which has achieved impressive results in image classification. Recent advances in hardware technology have allowed the network to become deeper and wider, such as ResNet [1] and Inception-ResNet [2] which also achieved excellent accuracy on ImageNet [3] challenge. However, these excellent network structures are carefully designed by experts, which also limits the promotion of convolutional neural networks to some extent. Recently, researchers are increasingly interested in the automation of both machine learning and deep learning, which inevitably promotes the development of neural network optimization and automation. It turns out that the choice of structure of the neural network is very significant, and the development of many deep neural networks also comes from the direct improvement of the network structure. Because of the high demand for computational resources and expert knowledge, automating the design process of neural networks can help researchers and practitioners benefit from deep learning faster [4].

2 Relate Works

Before deep learning is applied to the computer vision field, some handcrafted methods of extracting image features are usually performed in image recognition, such as Histogram of Oriented Gradients (HOG) [5] and Scale-Invariant Feature Transform (SIFT) [6], which are usually combined with Adaptive Boosting (AdaBoost) [7] or Support Vector Machines (SVM) [8]. Based on features of high dimension and redundancy of hyperspectral imaging data, Ziyi [9] constructed a stack sparse auto-encoder (SSAE) for the effective representation of sparsity information at different infestation stages of pests. Since the feature extraction of the traditional recognition method is usually done manually, the generalization ability is weak. It is difficult to extract the high-level features of an image, so a large number of experiments are required to select appropriate feature parameters. Moreover, it can only deal with small-scale category recognition, and its performance is poor when in the task of a wide range of category recognition.

Compared with traditional handcrafted methods of extracting features, deep neural networks not only have excellent performance but also can be trained to an end-to-end model, which does not need to be separated into multiple processes. Juan et al. [10] improved the Residual Network (ResNet) by increasing the number of convolution layers and the number of channels and optimizing the hyperparameters using Bayesian methods. Compared to the traditional methods (Bayesian neural network, convolutional neural network, and support vector machines), it improved the accuracy of classification by an average of 9.6%. In addition, Bian et al. [11] also did similar work on the image recognition of stored-gain insects. AlexNet [12] was used to recognize stored-gain insects, which accuracy reached 97.62%. Qiang et al. [13] designed a real-time monitoring system for stored-grain insects through deep learning and the Internet of Things (IoT) technologies. The pest trap captures the image and uploads the image to the server which uses VGGNet [14] to recognize the pests and returns the result to the mobile client. It provides a new idea for the intelligent detection system of agricultural pests.

3 Methodology

3.1 Tea Pests Classification

As mentioned above, due to the increase in computational resources, the convolutional neural network has flourished, and the traditional handcrafted methods of feature extraction have been gradually replaced by it because of excellent performance. Within the last few years, researchers in the machine learning community make an effort to automate the process of machine learning or deep learning and have achieved amazing results. We propose to use the convolutional neural network that obtained by progressive neural architecture search which is an automatic machine learning method to build a tea pests recognization system and choose the current popular neural networks (ResNet and its variant models, such as ResNeXt [15] and Inception-ResNet) for comparison.

PNASNet: PNASNet proposed by Liu et al. [16], evolves from NASNet proposed by Zoph et al. [17] and explores the architectures by progressively increasing the number of block in a cell in the same search space with NASNet, where the number of blocks and cells are hyperparameters. They use RNN with LSTM [18] to build a predictor for the searching process. The overall process of progressive neural architecture search is as follows. The initial candidate set is composed of cells with one block. These cells will be trained and use the validation results to update the predictor. Each cell is expanded in the next iteration to generate a new set of candidate cells, then the predictor scores them and chooses the top - K cells to update the candidate set. Moreover, train and validate candidate cells on the validation set to update the predictor.

ResNet: Different levels of convolutional layers in neural networks can effectively extract features of different levels in the image. The deeper convolutional layer can extract more abstract features and semantic information from images. However, improving network performance by simply stacking the convolutional layer may not work. This inconsistency may be due to the vanishing gradient and exploding gradient. Although it can be solved by using Batch Normalization [19], it still faces the degradation problem. He et al. [1] proposed a Residual Learning method to solve the above problems and built a Residual Network (ResNet). The core of the residual network is the identity mapping, which makes network optimization easier by using shortcut a connection to add their outputs to the outputs of the stacked layers.

ResNeXt: Xie et al. [15] improved on ResNet and proposed ResNeXt. The innovation of ResNeXt lies in aggregated transformations. The block of ResNeXt is composed of multiple branches of the same topology at the same level, replacing the block of the 3 convolution layers in ResNet, which improves the accuracy of the network without increasing the parameter amount.

Inception-ResNet: Since the residual network has a significant effect in accelerating the network training process and improving network performance (preventing vanishing/exploding gradient), the Inception [20] architecture is good at extracting sparse and non-sparse features in the same level layer. Szegedy et al. [2] were inspired by this and combined ResNet and Inception and proposed Inception-ResNet architecture.

3.2 Data Augmentation

Compared with traditional methods, although deep neural network shows powerful performance, it requires a large amount of data to feed, otherwise, it will encounter overfitting problems. We have used several methods to reduce the impact of overfittings, such as resize, crop, flipping, rotation, contrast enhancement, brightness enhancement, and saturation enhancement.

4 Experimental Results

4.1 Tea Pests Datasets

The datasets we used in our research were constructed from Internet images that were obtained by a search engine, such as Baidu, Google, Flickr, Fresheye, etc. 22 types of tea pests are included in the dataset. Considering that some pests' larval and adult stages are very different in appearance, their larvae and adults are divided into two separate categories. So, the final dataset contains 29 categories of pests, as shown in Table 1.

We manually de-duplicated and assign labels to every image, after that, we got a total of 1752 images of the original dataset. The images downloaded from the Internet cannot guarantee the balance of different categories. Meanwhile, the size of the original dataset is so small that it would cause overfitting problem. In order to avoid overfitting and data imbalance, we use data augmentation methods to expand the original dataset, which expanded to include 34,938 images and each category was large and balance enough. Moreover, 80% of the samples were randomly selected from each category of image to construct a training set, and another 20% of the samples were used to construct a validation set. Figure 1 shows a partial sample of the training set. The input images also need to be preprocessed during each iteration of training to enhance the generalization capabilities of the model. For example, random crop, random flip, and random color transformations (brightness, contrast, and saturation).

4.2 Quantitative Results

We used 4 NVIDIA Tesla V100 GPUs to train our convolutional neural network. In all experiments, we used Stochastic Gradient Descent (SGD) optimizer, momentum of 0.9, weight decay of 10^{-4}, initial learning rate of 0.001, and batch size of 16. The total training epochs is 60. Input image size is 224×224.

Figure 2a–c shows the loss curves and the top-1 error and top-5 error of different models during training. From the loss curves in Fig. 2, it can be seen that all four models can converge well, but PNASNet-5 is faster than ResNet-50, ResNeXt101 32x4d, and Inception-ResNet-V2. In addition, PNANet-5 achieved a minimum Top-1 error of 5.06% on the validation set which reflects better performance in classification tasks. The Top-1 classification errors of ResNet-50, ResNeXt101 32x4d and Inception-ResNet-V2 are: 11.03%, 6.65%, and 2.79%, respectively, as shown in Table 2.

We validate the model on the validation set after each training epoch. Figure 2d–f shows the loss curves and top-1 and top-5 of the different models on the validation set after each epoch. The curves of the validation set show a similar trend to the curves of

Table 1 The overview of tea pests dataset

Category	After data augmentation	Training set	Validation set
Zeuzera coffeae nietner imago	1248	999	249
Zeuzera coffeae nietner larva	1204	964	240
Mesosa perplexa	1170	936	234
Jacobiasca formosana	1170	936	234
Geisha distinctissima	1180	944	249
Euproctis pseudoconspersa imago	1204	964	240
Euproctis pseudoconspersa larva	1224	980	244
Arctornis alba	1230	984	246
Measuring worm	1136	909	227
Tortricidae imago	1170	936	234
Tortricidae larva	1224	980	244
Thosea sinensis imago	1224	980	244
Thosea sinensis larva	1200	960	240
Monema flavescens imago	1184	948	236
Monema flavescens larva	1224	980	244
Hypomeces squamosus Fabricius	1224	980	244
Setora postornata imago	1188	951	237
Setora postornata larva	1224	980	244
Agrilus sp.	1232	986	246
Amata germana	1368	1095	273
Helopeltis sp.	1136	909	227
Tanaoctenia haliaria	1190	952	238
Atractomorpha	1260	1008	252
Euricania ocellus	1164	932	232
Tridrepana unispina	1224	980	244
Black Vine Weevi	1160	928	232
Caloptilia theivora	1224	980	244
Prasa consocia Walker imago	1176	941	235
Prasa consocia Walker larva	1176	941	235
Total	34,938	27,963	6975

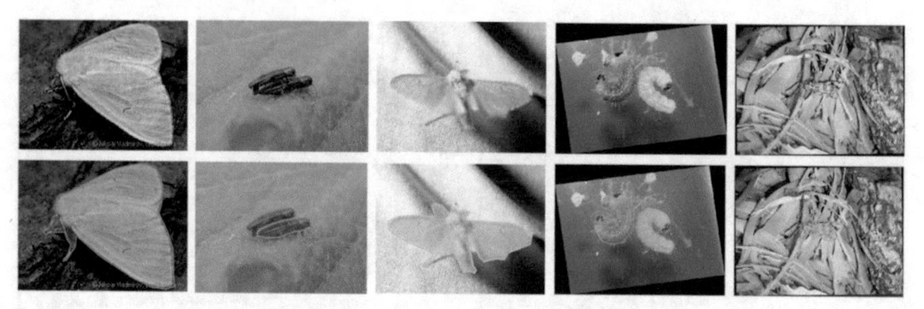

Fig. 1 An overview of the training set. The first line of the images are the training images, and the second line of the images are the visualization of segmentation annotations

training set. We can see that PNASNet not only converges faster during training, but also achieves the smallest top-1 error and top-5 error on the verification set, which indicates PNASNet obtained by progressive neural architecture search has better performance than handcrafted models of ResNet, ResNeXt, and Inception-ResNet. Table 3 shows the classification accuracy of the four models in each category, from which it can be clearly seen that the PNASNet model achieves the highest recognition accuracy in most categories.

5 Conclusions

Due to the wide variety of tea pests, their shape and behavior will change with the environment, and the recognization model needs to be improved according to the actual situation. In this situation, constructing a pest recognization system based on deep learning and optimizing the system from the aspect of the algorithm is the development trend of automatic pest identification in the complex environment of the field. The evaluation of traditional deep neural network relies on experts and their experience. Recently, automatic machine learning technology has attracted the interest of researchers with its amazing performance. Its rapid development has provided a new development direction for machine learning and deep learning community. We propose a convolutional neural network based on PNASNet which is obtained by automatic machine learning method called progressive neural architecture search to recognize tea pests. Experimental results show that PNASNet has higher accuracy than a traditional handcrafted neural network, and its top-1 error reaches 5.0609%. Future studies will focus on the combination of deep learning and sensors to realize real-time monitoring of tea pests automatically and intelligently. Moreover, we can integrate data and build an agricultural big data platform through IoT technology. We believe that an Internet-based agricultural information service platform will be established to realize smart agriculture.

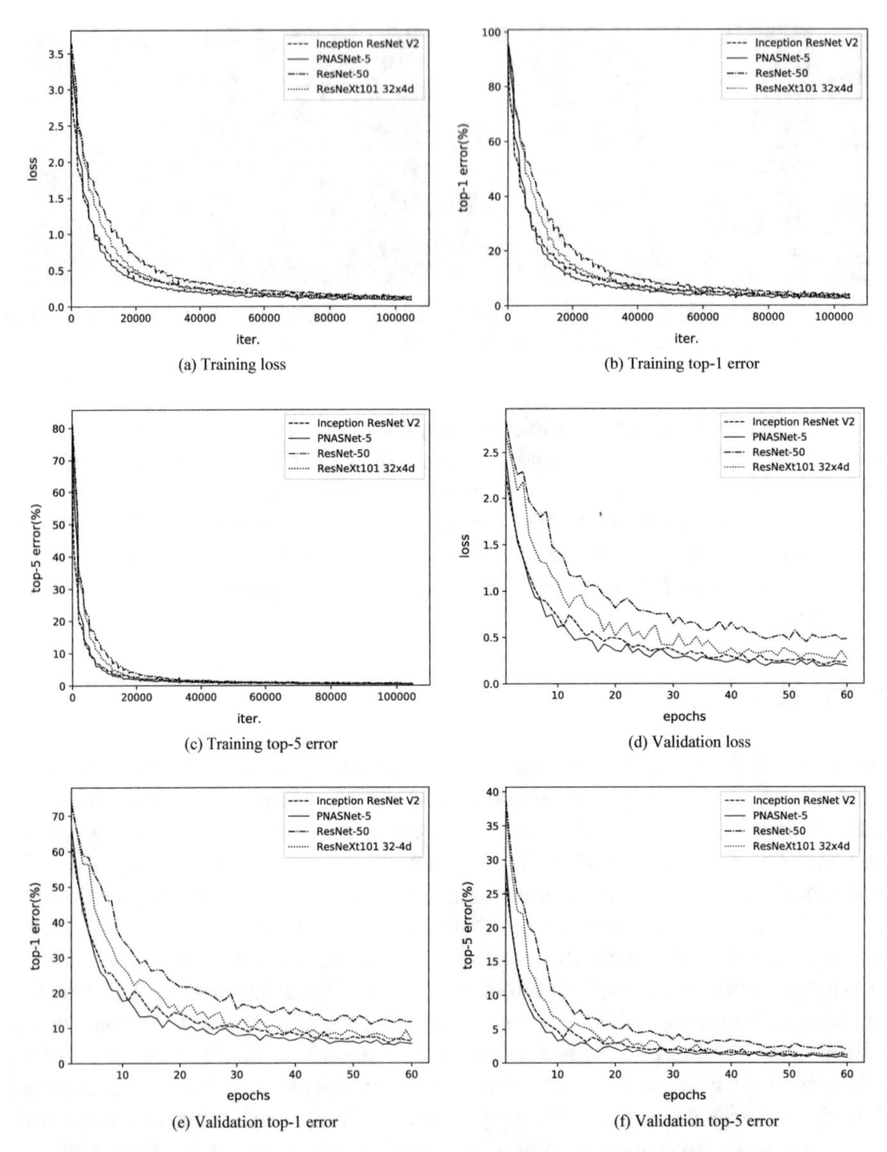

(a) Training loss

(b) Training top-1 error

(c) Training top-5 error

(d) Validation loss

(e) Validation top-1 error

(f) Validation top-5 error

Fig. 2 Loss curves and top-1 and top-5 of the different models on the validation set

Table 2 The error of different classification models in tea tree pest classification tasks

Architecture	Top-1 error (%)	Top-5 error (%)
ResNet-50	11.0251	1.9642
ResNeXt101 32x4d	6.6523	0.9749
Inception-ResNet-V2	5.7921	0.9892
PNASNet-5	5.0609	0.8029

Table 3 Classification accuracy of various categories of tea pests

Category	ResNet-50	ResNeXt-101 32x4d	Inception-ResNet V2	PNASNet-5
Agrilus sp.	0.96	0.93	0.96	**0.98**
Amata germana	0.9	0.9	0.95	**0.96**
Arctornis alba	0.95	0.98	0.97	**1**
Atractomorpha	0.84	**0.93**	0.92	0.88
Black Vine weevil	0.97	0.98	0.98	**0.99**
Caloptilia theivora	0.9	0.95	0.97	**0.97**
Euproctis pseudoconspersa imago	0.9	0.93	0.96	**0.98**
Euproctis pseudoconspersa larva	0.96	0.96	0.95	**0.99**
Euricania ocellus	0.9	0.93	**0.97**	0.95
Geisha distinctissima	0.92	0.97	0.96	**0.99**
Helopeltis sp.	0.96	0.96	0.97	**0.98**
Hypomeces squamosus Fabricius	0.9	0.94	0.97	**0.97**
Jacobiasca formosana	0.9	0.96	0.97	**0.97**
Measuring worm	0.85	0.91	0.96	**0.98**
Mesosa perplexa	0.95	0.98	**0.99**	0.97
Monema flavescens imago	0.91	0.92	0.92	**0.98**
Monema flavescens larva	0.96	**0.99**	0.96	0.96
Prasa consocia Walker imago	0.89	0.95	0.93	**0.97**
Prasa consocia Walker larva	0.96	0.97	0.97	**0.99**
Setora postornata imago	0.94	0.97	0.97	**0.99**
Setora postornata larva	0.97	0.97	**0.98**	0.97

(*continued*)

Table 3 (*continued*)

Category	ResNet-50	ResNeXt-101 32x4d	Inception-ResNet V2	PNASNet-5
Tanaoctenia haliaria	0.96	0.97	0.98	**0.98**
Thosea sinensis imago	0.92	0.95	0.94	**0.95**
Thosea sinensis larva	0.94	0.97	0.97	**0.98**
Tortricidae imago	**0.96**	0.94	0.91	0.95
Tortricidae larva	0.93	0.95	0.97	**0.97**
Tridrepana unispina	0.92	0.95	0.96	**0.97**
Zeuzera coffeae nietner imago	0.97	0.98	0.98	**0.98**
Zeuzera coffeae nietner larva	0.96	0.95	0.97	**0.97**

Acknowledgements. This work is supported by the Key Special Project National Key R&D Program of China (2018YFC1604000), Natural Science Foundation of Hubei Province of China (Program No. 2019CFB547), and the Fundamental Research Funds for the Central Universities, Huazhong Agricultural University (No.: 2662017PY119, 2662020XXPY05).

References

1. He KM et al (2016) Deep residual learning for image recognition. In: 2016 IEEE conference on computer vision and pattern recognition (CVPR), pp 770–778
2. Szegedy C et al (2017) Inception-v4, inception-resnet and the impact of residual connections on learning. In: Thirty-First AAAI conference on artificial intelligence
3. Deng J et al (2009) Imagenet: A large-scale hierarchical image database. In: 2009 IEEE conference on computer vision and pattern recognition. IEEE, New York
4. Wistuba M, Rawat A, Pedapati T (2019) A survey on neural architecture search. arXiv e-prints
5. Dalal N, Triggs B (2005) Histograms of oriented gradients for human detection
6. Lowe DG (2004) Distinctive image features from scale-invariant keypoints. Int J Comput Vision 60(2):91–110
7. Schapire RE (1999) A brief introduction to boosting. In: Proceedings of the sixteenth international joint conference on artificial intelligence (IJCAI-99), vols. 1 & 2, pp 1401–1406
8. Cortes C, Vapnik V (1995) Support-vector networks. Mach Learn 20(3):273–297
9. Ziyi L (2017) Detection of agricultural pest insects based on imaging and spectral feature analysis. Zhejiang University
10. Juan C et al (2019) Pest recognition based on improved residual network. Trans Chin Society Agric Mach 50(05):187–195
11. Bian G, Yu-hua Z, Tong Z (2018) Application of convolutional neural network in image recognition of stored grain insects. Sci Tech Cereals Oils Foods 26(06):73–76

12. Krizhevsky A, Sutskever I, Hinton GE (2012) Imagenet classification with deep convolutional neural networks. In: Advances in neural information processing systems
13. Qiang L, Ruilan H, Yi Z (2019) Real-time monitoring and prewarning system for grain storehouse pests based on deep learning. J Jiangsu Univ (Natural Science Edition) 40(02):203–208
14. Simonyan K, Zisserman A (2014) Very deep convolutional networks for large-scale image recognition. arXiv preprint arXiv:1409.1556
15. Xie SN et al (2017) Aggregated residual transformations for deep neural networks. In: 30th IEEE conference on computer vision and pattern recognition (CVPR 2017), pp 5987–5995
16. Liu C et al (2018) Progressive neural architecture search. In: Proceedings of the European conference on computer vision (ECCV)
17. Zoph B et al (2018) Learning transferable architectures for scalable image recognition. In: 2018 IEEE/CVF conference on computer vision and pattern recognition (CVPR), pp 8697–8710
18. Hochreiter S, Schmidhuber J (1997) Long short-term memory. Neural Comput 9(8):1735–1780
19. Ioffe S, Szegedy C (2015) Batch normalization: accelerating deep network training by reducing internal covariate shift. arXiv preprint arXiv:1502.03167
20. Szegedy C et al (2015) Going deeper with convolutions. In: 2015 IEEE conference on computer vision and pattern recognition (CVPR), pp 1–9

Thunderstorm Service and Decision Support Technology Based on Composite Reflectivity Information

Yao Shan(✉), Yuxin Hu, and Yungang Tian

State Key Laboratory of Air Traffic Management System and Technology, Nanjing 210007, China
shanyao923@163.com

Abstract. Thunderstorm is a great threat to civil aviation flight safety. The composite reflectivity data can directly reflect the intensity of thunderstorm and used as a means of thunderstorm prediction and analysis in aviation meteorology. First, the applications of convection prediction products and their aided decision making are investigated. Also, the processing of composite reflectivity data and the generation algorithm of hazardous weather avoidance field are researched. Finally, this paper designs a civil thunderstorm information processing system and publishes the thunderstorm through web service. When the user inquires, he can obtain the thunderstorm danger area information during the time period, and display it on the situation map, providing the decision-maker with auxiliary information.

Keywords: Composite reflectivity · Thunderstorm service · Decision support · Weather avoidance field

The appearance of thunderstorms, often accompanied by gusts of wind and heavy precipitation, is extreme weather that threatens the safety of civil aviation flights. In this context, in addition to the high requirements for the accuracy and timeliness of thunderstorm meteorological services, it puts forward urgent needs for the intensification of information resources, the integration of multiple information sources, and the decision-making assistance capabilities of meteorological services. The composite reflectivity strength is often used as a means of thunderstorm forecast analysis in aviation [1]. Traditional civil aviation thunderstorm forecast a wide range of products, available from the different data sources to provide thunderstorm prediction results on different time scales, etc. However, at the same time, there are problems such as large data volume and inconsistent format, and failure to directly access the user subsystem, which brings great inconvenience to users.

As the demand for the integrated operation of aeronautical meteorology and air traffic control becomes more and more urgent, how to realize the integration and interaction of meteorological information and control has become a problem that experts in the industry think deeply and actively explore. Taking thunderstorm forecast products as an example, in order to create a convenient and intuitive platform for users, there is an urgent need for a civil aviation thunderstorm service and auxiliary decision-making system, that

Q. Liang et al. (eds.), *Artificial Intelligence in China*, Lecture Notes
in Electrical Engineering 653, https://doi.org/10.1007/978-981-15-8599-9_36

is, real-time analysis of thunderstorm forecast product data, an intuitive query, display platform, and to provide decision-makers with reference to assist in decision-making.

1 Research Status at Home and Abroad

1.1 Research Status Abroad

In the field of aeronautical meteorological information services, scholars from various countries have conducted many studies and gradually applied them to the first-line production scenarios of civil aviation. Representative thunderstorms products include American Airlines Meteorological Center (Aviation Weather Center, AWC collaborative convection weather prediction products) released (Collaborative Convective Forecast Product, CCFP) [2, 3]. In the previous test, AWC generated two daily forecast products on the impact of convection on aviation: the "aeronautical meteorological impact" and "out-of-probability" forecast products. The "Aeronautical Meteorological Impact" product mainly describes the characteristics of convective weather, while the "over-probability" product focuses on forecasting the 30% and 60% probability areas of the combined reflectivity factor exceeding 40 dBz and the echo top height exceeding 37,000 feet MSL (mean sea level). In the later period, AWC officially renamed the product as CCFP to forecast convective weather within 8 h [4], the system shown in Fig. 1a. The product mainly serves the national traffic management center, the route control center, etc.

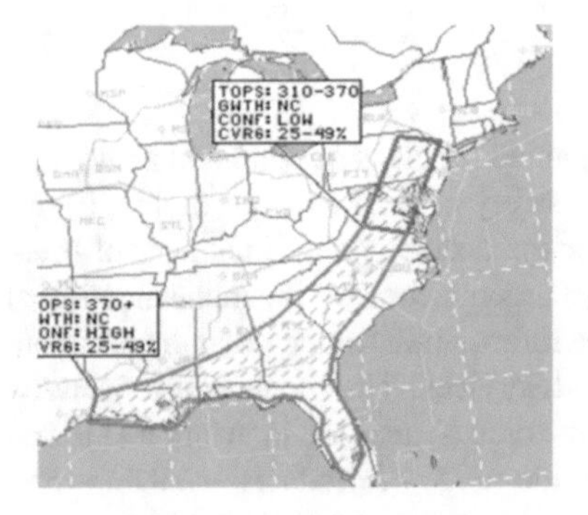

(a) US CCFP system

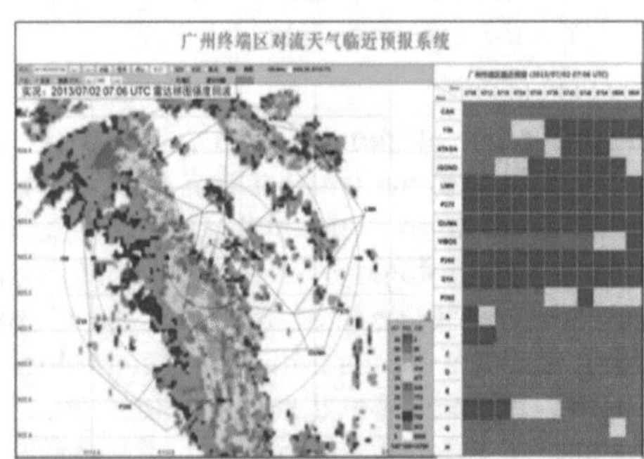

(b) Convective weather nowcasting system in Guangzhou terminal area

Fig. 1 Convective weather forecast system in domestic and international aviation weather

1.2 Initial Domestic Exploration

A series of domestic explorations have also been made on the impact of convective weather on the auxiliary decision-making of air traffic management. South ATMB from their own land meteorological feature field of view, the development of Guangzhou terminal convective weather nowcasting system [5], and put into use in the relevant regulatory authorities. The forecast system for convection within the scope of the terminal region of Guangzhou activities forecast, and gradually to extend the effective time of 2 h, the system shown in Fig. 1b. And the system can now measure the degree of influence of convective weather on each key point. Based on the study of the system on the South ATMB for convective weather, a study on the impact of the Pearl River Delta region and airspace capacity, select 15 typical convective weather day from March to August, on the flight actually about to fly through the normal convective clouds where group analysis, weather, and extracts feature parameters, using a correlation between the two methods of Gaussian classification, avoid establishing convective zone (Weather Avoidance Field, the WAF) mode type, and with minimal cutting maximum flow technique [6], to calculate the capacity value of the terminal area. At present, the subject is still in the research stage and has not been put into use.

2 Composite Reflectivity Data Processing Technology

2.1 Composite Reflectivity Data Preprocessing

Composite reflectivity can reflect the intensity and location of precipitation and convective weather more clearly. The composite reflectivity data usually includes the original data file and its supporting description file. The data file is the combined reflectivity binary data within the forecast range and time period. The description file records the latitude and longitude of the center point of the radar, the scan radius, the scan time, and the time interval of a single scan. Using GrADS software, and binding profile maybe of composite reflectivity preliminary analysis data.

2.2 Contour Extraction Technology

After the two-dimensional array of combined reflectivity is obtained, the contour is extracted by this system. The contour map is implemented using a rule-based grid tracking method [7]. The basic principle is: starting from the boundary of the grid area, using linear interpolation to traverse each grid point, Trace each contour line, get the coordinate value of the contour line on the edge of the grid it passes through, record the coordinates in the dynamic array, and finally connect these points to get a smooth and continuous contour line information.

The specific method is: first, find an equivalent point from the edge of the drawing grid area or the edge of the internal grid, and use the position of this point as the starting point of the contour; then, from this point, judge The coordinates of an equivalent point until the next equivalent point falls on the boundary of the grid area, or coincides with the starting point. At this point, the tracking of a contour is completed. Store all the contour information to be tracked in the file, which can be used for subsequent situation display and calculation of dangerous weather avoidance zone.

3 Dangerous Weather Avoidance Zone Calculation Technology

Through the above basic combined reflectance data processing technology, the real-time processing and situation display functions of the combined reflectivity of the meteorological radar are realized, which can be used as a basic thunderstorm forecast product for users such as control and flow. However, the display of the combined reflectivity on the situation map is based on the multi-layer overlay of the reflectance intensity, which makes it difficult for non-meteorological professionals to consult and understand. Therefore, this paper proposes a concept of dangerous weather avoidance zone, which expands and merges the area with a large combined reflectance intensity and provides it to the user as a reference data for user decision-making.

The calculation idea of the dangerous weather avoidance zone can be summarized as setting the dangerous weather reflectance threshold, expanding the reflectance contours greater than or equal to the threshold by a certain distance, and fusing the polygons obtained after the expansion, and after fusion. The formed polygon boundary is regarded as the dangerous weather avoidance zone of the reflectance threshold.

Specific embodiments may be designed as follows: First of all, the contour of 35 dBz exceeding the preset threshold is externally expanded. The external expansion distance refers to the actual thunderstorm flying distance and is selected to be 20 km. That is, the 20 km area outside the area where the reflectivity exceeds 35 dBz still is not safe for flight. Then, the areas covered by the contour lines after the outer expansion are merged, that is, all the areas where there may be dangerous weather are merged to obtain the information of the dangerous weather avoidance zone.

The extended fusion thunderstorms boundary contour geometry calculation, respectively, open-source library Geometry of the buffer and the union function, which functions as a schematic diagram in Fig. 2. The Buffer function realizes the function of expanding the internal polygon in (a) graph with a fixed width, while the union function realizes the function of fusing the two polygons in (b) graph to obtain the outer edge.

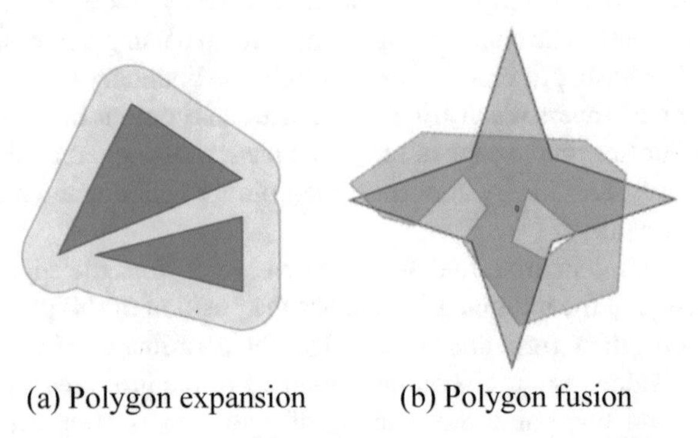

(a) Polygon expansion (b) Polygon fusion

Fig. 2 Schematic diagram of polygon expansion and fusion

When the number of polygons is large, polygon expansion and fusion takes a long time, which does not meet the timeliness requirements of the system for meteorological processing. Therefore, the above algorithm is improved to speed up the calculation of dangerous weather avoidance zones. A plurality of dangerous weather reflectivity threshold contour outline configuration of a multi-polygon (Multi-Polygon) as a reference, the other reflectivity profile contour polygons (Polygon) therewith a comparison, if the multi-polygon is within the coverage area, the polygon is added to it; otherwise, it is discarded. The resulting multi-polygon extended and fusion bonding, to give a final contour avoid hazardous weather information area, can effectively save the time required for calculation.

4 Application System Design

For civil air traffic management, aviation demand access to meteorological information, and decision support recommendations, based on the study of the above techniques, design weather information processing, services, and decision support systems, system architecture as Fig. 3 shown. The system includes three functional modules combined reflectivity data processing, thunderstorm information service, and thunderstorm information situation display and auxiliary decision-making.

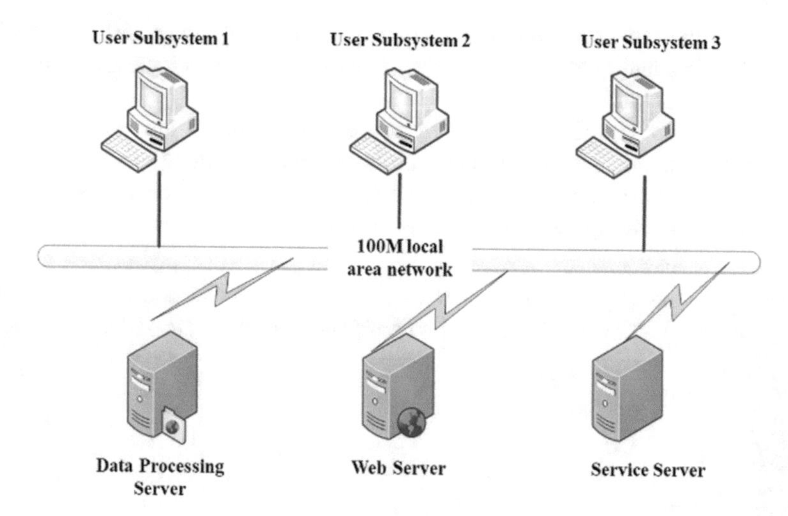

Fig. 3 System architecture diagram

Data flow diagram of the system is shown in Fig. 4. After the combined raw reflectivity data is connected by the Meteorological Center, the thunderstorm data processing module processes the data to obtain the reflectivity grid data under a single time slice, and extracts the contour information in the grid data, which is stored in the memory structure, and send to the topic corresponding to thunderstorm forecast information through active mq [8]. In addition, the thunderstorm data processing module also calculated the results of contour expansion and fusion and sent it to the theme of dangerous weather avoidance zone with a new interface.

In the thunderstorm information service module, the system analyzes the received thunderstorm data and converts the thunderstorm information into standardized XML

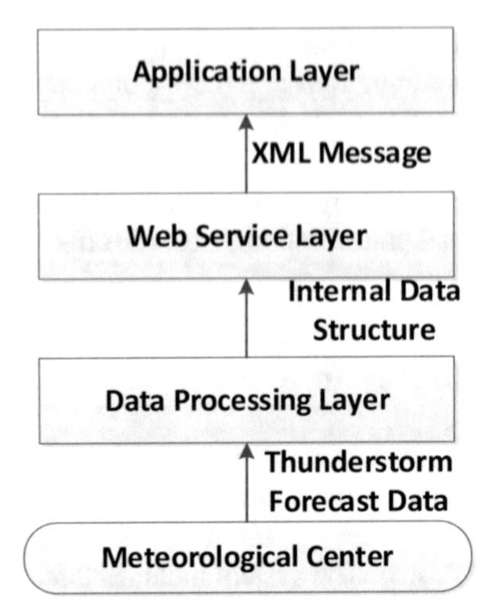

Fig. 4 Data flow diagram

messages. At the same time, the module manages data according to different regions, times and topics, and provides sub-user systems with query and subscription service interfaces based on different topics.

Thunderstorms information service website achieve as Fig. 5. Select one flight plan, you can query the thunderstorm forecast information during flight, and draw it on the situation map. Users can access the site to drag on the timeline, see the thunderstorms forecast, and avoid dangerous weather zone, as Fig. 6 shown.

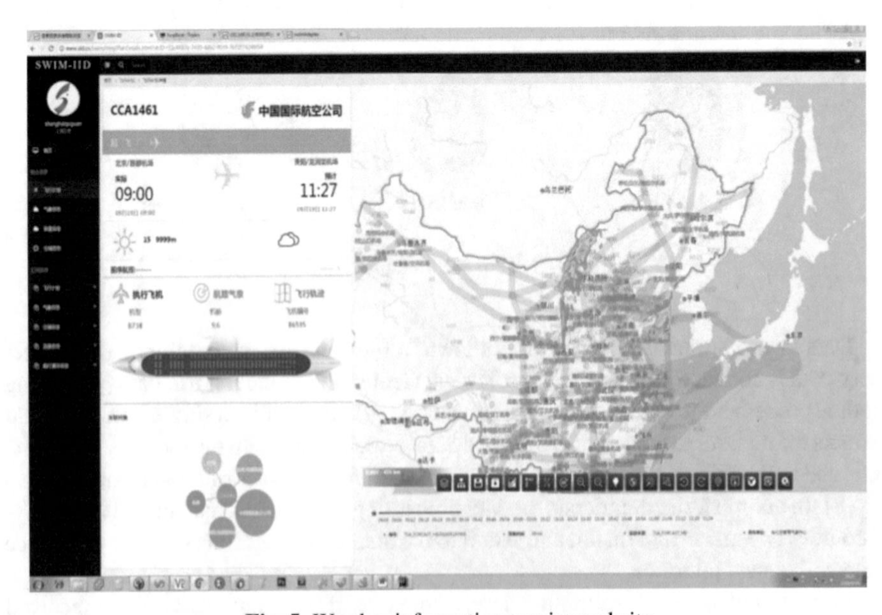

Fig. 5 Weather information service website

Fig. 6 Dangerous weather avoidance zone

Thunderstorm information situation display and auxiliary decision-making are mainly reflected in the application layer. According to their respective needs, each subsystem of the user queries or subscribes to the thunderstorm information on different topics according to time period, region, etc. to obtain the reflectivity and equivalent value returned by the application server. After parsing the XML message, the data such as line coordinates is displayed on the situation map.

5 Conclusion

In response to the dangerous extreme weather phenomenon of a thunderstorm, this paper mainly introduces a thunderstorm service and decision-making technology based on comprehensive reflectance data. In this paper, the domestic and international research status of the integrated operation of aeronautical meteorology and control is first described, and the processing method of composite reflectivity data is studied. On this basis, a method for calculating the dangerous weather avoidance zone is proposed, which expands the area with composite reflectivity greater than 30 dBz by 20 km and merges it. Finally, this paper designs and implements a set of thunderstorm service and auxiliary decision-making system based on combined reflectivity.

This article is designed composite reflectivity treatment options with easy and intuitive, information integration, it will provide a source of information to assist decision-makers of air traffic management. However, other convective weather information has not been considered comprehensively, the integration of multiple weather information has not been achieved, and the calculation scheme of the dangerous weather avoidance zone is too simple, without reference to the actual operation case, there is still room for improvement in the future.

Acknowledgements. This study was supported by National Key Research and Development Program of China (2018YFE0208700).

References

1. Jianhua Z, Zhongfeng Z, Weifang Z et al (2011) Aviation meteorological business. Meteorological Press, Beijing
2. Huberdeau M, Gentry J (2004) Use of the collaborative convective forecast product in the air traffic control strategic planning process. J Air Traffic Control 2:9–14
3. Hudson HR, Foss FP (2002) The collaborative convective forecast product from the aviation weather center's perspective. In: 10th conference on aviation, range, and aerospace meteorology/13th conference on applied climatology, Portland, OR
4. Yongguang Z, Ming X, Jue TZ (2015) US NOAA test platform and spring forecasts test summary. Meteorology 41(05):598–612
5. Gang W (2012) Preliminary study on airport terminal convective weather nowcasting. Guangdong Meteorological Society. Guangdong Provincial Meteorological Society 2012 Annual Conference Abstract Proceedings. Guangdong Meteorological Institute; Guangdong Provincial Science and Technology Association of scientific communications, 2012: 1
6. Rong X, Zheng C, Bo K, Ying P (2018) A preliminary study on the impact of convective weather on the airspace capacity of the Pearl River Delta region. Air Transp Bus (04): 66–68+72
7. Feng Y, Zhipei C (2011) Kuo novelty, strict construction steel. Based on VC++ track and contour filling algorithm. Industry Automation 30(04): 81–84
8. Caide X (2016) Design and implementation of civil aviation meteorological information service system. Shaanxi Meteorol 5:29–31

Research on Intrusion Detection Method Based on PGoogLeNet-IDS Model

Min Sun(✉), Xue Hao, and Wenbin Li

School of Computer and Information Technology, Shanxi University, Taiyuan 030006, China
476957266@qq.com

Abstract. The current network environment generally presents a situation of high latitude and a massive amount of information. Intrusion Detection Systems not only need to improve detection rate but also enhance detection speed. Therefore, this paper applies the PGoogLeNet-IDS model to the Intrusion Detection System, which can effectively balance the relationship between detection rate and detection speed. First, the PGooglLeNet-IDS need to replace the Inception structure of GoogLeNet with the SE_DSC structure. Second, this model uses a cross-layer connection to the network layer of the model, which can reduce the loss of features in the training process. Finally, this model uses the Focus Loss Function to the model to solve the uneven distribution of sample data. This paper uses the NSL-KDD benchmark data set to verify the performance of the model. The experimental results show that PGoogLeNet IDS has been significantly improved in various performance evaluation indicators, and the training time of the model is also significantly shortened. The model proposed in this paper can be effectively detected in the current complex network environment.

Keywords: GoogLeNet · Deeply separable convolution · Across the layer connection · SENet · Focus loss function · Intrusion detection · NSL-KDD

1 Introduction

The rapid development of the Internet has provided convenience and services for all aspects of human life. At the same time, information flooding has also brought about various events that seriously threaten network security, such as virus intrusion, malicious attacks, and information tampering. Various hackers made illegal intrusions into the network, which caused important information of individuals, enterprises and the country to be leaked, and also caused huge losses. Therefore, how to effectively defend network security has become the focus of current research. At present, the mainstream defense measures contain Network Firewall, Data Encryption, Access Control Technology, Virtual Network Technology, and Intrusion Detection Technology. The first three are passive defense methods, while the latter two are active defense methods. Compared with passive defense, the active defense can not only locate and track external attacks more accurately but also provide better defense effects against internal attacks. When

Q. Liang et al. (eds.), *Artificial Intelligence in China*, Lecture Notes
in Electrical Engineering 653, https://doi.org/10.1007/978-981-15-8599-9_37

it comes to the Intrusion Detection System [1] has become a popular active defense method.

In essence, the Intrusion Detection System is a classification problem of distinguishing normal data flow and abnormal attack categories on the network data flow. The machine learning algorithm is the mainstream algorithm to deal with the classification problem. In the early days of the development of intrusion detection systems, shallow learning algorithms [2–4] has widely used in this field. However, with the rapid development of the network, this way of manually extracting features still has some limitations when dealing with massive non-linear problems. For example, the detection process is slow, and the detection result is not accurate enough. With the prevalence of deep learning, many scholars apply models of deep learning algorithms to Intrusion Detection Systems, and they provide a new research direction for the development of Intrusion Detection Technology. Yin et al. [5] proposed to apply a recurrent neural network to the Intrusion Detection System. They use the NSL-KDD data set to evaluate the performance of the system. Xiao et al. [6] proposed an intrusion detection system that combines self-encoding and convolutional neural networks. Simultaneously, they convert the KDD99 data set into an image data format, which is convenient for convolutional neural networks to obtain the characteristics of net traffic data. Wang et al. [7] used a combination of deep belief networks and genetic algorithms to improve the detection rate significantly, but the use of genetic algorithms to adjust the parameters of the network resulted in the longer training of the model. Khan et al. [8] used the GoogLeNet model and the ResNet model to detect malware. Although the detection rate of the ResNet model is high, the model training is slow due to the deep network structure used by its network. To solve those problems, this paper proposes an Intrusion Detection System based on the PGoogLeNet-IDS model, which can solve those problems, such as incomplete feature extraction, uneven distribution of data sets, and slow model training.

2 Proposed Methodologies

GoogLeNet [9] is a high-performance deep learning structure launched by Google. The entire structure was assembled from multiple Inception modules. Inception Module leverages the computational resources of a network model. And the width and depth of the network are increased without increasing the computational load. In this way, the problems of gradient disappearance, gradient explosion, and overfitting caused by simply increasing the network structure are avoided to some extent. The entire model uses fewer parameters but learns more abundant features. Therefore, the GoogLeNet network is used as the backbone framework of this paper to meet the detection speed requirements of intrusion detection.

2.1 PGoongLeNet-IDS Network Structure

The PGoogLeNet-IDS structure proposed in this paper uses the GoogLeNet network model as the backbone network structure and uses the SE_DSC structure to replace the inception structure in the model. According to the needs of input data, PGooglenet-IDS changed the convolution kernel size in the GoogLeNet model and removed some structures of GoogLeNet, such as the auxiliary classifier, Dropout layer, and full connection layer. The model connected the output data of SE_DSC (3B), SE_DSC (4E), and SE_DSC (5b) structures by across channels. The penultimate layer of the model uses global average pooling [10] to reduce the number of parameters to prevent overfitting. Use the focus loss function in the Softmax layer to solve the problem that the uneven distribution of the number of samples affects the classification performance of the model (Fig. 1).

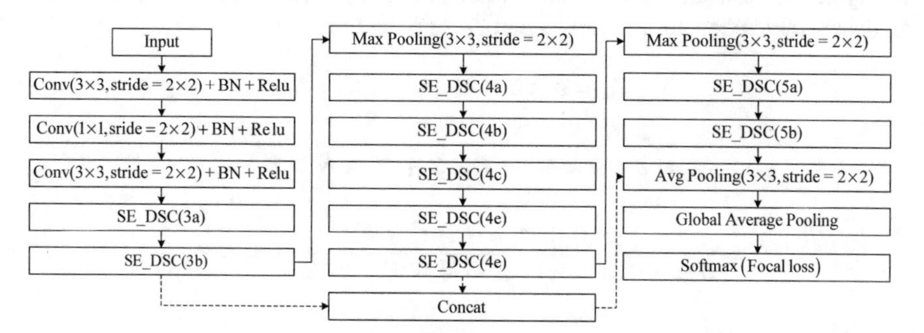

Fig. 1. PGoogLeNet-IDS model

2.1.1 SE_DSC Structure

To better balance the relationship between the detection speed and detection accuracy of the model, this paper presents the SE_DSC structure (Fig. 2).

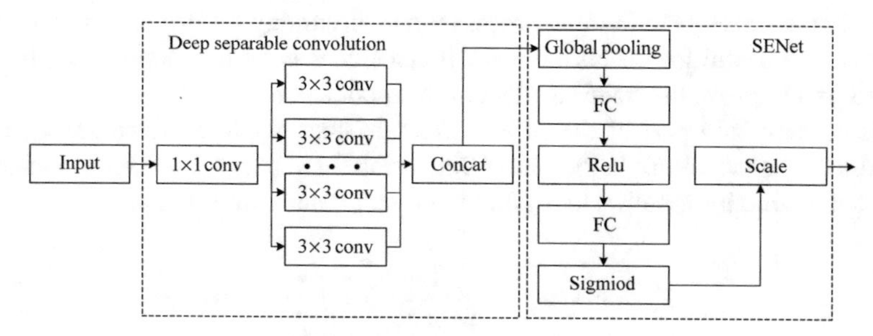

Fig. 2. SE_DSC structure

A. *Depth Separable Convolution*

The ordinary spatial convolution operation needs to consider both spatial and channel features, making the model have huge calculation problems during the training process, resulting in a slow model detection speed. The depth of the convolution is the ultimate inception structure [11]. The structure first uses a 3 × 3 convolution kernel to map each channel of the feature map to a new space. Then convolution through a 1×1 convolution kernel, while learning spatial correlation and correlation between channels. Introducing it into the GoogLeNet model, the channel correlation and spatial correlation decoupling method can be used to effectively reduce the spatial convolution and reduce the required parameters, optimize the network structure, improve the training speed of the network model, and further improve The width of the network is retained, and more feature information is retained to propagate in the network, which enhances the reconstruction quality of the network. The calculation amount of standard convolution operation and depth separable convolutions are Z_1 and Z_2, and the ratio is L:

$$Z_1 = b \times b \times M \times N \times K \times K \tag{1}$$

$$Z_2 = b \times b \times M \times K \times K + M \times N \times K \times K \tag{2}$$

$$L = \frac{b \times b \times M \times N \times K \times K}{b \times b \times M \times K \times K + M \times N \times K \times K} = N + \frac{1}{b^2} \tag{3}$$

B. *SENet Structure*

To reduce the impact of interference information on the performance of the model, focus on global features, and obtain useful feature information. Therefore, this paper introduces SENet [11], and combines it with deep separable convolution to form the

SE_DSC structure. SENet mainly adaptively calibrates the weight of each feature channel based on the loss value. It makes the model selectively focus on the feature information. In other words, the model can focus on the feature information that is useful for the current task according to the weight. The network can fully learn these features in the subsequent training process while discarding feature information that is useless or less useful for the current task. It can enhance the interdependence between channels and improve the training effect of the model.

The compression part of the SENet structure causes Global Average Pooling to compress the input features in the spatial dimension. In this way, the feature channel can use global feature information to capture important feature information.

$$Z_c = F_{sq}(u_c) = \frac{1}{W \times H} \sum_{i=1}^{W} \sum_{j=1}^{H} u_c(i, j) \tag{4}$$

The Excitation of the SENet structure uses two fully connected layers to fuse feature information in each channel. The first fully connected layer is to reduce the dimension of the compressed features, while the second fully connected layer will go through Relu function, after the function is activated, the features are upgraded. Finally, the sigmoid function is used to obtain the final weight.

$$S = F_{ex}(z, W) = \sigma(g(z, W)) = \sigma(W_2 \delta(W_1, z)) \tag{5}$$

Finally, it is necessary to weight the channels of the original feature information of the channels of the output data of the compression and excitation operations to form new features.

$$\widetilde{x}_c = F_{scale}(u_c, s_c) = s_c \cdot u_c \tag{6}$$

2.1.2 Cross-Layer Connection

Shallow feature maps often require fewer convolution operations and sampling operations, so they have complete feature information, and high-level feature maps often contain more abstract semantic information. If low-level feature information is integrated into the higher layer, the deeper network layer has more luxurious features, which help improve the classification accuracy of the model. Therefore, referring to the connection mode of SSD model [12], connect SE_DSC (3B), SE_DSC (4E), and SE_DSC (5b) in the model, adopt the output features of the 1×1point convolution operation fusion structure, and conduct weighted average pooling as the input data.

2.1.3 Focus Loss Function

In KDDTran+, the number of U2R and R2L attacks is much smaller than in other categories. Therefore, the model cannot fully learn the characteristic information of samples, and the model cannot recognize the new intrusion in KDDTest+, resulting in the low detection rate of these two types. On the contrary, samples with large data volume appear more frequently, and the model can learn the characteristics of this type well.

Lin et al. [13] propose the Focus Loss function. The function can dynamically adjust the cross-entropy function ratio, which can show a better experimental effect in the training process of a few samples. The Focal Loss function can increase the attention to the attack types of the minority class, and reduce the attention to the samples of the majority class. This way can improve the classification performance of the model.

The focus loss function is improved on the basis of the cross-entropy function. On the one hand, the focusing parameter γ is introduced to set the weights of the difficult-to-classify samples and the easy-to-classify samples. On the other hand, the weighting factor is introduced to adjust the weights of minority samples and majority samples, and the method of increasing the weight of minority samples and reducing the weight of majority samples is used to balance the data distribution. The final form of the focus loss function is:

$$FL(p_t) = -\alpha_t (1 - p_t)^{\gamma} \log(p_t) \tag{7}$$

3 Experiments and Results

This article will use the NSL-KDD benchmark data set to verify the effectiveness of the PGoogLeNet-IDS model. The operating system is Ubuntu16.04, the compilation environment is Python3.6, and the deep learning framework of TensorFlow-Gpu1.5 is downloaded and installed. The GPU framework of the CUDA 9.0 version.

3.1 Data Set

The NSL-KDD data set removes the redundant data in the KDDCUP99 data set and divides it into KDDTrain+ and KDDTest+ reasonably. The data set contains five types of tags, and it contains Normal, DOS, Probe, U2R, and R2L (Table 1).

Table 1. Classification of sample labels

Category	Normal	Dos	Probe	U2R	R2L	Total
$KDDTrain^{+}$	67,343	45,927	11,656	995	52	125,973
$KDDTest^{+}$	9711	7458	2421	2754	200	22,544

3.2 Data Set Sample Preprocessing

In order to form the standard data conforming to the model input, a series of preprocessing work is carried out on the data set. First, a single thermal encoder is used to digitize the non-digital features in the data set to form 122 dimension. Secondly, the minimum-maximum criterion is used to normalize the data and map the interval of data features [0, 1] to reduce the difference between data features. Finally, the Denosing AutoEncoder is used to reduce the 122 dimensions to 121 dimensions and convert it into 11×11 matrix form.

3.3 Evaluation Indicators

This article uses Accuracy, Recall, Precision, and $F1$-Score to evaluate the performance of the model.

$$\text{Accuracy} = \frac{\text{TP} + \text{TN}}{\text{TP} + \text{TN} + \text{FP} + \text{FN}} \tag{8}$$

$$\text{Recall} = \frac{\text{TP}}{\text{FN} + \text{TP}} \tag{9}$$

$$\text{Precision} = \frac{\text{TP}}{\text{FP} + \text{TP}} \tag{10}$$

$$F1\text{–Score} = \frac{2(\text{Recall} * \text{Rrecesion})}{\text{Recall} + \text{Precesion}} \tag{11}$$

TP means that normal traffic is correctly classified as normal traffic. TP means that abnormal traffic is correctly classified into corresponding categories. FN means that abnormal traffic error is classified as normal traffic; FP means that normal traffic error is classified as abnormal traffic.

3.4 Comparative Experiment and Result Analysis

The effectiveness of the PGoogLenet-IDS intrusion detection model will be verified through the following three groups of experiments.

Experiment 1: verify the effect of using SE_DSC, cross-layer connection, and focus loss function on the model accuracy and training time in the model.

As shown in Table 2, the pgooglenet-IDS (V1) model is formed by using the SE_DSC structure in the modified GoogLeNet model. The classification accuracy of the model was significantly improved, the training time was reduced by about half (V1), and the pgooglenet-IDS (V2) model was formed by cross-layer connection. The fusion of high and low-level information increases the training time of the model but also improves the accuracy of the model. Based on Pgooglenet-IDS (V2), a focus loss function is used to form Pgooglenet-IDS. This function is used to solve the problem of uneven data distribution. The experimental results show that the model proposed in this paper is accurate or trained, and the PGoogLeNet model is superior to other models.

Table 2. Comparison of accuracy and training time of different models

Model	Accuracy (%)	Time (s)
GoogLeNet	78.81	6835.74
PGoogLeNet-IDS (v1)	81.72	3725.18
PGoogLeNet-IDS(v2)	84.36	3925.09
PGoogLeNet-IDS	87.16	4137.83
RNN [14]	83.28	5516
CNN-BiLSTM [15]	83.58	4369.99

Experiment 2: In order to further verify the impact of focus Loss function on the performance of U2R and R2L attack types, this paper USES F1-Score index to evaluate the performance of each attack type (Table 3).

Table 3. Comparison of various models $F1$-Score of different models

Model	Normal (%)	Dos (%)	Probe (%)	U2R (%)	R2L (%)
GoogLeNet	82.96	87.88	70.74	13.25	36.36
RNN [14]	83.21	89.34	73.61	19.57	38.02
PGoogLeNet-IDS	89.21	89.07	79.13	62.37	77.17

In the RNN model and GoogLeNet model, and PGoogLeNet-IDS, the detection rate of R2L and U2R is very low, because the intrusion detection system can easily identify numerous categories, and it is more difficult to detect the occurrence Less frequent categories. After the focus loss function was added into the model, the $F1$-score of U2R and R2L was significantly improved.

Experiment 3: Compare the accuracy rate, recall rate, and $F1$ score of the model proposed in this paper with other intrusion detection models.

For data sets with uneven distribution of data samples, model performance cannot be measured by accuracy alone. Because the classifier of the model tends to the majority class samples Even if the number of minority samples is small, the overall accuracy of the model will not be significantly affected. Therefore, the accuracy rate cannot comprehensively evaluate the performance of the model. As shown in Fig. 3, the PGoogLeNet-IDS model is better than other classification models on the KDDTest + . Therefore, the performance of the model proposed in this paper is better than other models, so the PGoogLeNet model can carry out efficient and accurate defense when attacking, which has stronger practicability.

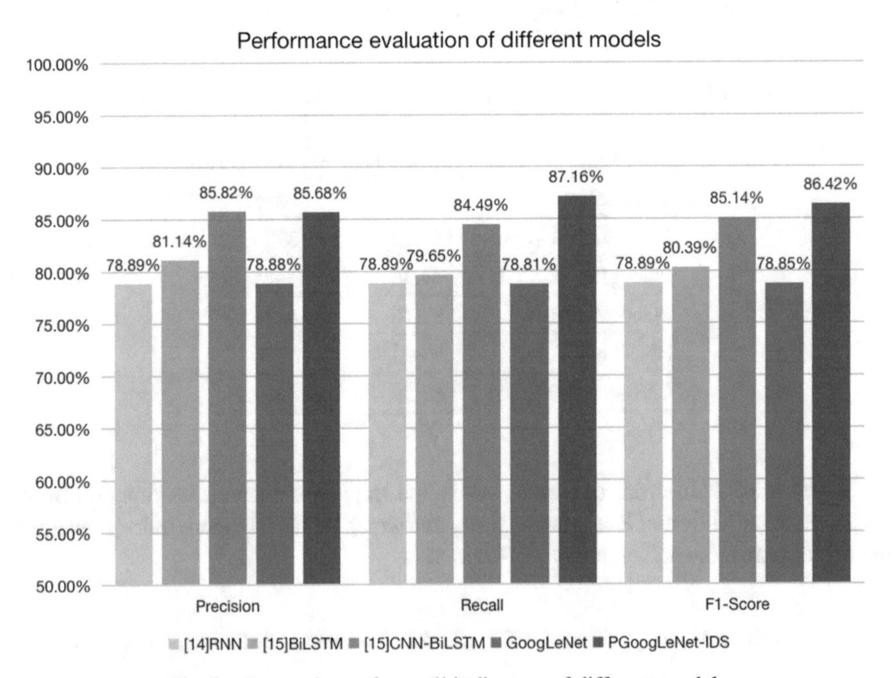

Fig. 3. Comparison of overall indicators of different models

4 Conclusion

To improve the accuracy and training speed of model detection, this paper proposes an intrusion detection system based on the PGoogLeNet-IDS model. The GoogLeNet model USES a fusion module that combines deeply separable convolution with Senet to make the network more lightweight. This method can effectively reduce the training time of the model. At the same time, the performance of the model is improved by increasing the focus on useful features. Cross-layer fusion is used to fuse the important features of the bottom layer with the features of the high layer, which can make the model to fully extract feature information. In addition, the model uses the Focus Loss Function, which not only can increase the weight of a small number of samples but improve the accuracy of a small number of samples. Experimental results show that the PGoogLeNet-IDS model has better performance than other classification models.

Acknowledgements. The work described in this paper is supported by Shanxi Natural Science Foundation (No. 201701D121052).

References

1. Denning DE (1987) An intrusion-detection model. J IEEE Trans Software Eng 2:222–232
2. Resende PAA, Drummond AC (2018) A survey of random forest based methods for intrusion detection systems. J ACM Comput Surv (CSUR) 51(3):1–36
3. Alhakami W, ALharbi A, Bourouis S et al (2019) Network anomaly intrusion detection using a nonparametric Bayesian approach and feature selection. J IEEE Access 7:52,181–52,190
4. Gao X, Shan C, Hu C et al (2019) An adaptive ensemble machine learning model for intrusion detection. J IEEE Access 7:82512–82521
5. Yin C, Zhu Y, Fei J, et al (2017) A deep learning approach for intrusion detection using recurrent neural networks. J IEEE Access 5: 21,954–21,961
6. Xiao Y, Xing C, Zhang T et al (2019) An intrusion detection model based on feature reduction and convolutional neural networks. J IEEE Access 7:42210–42219
7. Zhang Y, Li P, Wang X (2019) Intrusion detection for IoT based on improved genetic algorithm and deep belief network. J IEEE Access 7:31711–31722
8. Khan RU, Zhang X, Kumar R (2019) Analysis of ResNet and GoogleNet models for malware detection. J Comput Virol Hacking Tech 15(1):29–37
9. Szegedy C, Liu W, Jia Y et al (2014) Going Deeper with Convolutions. J
10. Chollet F (2017) Xception: deep learning with depthwise separable convolutions. In: Proceedings of the IEEE conference on computer vision and pattern recognition, pp 1251–1258
11. Hu J, Shen L, Sun G (2018) Squeeze-and-excitation networks. In: Proceedings of the IEEE conference on computer vision and pattern recognition, pp 7132–7141
12. Liu W, Anguelov D, Erhan D et al (2016) SSD: single Shot MultiBox Detector. J
13. Lin TY, Goyal P, Girshick R et al (2019) Focal loss for dense object detection. In: Proceedings of the IEEE international conference on computer vision, pp 2980–2988
14. Chuan LY, Yue-Fei Z, Jin LF et al (2019) A deep learning approach for intrusion detection using recurrent neural networks. J IEEE Access 99:1–1
15. Jiang K, Wang W, Wang A, H Wu et al (2020) Network intrusion detection combined hybrid sampling with deep hierarchical network. J IEEE Access 8:132,464–32,476

Research on ADS-B Interference Principle and Suppression Method

Yi Yang(✉), Ruheng Xie, and Yang Ding

State Key Laboratory of Air Traffic Management System and Technology, Nanjing 210007, China
nuaa_yang@nuaa.edu.cn

Abstract. As the core technology of next-generation air traffic management system for various countries, the security problem of automatic dependent surveillance-broadcast (ADS-B) is constraint to its application. In this paper, the various principles of ADS-B interferences are analyzed. The state-of-the-art interference suppression methods are classified as broadcast authorization and location verification. These mentioned methods are reviewed in-depth and compared with each other. Finally, an available route for ADS-B anti-interference is provided.

Keywords: Air traffic surveillance · Automatic dependent surveillance-broadcast · Interference · Suppression

1 Introduction

In order to cope with the contradiction between the growing aviation demand and the relatively backward air traffic control, ICAO has proposed an automatic dependent surveillance-broadcast technology [1] under a new navigation system. Compared with traditional radar surveillance systems, ADS-B can obtain higher-precision position and altitude information of the target, and breakthrough the problem that it cannot cover desert and ocean areas and has limited data accuracy [2]. By providing pilots with faster flight data transmission services, the use of ADS-B can optimize route settings, reduce separation standards, and make the airspace more fully utilized [3]. In addition, ADS-B can also provide aircraft with more comprehensive weather, terrain, and surrounding traffic flight information, so that the crew can better understand the situation around the aircraft, improve situational awareness, and thus achieve integrated space, air, and ground information sharing and collaborative monitoring [4]. Therefore, ADS-B is considered to be the future of air traffic control.

The international aviation community has also been actively promoting the promotion and application of ADS-B in recent years, in order to enable the conversion of new forms of ATM to meet the needs of the next 30 years. At present, the more representative and influential ADS-B technology research and application projects mainly include the CAPSTONE plan in the United States, the CASCADE plan in Europe, and the UAP

Q. Liang et al. (eds.), *Artificial Intelligence in China*, Lecture Notes
in Electrical Engineering 653, https://doi.org/10.1007/978-981-15-8599-9_38

project in the high altitude airspace plan in Australia [5, 6]. The Civil Aviation Administration of China formulated the "ADS-B Implementation Plan for Civil Aviation of China" in 2012, which clarified the overall objectives, stage planning, and technical plan for the implementation of ADS-B, aiming to build an integrated ADS-B operation of space, air, and ground System, and actively promote the construction and operation of ADS-B, so that China will move from a big civil aviation power to a strong civil aviation power [7, 8].

As an open surveillance method, the safety problem of ADS-B becomes the biggest obstacle to its widespread use or even replacement of the existing secondary surveillance radar. Among the many security weaknesses that appear, the most prominent issue in the security problems of the ADS-B surveillance system is man-made active interference, which includes interference to a single node (ground station or aircraft) or malicious transmission of S-mode high power 1090 MHz frequency by many participants in an organized area and makes the original receiver unable to send/receive messages [9].

The structure of this paper is as follows: First, the mechanism of ADS-B suppression, deception, and multipath reflection interference is analyzed according to the signal transmission theory. Secondly, the existing ADS-B interference countermeasure technology is divided into two categories: broadcast authentication and location confirmation for detailed analysis. Finally, the advantages and disadvantages of the anti-interference methods involved in this article are compared, and the future research ideas in this direction have prospected.

2 ADS-B Interference Mechanism Analysis

For an ADS-B receiver, the interference signal ratio p_{Ji}/p_{Si} of the ADS-B signal transmitted by an aircraft can be described as

$$\frac{P_{Ji}}{P_{Si}} = \left(\frac{P_{TJ}}{P_{TS}}\right)\left(\frac{G_{TJ}}{G_{TS}}\right)\left(\frac{G_{RJ}}{G_{RS}}\right)\left(\frac{L_S}{P_J}\right)\left(\frac{1}{L_f L_t L_p}\right) \tag{1}$$

Among them, P_{TS} is the target signal transmission power, G_{TS} is the transmit antenna gain, G_{RS} is the receive antenna gain, and L_S is the signal loss; L_f, L_t, L_p is the target signal transmission loss in the frequency domain, time domain, and polarization. Based on the above model, the following interference methods currently exist:

1. Suppressive interference: This interference method is to transmit a high-power signal to the ADS-B receiver to suppress the ADS-B receiver from receiving the real signal;
2. Deceptive interference: This interference method is to send information to the ADS-B receiver through a false information source, or send false information to the ADS-B;
3. Multi-path reflected interference: This interference refers to the interference caused by the superposition of multiple signals received by the ADS-B receiver on the real signal.

3 ADS-B Interference Suppression Method

As the deadline for the implementation of ADS-B in various countries approaches, the need to solve the interference problems involved is becoming increasingly urgent [10].

Therefore, in recent years, the related theoretical and practical research work of ADS-B anti-jamming has become a hot issue in civil aviation communication, navigation, and surveillance technology. The research results of ADS-B anti-interference are mainly divided into two ways: broadcast certification and location confirmation [11].

3.1 Broadcast Certification

Compared with peer-to-peer communication, information authentication on broadcast media is more difficult. The reason for this is that symmetric attributes are only useful in peer-to-peer authentication where both parties trust the other. Therefore, ADS-B's broadcast authentication technology requires an asymmetric mechanism inside the receiver to verify the information rather than generating real information itself [12]. Its purpose is to maintain the integrity of ADS-B features where suspicious conditions are detected in global or local areas when providing a potential authentication mechanism.

In addition, there are differences between broadcast schemes, some based on users, some based on nodes, and some based on both. Node-based (namely host) programs need to determine the authenticity of the nodes, namely hardware; the user-based scheme ignores the hardware, but focuses on authenticating users. This section will summarize the various methods mentioned in the literature and analyze their feasibility on ADS-B.

3.1.1 Physical Layer Scheme

The non-password scheme is similar to fingerprint identification, based on the incompleteness of the wireless channel, and is used for wireless user authentication and device technology authentication. The purpose is to identify suspicious activity on the network. Finding signals from legal beacons on the network may be able to distinguish between ground stations and aircraft, identify aircraft types, and even enable individual machines to provide data that can help develop intrusion detection systems [13]. If specific differences between legal and illegal packets are found on the physical layer, machine learning techniques will build a model to predict the statistical thresholds of normal and suspicious activity. This is of great value in detecting intruders, even if only the type of equipment can be identified, not an abnormal participant. However, fingerprint authentication cannot guarantee accuracy for the time being.

3.1.2 Public Key Cryptographic Scheme

The simplest method of a cryptographic scheme is to distribute the same key to all ADS-B participants in the world, or to aircraft and ground stations in at least a certain area. However, this large-scale encryption mechanism, even including general aviation, is extremely insecure whether it is an internal or external attack. Finke et al. tested a variety of encryption mechanisms, including key management for symmetric encryption of different channels, such as controller-pilot data link communication (CPDLC) [14]. Through analysis of the security, symmetric and asymmetric practicality, and format retention encryption technology, they recommended the use of FFX algorithm for symmetric ciphers, which can encrypt the non-standard size data block (112-bit ADS-B message) with sufficient entropy.

In order to make the broadcast authentication add an asymmetric attribute to the ADS-B monitoring system, it can be done by public-key encryption. Samuelson and Valovage use hashing to create an information authentication code (MAC). Sensis uses the challenge query/response format of the ground station authenticator to authenticate each participant within its scope and notifies the superior authority/all other failed participants. This concept requires a ground station connected to the world-wide security key database, which not only makes it difficult to maintain globality but also has security holes. Because such a system can not only be used to identify ADS-B participants, pilots and other individuals/systems can also be added to the flight procedures. In addition, ADS-S requires a changing modulation and a complete overhaul of the message communication system, which makes it incompatible with ADS-B and costly.

The PKI scheme is essentially equivalent to the traceback part of a key: Aircraft A sends N signals distributed in ADS-B messages so that after every N message, nearby participants have received the signal of A. The recipient will keep these messages until the signal is completely sent, at which time the buffered messages can be authenticated.

There are some difficulties in using encryption technology that is difficult to overcome, as described below:

(1) To despise data frames, management and control frames are not protected by encryption.
(2) Not compatible with the installation base.
(3) Non-centralized key exchange in a self-organizing network is very difficult. It often moves too fast and needs to be modified frequently which leads to the message excessive and too large.
(4) The characteristics of ADS-B opening is that compared with ADS-S, the installed cryptosystem has no public broadcast communication.
(5) One-time signature is not feasible, because it can only sign 60 bytes, but the requirements exceed 80 bytes or more.

3.1.3 Backtracking Key Release

A variation of the traditional asymmetric encryption technology is to use the transmitter to retroactively publish its key to the receiver and later authenticate the broadcast message. This method has been mentioned in various fields [15]. The broadcast entity generates an encrypted message authentication code when sending each message. After a period of time or a group of messages, the key to decrypt the MAC will be released. All listeners who have buffered the previous message can decrypt the message at this time and determine the continuity of the sender.

The Timed Effective Streaming Loss Authentication (TESLA) protocol [16] uses RFC4082 as the standard, can provide effective broadcast authentication on a large scale, and can also handle packet loss and apply it in real-time. The μ TESLA broadcast certification agreement is a modified version of TESLA for wireless sensor networks.

TESLA and μ TESLA use a one-way keychain. Because of the asymmetric nature of TESLA, μ TESLA can be used to adapt to ADS-B in time, and accurate time calibration can be obtained through GPS. Although ADS-B needs to be added to identify source integrity, μ TESLA does not need to determine the continuity of the sender to maintain open broadcastability and complex PKI infrastructure (such as separate well-connected ground stations), so μ TESLA has an advantage. Moreover, it can enable participants to resist camouflage attacks, in areas covered by ground stations with good coverage, and once the continuity is interrupted, an alarm is triggered.

Another advantage of μ TESLA is to ensure that data packets will not be lost on the congested 1090 MHz frequency. Moreover, the additional communication and modification of the ADS-B protocol required are also much less than traditional asymmetric encryption methods.

On the other hand, when used in AANETs, μ TESLA also has shortcomings: it needs to be restarted and storage paralysis is easy (according to receiver settings). To overcome these, Eldefrawy et al. proposed to use forward hashing, using two different built-in hashes and the Chinese remainder theorem, so that the system does not need to be restarted.

3.2 Location Confirmation

In addition to protecting ADS-B communication data, there are other ways to ensure the integrity of air traffic management. The concept of safe location certification is that the aircraft and other ADS-B participants double-check the authenticity of the position report. This is different from broadcast authentication [17]. The method is to establish a way to find the exact location of the sender and to provide redundancy and the ability to double-check reports. The advantage is that it will generate more positioning data, combined with ADS-B and radar as a backup system to prevent failure of the dominant system or GPS.

3.2.1 Multi-point Positioning

Multi-point positioning is a relatively popular and mature collaborative monitoring method. The core idea is that by establishing the exact distance between 4 or more targets with known positions and 1 target with unknown positions, the positioning problem can be transformed into a geometric problem. Based on multiple antennas distributed in different locations, receiving the same signal at different times, using the time difference of arrival to calculate the position of the aircraft.

However, the multi-point positioning itself has the following defects:

(1) Easily affected by multi-path propagation.
(2) A relatively large number of receiving stations are required to accurately detect the signal to be verified, which brings a certain complexity in synchronization, on the other hand, it greatly increases the total cost of its equipment.
(3) The central processing station needs to link with all receiving stations.

Due to limitations of cost and logic, it is difficult to install more multi-point positioning base stations in remote or difficult-to-reach areas.

3.2.2 Organization Verification

Organization verification is another concept that alleviates ADS-B security and privacy issues [18]. Its purpose is to protect the air ADS-B In communication with the multi-point positioning of an organization to verify the position report of non-organization members in flight. Just like ground stations, 4 or more certified aircraft establish mutual trust and communicate to use multipoint positioning (based on TDOA or RSS). If a false position report is detected, starting from the position of the aircraft that cannot be accurately located, the avoidance circle is enlarged, so that safety can be guaranteed.

Organizational verification has some disadvantages. First, it requires a lot of additional information to carry out verification and trust procedures. And ADS-B only has one-way broadcasting, so the concept of related organizations needs a new protocol to support, such as L-band digital aviation communication system (L-DACS). Eurocontrol has developed L-DACS as the future IP technology to cope with air-to-air communications, but there is no detailed plan yet. In addition, to successfully use such a protocol, the first problem to be solved is how to manage the security certification of members. It is very complicated to establish mutual trust and avoid malicious attacks in the new MANET organization. In addition, the performance of the system in response to intelligent interference in communications needs to be considered.

3.2.3 Distance Boundary Verification

In wireless networks, the distance boundary to a certain extent can also locate other participants and determine secure transactions, that is, RFID communication [19]. The distance boundary is to establish an encryption protocol, use a calibrator P to guide a proofreader V, and P is within a certain physical distance. Electromagnetic waves propagate at a speed no higher than the speed of light c, which forms the basis of the distance boundary protocol. In this way, the distance can be calculated based on the time between the calibration machine's query and the calibrator's response. The determined distance is the upper boundary. As an additional piece of information, the authenticity of the sent information will be checked afterwards to verify the node. When various trusted entities implement distance boundaries, they can collaborate and find the actual position of the calibrator through trilateration.

The safety distance boundary of VANETs is divided into three steps:

(1) Use the traditional distance boundary to find the lower boundary between V and P. P can only increase the time to respond to V's query, which will be more and more than the real.
(2) V checks the authenticity of the location reported by P:

    (a) Confirmation based on distance: The actual maximum distance of wireless transmission is limited. The trial operation can find the actual maximum value for ADS-B users at a certain place/route.
    (b) Confirmation based on speed: Considering that the speed of the aircraft at different stages of flight is known (the actual minimum and maximum speeds possible), continuous position reports must be made in the given window.

(3) In order to improve security, after passing all the authenticity checks, the proofreader selects a neighbor B, and then B gives P its estimated position E. Once the estimate exceeds the error tolerance, B knows that P has increased the distance.

In addition to being unsuitable for long-distance and high-speed, currently in air traffic control, another major disadvantage of the distance boundary is that it itself requires the calibrator to respond to queries from the proofreader. In this way, from the perspective of ADS-B, a completely new agreement is needed.

3.2.4 Data Fusion and Trust Management

Data fusion is quickly becoming the cornerstone of modern intelligent transportation systems (ITS). This hybrid estimation algorithm often uses multiple monitoring technologies (PSR, SSR, multi-point positioning) and flight plan information to fuse multiple sensors with the purpose of comprehensive fault detection rather than specific security. Then determine whether the relevant system is operating normally and determine whether there is a malicious attack. This process includes analysis of data reliability (according to whether the system is vulnerable to interference attacks) and the accuracy/measurement uncertainty of each technology [20, 21]. According to the cosine similarity between the reported position and the estimated position, the credibility of the participant's report is judged and the trust information of the historical beacon is maintained. Then calculate the weighted credibility of the beacon based on time.

The advantage of data fusion is to retain the compatibility of existing systems, including no need to modify the ADS-B protocol to provide other features, the disadvantage is to provide the necessary redundancy, the cost of additional systems will increase.

3.2.5 Purpose Verification Based on Kalman Filtering

Kalman filtering was first used in automatic traffic control (ATC) systems to filter and smooth GPS position data in messages [22–24]. Kalman filtering is used to observe noisy time series in the measurement, and to determine the future state of variables in the best-predicted underlying system.

Kalman filtering is used in the ground system to filter and verify the status vectors and trajectory changes reported by ADS-B aircraft, and to check the authenticity of these data, and to verify the purpose of the aircraft by defining local and global correlation functions to evaluate the correlation between aircraft movement and ADS-B objectives. Then calculate the geometric consistency, that is, whether the aircraft is within the given horizontal and vertical limits, and the purpose consistency, that is, to analyze the aircraft movement and compare it with a reasonable destination model (such as horizontal and vertical speed).

However, for an attacker, Kalman filtering can be countered by a so-called boiled frog attack: it interferes with the correct signal when it continues to send slightly modified positions. If the completion is slow enough, Kalman filtering will treat the injected data as a legitimate trajectory change. This exposes a major weakness of Kalman filtering: the historical data on which it is based is relatively scarce. However, it is still very useful, because obvious false maneuvers, speeds, and characteristics can still be detected, and it also greatly increases the complexity of the attack.

4 Comparison and Suggestions of Interference Suppression Methods

4.1 Comparison of Suppression Methods

We compared the advantages and disadvantages of various ADS-B interference suppression strategies involved in the previous section. The feasibility analysis is shown in Table 1.

Table 1 Availability issue comparison on different ADS-B interference suppression methods

	Difficulty	Cost	Expansibility	Compatibility
Physical layer verification	Mutative	Mutative	Mutative	Additional hardware and software are required. No need to modify ADS-B protocol
Uncoordinated spread spectrum	Medium	Medium	Medium	New hardware and physical layer are required
PKI	High	High	Medium	Need to allocate infrastructure, modify protocols and message processing
μTESLA	Medium	Medium	High	A new message category is required to publish the key. Join MAC
Wide area multi-point positioning	Low	Medium	Medium	No need to change ADS-B. Independent hardware system
Distance boundary	High	Medium	Low	New messages and protocols are required
Kalman filtering	Low	Low	High	Additional messages are required. Independent hardware system
Organization certification	High	Medium	Low	New messages and protocols are required
Data fusion	Low	Medium-high	Medium	No need to change ADS-B. Independent system

4.2 ADS-B Anti-interference Feasibility Suggestions

Summarizing the existing ADS-B interference suppression and countermeasure technologies, comprehensively considering the distribution and use of my country's existing air traffic control equipment, the following feasibility suggestions are obtained:

First, through the spread spectrum and encryption algorithms in the traditional electronic countermeasure technology, the two-way authentication mechanism of ADS-B transmission and acceptance is established, which is directly applied to the method of ADS-B transmission equipment. There are difficulties in adding software and hardware modules. Considering that the manufacturers of my country's civil aviation operating fleet are Boeing and Airbus, such conversions that meet my country's special needs are relatively difficult. In addition, such technologies make it difficult for the aircraft to meet the criteria of free flight later.
Second, the method of enhancing the accuracy of ADS-B information through fusion with secondary radar monitoring results information or multi-point positioning auxiliary verification, although such methods redundant ADS-B receiving systems, reducing the inherent ADS-B system the advantage of low cost, but considering the number and coverage of existing radar base stations in my country, this kind of method is feasible, especially in the eastern region.
Third, in order to speed up the monitoring and support capability of the airspace in the western region and the ocean region of my country, it is necessary to improve the anti-interference ability of the ADS-B receiver. It is feasible to consider ADS-B transmitters and receivers with array antennas. On the one hand, by estimating the direction of incoming waves, the authenticity of the ADS-B signal can be judged; on the other hand, the ADS-B transmitter can also prevent its transmitted signal from being intercepted where it is judged to interfere.

5 Conclusion

This paper analyzes the mechanism of repressive, deceptive, and multi-path reflection interference existing in the process of ADS-B signal propagation. The existing ADS-B interference countermeasures are divided into two categories: broadcast certification and location safety confirmation, and are discussed in detail. The advantages and disadvantages of the various methods listed in this article are compared, and possible suggestions are proposed for my country's ADS-B interference suppression research.

References

1. FAA. FAA aerospace forecast: fiscal years 2012–2032. FAA Aerospace Forecast: Fiscal Years 2012–2032, 2012, pp 1–115
2. Rekkas C, Rees M (2008) Towards ADS-B implementation in Europe. In: International workshop on digital communications-enhanced surveillance of aircraft and vehicles, September 2008, pp 1–4
3. China Civil Aviation Administration (2014) 2013 Statistical Bulletin of Civil Aviation Industry Development, pp 1–7

4. Erzberger H, Paielli RA (2002) Concept for next generation air traffic control system. Air Traffic Control Q 10(4):355–378
5. Vidal L (2013) ADS-B out and in airbus status. ADS-B Taskforce-KOLKATA, April 2013
6. ICAO. Status of ADS-B avionics equipage along ATS routes L642/M771 for harmonized ADS-B implementation. ADS-B Seminar and 11th Meeting of ADS-B Study and Implementation Task Force, Apr. 2012
7. China Civil Aviation Administration (2012) China Civil Aviation ADS-B Implementation Plan
8. Jing X, Tao S, Chen Yu (2005) Discussion on the implementation of ADS-B technology in my country. China Civil Aviation 58(10):25–29
9. McCallie D, Butts J, Mills R (2011) Security analysis of the ADSB implementation in the next generation air transportation system. Int J Crit Infrastruct Prot 4(2):78–87
10. Schafer M, Storhmeier M, Lenders V et al (2014) Bringing up opensky: a large-scale ADS-B sensor network for research. In: 13th International symposium on information processing in sensor networks, 2014, pp 83–94
11. Storhmeier M, V. Lenders, I. Martinovic, On the security of the automatic dependent surveillance-broadcast protocol. IEEE Commun Surveys Tutorials. https://doi.org/10.1109/comst.2014.2365951
12. Strasser M, Pöpper C, Capkun S et al (2008) Jamming-resistant key establishment using uncoordinated frequency hopping. In: IEEE symposium on security and privacy, May 2008, pp 64–78
13. Pöpper C, Strasser M, Capkun S (2009) Jamming-resistant broadcast communication without shared keys. In: USENIX Security Symposium, pp 231–247
14. Robinson RV, Li M, Lintelman SA et al (2007) Impact of public key enabled applications on the operation and maintenance of commercial airplanes. In: AIAA aviation technology integration, and operations (ATIO) conference, 2007, pp 750–759
15. Viggiano M, Valovage E, Samuelson K, Hall D et al (2010) Secure ADSB authentication system and method. US Patent, 7730307
16. Perrig A, Szewczyk R, Tygar JD et al (2002) SPINS: security protocols for sensor networks. Wireless Netw 8(5):521–534
17. Smith A, Cassell R, Breen T et al (2006) Methods to provide system-wide ADS-B back-up, validation and security. In: 25th digital avionics systems conference, 2006, pp 1–7
18. Sampigethaya K, Poovendran R (2011) Security and privacy of future aircraft wireless communications with offboard systems. In: IEEE international conference on communication systems and networks (COMSNETS 2011), Jan. 2011, pp 1–6
19. Kovell B, Mellish B, Newman T et al (2012) Comparative analysis of ADS-B verification techniques. In: IEEE integrated communications, navigation and surveillance conference (ICNS), 2012, pp K3-1–K3-9
20. Krozel J, Andrisani D, Ayoubi MA et al (2004) Aircraft ADS-B data integrity check. In: AIAA 4th aviation technology, integration and operations (ATIO) Forum, 2004, pp 1–11
21. Wei Y-C, Chen Y-M, Shan H-L (2011) Beacon-based trust management for location privacy enhancement VANETs. In: 13th Asia-Pacific network operations and management symposium (APNOMS), Sept. 2011, pp 1–8
22. Chan-Tin E, Heorhiadi V, Hopper N et al (2011) The frog-boiling attack: limitations of secure network coordinate systems. ACM Trans Inf Syst Security 14(3):27
23. Meng X, Gao Y (2004) Electric systems analysis. Higher Education Press, Beijing, pp 3–21
24. Li Y, Liu J (2007) Mechanism and improvement of direct anonymous attestation scheme. J Henan Univ 37(2):195–197

A Multispectral Image Enhancement Algorithm Based on Frame Accumulation and LOG Detection Operator

FengJuan Wang, BaoJu Zhang(✉), CuiPing Zhang, ChengCheng Zhang, and Man Wang

Beijing, China
wangfengjuan327@163.com

Abstract. In multispectral transmission image, due to the strong scattering and strong absorption characteristics of biological tissue, the weak image signal, low signal-to-noise ratio, and blurred edges of the transmission image greatly hinder the detection of heterogeneity of transmission tissue. And image preprocessing is crucial to the complex image algorithm in the later experiments. In order to improve the quality of low-resolution multispectral transmission images, this paper proposes an enhancement algorithm for transmission tissue images through the simulation experiment of multispectral transmission images: Single-channel frame algorithm that combines with LOG detection operators. The experimental results show that the peak signal-to-noise ratio (PSNR) of the image processed by the enhanced algorithm in this paper is increased to 52.289 dB, which is 1.515 dB higher than the original image and is 1.386 dB higher than the only filtered image. The multispectral image processed by this algorithm shows that the contrast ratio at the edge of the image has been improved, and the image quality has been improved to a certain extent. Finally, through a comparative experiment, the effectiveness and practicability of the enhancement algorithm for improving the quality of multispectral transmission images are verified.

Keywords: Multispectral transmission image · Frame accumulation · Bilateral filtering · LOG detection operator · Image enhancement

1 Introduction

In multispectral transmission image, the strong scattering and strong absorption characteristics of the tissue cause the imaging to be blurred, which brings greater difficulty to the subsequent detection of heterogeneity, and it is difficult to achieve more accurate detection of early breast lesions. Frame accumulation technology can greatly improve the signal-to-noise ratio of the image and is an effective method to improve the image quality [1, 2]. Experiments show that this technology plays an important role in improving the quality of multispectral transmission images [3, 4]. Taking breast tumors as an example, the difference in gray scale between normal breast tissue and the tissue at the

Q. Liang et al. (eds.), *Artificial Intelligence in China*, Lecture Notes
in Electrical Engineering 653, https://doi.org/10.1007/978-981-15-8599-9_39

beginning of the lesion is small. If the image signal-to-noise ratio is improved and the image edge details are enhanced, the edge of the diseased tissue part can be extracted more accurately and make early diagnosis more accurately [5]. Image enhancement technology plays an important role in digital image processing, and some of these methods are improved and applied to the transmission tissue image processing, which can improve the image contrast ratio, which is conducive to the improvement of the later heterogeneity detection effect.

Frame accumulation technology is one of the most effective methods successfully applied to various low-light-level image detection devices and weak image signal detection and has been proven to improve gray scale resolution to a certain extent. Gang et al. [6, 7] used a combination of frame accumulation and shape function signal modulation technology to greatly improve the signal-to-noise ratio and gray-scale resolution of the image, thereby improving the detection sensitivity of low-light transmission images. Image enhancement is an image enhancement technology that can highlight the edges of smoothed object contours in the image to accurately perform subsequent target detection and recognition [8, 9]. Enhancement technology is one of the simple and effective methods that can clear the edge lines and improve the image contrast ratio. It has been widely used in various low contrast ratio image enhancements. They are combined in the multispectral transmission image of biological tissue reinforcement has good application prospects [10].

2 Experimental

The structure of the experimental device: LED light source (0.5 W, blue light (465–470 nm), and yellow light (577–597 nm) two wavelengths of synthetic LED light source), phantom, mobile phone (using its camera to collect images, model: HUAWEI mate9, Frame rate: 59 fps, image resolution:), cputer (connected to mobile phone, transfer image to computer for processing), shading cloth (used to isolate ambient light), using constant current power to drive the light source. Put carrot and pork cut pieces of different sizes and shapes in the simulated tissue fluid of the phantom to obtain the original milk solution and heterogeneous image (Fig. 1).

Experimental data acquisition:

(1) Adjust the distance between the phantom and the camera and light source, turn on the light source, cover the shading cloth, and turn on the camera to record phantom video.
(2) Transfer the video in (1) to the computer and extract images excluding the gross errors (such as Images with large jitter and smearing) from it.
(3) Cut the non-interesting regions (non-ROI, mostly unrelated edge parts) with noise in the image appropriately. By calculating the signal-to-noise ratio of the original image and the image with the non-ROI region cut, it is found that the image noise after cutting is reduced compared with the original image, and the signal-to-noise ratio is increased.
(4) Separate the RGB images from the cut image and obtain three single-channel grayscale images from each original image to prepare for the subsequent single-channel frame accumulation processing.

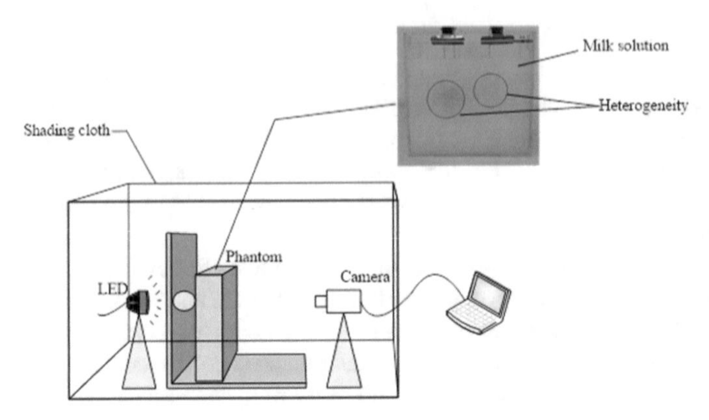

Fig. 1 Experimental device

3 Frame Accumulation Technology

Gray scale is the core of image accuracy and heterogeneous body detection sensitivity. The higher the gray scale resolution of the image, the richer the image information is, and the more favorable it is for the classification and analysis of tissues. Frame accumulation can increase the gray level and improve the gray resolution to a certain extent. The frame accumulation technique used in this paper is to accumulate the averaged multi-frame images, add the pixel gray levels of the corresponding points between two or more frames of static images at different acquisition times, and then calculate them at these times. The average image can increase the signal-to-noise ratio of the image exponentially, and at the same time avoid the edge loss caused by the filtering method.

Use the experimental device 1 to collect transmission images, separate the original image channels to obtain a single-channel grayscale image, and then accumulate the single-channel images to increase the grayscale of the image and increase the signal-to-noise ratio.

For single-channel images, cumulative average of each frame, there is:

$$\begin{aligned}\sum_{i,j=1}^{i+N-1} x_{i,j} &= x_{i,1} + x_{i+1,1} + \cdots + x_{i+N,1} \\ \sum_{i,j=2}^{i+N-1} x_{i,j} &= x_{i,2} + x_{i+1,2} + \cdots + x_{i+N,2} \\ \sum_{i,j=3}^{i+N-1} x_{i,j} &= x_{i,3} + x_{i+1,3} + \cdots + x_{i+N,3}\end{aligned} \tag{3.1}$$

Calculate the single-channel average after accumulation:

$$\overline{x}_{t,j} = \frac{1}{N}\sum_{i}^{i+N-1} x_{i,j} \quad t = 1, 2, \ldots, j-1, 2, 3 \tag{3.2}$$

During the experiment, considering the improvement of image quality and operation speed, when $N = 100$, the effect is the best. The following is the result of single-channel frame accumulation experiment (Figs. 2, 3, 4, and 5).

Fig. 2 Channel B

Fig. 3 Channel G

Fig. 4 Channel R

Through the experimental results, it can be seen that after the frame accumulation of the single-channel gray-scale image, the image quality is significantly improved, and the contrast ratio of the image is more obvious. After the frame accumulation processing of the multispectral transmission image, the signal-to-noise ratio of the image is improved, and the image quality and contrast ratio are a significant improvement.

It can be seen from the visual angle that the image quality has been significantly improved, and the following can be analyzed by the evaluation index of the image quality.

Fig. 5 Original color image and color image after frame accumulation

PSNR is an objective measurement method used most frequently to measure image quality. The larger the PSNR value, the smaller the image distortion. The expression is as follows:

$$\mathrm{MSE} = \frac{\sum_{0 \le i \le M} \sum_{0 \le J \le N} \left(f_{ij} - f_{ij}\right)^2}{M \times N} \tag{3.3}$$

$$\mathrm{PSNR} = 10 \log_{10} \frac{(2^{\mathrm{bits}} - 1)^2}{\mathrm{MSE}} \tag{3.4}$$

Among them, f_{ij} represents the pixels of the original blurred image, $f_{ij}^{'}$ represents the pixels of the enhanced image, the image size is $M \times N$, and MSE is the mean square error. After calculation, the peak signal-to-noise ratio (PSNR) of the single-channel grayscale image before and after frame accumulation is shown in Table 1.

Table 1 Frame accumulation result

Channel B image	PSNR
Single channel	50.742
100 frame accumulation	**52.076**

4 Principle of Bilateral Filtering

When collecting images in an experiment, continuous image sensor shooting for several hours will cause its temperature to be too high, and multiple components in the sensor circuit will also affect each other. These factors will bring a lot of Gaussian noise. Therefore, the averaged gray image is further denoised using bilateral filtering and is also prepared for the edge enhancement of the LOG operator later.

Bilateral filter is a nonlinear filtering method with two weights: spatial weight and similar weight. Compared with Gaussian filtering, the main advantage is edge-preserving and denoising, while smoothing the filter, it retains more edge information.

Spatial weight (fuzzy denoising): Related to the position of the pixel, the distance between the pixels (Euclidean distance, spatial metric), so it can be defined as a global variable placed outside the loop, usually defined as:

$$c(\xi, x) = e^{-\frac{1}{2}\left(\frac{d(\xi,x)}{\sigma_d}\right)^2}$$
$$d(\xi, x) = d(\xi - x) = \|\xi - x\| \tag{4.1}$$

where represents the distance between two pixels (Euclidean distance), which is filtered as follows:

$$h(x) = k_d^{-1}(x) \int_{-\infty}^{\infty} \int_{-\infty}^{\infty} f(\xi)c(\xi, x)\mathrm{d}\xi \tag{4.2}$$

Weights are:

$$k_d(x) = \int_{-\infty}^{\infty} \int_{-\infty}^{\infty} c(\xi, x)\mathrm{d}\xi \tag{4.3}$$

Similarity weight (protection edge): It is related to the size of the pixel value, which is the distance between the pixel values, which varies according to the pixel value, and needs to be placed in the loop, usually defined as:

$$s(f(\xi), f(x)) = e^{-\frac{1}{2}\left(\frac{\sigma(f(\xi),f(x))}{\sigma_\tau}\right)^2}$$
$$\sigma(\xi, x) = \sigma(\xi - x) = \|\xi - x\| \tag{4.4}$$

where $f(\xi), f(x)$ represents the distance between two pixel values; the process is filtered as follows:

$$h(x) = k_\tau^{-1}(x) \int_{-\infty}^{\infty} \int_{-\infty}^{\infty} f(\xi)s(f(\xi), f(x))\mathrm{d}\xi \tag{4.5}$$

Weights are:

$$k_\tau(x) = \int_{-\infty}^{\infty} \int_{-\infty}^{\infty} s(f(\xi), (x))\mathrm{d}\xi \tag{4.6}$$

The combination of the two can obtain bilateral filtering based on the spatial distance, and the overall degree of similarity as follows:

$$h(x) = k^{-1}(x) \int_{-\infty}^{\infty} \int_{-\infty}^{\infty} f(\xi)c(\xi, x)s(f(\xi), f(x))\mathrm{d}\xi \tag{4.7}$$

Weights are:

$$k(x) = \int_{-\infty}^{\infty} \int_{-\infty}^{\infty} c(\xi, x)s(f(\xi), (x))d\xi \tag{4.8}$$

Comparison of experimental results of Gaussian filtering and bilateral filtering for multispectral images after frame accumulation (Fig. 6).

Fig. 6 Comparison of experimental results

The multispectral grayscale image after Gaussian filtering and bilateral filtering changes slightly visually, and its peak signal-to-noise ratio is 1.096 dB higher than that of Gaussian filtering, as can be seen from following Table 2.

Table 2 Filter effect comparison

Filtering method	PSNR (dB)
Gaussian filtering	51.193
Bilateral filtering	**52.289**

5 LOG Detection Operator Theory

The basis of the Laplacian of Gaussain (LOG) operator is the Laplacian operator, and the Laplace operator is the simplest isotropic differential operator with rotation invariance. The Laplace transform of a two-dimensional image function is the isotropic second derivative, defined as:

$$\nabla^2 f(x, y) = \frac{\partial^2 f}{\partial x^2} + \frac{\partial^2 f}{\partial y^2} \tag{5.1}$$

In multispectral images express this equation as a discrete form (two dimensional):

$$\frac{\partial^2 f}{\partial x^2} = f(x+1, y) + f(x-1, y) - 2f(x, y)$$

$$\frac{\partial^2 f}{\partial y^2} = f(x, y + 1) + f(x, y - 1) - 2f(x, y) \tag{5.2}$$

The above is the Laplace operator in mathematical expression, expressing it as a template form:

$$\nabla^2 = [f(x+1, y) + f(x-1, y) + f(x, y+1) + f(x, y-1)] - 4f(x, y) \tag{5.3}$$

The basic method of Laplace sharpening can be expressed by the following formula:

$$g(x, y) = \begin{cases} f(x, y) - \nabla^2 f(x, y), & k > 0 \\ f(x, y) + \nabla^2 f(x, y), & k < 0 \end{cases} \tag{5.4}$$

where k is the center coefficient of the Laplace mask. This simple sharpening method can not only produce the Laplacian sharpening effect, but also retain the background information and superimpose the original image to the Laplacian transformation processing result, so that each gray in the image degree value is preserved, so that the contrast ratio at the abrupt change of gray level is enhanced, and the final result is to highlight small details in the image while preserving the background of the image. However, we found that the Laplacian operator does not have a smoothing process like Sobel or Prewitt when performing edge detection. It is very sensitive to noise and cannot obtain edges in the horizontal direction, vertical direction, or other fixed directions, respectively. Here, it can perform Gaussian smoothing on the image first and then convolve with the Laplacian operator. This operation method is the Laplacian of Gauss (Log).

Constructing a convolution template from the expression of the Log operator, you can first smooth the image with Gaussian and finally find the Laplacian of the result. Because convolution is a linear operation, the expression of the log operator can be written as:

$$g(x, y) = \nabla^2 \left[G(x, y) \otimes f(x, y) \right] \tag{5.5}$$

Analysis of comparative experiment results (Fig. 7).

In order to verify the validity of the experimental results, the LOG detection operator is compared with other detection operators such as Sobel, Prewitt. By comparing the experimental results, it can be seen that the multispectral image processed by the Sobel and Roberts detection operators on the frame accumulates bilateral filtering. The detection effect is very poor; almost no heterogeneity and the surrounding edges are detected. Canny can detect some edges, but the detection effect of the heterogeneity bodies in the solution is still very poor. Under the same conditions, the detection effect of the LOG detection operator is better. Not only the position of the heterogeneity is detected, but also the details of its surroundings are reflected. This shows that the image enhancement algorithm combining the frame accumulation and log detection operator proposed in this paper is effective.

6 Conclusion

Based on the characteristics of biological tissues and the problems of low signal-to-noise ratio and low contrast ratio in transmission imaging, this paper proposes a multispectral

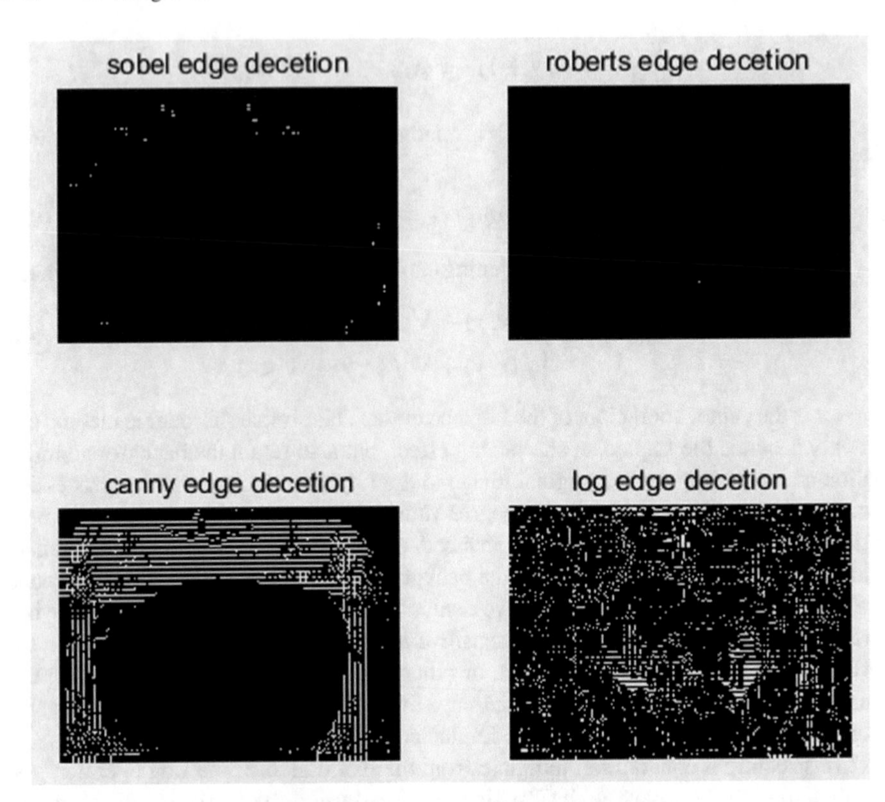

Fig. 7 Comparison of experimental results

transmission image enhancement algorithm combining frame accumulation and LOG edge enhancement. After processing by this algorithm, a higher quality transmission image suitable for heterogeneous body detection is obtained. The experimental results prove that the peak signal-to-noise ratio of the image after the enhancement algorithm in this paper is increased to 52.257 dB, the signal-to-noise ratio of the image is increased by 1.515 dB, and the edges are retained, the contrast ratio is greatly improved, and the quality of the multispectral image is obtained. Then, the image quality evaluation indicators such as image signal-to-noise ratio and contrast ratio are calculated, which proves the effectiveness of the enhanced algorithm in this paper. Finally, through comparison experiments, it is found that the image processed by the enhanced algorithm in this paper increases the signal-to-noise ratio more, and the gray scale resolution and edge contrast ratio have been improved to a certain extent. This improves the detection accuracy of non-uniformity to a certain extent and also provides a new way to improve the image quality of multispectral tissues.

References

1. Huaqing H, Ming L, Peng X (2019) Multi-lead model-based ECG signal denoising by guided filter. Eng Appl Artif Intell 79:34–44

2. Zhang C, Zhang B, Li G, Lin L, Zhang C, Wang F (2019) A preprocessing algorithm based on heterogeneity detection for transmitted tissue image. EURASIP J Wireless Commun Networking 1:99
3. Jiaxin L (2017) Research and implementation of image contrast enhancement algorithm based on Matlab
4. Fei L, Gang L, Wen Y et al (2019) Improving heterogeneous classification accuracy based on the MDFAT and the combination feature information of multi spectral transmission images. Infrared Phys Technol 102:102992
5. Yi W, Gang L, Wen Y et al (2019) Heterogeneity detection method for transmission multispectral imaging based on contour and spectral features. Sensors 19(24):5369
6. Baoju Z, Chengcheng Z, Gang L (2019) Multispectral heterogeneity detection based on frame accumulation and deep learning. IEEE Access 7(1):29277–29284
7. Xuefeng Z, Hui Y (2020) Image denoising and enhancement algorithm based on median filtering and fractional filtering 41:4
8. Liwen H, Bo W (2020) Song Tao. Research on low-light color image enhancement algorithm 1:34
9. Siok K, Jenerowicz A, Ewiak I (2020), A simulation approach to the spectral quality of multispectral images enhancement 174
10. Yuebin Z, Jianmin Y, Yujin D, Rashid SF (2020) New improved optimized method for medical image enhancement based on modified shark smell optimization algorithm 21:20

Research on Intelligent Release Strategy of Air Traffic Control Automation System Based on Control Experience

Liu Yan(✉)

State Key Laboratory of Air Traffic Management System and Technology, Nanjing 210007, China
16108456@qq.com

Abstract. The core of air traffic control is interval management. Our running status quo controllers based on experience to adjust the aircraft's altitude, heading and other flight parameters and the planned route, notification through voice communication of flight staff. When the aircraft between occurs when dangerously close automated systems have warning prompt notice, but the system lacks real-time the intelligence interval management decision support tool. Aiming at this problem, this article introduces conflict detection methods and studies of the release strategy based on regulatory experience. The technology used in intervals to maintain aid decision-making tool designed With this tool, the controllers in the system under the condition of intelligent assistance, more reasonable control instructions are issued to reduce the burden on the controller. Solve the problem of excessive interval caused by completely manual mode and improve the efficiency of airspace use.

Keywords: Air traffic control · Control experience · Intelligent conflict relief

1 Introduction

With the continuous growth of the national economy, China's civil aviation industry has developed rapidly. However, with the development of the civil aviation industry and the increase of military aircraft training tasks, the contradiction between the limited airspace resources and the increasing traffic flow has become increasingly prominent. The surge in flights has greatly increased the likelihood of flight conflicts between aircraft, which undoubtedly increased the workload of controllers. It puts forward higher requirements for improving the efficiency of airspace use.

Currently, most parts of China are using radar control, and the flight controllers are monitoring the situation and give control instructions based on experience, to achieve spacing. Controllers often put safety first in the process of being on duty. Due to the lack of scientific and reliable system strategies and support tools, it often leads to enlargement of flight intervals. This extensive interval management model has been unable to adapt to the current operating environment. Therefore, the key to intelligent air traffic management

Q. Liang et al. (eds.), *Artificial Intelligence in China*, Lecture Notes in Electrical Engineering 653, https://doi.org/10.1007/978-981-15-8599-9_40

is to be able to provide decision support tools for controllers, to improve China's air traffic control level, and to alleviate the current situation of airspace congestion, which are important.

NASA in 2014 is to carry out verification airspace technology (Airspace Technology Demonstration, ATD) project, in which the first stage is to focus on running the terminal area stage, through the development of airborne separation management tools (Flight Deck Interval Management, FIM) and ATC interval management tool (the Controller Managed Spacing, CMS), etc., to enhance the operating efficiency of flights from cruise to landing period, increasing airspace capacity. Among them, the controller interval management tool CMS is used to maintain the interval of the terminal area and can provide speed and height adjustment suggestions according to the setting of the interval standard.

Through the analysis of actual operation requirements and foreign research progress, this article focuses on the conflict detection technology and design of auxiliary tools in interval management. First, carding aviation operations flight separation standards introduce interval management functions of the current air traffic control automation system, based on research 4D trajectory 10 collision detection within minutes method. Finally, design air traffic control system tools to assist controllers in their daily control.

2 Flight Interval Standard

2.1 Main Factors Influencing the Separation Standards and Principles to Determine the Minimum Safety Distance

The factors that determine the separation standard include safety, capacity, the fuel-efficient route that the airline expects to fly, the performance of the communication navigation monitoring system, weather, etc. which are all factors that affect the development of the separation standard. With the improvement of aircraft performance and communication navigation monitoring performance, the separation standard can be appropriately reduced. The more complex the airspace structure, the worse the flight environment, and the separation standard should be appropriately increased. The factors for setting the separation standards for each country are different according to their own circumstances.

The optimal value of the interval standard is a comprehensive consideration of the deviation of the route. The implementation of the interval standard may lead to delay losses, and the loss caused by possible collisions. The minimum interval reached by consumption. If the interval standard is set too large, the capacity will be reduced, and if the interval standard is set too small, the risk of collision will increase. The sum of the delay cost and the collision cost is called the total cost. The total cost will increase with the increase of the interval standard and will increase to an inflection point. This inflection point is the optimal value of the interval standard.

2.2 General Interval Standards

(a) Horizontal interval

The standard for longitudinal (front-to-rear) separation between aircraft is defined as 10 min (approximately 150 km) under procedural control conditions, and the standard for longitudinal (front-to-back) separation under radar surveillance is 75 km and 10 km under radar control conditions The horizontal interval under international radar control conditions is 5 nautical miles.

(b) Vertical interval

In November 2007, China began to reduce the vertical separation in high-altitude control areas. Detailed vertical separation of 600–8400 yards is in a range of 300 m, at 8400–8900 m in a range of 500 m, at 8900–12,500 m in a range of 300 m, 12,500 m or more to 600 m.

2.3 Intervals of China Issued Regulations

According to the "Flight Separation Regulations" issued by the Air Traffic Control Commission of the State Council of my country, my country's flight separation standards include vertical separation standards, visual flight horizontal separation standards, instrument flight horizontal separation standards, radar separation standards and wake separation standards. It is stipulated that when implementing radar control, the approach control range shall not be less than 6 km, and the regional control range shall not be less than 10 km [1].

The Basic Flight Rules of the People's Republic of China (2007) describes that the flight interval is the minimum safety distance that should be maintained between aircraft specified in order to prevent flight conflicts, ensure flight safety and improve flight space and time utilization [2].

3 Interval Management Function of ATC Automation System

Air traffic control automation system can be considered as interval management function, including short-term conflict alert, flight plan mid-term conflict detection warning, approach and departure ranking.

3.1 Short-Term Conflict Warning

The system can detect the possibility of dangerous approach within a short period (2–5 min) of the track located in the conflict warning zone in real time. If the current horizontal and vertical distances between the tracks are lower than the warning interval at the same time, or if they will be lower than the warning interval at a certain time in the future, a conflict warning will be issued. The set parameter values include alarm zone, horizontal interval, vertical interval, look-ahead time and warning time [3].

3.2 Mid-Term Conflict Detection and Early Warning of Flight Plan

The mid-term conflict early warning of the flight plan is to speculate on all plans related to the trajectory (15–25 min) [4], using the warning area parameters where the current location of the relevant trajectory is located. Based on the predicted flight plan trajectory, the system calculates the vertical and horizontal distances of the two aircraft in the forward-looking time and checks whether they will be less than the flight plan warning standard at the same time. The parameter settings and short-term conflicts are basically the same.

4 Conflict Detection Technology

4.1 Flight Conflict Screening

In order to improve the computational efficiency, a 4D grid-based flight conflict preliminary screening algorithm is proposed. Taking en route flight as an example, this method first uses a four-dimensional space-time grid to discretize the entire flight space, and the size of each grid unit is based on flight safety. The standard interval setting is 5 nautical miles in length and width and 1000 feet in height. Distribute the discrete track points of the aircraft into the corresponding 4D grid. By checking each non-empty adjacent grid, potential flight conflict points can be detected. Generally, if there are co-existing track points of different flights in the 4D grid or there are track points of different flights in the adjacent grid, it can be judged that there is a potential flight conflict.

Set (x_0, y_0, z_0, t_0) as the coordinate origin 4D grid temporal position, the time t_0 variation range $[0, +\infty)$, although the time axis is continuously changed, but in the actual operation of the conflict determination impossible at any given time, only certain time intervals, to achieve discrete time period $\Delta\tau_0$ (small enough) conflict judgment.

In order to realize the preliminary screen of flight conflict based on 4D grid, the 4D coordinate of any reference aircraft is set as, where is the track point number, and it is defined that it falls in the airspace grid cell in the time period. $i(x_{ij}, y_{ij}, z_{ij}, t_{ij})j\ t_n = t_{ij}A_{ij}^0$. In order to judge whether the flight path of the aircraft is in potential conflict danger, it is necessary to determine whether the 4D grid or the neighborhood grid corresponding to each time period has the coexistence of flight path points of other aircraft. The "three-dimensional" matrix formed by the grid and the grids in its neighborhood is defined as, where: $A_{ij}^0\ 3^3 - 1 = 26\ A_{ij} = \left[A_{ij}^1\ A_{ij}^2\ A_{ij}^3\right]$

$$A_{ij}^1 = \begin{bmatrix} A_{ij}^{111} & A_{ij}^{112} & A_{ij}^{113} \\ A_{ij}^{121} & A_{ij}^{122} & A_{ij}^{123} \\ A_{ij}^{131} & A_{ij}^{132} & A_{ij}^{133} \end{bmatrix}_{3\times3}, A_{ij}^2 = \begin{bmatrix} A_{ij}^{211} & A_{ij}^{212} & A_{ij}^{213} \\ A_{ij}^{221} & A_{ij}^{222} & A_{ij}^{223} \\ A_{ij}^{231} & A_{ij}^{232} & A_{ij}^{233} \end{bmatrix}_{3\times3}, A_{ij}^3 = \begin{bmatrix} A_{ij}^{311} & A_{ij}^{312} & A_{ij}^{313} \\ A_{ij}^{321} & A_{ij}^{322} & A_{ij}^{323} \\ A_{ij}^{331} & A_{ij}^{332} & A_{ij}^{333} \end{bmatrix}_{3\times3}. \tag{3.1}$$

The matrix represents the nine mesh neighborhoods of the upper layer and the lower layer, respectively $A_{ij}^1, A_{ij}^3 A_{ij}^0 A_{ij}^2$. Represents the nine grid neighborhoods of the grid layer, grid and is the same grid $A_{ij}^0 A_{ij}^{222} A_{ij}^0$.

It is defined that when there are other aircraft track points in any grid of and neighborhood, that is, there are any elements in the matrix, where, it indicates that there is a potential flight conflict $A_{ij}^{0}A_{ij}A_{ij}^{mnk} = 1m, n, k \in \{1, 2, 3\}$. Otherwise, when all elements are zero, formula (3.2) is satisfied, indicating that no conflict occurs.

$$\sum_{m=1}^{3}\sum_{n=1}^{3}\sum_{k=1}^{3} A_{ij}^{mnk} = 0. \tag{3.2}$$

This potential conflict detection scheme is implemented using a hash table data structure. For a given discrete 4D track, each sampling point is mapped to each 4D grid unit, and a series of flight identification information. There is no need to store 4D coordinates in the data structure, which greatly reduces the required memory space. After the initial flight release time is modified, the total number of potential conflicts can also be easily updated [5].

4.2 Based on Geometric Method of Flight Conflict Detection

Since it is only determined based on the divided 27 grids whether there is a flight conflict, the safety separation standard has been expanded three times invisible, and it is easy to cause too many false alarms. Therefore, the potential flight conflict situation judged by the 4D grid detection method is only a preliminary screening process. To obtain accurate conflict detection, further flight conflict determination is required. Mainly for the predicted flight trajectory 4D conflict detection, i.e., in line with geometric deterministic algorithm [6], by predicting the 4D trajectory inferred track point vector difference between the aircraft is less than the minimum safe separation standards to achieve flight conflict detection.

Assuming that the preliminary screening algorithm for flight conflict based on 4D grid has determined that there is a potential flight conflict between track points on the predicted flight tracks of two aircraft (numbered as respectively) in the time period, the corresponding THREE-DIMENSIONAL space coordinates are denoted as and respectively, then the relative position vector between aircraft is expressed as t' i, j $p_i^k(x_i^k, y_i^k, z_i^k)$ $p_j^l(x_j^l, y_j^l, z_j^l)$ $\vec{p}_{i,j}^{\,k,l} = (\Delta x_{i,j}^{k,l}, \Delta y_{i,j}^{k,l}, \Delta z_{i,j}^{k,l}) = p_A - p_B$.

When there is flight conflict between reference aircraft A and test aircraft B, the following equations can be met:

$$\begin{cases} \sqrt{(\Delta x_{i,j}^{k,l})^2 + (\Delta y_{i,j}^{k,l})^2} < s \\ |\Delta z_{i,j}^{k,l}| < H \end{cases}, \tag{3.3}$$

where, is the minimum vertical safety interval of the flight protection area, that is, the height of the cylinder protection area; $H = 304.8$ m (1 ft $= 0.3048$ m) $s = 9260$ m (1 NM $= 1852$ m). Is the minimum horizontal safety interval of the flight protection zone, namely the radius of the cylindrical protection zone, which can be set flexibly through the configuration file.

5 Conflict Relief Strategy Based on Regulatory Experience

At present, the main method for controlling units to resolve conflicts is to change the course, altitude, speed and ascent/descent rate of the aircraft. Among them, the fastest and most effective method is to change the course and altitude of the aircraft. According to the different types of conflicts, conflict relief methods can be divided into horizontal avoidance and vertical avoidance. Wherein, in accordance with the positional relationship of the different aircraft conflict occurs, the level of the relief is divided into forward flight, the flight and cross reverse flight, the flight but also cross points of intersection an acute, obtuse angle, intersect at right angles and the like.

The method of conflict resolution on the front line is usually a "coarse" classification, which can only cover most conflict situations. Therefore, in the following classification of control conflicts and methods of control conflict relief, there may be some deviations in the control suggestions provided for specific situations, so we combined the airspace operation in the algorithm to determine the best solution for conflict relief, reducing the number of controllers. Reconfirm the type of conflict load after receiving regulatory advice. The control suggestion database can analyze the history of flight conflicts through big data analysis based on historical data, continuously expand its content and optimize its method of conflict relief. The release strategies for different positional relationships are described in detail below [7].

5.1 Horizontal Avoidance

As shown in Fig. 1, the front-to-back aircraft is flying forward, and there is catchup, which may cause the front-to-back aircraft interval to not meet the horizontal interval in the future. Conflict resolution is as follows:

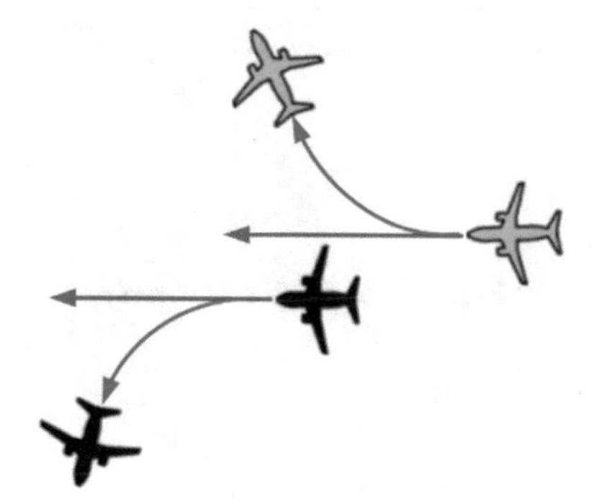

Fig. 1 Forward flight

(1) First direct the second plane to turn right (or left), and then direct the front plane to turn in the opposite direction;
(2) If the two aircraft have a certain lateral separation, then command each to evade away from each other.

(a) **Reverse flight**

As shown in Figs. 2 and 3, the front and rear aircraft is flying head to head, and the relative speed between the two aircraft is very fast. If they are not deployed in time, it is likely to cause conflict. Conflict resolution is as follows:

(1) When there is no lateral separation between the two aircraft, the two aircraft are usually instructed to turn right and avoid.
(2) If the two aircraft have a certain lateral separation, then command each to avoid the other side.

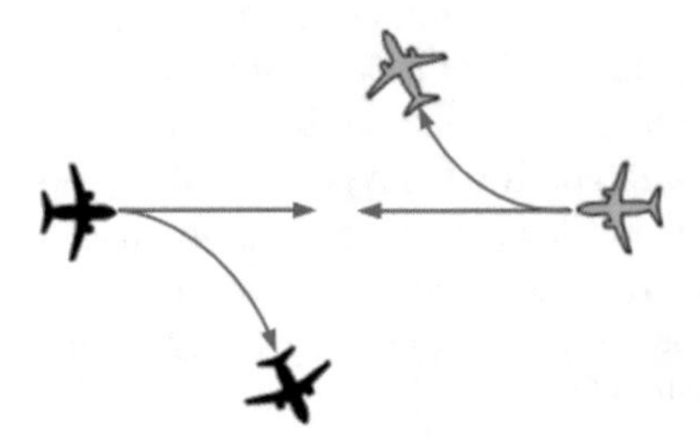

Fig. 2 Reverse flight

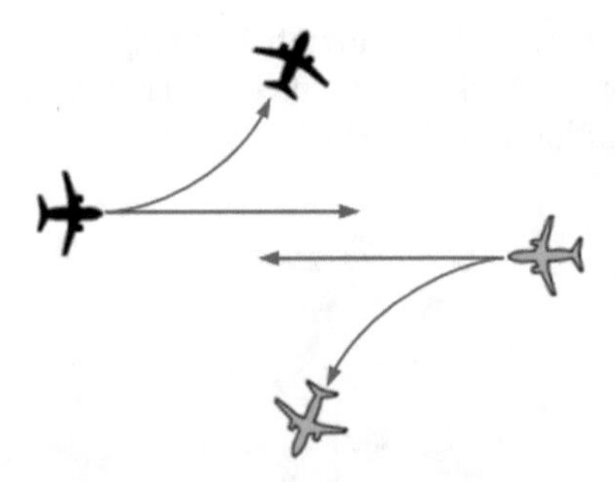

Fig. 3 Lateral interval reverse

(b) **Cross flight**

Due to the high speed and close approach of the aircraft in the area of regional control, and the maneuverability is not as good as that of low-altitude flight, coupled with the efficiency of conflict release commands, aircraft pilot operations, and aircraft response are delayed, the conflict release of cross flight is more complicated and indeterminate (Figs. 4, 5, 6 and 7).

1. **Acute Angle of cross**

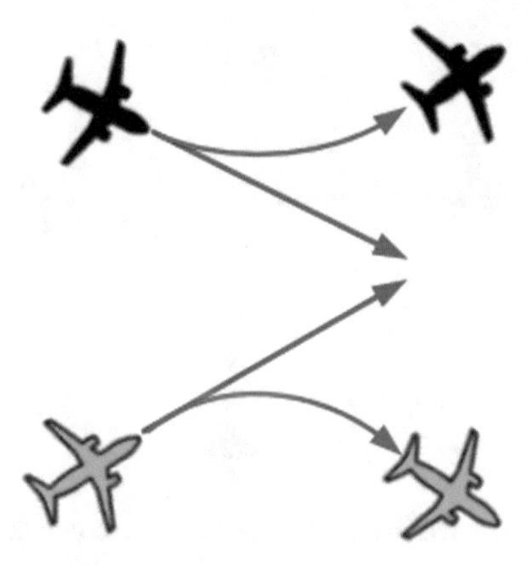

Fig. 4 Intersection of acute angles

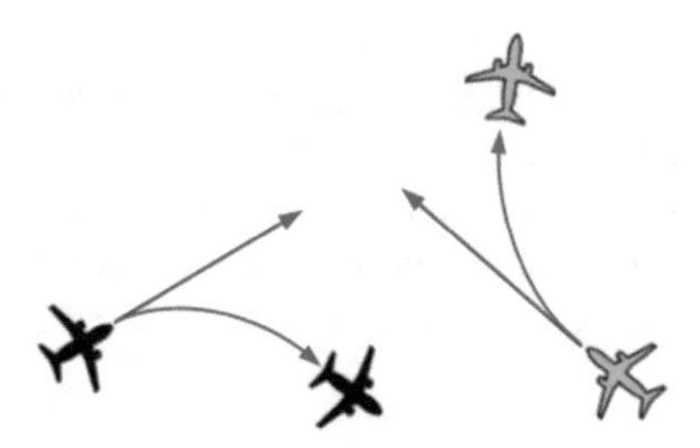

Fig. 5 Intersection of obtuse angles

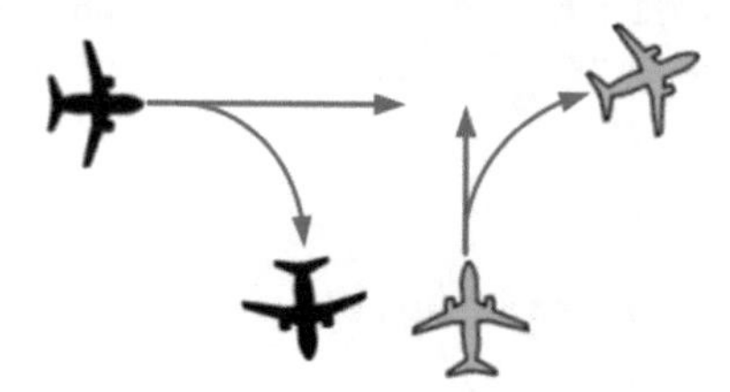

Fig. 6 When two machines are far from the intersection

Conflict resolution: The intersection angle of the two machines is between 0° and 90°. It is suggested that the two machines turn away from each other.

2. **Obtuse Angle of cross**

Conflict resolution: The intersection angle is between 90° and 180°. It is recommended that both machines turn to the right or to the left at the same time to avoid the large angle.

3. **Rectangular cross**

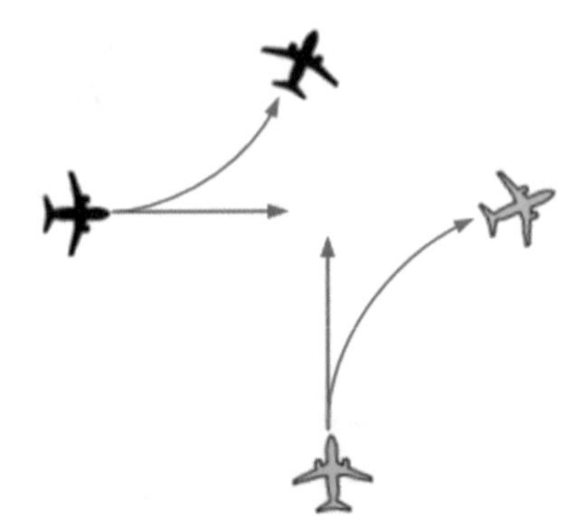

Fig. 7 When the two machines are close to the intersection

Conflict resolution is as follows:

(1) When the two planes are far from the intersection (more than 1 min), it is suggested to turn away from each other.
(2) When the two aircraft are close to the intersection (less than 1 min), it is suggested to direct the aircraft far from the intersection to make a big angle turn to avoid the other side and at the same time direct the other aircraft to avoid the direction far from the other side.

If it is impossible to judge the distance between the two aircraft and the intersection point, the aircraft with good maneuverability (slow speed, small aircraft) should be given priority to steer away from the other side at a large angle and at the same time command the other aircraft to steer away from the other side.

5.2 Vertical Avoidance

(a) The vertical avoidance of forward flight

Considering that there is an altitude difference of at least 600 meters between the forward-flying aircraft and that the automatic at C equipment has a relatively complete altitude breakthrough alarm function at present, as long as the controller can identify and deal with the altitude breakthrough alarm in the first time, it is generally not easy to have the problem of secondary crossing. Therefore, it is suggested that the first choice is to have the aircraft stop ascending/descending immediately and then instruct another aircraft to accelerate ascending/descending to avoid the other side.

(b) Vertical avoidance in cross or reverse flight

In this type of conflict, the principle of vertical avoidance cannot be simply determined. The general principle is to avoid the occurrence of secondary crossing. Depending on the specific situation, the termination of the crossing can be judged according to the vertical approach rate and the height difference between the two planes when the conflict occurs, so that the aircraft will immediately stop ascending/descending or continue the current movement trend of the aircraft to accelerate the crossing.

6 Height Adjustment Aid Decision Tool Design

Controller specified aircraft flying height, arrival time, flight routes, fails to notify the pilot before the 4 d track forecast system for aircraft (in this paper, the prediction method is not detailed description), on the basis of prediction, using a third chapter algorithm inference will happen between track flight conflict or danger close, namely, to judge the internal between aircraft is less than the minimum interval flight safety standards, will risk information feedback to the controller, is used to make decisions. In this paper, the height adjustment in the control process is considered, and the tools for specifying the height and for deducing the access to the control area are designed.

(a) Specified height tool

The controller adjusts the flight altitude of CCA0623. Click the current altitude and the height level that may be adjusted, and the corresponding climbing and descending rate will pop up. The system calculates that those with possible conflicts in the next 10 min will be red, while those without conflicts will be green (Figs. 8 and 9).

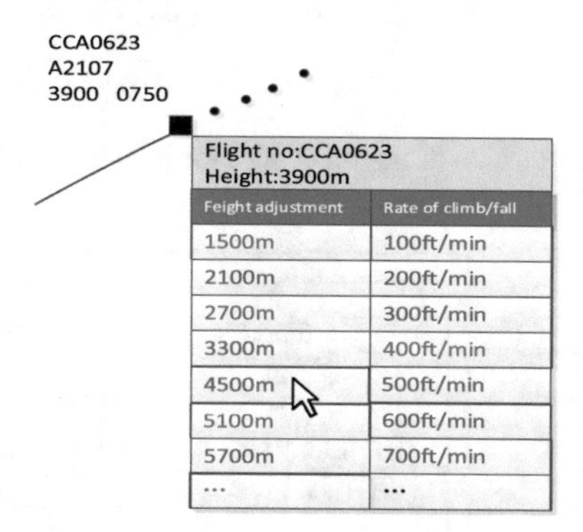

Fig. 8 Specifies the height tool design interface

(b) Height deduction tool for access control area

The controller conducts the height of entering and leaving the control area for the target about to enter or fly out of the control area (flight height is generally used between sectors) and sets the height of entering and leaving the area. The system deducts whether there is a conflict in the next 10 min and prompts the controller to adjust if there is a conflict, as shown in Fig. 10, and the system processing process is shown in Fig. 11.

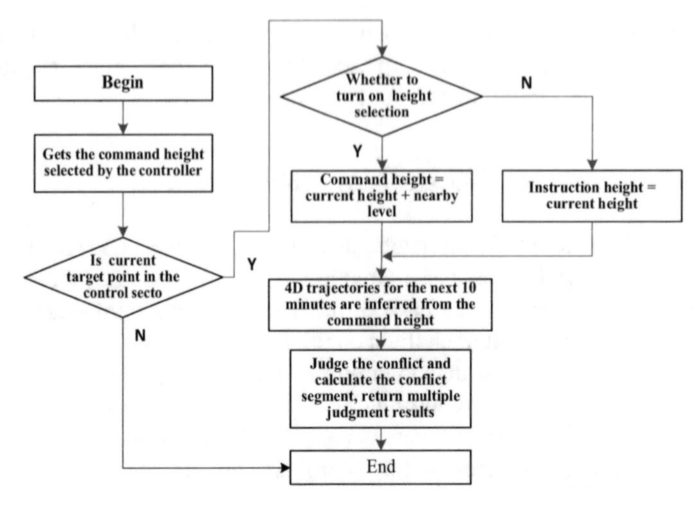

Fig. 9 Specifies the height conflict detection processing flow

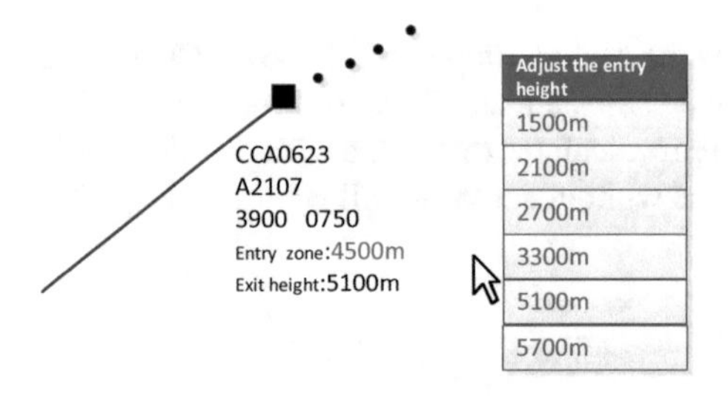

Fig. 10 Design interface of inference tool for inlet and outlet control area

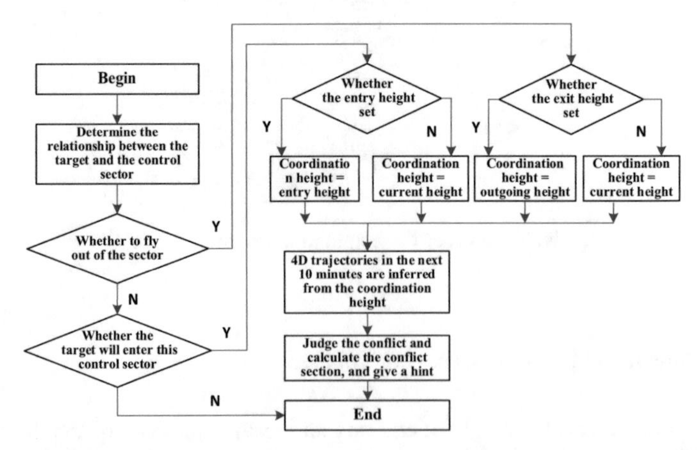

Fig. 11 Processing flow of high conflict detection for inlet and outlet area

7 Conclusion

This article sorts out China's flight separation standards, explains the current interval management functions of the ATC automation system and introduces flight conflict detection. Combined with the control experience, the intelligent conflict resolution strategy is studied, and an auxiliary decision-making tool is designed, which can assist the controller to adjust the aircraft flying height reasonably. But if we want to achieve the goal of maintaining reasonable intervals and improving the efficiency of airspace use, we also need to provide auxiliary decision-making tools such as heading and arrival time adjustment. This is a problem that we need to study together.

References

1. Regulations on Flight Separation of the Air Traffic Control Commission of the Central Military Commission of the State Council. Air Traffic Control Commission of the Central Military Commission of the State Council, Beijing, May 31, 2002
2. State Council Order No. 312. Basic Flight Regulations of the People's Republic of China
3. Hongyuan Z (1998) Study on the model of the number of dangerous collisions between aircrafts on two crossing routes. Syst Eng Electron Technol 20(5):6–9
4. Mining algorithm of radar track control intention. Command Information System and Technology, 2014, 1(5):33–36
5. Du solid, Haibo, the transfer of a multi-sector control separation Management. Haibo 20 16, 13 (16):130–137
6. Chunjin L, Yingxun W (2001) Mathematical model of the collision hazard of planes on parallel routes. J Civil Aviation China 23(5):31–34
7. White AL (2010) An aircraft separation algorithm with feedback and perturbation. NASA Langley Research Center USA

A Method Based on Deep Reinforcement Learning to Generate Control Strategy for Aircrafts in Terminal Sector

Qiucheng Xu[1,2(✉)], Jinglei Huang[1], Zeyuan Liu[2], and Hui Ding[2]

[1] College of Civil Aviation, Nanjing University of Aeronautics and Astronautics, Nanjing 211106, China
xuqiucheng1987@163.com

[2] State Key Laboratory of Air Traffic Management System and Technology, The 28th Research Institute of China Electronics Technology Group Corporation, Nanjing 210007, China

Abstract. With the increasing air traffic density and complexity in terminal sector, air traffic controllers will face more challenges and pressures to ensure the safe and efficient operation of air traffic. In this work, an artificial intelligence (AI) agent based on deep reinforcement learning is built to mimic air traffic controllers, such that the dense, complex and dynamic air traffic flows in terminal airspace can be handled sequentially and separated. To simplify the problem, the complex three-dimensional terminal sector is projected onto the vertical plane by dispersing state space and action space. And then, the typical reinforcement learning algorithm, double deep Q-network, is taken to realize the AI agent. Results show that the built AI agent can guide 6 aircrafts safely and efficiently through Sector 01 of Nanjing Terminal, simultaneously.

Keywords: Air traffic management · Terminal sector · Artificial intelligence agent · Deep reinforcement learning

1 Introduction

With the rapid development of China's air transport industry, the number of flights is increasing at an average annual rate of over 10% in recent years [1]. What is more, by 2030, there will be more than 450 civil transport airports, and the volume of passenger traffic will reach 1.8 billion [2]. However, the rapid development of air transport would lead to the increasing challenges and pressures [3] for air traffic controllers to ensure the safe and efficient operation of air transport.

To deal with the challenges of current and future air traffic demands, Civil Aviation Administration of China (CAAC) has proposed the idea of

Q. Liang et al. (eds.), *Artificial Intelligence in China*, Lecture Notes in Electrical Engineering 653, https://doi.org/10.1007/978-981-15-8599-9_41

Four Enhanced Air Traffic Management (ATM) Solutions in 2018 [4], including enhanced security, enhanced efficiency, enhanced intelligence and enhanced collaboration, where the enhanced intelligence is to suggest that the ATM should take new technologies, such as big data, blockchain, artificial intelligence (AI) and so on, to promote the operational effectiveness significantly. Therefore, some researchers are working on how to apply these new technologies to the aviation.

Deep Reinforcement Learning (DRL) framework and algorithm is one of the most famous AI technologies, which has a great ability of dealing with continuous sequential decision-making problems [5,6], and have been demonstrated to perform high level tasks and learn complex strategies, such as play the games of AlphaGo [7] and StarCraft-II [8].

Inspired by this, many researchers try to solve many difficult decision-making problems in ATM by using the deep-reinforcement Learning Algorithm.

In [9], the authors adopted a reinforcement learning method to predict the taxi-out time of the flight, and the predicted taxi-out time result is then compared with the actual taxi-out time to reduce the taxi-out time error. In [10], a Multi-agent system using Reinforcement Learning is developed for both simulation and daily operations to support human decisions, where two types of reward functions are proposed for air traffic flow management (ATFM) decision making to control safety separation of Ground Holding Problem (GHP) and Air Holding Problem (AHP).

In Tumer and et. al [12], proposed a multi-agent algorithm based on reinforcement learning for traffic flow management, where each agent is associated with a fix location and its goal is to set separation and speed up or slow down traffic flows to manage congestion. At last, the proposed method is tested on an air traffic flow simulator, FACET. In their following work [11], the authors proposed a distributed agent based solution where agents provide suggestions to human controllers, and an agent reward structure is designed well to allow agents to learn good actions in the indirect environment, such that the "Human-in-the-Loop" solution can be achieved.

Brittain and Wei [13] proposed a hierarchical deep reinforcement learning algorithm to build an AI agent, which takes the NASA Sector 33 app as the simulator. The well-trained AI agent can guide aircraft safely and efficiently through "Sector 33" and achieve required separation at the metering fix. And then, the authors [14] also proposed a deep multi-agent reinforcement learning framework to identify and resolve conflicts between aircrafts in a high-density, stochastic, and dynamic en route sector with multiple intersections. However, these work only considered the horizontal space separation and ignored the vertical separation of airspace.

In this work, an artificial intelligence (AI) agent based on deep reinforcement learning is built to mimic air traffic controllers, such that the dense, complex and dynamic air traffic flows in terminal airspace can be handled sequentially and separated. To simplify the problem, the complex three-dimensional terminal airspace is projected onto the vertical plane by dispersing state space and action space. And then, the typical reinforcement learning algorithm, double deep Q-network, is taken to realize the AI agent. Results show that the built AI agent can guide 6 aircrafts safely and efficiently through Sector 01 of Nanjing Terminal, simultaneously.

The remainder of the paper is organized as follows. Section 2 describes the problem definition. Section 3 shows the proposed DDQN-based method to generate control strategy. Experimental results and conclusions are discussed in Sects. 4 and 5, respectively.

2 Problem Definition

In terminal sectors, air traffic controllers are responsible to ensure safe separation among all aircrafts by sending sequential instructions, such as height adjustment and speed adjustment. However, it will be a challenge problem for controllers to ensure the safe and efficient operation of air transport with the increasing air traffic density and complexity in terminal airspace.

Reinforcement learning is one type of sequential decision making, which is do well in learning how to act optimally in a given environment. Therefore, this air traffic control decision-making problem can be formulated as a reinforcement learning model, which is consistent with Markov decision process and Fig. 1 shows an example of reinforcement learning problem interacting with the environment through trial and error. As shown in Fig. 1, we can get the decisions-making sequence $s_t \rightarrow a_t \rightarrow s_{t+1} \rightarrow a_{t+1} \cdots$, and the decisions sequence from the first stage to the last stage form one strategy. In addition, the AI agent would receive a reward r from each transition.

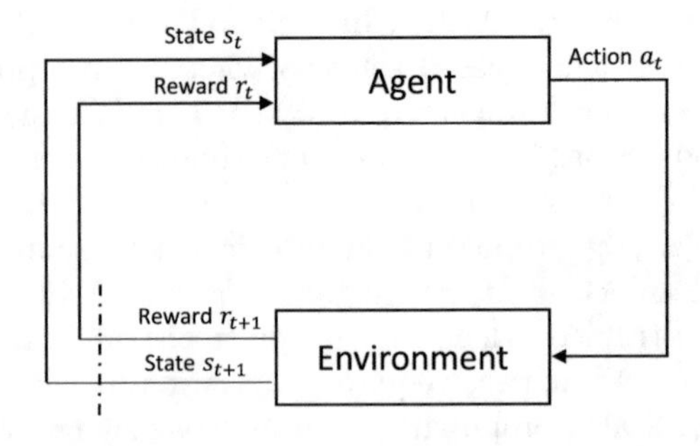

Fig. 1. An example of reinforcement learning problem interacting with environment

Learning to AI agent directly from three-dimensional terminal sector environment is a big challenge of reinforcement learning. In this work, to simplify the problem, we only adjust the height of aircrafts, thus the complex three-dimensional terminal airspace can be projected onto the vertical plane by dispersing state space and action space.

Generally, a reinforcement learning problem involves state space, action space, reward function and objective function, which are defined as follows for the built AI agent.

(1) State space S, which contains all information about the environment and each element $s_t \in S$ can be considered a snapshot of the environment at time t, where aircraft positions (x_i, y_i) and height h_i for each aircraft i are included.

Figure 2 gives a state at time t of the environment that is dispersed by the rectangles with size of h x d, where h is the height that aircraft can raise or descend in four seconds, d is the distance that aircraft moves horizontally in four seconds.

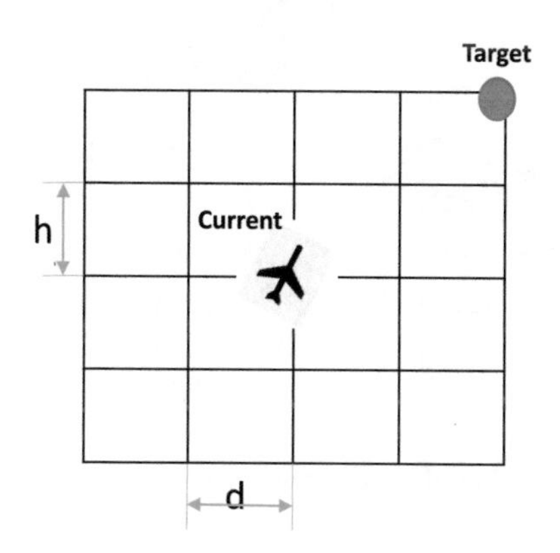

Fig. 2. The dispersed state space on the vertical plane

(2) Action space A, which is the set of all actions that AI agent could select in the environment. In this work, the built AI agent can take one action every four seconds to change the height of aircrafts, and the actions include maintain height, raise height and descend height.

(3) State transition: In each state, the AI agent can choose one decision a from a set of feasible decision options A. Corresponding to a decision a, we can get the transition from a state s_i to another state s_j.

Figure 3 gives a simple example to show the state transition from one state to another by taking different actions. As shown in Fig. 3, given the current state of Fig. 3a, we can get the next environment state as shown in Fig. 3b–d based on different actions, such as maintain height, raise height and descend height.

(4) Objective: The goal of the agent is to interact with the emulator by selecting actions in a way that maximises future rewards, where the selected action can maintain safe separation between aircraft and resolve conflicts for all aircraft in the sector by providing height adjustment, and all aircrafts arrival at the target positions with maximization of the cumulative reward from each transition.

3 Deep Reinforcement Learning Based Method

In this work, we consider a typical reinforcement learning algorithm, double deep Q-network (DDQN), to generate strategy for AI agent. In this work, the

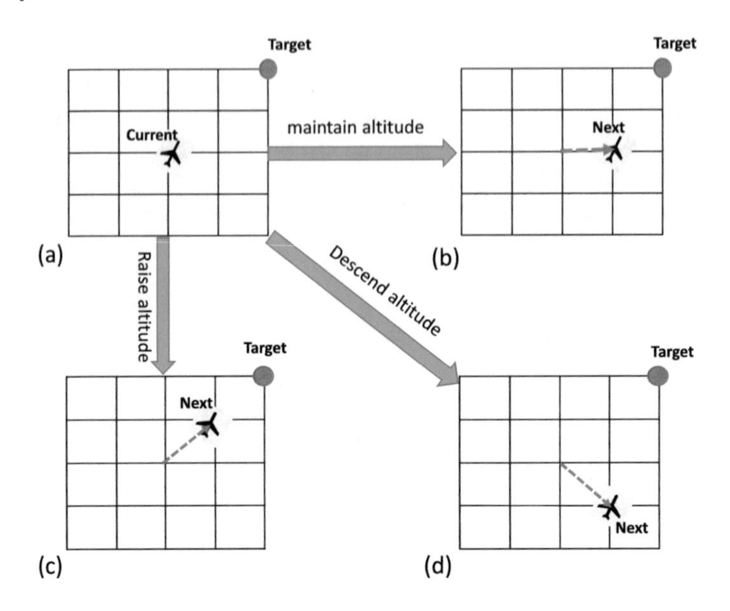

Fig. 3. An example of state transition based on different actions

designed architecture of DDQN contains three fully-connected hidden layers, and the number of nodes in each layer are 512, 256 and 256, respectively. **Algorithm 1** shows the overall flow of the proposed DDQN-based method.

During the training process, ϵ-greedy search strategy is taken, and ϵ is decayed from 1.0 to 0.01. In the experiment, the $MaxBuffer$ is set to 2000.

In step 7, when the selected action a_t is taken on the environment, we can get the next state of environment s_{t+1}, reward r_t and the flag is_end that represents whether the episode is terminated.

The reward function r_t should be designed to reflect the goal of AI agent, which is defined as follows:

$$r_t = \begin{cases} 1000 \cdot \alpha/(dis+1), & \text{if action is raise height} \\ 0.2, & \text{if action is maintain height} \\ -1000 \cdot \alpha/(dis+1). & \text{if action is descend height} \end{cases} \tag{1}$$

where dis represents the distance from the current position to the target position, and α is a flag that represents whether the aircraft is approaching or departing. If the aircraft is approaching, we set $\alpha = -1$, otherwise, α=1. Reward function r_t is calculated at each time-step. And, once safe separation are not satisfied, or aircraft overpasses the terminal sector boundary, or sector handover condition is not satisfied, reward r_t will minus 10. If all aircrafts reached their corresponding target positions, reward r_t will add 10. With the reward definition, the AI agent would prioritize control the aircrafts nearest the corresponding target position to land or climb.

Algorithm 1 DDQN-based method

```
1: Initialize Evaluation network Q and Target network Q';
2: Initialize a queue ReplayBuffer to empty;
3: score = 0;
4: for i = 1 to n do
5:     while is_end is not true do
6:         a_t = Q.ChooseAction(s_t);
7:         s_{t+1}, r_t, is_end = Environment.ExecuteAction(a_t);
8:         score += r_t;
9:         ReplayBuffer.push(s_t, a_t, s_{t+1}, r_t, is_end);
10:        if ReplayBuffer.size() ≥ MaxBuffer then
11:            Randomly drawing samples from ReplayBuffer;
12:            Calculate the target Q value;
13:            Update the network parameters;
14:        end if
15:    end while
16: end for
```

In this work, the learning process in one episode is terminated when one of the following four situation is satisfied:

(1) All aircrafts reached their corresponding target positions $(x^i_{target}, y^i_{target}, h^i_{target})$ without collision, that is,

$$\sqrt{(x_i - x^i_{target})^2 + (y_i - y^i_{target})^2 + (h_i - h^i_{target})^2} = 0, \ \forall i. \tag{2}$$

(2) An aircraft overpasses the terminal sector boundary;
(3) Sector handover condition is not satisfied, that is,

$$\sqrt{(x_i - x^i_{target})^2 + (y_i - y^i_{target})^2} = 0 \quad \text{and} \quad h_i - h^i_{target} \neq 0, \ \forall i \tag{3}$$

(4) Collision is occurred between aircrafts, that is,

$$\sqrt{(x_i - x^i_{target})^2 + (y_i - y^i_{target})^2 + (h_i - h^i_{target})^2} < \delta, \ \forall i \neq j \tag{4}$$

At last, by training the proposed DDQN model until it converges, we can obtain the optimal control strategy for aircrafts in terminal sectors.

4 Experimental Results

The proposed AI agent construction method have been implemented in Python-language on a 64-bit workstation (Intel 2.4 GHz, 256 GB RAM).

In this work, the simulator based on Sector 01 of Nanjing terminal is constructed as our air traffic control environment. Figure 4 gives an example of the constructed simulator environment, where there are 10 approach routes and 15 departure routes.

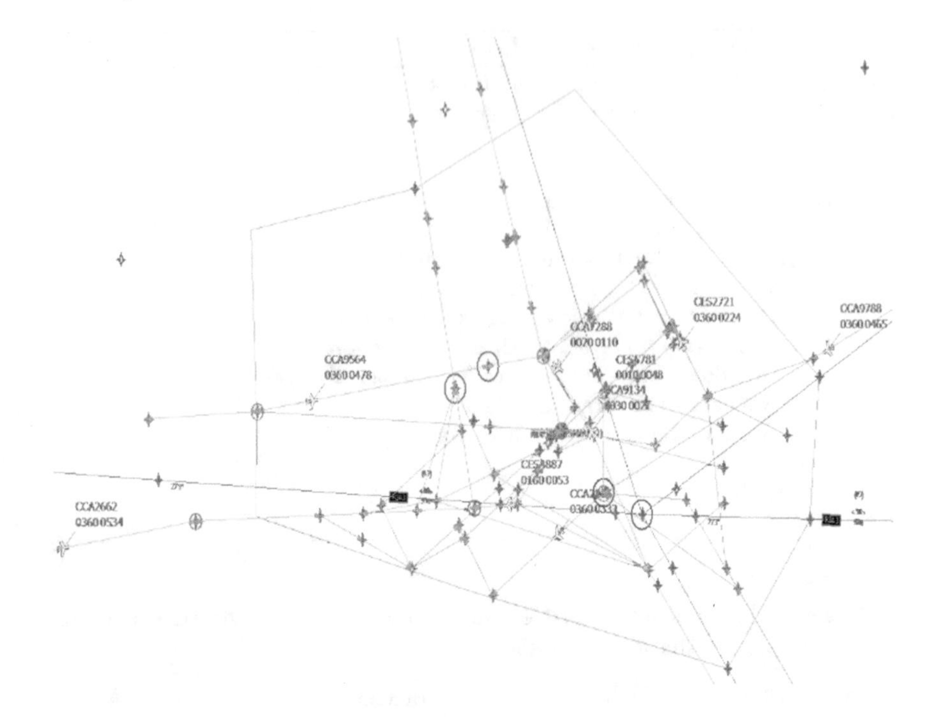

Fig. 4. The simulator environment based on Sector 01 of Nanjing terminal

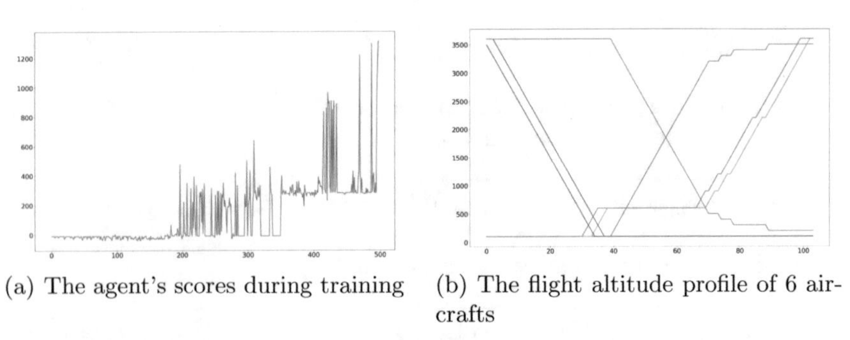

(a) The agent's scores during training

(b) The flight altitude profile of 6 aircrafts

Fig. 5. The experiment results on Sector 01 of Nanjing terminal

In the experiments, we considered 6 aircrafts with control to evaluate the performance of our reinforcement learning framework. By training the AI agent on around 1000 episodes and choosing a time-step of four seconds, we can obtain the optimal solution for this problem and Fig. 5 shows the experiment results. Figure 5a shows the agent's scores during training, which shows that the score increases with the number of training episodes. Figure 5b gives the flight altitude profile of these 6 aircrafts, and we can see that, these 6 aircrafts can successfully reach their target positions using the control strategy generated by AI agent, which demonstrates the effectiveness of the proposed method.

5 Conclusion

In this work, an artificial intelligence (AI) agent based on deep reinforcement learning is built to mimic air traffic controllers, such that the dense, complex and dynamic air traffic flows in terminal airspace can be handled sequentially and separated. To simplify the problem, the complex three-dimensional terminal airspace is projected onto the vertical plane by dispersing state space and action space. And then, the typical reinforcement learning algorithm, double deep Q-network, is taken to realize the AI agent. Results show that the built AI agent can guide 6 aircrafts safely and efficiently through Sector 01 of Nanjing Terminal, simultaneously.

References

1. Zhao W (2017) The opportunities, challenges and obligations in internationalization of China civil aviation. Civil Aviation Manage 09:6–11
2. Yan Y, Cao G (2018) Operational concepts and key technologies of next generation air traffic management system. Command Inf Syst Technol 9(3):8–17
3. Ma X, Xu X, Yan Y et al (2018) Correlation analysis on delay propagation in aviation network. Command Inf Syst Technol 9(4):23–28
4. http://www.caac.gov.cn/XWZX/MHYW/201803/t20180315_55771.html
5. Sutton RS, Barto AG (2011) Reinforcement learning: an introduction
6. Mnih V, Kavukcuoglu K, Silver D, Rusu AA, Veness J, Bellemare MG, Graves A, Riedmiller M, Fidjeland AK, Ostrovski G et al (2015) Human-level control through deep reinforcement learning. Nature 518(7540):529–533
7. Deepmind. Alphago at the future of go summit, pp. 23–27, May 2017. http://deepmind.com/research/alphago/alphago-china/
8. Vinyals O, Ewalds T, Bartunov S, Georgiev P, Vezhnevets AS, Yeo M, Makhzani A, Küttler H, Agapiou J, Schrittwieser J et al (2017) Starcraft II: A new challenge for reinforcement learning. arXiv preprint arXiv:1708.04782
9. George E, Khan SS (2015) Reinforcement learning for taxi-out time prediction: an improved q-learning approach. In: 2015 international conference on computing and network communications (CoCoNet). IEEE, New York, pp 757–764
10. Cruciol LL, de Arruda Jr AC, Weigang L, Li L, Crespo AM (2013) Reward functions for learning to control in air traffic flow management. Transp Res Part C: Emerging Technol 35:141–155
11. Agogino A, Tumer K (2009) Learning indirect actions in complex domains: action suggestions for air traffic control. Adv Complex Syst 12(04–05):493–512
12. Tumer K, Agogino A (2007) Distributed agent-based air traffic flow management. In: Proceedings of the 6th international joint conference on autonomous agents and multiagent systems, pp 1–8
13. Brittain M, Wei P (2018) Autonomous aircraft sequencing and separation with hierarchical deep reinforcement learning. In: Proceedings of the international conference for research in air transportation
14. Brittain M, Wei P (2019) Autonomous separation assurance in an high-density enroute sector: a deep multi-agent reinforcement learning approach. In: IEEE intelligent transportation systems conference (ITSC). IEEE, New York, pp 3256–3262

Sand Table Design of Virtual Reality Psychotherapy

Chaoran Bi[1], Hai Wang[2](✉), and Rui Dong[1]

[1] College of Media Design, Modern Vocational Technology College, Tianjin 300350, China
[2] Teaching Center for Experimental Electronic Information, College of Electronic Information and Optical Engineering, Nankai University, Tianjin 300350, China
wanghai@nankai.edu.cn

Abstract. With the rapid pace of life and social pressure, the number of patients with mental disorders has increased, and the demand for psychological counseling has increased. The sand table therapy is a popular psychotherapy method on the market, which is suitable for a wide range of people, based on the analysis of the development and current situation of the sand-play therapy in China, combining the Internet resources and the rapid development of virtual reality technology, integrating the new media technology and the limited interactive computer database into the psychological sand-play therapy, it can not only accurately and vividly complete the subconscious inner expression and spiritual release of patients, but can also reduce the work pressure of psychiatrists, and this design is an expansion and innovation of the traditional sand table game mode.

Keywords: Virtual simulation · Sand-play therapy · Online simulation · Natural human–computer interaction

1 Prospect and Current Situation

The important theoretical basis of sand-play is based on Jung's analytical psychology, world techniques and oriental philosophy a psychotherapy technique founded by Dora Calve, a Swiss analytical psychologist, on the basis of world techniques, analytical psychology and oriental philosophy [1]. This kind of game is operated by the help-seeking person after the psychological consultant's explanation, it also requires a good relationship between the psychologist and the patient, and let the patient remove all psychological defenses, use all kinds of sand tools in the context of extreme trust and relaxation, make a scene to show the player's subconscious, and promote the communication and integration of consciousness and subconscious; then the patients can be cured in the game, even self-healing [2].

Virtual reality (VR) is a new and practical high-tech imaging technology rising in the early 1990s, and it makes comprehensive use of multimedia imaging, simulation technology, network technology and computer technology [3, 4]. By creating a realistic three-dimensional visual experience space and model by computer, users can input

Q. Liang et al. (eds.), *Artificial Intelligence in China*, Lecture Notes in Electrical Engineering 653, https://doi.org/10.1007/978-981-15-8599-9_42

information and get feedback through many types of sensors, so as to achieve immersive sense of reality and achieve the goal of human–computer interaction by using new interdisciplinary technology.

In this design, new virtual reality and interactive technology are used in traditional sand table games to improve the effect of sand table therapy, and it can promote and popularize the therapy in a wider range, reduce the cost of sand tools, save space resources and reduce the work intensity of psychological consultants.

Analysis on the problems of traditional sand table: A small amounts of models may not be enough to express the psychological image of the player, affect the game and evaluation results [5]. A lot of models take a lot of time to browse while occupying a large space at the counseling rooms. At the same time, players may have different degrees of obsessive-compulsive disorder and spend a lot of time looking for models because of the excessive pursuit of the image in the heart; thus, excessive action will lead to fatigue and affecting the consistency of game progress. The effect and evaluation will be affected. It takes too much time for psychological consultants [6]. The expression of sand table is mostly outdoor fine sand which is more like water than sand and blue organic glass, but it is not enough to describe the weather, and the abstract elements of wind, rain and lightning are not well represented from the production process. The sand table space is limited. If players express too much thought, they cannot do it any more when they are full of sand table. At the same time, the consultant will spend a lot of time in communicating with the players and evaluating the sand table with more content. Sand table is lack of indoor space, and the performance mode is insufficient. It cannot enter the building to show the scenes in the hearts of players, such as theater stage, church, gymnasium and so on.

2 The Conception and Feasibility Analysis of Virtual Reality Psychological Sand-Play

2.1 Improvement of Hardware Data Storage and Calculation

As we all know, the amount of information in high-density disk is huge, the current pervasive hard disk capacity can reach 500 g to 2 TB, and it can store a lot of data information which make up for the limited short board of traditional sand table model [7]. For example, we talk about some abstract expressions of "wind, rain, thunder and electricity", and players can find more accurate expression of images in line with their inner intention. And no longer need to occupy a large number of consulting room space, which can effectively save resources.

The data processing resource library can distribute the prompt of submenu more orderly, and players can find the corresponding classification more accurately and in detail. It is no longer like the traditional sand table, which takes up a lot of searching time, making up for the inconvenience of traditional browsing. At the same time, it saves the time of counselors and games, making the treatment more coherent and the effect will be more significant.

Data records of players, including demographic data, personal growth history, mental state, physical state, social function records and information preservation, can be saved in large quantities, so as to generate rough evaluation results in later interaction.

2.2 Immersive Art Expression Function of Virtual Reality Technology

The expression of immersive art is similar to the atmosphere of the church, which makes the participants more quieted and engaged, and can enter the game state more effectively and quickly. Compared with the traditional sand table, it can create a safe and free space more quickly. Under the influence of immersive atmosphere, the players can express their inner life more unconsciously [3].

Compared with the traditional sand tools, immersive new media art can be a more real dynamic three-dimensional model and can be equipped with sound effects to express the players' consciousness from multiple dimensions.

Scene selection. Instead of the traditional sand table outdoor boundless mode, players can choose the scene first, which can be "hospital, station, theater, arena" and other indoor scene. In particular, the concept of "home" is presented, which includes "family guidance sandbox" in the traditional treatment of mental sandbox. It is a very important and common mode in mental sandbox game, mainly used to show and adjust the relationship between family members. The general method is to make a sandbox by combining families and placing their own image in it, so it is necessary to involve indoor "home" scene [8].

At the same time, immersive artistic expression can also be such an anti-interference factor. Each independent scene can adjust the brightness, color **and** temperature of the light achieving the best display and experience effect and more accurately express the consciousness of the players. And then the evaluation results are more accurate.

The control of the model proportion and the proportion of some objects and images can be as large as that of reality with the reality, which can increase the details that many players want to express in their consciousness; for example, it can only be "a table full of their favorite food", which increases the optional choice of the scene of the content of game, and the sense of substitution and reality is stronger [9].

The immersive sand table game makes the players more engaged, and the players have the experience of being in the real world. In the process of the players, it reduces the time and labor intensity of the consultants, realizes the substitution of technology for human labor, reduces the excessive contact between doctors and patients and avoids the unnecessary transference and countertransference between many doctors and patients. In the traditional sand table game, the counselor needs to let the player remove the psychological defense, fully trust the counselor and project all his dark negative emotions. However, the consultant's handling of transference is very time-consuming and energy-consuming, and if the consultant does not control "transference" and "countertransference", it may cause damage to professional theoretical norms, resulting in negative results subconscious images into the sand table, which is empathy. Freud later admitted that transference is inevitable. It is an important way in psychoanalysis. In order to treat some patients and even to intentionally induce transference, many interventions are also the embodiment of countertransference. The emergence of virtual sand table games can let players Project Empathy into the game, increase the effect of the game and reduce the degree and time of empathy with consultants [10, 11].

3 Design and Implementation of Virtual Reality Sand Table Game

3.1 Design of Rules of Virtual Reality Psychological Sand Table Game

Virtual reality psychological sand-play is the imagination and experiment of applying virtual reality imaging technology in modern science and technology to psychological therapy [12]. Virtual reality technology has the characteristics of multi-perception, immersion, interaction and imagination. Immersion and multi-perception can enhance the authenticity of the whole game process; interactivity enables visitors to interact naturally, instead of watching from the perspective of "God" in an objective environment, especially imagination, which broadens the thinking of players, the diversity of game types and scenes, and enriches the performance of self-consciousness.

The characteristics of virtual reality applied to the traditional sand-play psychotherapy can enrich the traditional treatment means and enhance the treatment effect. This design is based on the original traditional psychological sand table game based on Jung's analytical psychology, using the characteristics of virtual reality, and improving and adjusting the original imaging technology and operation process. As for the rules of the game, it is still according to the traditional game rules that players make use of the sand table model to present the unconscious image of the heart, and psychological consultants analyze and guide the treatment.

3.2 Interactive Design of Virtual Reality Psychological Sand-Play

The interaction mode of virtual reality psychological sand-play is the most basic expression mode in the game, in which the player transforms in the virtual sand-play through hardware input [13]. According to the rules of the game, the player adapts to the concept understanding of the first-person subjective perspective and carries out immersive virtual visual experience operation. The operation behavior can be to click the remote control hardware handle or make specific gestures to the sensing device. As for the interactive hardware, it will be detailed in the next chapter. The interaction of virtual reality psychological sand table game aims to enhance the authenticity of experience effect and operation fun, so as to achieve a real sense of immersion, let players relax and provide a free and safe space for them. Through the three-dimensional virtual model matching with sound effect, some objects even have dynamics, which can meet the needs of players to show their inner world unconsciously and meet their self-knowledge. They are eager for consultants to understand them and help them to understand their own needs. They have empathy for the virtual sand table game, which extends from the interaction of operation to the interaction of mind. Such a human–computer interaction mode can be described as one access, access to the patient's heart heal.

3.3 Operation Interface and Environment Design of Virtual Reality Psychological Sand Table Game

The first thing we need to touch is the guidance interface. The design of the guidance interface is directly related to the players' good impression and acceptance of the virtual reality sand table game and then to the trust of the game process and treatment effect.

3.3.1 Visual Design of Guidance Interface

According to visual communication design, Gestalt psychology and color psychology, the guide interface is guided by shape, shape size, color brightness, saturation, etc. Psychotherapy for the purpose of the game does not need to be as profound as the public game interface, just simple and easy to understand and color chosen should be bright and soft. In the design, the same type of elements can be used in the same shape and color. From the perspective of color psychology, the main color of the interface menu design can consider the use of light yellow, light blue or light green. The high brightness and low saturation hue make people relaxed, eye-catching and non-stimulating. From the beginning, it creates a comfortable and relaxed operating environment, which makes players easy to relax and can be put into the game as soon as possible. Or the game can provide a variety of different interface design styles, for different players, the players themselves or consultants choose the appropriate interface style.

3.3.2 Function Design of Guidance Interface

The process of menu should not be too complex; otherwise, players are easily lost in the process, resulting in pressure or even giving up tasks.

The inevitable task operation requires multi-step operation across multiple pages. You can use wizard control to guide users to complete multi-step operation. The design follows people's daily life habits, from left to right, from top to bottom. At the same time, when necessary, we can use significant prompt mode to guide the user to operate. For example, motion guidance is a good way, because the human subconscious likes to capture moving objects. This movement is not only the movement of position, but also the change of size, flicker or transparency to attract the visual center of the player. There is also a hint to tell the players how many steps they need in total, which stage they are going to go through now, and how many are left to be completed. To sum up, we should guide the players in the interactive interface, try to be concise and concise, and feed back information through visual elements, so that the players can better and faster enter the game.

3.3.3 Initial Game Environment Scene Design

Considering the space design based on the traditional psychological consulting room, its purpose is to make the consultants feel warm and safe, and the first impression of the three-dimensional space of the scene of random game interface is moderate. That is to say, it should not be too large, which cannot make players feel empty and thus bring insecurity, or even fear because of the infinite and vast virtual space. It cannot be too small, which will make the players feel depressed and claustrophobic. According to the reality of traditional consulting room as a reference, it can give people a feeling of 15 km^2 of white wall space. Subsequently, players can adjust and control the expansion or reduction of space according to their own wishes. Select the relevant indoor or outdoor scene according to the menu and submenu prompts.

3.4 Construction of Sand Tool Model in Virtual Reality Sand Table Game

The principle of virtual reality sand-play is equivalent to the psychological projection of the traditional psychological sand-play model. The game rules also allow players to place sand devices freely in unconscious condition, simultaneous interpreting their intentions.

Virtual reality technology presents sand model. Small objects can be scanned by ordinary 3D camera in close range to present visual image similar to human perspective. The principle is to build a virtual three-dimensional model through dot grating imaging or diamond grating holographic imaging technology (shown in Fig. 1). This technology is suitable for small models with real objects and has a high degree of reduction [14].

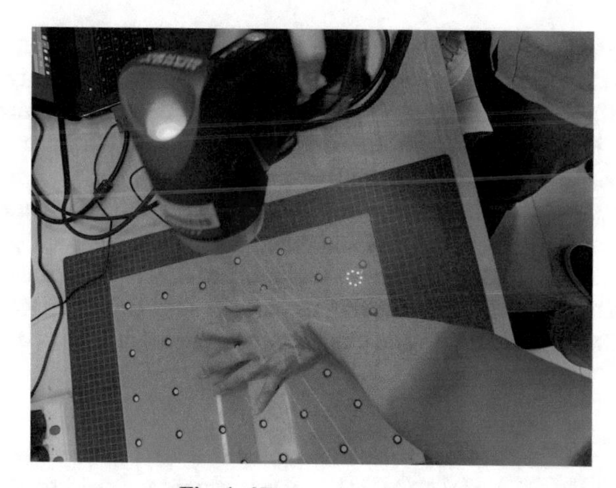

Fig. 1 3D camera scanning

Large objects or macro scenes can be imaged using either 3D software (3dmax Maya C4D) or computer-aided design space software (CAD), which uses equations to generate lines and shapes, generating an image of an object on a screen or in virtual reality. The production process is divided into two parts: modeling and rendering.

3.5 The Utilization of Human–Computer Interaction Technology

In the process of the game, players use gestures to control the process of the game. Players can participate in it more truly. As feedback, the game can feed back information to players through sound effects and dynamics. Many elements represent complicated connotation. Through database information, a rough evaluation result is generated to save doctors' labor. At the same time, using the previous player's data analysis and more accurate to the player's consciousness image positioning, for example: the horse symbolizes grassland, freedom and sex at the same time, different information is easier to locate the meaning of the object image.

3.6 Hardware Equipment

In order to achieve a near-real visual experience, we need a VR headset to complete the user's visual operation. In order to popularize widely, the technology should be compatible with most of the hardware devices on the market. Common technologies on the market can be divided into PC end and mobile end. The PC end represents oculus rift, HTC Vive, Yishi helmet, Uglasses and PS VR. The mobile end includes Google Cardboard, Samsung Gear VR, Microsoft HoloLens, Joycher, etc., which can be matched.

4 Conclusion

After studying the current situation and background of psychological sand-play and spatial augmented reality technology, this paper finds out the possibility of combining them and puts forward the game form of spatial augmented reality psychological sand-play, so that it is not only a sand-play, but also an evaluation system. Based on the discussion of the mechanism of traditional psychological sand-play and the manifestation of spatial augmented reality technology, this paper has worked out the content, interactive behavior, role composition, game task and other elements of spatial augmented reality psychological sand-play and analyzed the feasibility of the technical route. Based on the traditional sand table game, it is improved. Its purpose and significance are to let players better release their consciousness and emotions, so as to achieve self-healing.

References

1. Lee CG, Dunn GL, Oakley I et al (2017) Visual guidance for a spatial discrepancy problem of in encountered-type haptic display[J]. IEEE Trans Syst Man Cybern Syst 50(4):1384–1394
2. Kim J, Nakamura T, Kikuchi H et al (2015) Covariation of depressive mood and spontaneous physical activity in major depressive disorder: toward continuous monitoring of depressive mood[J]. IEEE Biomed Health Inf 19(4):1347–1355
3. Gaffary Y, Benoît LG, Marchal M et al (2017) AR feels "Softer" than VR: haptic perception of stiffness in augmented versus virtual reality[J]. IEEE Trans Vis Comput Graph 23(11):2372–2377
4. André Z, Makhsadov A, Sören K et al (2020) Immersive process model exploration in virtual reality[J]. IEEE Trans Vis Comput Graph 26(5):2104–2114
5. Mallya GP, Hiremani A, Gajakosh G et al (2020) Predicting depression symptoms in an arabic psychological forum [J]. IEEE Access 57317 – 57334
6. Noble SL (2014) Control-theoretic scheduling of psychotherapy and pharmacotherapy for the treatment of post-traumatic stress disorder[J]. IET Control Theory Appl 8(13):1196–1206
7. Iyengar AK (2018) Enhanced clients for data stores and cloud services[J]. IEEE Trans Knowl Data Eng 31(10):1969–1983
8. Lu ZK, Lin W, Yang X et al (2005) Modeling visual attention's modulatory aftereffects on visual sensitivity and quality evaluation[J]. IEEE Trans Image Process 14(11):1928–1942
9. Wu JJ, Lin WS, Shi GM et al (2015) Visual orientation selectivity based structure description[J]. IEEE Trans Image Process A Publ IEEE Signal Process Soc 24(11):4602–4613
10. Manschreck TC (1986) Schizophrenic disorders and reading disturbance[J]. IEEE Prof Commun IEEE Trans 29(2):27–30

11. Zhang WR, Pandurangi AK, Peace KE, Yang Y (2007) Dynamic neurobiological modeling and diagnostic analysis of major depressive and bipolar disorders[J]. IEEE Trans Biomed Eng 54(10):1729–1739
12. Yeh SC, Li YY, Chu Z et al (2018) Effects of virtual reality and augmented reality on induced anxiety[J]. IEEE Trans Neural Syst Rehab Eng 26(7):1345–1352
13. Katja Z, Elena K, Rachel M (2018) The effect of realistic appearance of virtual characters in immersive environments-does the character's personality play a role [J]. IEEE Comput Soc 24(4):1681–1690
14. Lee B, Jang C, Kim D et al (2019) Single grating reflective digital holography with double field of view[J]. IEEE Trans Ind Inf 15(11):6155–6161

Massive Flights Real-Time Rerouting Planning Based on Parallel Discrete Potential Field Method

Yang Ding[1](✉), Zelin Li[1], Bingyu Li[2], and Ruheng Xie[1]

[1] State Key Laboratory of Air Traffic Management System and Technology, 210014 Nanjing, China
dyjeremy@126.com

[2] The 28th, Research Institute of China Electronics Technology Group Corporation, 210007 Nanjing, China

Abstract. Severe weather is an important reason that threatens the safety of air transportation. As the flights grow, path planning for massive aircraft is becoming an important problem for rerouting. This paper proposes a parallel computing method to carry out rerouting path planning. Compared with the traditional method, this method focuses on improving the computational speed, making it more suitable for the dynamic operating situation. Numerical experiment also proved the validity and efficiency of this method.

Keywords: Air traffic control · Rerouting · Potential field · Parallel computing

1 Introduction

Severe weather is an important reason that threatens the safety of air transportation. When certain airspace or routes are affected by severe weather and unavailable to flights, rerouting is a common means in the operation [1, 2]. Rerouting is to temporarily arrange the flights to an unaffected route so as to bypass the restricted airspace. Accordingly the path planning is a key issue for rerouting.

Current research on flight rerouting began in the 1990s. After about 30 years of research, a variety of related algorithms have been proposed, such as grid-based rerouting method [3], convex-polygon-based rerouting method [4], rule-based method [5], A*-based searching method [6] on the existing waypoints. Although the above methods can effectively carry out rerouting planning, most of them focus on a single flight with relatively low-calculation load. In actual operation, the rerouting planning problem of multiple flights or even a large number of flights is common and remains to be solved, which is the focus of this paper.

2 Problem Description

This paper focuses on the rerouting path planning for a large number of aircraft. In order to simplify the problem, the following assumptions are made [7].

Q. Liang et al. (eds.), *Artificial Intelligence in China*, Lecture Notes in Electrical Engineering 653, https://doi.org/10.1007/978-981-15-8599-9_43

(1) The aircraft is simplified to one particle.
(2) The starting point and target point of each aircraft are known.
(3) Regardless of vertical climb and descent, it is simplified to the problem of path planning in the two-dimensional space.
(4) The aircraft maintains its ideal velocity while there is no severe weather.

3 Application of Parallel Discrete Potential Field Method

The classic artificial potential field method is a virtual field method, which is widely used in the research of robot path planning. The basic idea of this method is to imagine the robot moving in an abstract artificial potential field [8–11]. The destination virtually generates attractive force to the mobile robot, whereas the obstacle generates repulsion force to it. The composition of forces finally controls the robot's motion. Due to its advantage of real-time calculation and relatively smooth path result, the method can also be applied to the research of rerouting of flights. That is to say, the aircraft is simplified to a particle which moves in a two-dimensional plane space. When the aircraft is moving toward the destination, it is simultaneously influenced by attractive force of the destination and the repulsion force of the obstacles, which include other aircraft and limited airspace.

The paralleled discrete potential field method proposed in this paper is based on the idea of classic artificial potential field, combining with parallel calculation algorithm [12], so as to realize the fast rerouting decision and simulation of massive flight flow. The flow chart of the algorithm is as follow (Fig. 1).

3.1 Scene Rasterization

The basis of discretization is to rasterize the scene and establishes a repulsive field based on the restricted area caused by severe weather. We divide the scene into square cells, convert the restricted area into polygons, and map them into the cells. Each cell can be identified by coordinates (i, j). Obviously, the size of the cell cannot be too small, otherwise it will affect the calculation speed, also the size cannot be too large, or it will reduce the accuracy of the calculation result. To ensure the real-time performance of the calculation, it is necessary to determine an appropriate size of one cell.

Each cell stores data such as density, velocity, potential value, where the repulsive force value represents the selection tendency for flying into the cell of a group of flights with the same destination. This paper assumes that the repulsion value is related to the flight density, namely the more flights in one cell, the higher the repulsion value of the cell is. When the density of flights is above certain threshold, the repulsion value is defined as positive infinity. If the repulsion value of a cell is positive infinity, then the cell represents completely impassable. In addition, the repulsion value of cells of flight restricted area is initialized to positive infinity. In addition, all flights to be rerouted are grouped according to their target cell. The calculation of the repulsion value is given by the following formula:

$$c_{i,j} = \begin{cases} \infty, & n_{i,j} > \varepsilon \\ n_{i,j}, & \text{else} \end{cases} \quad (1)$$

Here, $c_{i,j}$ indicates the repulsion value of one cell; $n_{i,j}$ indicates the number of flights in one cell; ε means the density threshold of one cell.

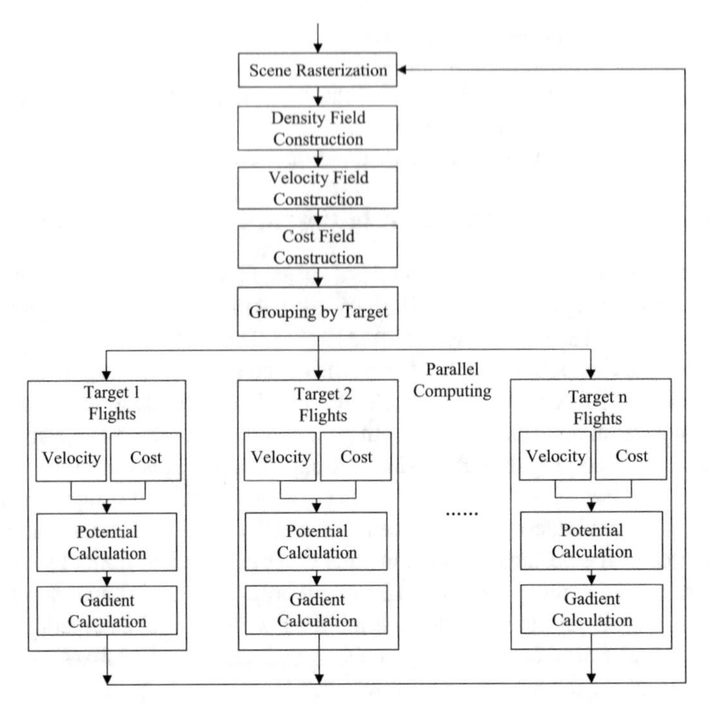

Fig. 1 Flow chart of parallel discrete potential field method

3.2 Velocity Field Construction

In order to get the maximum reachable velocity in certain cell (i, j), a flight density contribution function is to be established. By determining the density contribution value of each aircraft and using the value as the weight, the weighted average velocity of flights in one cell can be calculated, so as to obtain the maximum reachable velocity of one aircraft flying into the surrounding cell. The specific steps are as follows:

1. Establish the density contribution function

The paper supposes each flight that has a separate density field. The density field reaches the maximum value at the position of the aircraft and gradually decreases toward its surrounding direction. The contribution value to surrounding cells can be calculated by the following formula:

$$\begin{cases} \rho_A = \min(1 - \Delta x, \Delta y)^{\lambda} \\ \rho_B = \min(\Delta x, \Delta y)^{\lambda} \\ \rho_C = \min(1 - \Delta x, 1 - \Delta y)^{\lambda} \\ \rho_D = \min(\Delta x, 1 - \Delta y)^{\lambda} \end{cases} \tag{2}$$

Here, Δx indicates the position of the aircraft in the *X*-axis direction; Δy means the position of the aircraft in the *Y*-axis direction; ρ indicates contribution value to surrounding cells (*A*, *B*, *C*, *D*); λ is a constant standing for the density decay rate.

2. Calculate the density contribution value

The density contribution of one aircraft to the affected cells is calculated in step 1. By adding contribution values of all aircraft in the cell, the density value of this cell is obtained, as the following Formula shows.

$$\rho_{i,j} = \sum_{a \in A} \rho_a \tag{3}$$

Here, $\rho_{i,j}$ indicates the density value of the cell;ρ_a is the density contribution value of aircraft a.

3. Calculate the weighted average velocity

Take the density contribution value of certain aircraft as a weight and thus calculate the weighted average velocity in the cell (i, j), as the following formula shows.

$$\bar{v}_{i,j} = \frac{\sum_{a \in A} \rho_a v_a}{\rho_{i,j}} \tag{4}$$

Here, $\bar{v}_{i,j}$ indicates weighted average velocity of the flights in the cell (i, j); v_a means the velocity of aircraft a.

4. Calculate the maximum reachable velocity

When the density of flights in adjacent cells in the flight direction is greater than a certain threshold, the upper limit of the flight velocity of the flight in that direction is just the limited velocity of the flight flow in that direction. In addition, when it is lower than the threshold, it means that the flight in the direction can be undisturbed, and it can maintain its ideal velocity. If the density of adjacent cells in the flight direction is an intermediate value, the velocity is obtained by linear interpolation. Corresponding equation goes below.

$$v_{i,jd}^{top} = \begin{cases} v_{i,jd}^{flow}, & \rho_{i',j'} > \rho_{\max} \\ v_{\max} - \frac{\rho_{i',j'} - \rho_{\min}}{\rho_{\max} - \rho_{\min}} \cdot (v_{\max} - v_{i,jd}^{flow}), & \rho_{\min} \le \rho_{i',j'} \le \rho_{\max} \\ v_{\max}, & \rho_{i',j'} < \rho_{\min} \end{cases} \tag{5}$$

Here, $v_{i,j\,\mathrm{d}}^{top}$ indicates the maximum achievable velocity of the flights in the cell (i, j) flying into the direction d; $\mathrm{v}_{i,j\,\mathrm{d}}^{flow}$ means the velocity of the flight flow flying into the direction d; $v_{\max}$ means the ideal velocity of an aircraft.

3.3 Cost Function Establishment

In the rasterized environment, the flying path can be regarded as a sequence of ordered cells. A flight can enter its four adjacent cells. Therefore, the cost of a single path can be approximated by the sum of the unit cost entering to the adjacent cell. Namely the cost function can be expressed as follows:

$$E'_p = \sum_{d \in \{W,N,E,S\}} \mathrm{e}_{i,j_d}$$

$$\mathrm{e}_{i,j_d} = \frac{w_l \cdot v^{\mathrm{top}}_{i,j_d} + w_t + w_c \cdot c_{i',j'}}{v^{\mathrm{top}}_{i,j_d}} \tag{6}$$

Here, $\mathrm{e}_{i,jd}$ indicates the unit cost flying into the adjacent cell in the direction *d;* $c_{i',j'}$ is the repulsive force value in the corresponding adjacent cell.

3.4 Discrete Potential Field Calculation

By establishing the discrete potential field calculation function, a scalar field is constructed, of which the gradient descent curve is the optimal path moving from point *A* to point *B* for the aircraft. The discrete potential field calculation function goes below.

$$P_{i,j} = \begin{cases} \frac{P_x + P_y + \sqrt{2(\mathrm{e}_x+\mathrm{e}_y)^2 - (P_x - P_y)^2}}{2} & 2(\mathrm{e}_x+\mathrm{e}_y)^2 - (P_x - P_y)^2 > 0 \\ \min\{P_x, P_y\} + (\mathrm{e}_x + \mathrm{e}_y) & \text{other} \end{cases} \tag{7}$$

Here, P_x indicates the potential value of adjacent cell in *X*-axis;P_y indicates the potential value of adjacent cell in *Y*-axis; e_x and e_y are the unit cost entering the adjacent cell.

According to the definition of gradient, the direction of the gradient points to the direction of the fastest growth of the scalar field. To calculate the gradient of the cell (i, j), the paper adopts a one-sided difference equation for the discretization, which is shown as follows:

$$g_x = \begin{cases} P_{i,j} - P_{i-1,j} & P_{i+1,j} < P_{i,j} < P_{i-1,j} \vee P_{i,j} < P_{i+1,j} < P_{i-1,j} \\ P_{i+1,j} - P_{i,j} & P_{i-1,j} < P_{i,j} < P_{i+1,j} \vee P_{i,j} < P_{i-1,j} < P_{i+1,j} \\ P_{i+1,j} - P_{i,j} & P_{i+1,j} < P_{i-1,j} < P_{i,j} \\ P_{i,j} - P_{i-1,j} & P_{i-1,j} < P_{i+1,j} < P_{i,j} \\ 0 & \text{other} \end{cases} \tag{8}$$

3.5 Parallel Computing Algorithm

In this paper, a parallel calculation algorithm is used to numerically solve the potential field function. By using the partition calculation rule, all grids are divided into predefined tiles, in which the potential values are solved in parallel by CUDA of GPU. The algorithm is as follows:

1. Definition

(1) Set *t* to represent a single tile that divides the environment area into equal sizes;

(2) Set the operation list L to store the tiles that need to be updated;

2. Initialization

(1) All cells are evenly divided into tiles of the same size;
(2) Move all tiles containing target cells into L;

3. Loop Calculation

(1) Update each tile t in the operation list L.
(2) Update the four adjacent tile of block t. If t is convergent, then remove t from L;
(3) Move all non-convergent tiles into L, repeat above steps until L is empty.

4 Example Analysis

The study is carried out in Shanghai terminal area. In order to analyze the relationship between flight numbers and computing efficiency, different numbers of aircraft were simulated in and out of the Shanghai terminal area. The rule-based method (RBM) and theparallel iscrete potential field method (PDPFM) are both used for the analysis. RBM carries out conflict detection for every pair of flights and then resolve the conflict based on predefined rules to get the continuous rerouting path, which is similar to current practice of controllers. In PDPFM, the flights are grouped by their targets so as to reduce the computation complexity. In the experiment, the size of one cell is 10 km * 10 km. The results are shown as follows (Fig. 2) (Table 1).

As can be seen, the increase of flight numbers reduces computing performance of RBM, while it has little effect on the computing performance of PDPFM. Therefore, thanks to the advantage of parallel computing, PDPFM is effective for rerouting path planning for large number of flights.

5 Conclusion

This paper proposes a new method to carry out rerouting path planning for large number of aircraft. Compared with the traditional method, this method focuses on improving the computational speed, making it more suitable for the dynamic operating situation. The experiment shows that this method is able to handle large number of aircraft without compromising computational efficiency, thus, to make better use of airspace resources and improve operational efficiency.

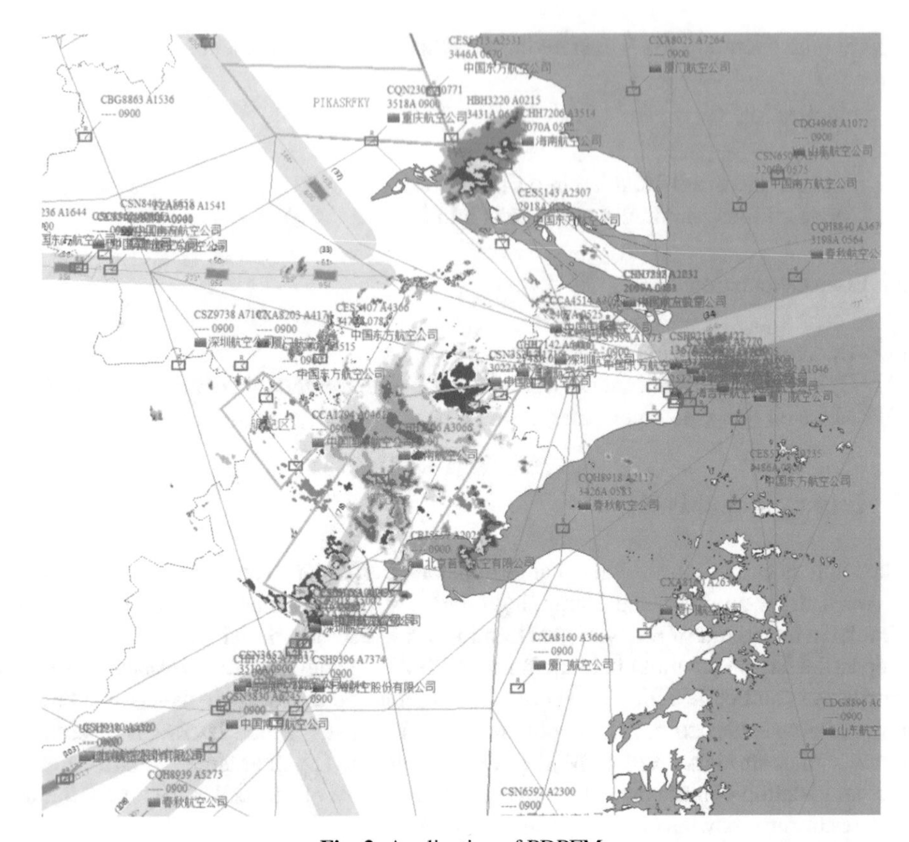

Fig. 2 Application of PDPFM

Table 1 Comparison of computing efficiency between two methods

Aircraft numbers	RBM	PDPFM
50	121 ms	493 ms
100	268 ms	548 ms
200	890 ms	623 ms
400	2719 ms	697 ms

References

1. Wang G, Wei W, Ding Y (2016) Transport aircraft route planning algorithm. Command Inf Syst Technol 7:43–46
2. Luo Q, Li R, Shao W, Yang M, Ben T (2018) Design of intelligent path planning for tactical enviroment. Command Inf Syst Technol 9:49–54
3. Dixon M, Weiner G (1993) Automated aircraft routing through weather-impacted airspace. In: Fifth international conference on aviation weather system, Vienna, pp 295–298

4. Tian Y, Song K, Gu Y (2008) The research on rerouting problem in air traffic flow management. Math Practice Theory 38(10):70–76
5. Li X, Xu X, Zhu C (2008) Air traffic reroute planning based on geometry algorithm. Syst Eng 26(8):37–40
6. Zhang XG, Mahadevan S (2016) Aircraft rerouting optimization and performance assessment under uncertainty. Decis Support Syst 96:67–82
7. Zhang Z, Zhao Z (2015) Multi-aircraft dynamic rerouting path planning based on improved artificial potential field algorithm. J Wuhan Univer Technol 37:44–49
8. Liu Y, Zhang Y (2006) Study of local path planning of mobile robot based on improved artificial potential field method. Modern Mach 6:48–49
9. Kuang F, Wang Y, Zhang H (2005) Real time path planning of mobile robot in dynamic world based on improved artificial potential field. Comput Appl 25(10):2415–2417
10. Wang M, Wang X, Li C (2008) Study of local path planning of mobile robot based on improved artificial potential field method. Comput Eng Des 29(6):1504–1506
11. Abdalla TY, Abed AA, Ahmed AA (2017) Mobile robot navigation using PSO-optimized fuzzy artificial potential field with fuzzy control. J Intell Fuzzy Syst 32(6):3893–3908
12. Mabrouk MH, Mcinnes CR (2008) Solving the potential field local minimum problem using internal agent states. Robot Auton Syst 56(12):1050–1060

Remote Sensing Image Detection Based on FasterRCNN

Shunmin Liu[1,2], Zhiming Ma[1,2], and Bingcai Chen[1,2](✉)

[1] School of Computer Science and Technology, Xinjiang Normal University, 830054 Xinjiang, China
cbc9@qq.com

[2] School of Computer Science and Technology, Dalian University of Technology, 116024 Liaoning, China

Abstract. The object detection algorithm can accurately detect the objects needed in remote sensing images. For the traditional object detection which adopts the sliding window method, both the detection accuracy and robustness are low and the generalization ability is weak. It is difficult to meet people's demand for production and application. With the establishment of large-scale image database, the development of deep convolutional network has been promoted, which greatly improves the accuracy of object detection in images. In the field of target object detection, the algorithm based on depth of deep learning is the most popular and effective ones. Through a large number of experiments on the NWPU VHR-10 data set, we verified the remote sensing image detection based on FasterRCNN. The experiment results showed that the average accuracy of the ten types of targets in the NWPU VHR-10 data set reached 81.59%, which therefore proved the effectiveness of deep learning algorithm in remote sensing image detection.

Keywords: Deep convolutional network · Remote sensing image · Object detection · FasterRCNN

1 Introduction

With the rapid development of remote sensing technology, it has been widely used in terrain mapping, land use survey, vegetation survey, and traffic survey. Remote sensing images have an open perspective so that more detailed ground information can be obtained and objects of interest in these remote sensing images can be accurately detected. Traditional object detection algorithms mainly adopt sliding window paradigm, HOG [1], and other manually designed features for object detection. Because it fails to utilize the features of deep semantics, it is only used to detect the given object, and the accuracy and robustness of the detection results are low and the generalization ability is weak. In addition, the ground environment is complex and changeable, and there are various types of targets. Different types of object have different object characteristics. In order to improve the efficiency of object detection, the mainstream object detection algorithms at present adopt deep learning scheme.

Q. Liang et al. (eds.), *Artificial Intelligence in China*, Lecture Notes in Electrical Engineering 653, https://doi.org/10.1007/978-981-15-8599-9_44

Deep learning deepens the level of convolutional network through weight sharing, which makes the convolutional network to have more powerful performance. With the establishment of large-scale image database, the development of convolutional network has been promoted. VGG16 [2], GoogleNet [3], and ResidualNet [4] promoted the convolutional network to a deeper level and greatly improved the performance of the network, which promoted the accuracy of large-scale image classification to a very high level. In terms of object detection, region-based convolutional neural network RCNN [5] successfully connects object detection with deep convolutional network, which improves the accuracy of object detection to a new level. Since RCNN is divided into three independent processes, the detection efficiency is very low. Based on this situation, scholars improved RCNN and fast region-based neural network FastRCNN [6]. It is not necessary to send all candidate windows into the network, but to send the image into the deep network once, and then map all candidate windows on a certain layer in the network, greatly improving the detection speed of the model. Region-based full convolutional network RFCN [7] is further improved on this basis. It is found that the network layer after ROI pooling no longer has translation invariance, and the number of layers after ROI pooling will directly affect the detection efficiency. Therefore, RFCN designs a position-sensitive ROI pooling layer to directly discriminate the results after pooling, greatly improving the detection efficiency. Faster region-based convolutional neural network FasterRCNN [8] uses candidate window network RPN to generate candidate windows and uses the same structure as FasterRCNN for classification and window regression. RPN and FastRCNN share the main deep network. FasterRCNN synthesizes object detection into a unified depth network framework.

In this paper, the NWPU VHR-10 [9] remote sensing data set was detected based on FasterRCNN. The 454 images in the NWPU VHR-10 data set were trained for 40,000 iterations, 196 images were used for testing, and then the experimental results were analyzed.

2 FasterRCNN Algorithm

FasterRCNN is the final version of RCNN series detection algorithms for regional neural network. RCNN is the first convolutional neural network object detection framework based on candidate regions. After the improvement of three subsequent versions of space pyramid pooling network SppgNET, FastRCNN, and FasterRCNN, the detection speed and accuracy are greatly improved. The common point of these detection frameworks is that they are divided into these three steps. First, the candidate region is extracted from the picture, and then the candidate region is sent to the trained convolutional network model to extract features, and finally the feature is used to classification and regression the candidate region. FasterRCNN proposes a regional recommendation network (RPN) that can generate candidate regions on the basis of FastRCNN, and integrates the two into a complete network that can be learned end-to-end, which not only ensures accuracy but also improves speed.

2.1 Network Framework

The FasterRCNN object detection network is divided into two steps. First, a series of candidate boxes are generated as samples, and then the samples are classified by the convolutional neural network. FasterRCNN framework in the input test images, the entire image input to the CNN for feature extraction, and image generated on the 300 proposals window each suggested the last layer of convolution window is mapped to a CNN characteristic diagram, ROI through the ROI pooling layer which generates the characteristics of the fixed figure, and finally through the classification and regression testing box for precise location. The whole process is shown in Fig. 1.

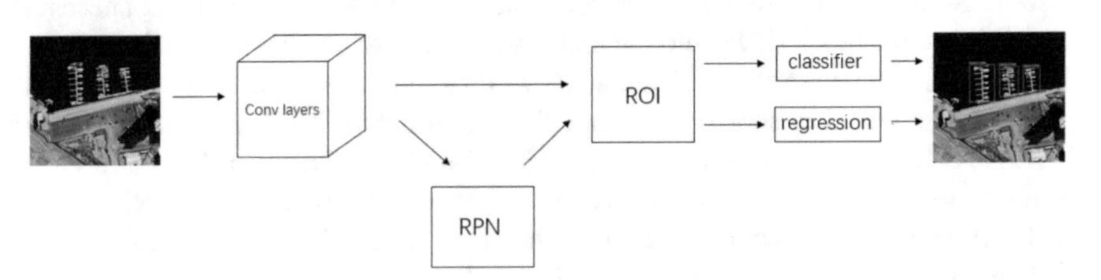

Fig. 1 Network structure of FasterRCNN

2.2 Network Training

When training RPN, specify each anchor with a binary tag to indicate whether the anchor is a target. Only specify anchor as positive sample in the following two cases, that is, anchor is a target:

(1) When an anchor has the highest IOU with one of the ground-truth boxes
(2) When the IOU of anchor and any ground-truth box is higher than 0.7.

A single ground-truth box may mark multiple anchors as positive. Normally, the second condition is sufficient to produce a suitable positive sample, but the first condition is added to prevent the second condition from occasionally finding a positive sample. If the IOU samples below 0.3 are assigned as negative samples, the rest anchors that are neither positive samples nor negative samples have no help for training, so they are all discarded.

For each anchor, a binary softmax is attached with two score outputs to represent the probability that it is an object and not an object (p_i). Then, with a bounding box, regressor output is representing the anchor's four coordinate positions (t_i), and RPN's loss function is shown below:

$$L(\{p_i\}, \{t_i\}) = \frac{1}{N_{\text{cls}}} \sum_i L_{\text{cls}}(p_i, p_i^*) + \lambda \frac{1}{N_{\text{reg}}} \sum_i p_i^* L_{\text{reg}}(t_i, t_i^*) \quad (1)$$

i represents the ith anchor, $p_i^* = 1$ when anchor is positive sample, and $= 0$ if anchor is negative sample. t_i^* represents a ground true box coordinate related to a positive sample anchor, and each positive sample anchor can only correspond to a ground true box: A positive sample anchor corresponds to a certain grand true box, and then the IOU of this anchor and ground true box is either the largest of all anchors, or greater than 0.7.

x, y, w, h, respectively, mean the central coordinates and width and height of box. x, x_a, x^*, respectively, mean predicted box, anchor box, and ground-truth box. t_i^* means the offset of box relative to anchor box. The formula is shown below:

$$t_x = (x - x_a)/w_a,\ t_y = (y - y_a)/h_a, \tag{2}$$

$$t_w = \log(w/w_a),\ t_h = \log(h/h_a), \tag{3}$$

$$t_x^* = \left(x^* - x_a\right)/w_a,\ t_y^* = (y - y_a)/h_a, \tag{4}$$

$$t_x^* = \log(w/w_a),\ t_y^* = \log(h/h_a) \tag{5}$$

where L_{reg} in the loss function of RPN:

$$\text{smooth}_{L1}(x) = \begin{cases} 0.5x^2 & |x| \leq 1 \\ |x| - 0.5 & \text{otherwise} \end{cases} \tag{6}$$

p_i^* denotes those loss pointers with regressor for positive samples, as the items using [Formula] $= 0$ were eliminated with negative samples. L_{cls} is about two kinds of log loss.

3 Experiment and Analysis

3.1 Data Set

The experimental data set NWPU VHR-10 was released by Northwestern Polytechnical University in 2014. The NWPU VHR-10 data set is a publicly available detection data set for ten categories of geospatial objects, which is only used for research purposes. These are aircraft, ships, storage tanks, baseball, tennis, basketball, ground runways, harbor bridges, and vehicles. The data set contains a total of 800 ultra-high resolution remote sensing images, including 650 images containing the target and 150 background images. These images were clipped from Google Earth and Vaihingen data sets and then manually annotated by an expert. In this experiment, 650 images containing targets were selected, among which 454 images were used as training set and 196 images were used as test set. The number of images containing various targets was shown in Table 1. The label of NWPU VHR-10 data set is in TXT file. The format of the data set suitable for operation is PASCAL VOC. Before word model training, the data set is first converted into PASCAL VOC format so that the program code can be trained.

Table 1 Contains the number of images for various targets

Type	Airplane	Ship	Storage	Ball park	Tennis court	Basket court	Track	Harbor	Bridge	Vehicle
number	88	65	26	158	74	69	143	36	59	85

3.2 Model Training

In this paper, the basic feature extraction network of the detection model remains consistent with FasterRCNN, and the VGG16 network is selected to initialize the pre-trained model on the ImageNet classification task. The VGG16 convolutional neural network obtained by pre-training of ImageNet classification is used to initialize the weight of feature extraction network convolutional layer. Parameters are weight decay was 0.0005, learning rate was 0.001, momentum was 0.9, batch size was 256, positive sample threshold was 0.7, negative sample threshold was 0.3, 40,000 and iterations were performed. The experimental hardware environment was Intel Core i5 CPU and Nvidia GeForce1050GPU. The experimental software environment was windows10 + tensorflow + cuda9.0 + python3.6.

3.3 Experimental Results and Analysis

Finally, FasterRCNN is based on the remote sensing image data collection trained by NWPU VHR-10. After 40,000 iterations, the final loss value drops from 1.090 to 0.096. The loss curve is shown in Fig. 2, and it gradually converges after 10,000 iterations in turbulence is bigger, came back he gradually began to convergence, but in the back of the time loss will produce some bigger shock.

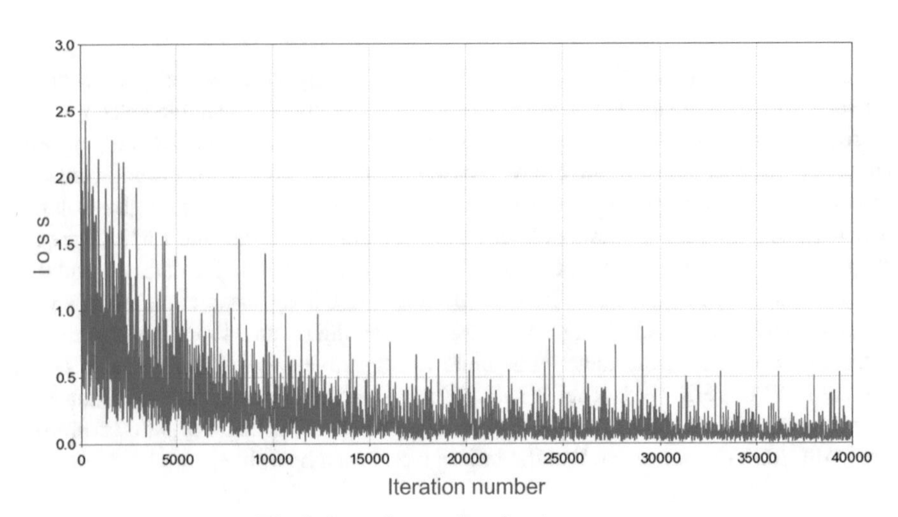

Fig. 2 Detection results of various targets

The accuracy of various targets in the data set after detection is shown in Table 2, with an average accuracy of 81.6%. Among the ten categories of target objects, aircraft and track had the highest accuracy of 90.9%, while bridge had the lowest accuracy of 62.7%. There are two reasons for this. The first is that the data set NWPU VHR-10 contains images of airplanes and track and field fields that are more abundant than those of bridges. Pictures of the bridge are only 59 images of the bridge. The data is too small. The trained model will not have enough samples to summarize the features, which will lead to poor generalization on the test data set. To solve this problem, besides adding images in the data set later, it can also be solved by cropping, rotating, panning or zooming the image. In addition, we can also add noise or change the color, brightness, sharpness, contrast, sharpness of the image and block part of the noise. Secondly, the bridge target in aerial photography is relatively small, it is easy to detect other targets as bridges, thereby reducing the accuracy of detection. This problem needs to improve the algorithm.

Table 2 Accuracy of various targets

Airplane	Ship	Storage	Ball park	Tennis court	Basket court	Track	Harbor	Bridge	Vehicle	Map
90.9%	81.4%	72.6%	90.1%	90.8%	80.3%	90.9%	68.6%	62.7%	87.7%	81.6%

3.4 Results

As shown in Fig. 3, aircraft, ships, tanks, baseball, tennis courts, basketball courts, ground track, ports, bridges, and vehicles were detected as a result. However, some small targets are dim and difficult to detect, which requires the later work to improve the performance of the algorithm to detect small targets. The detected result is in the upper left corner of the target box.

4 Conclusion

In this paper, ten types of targets in remote sensing images of NWPU VHR10 data set are detected based on FasterRCNN, which was proved to be able to provide higher effectiveness in remote sensing image detection. Our experimental results showed, when adopted to remote sensing image recognition, this model can achieve better training effect in shorter training time and thus significantly improve the efficiency of remote sensing image recognition.

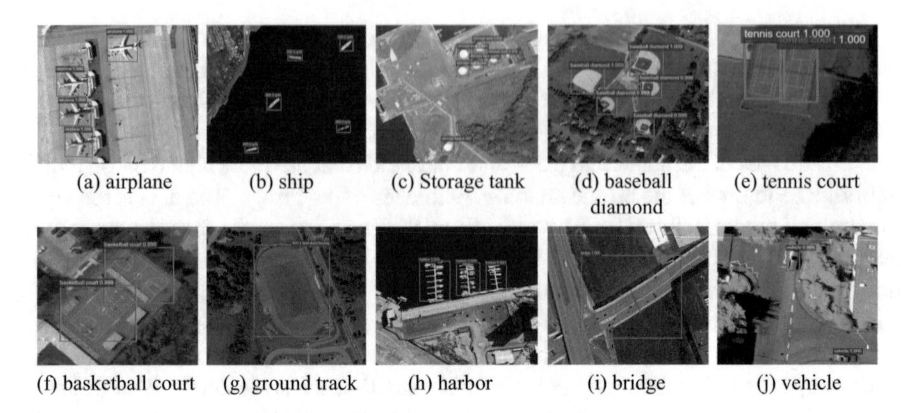

(a) airplane (b) ship (c) Storage tank (d) baseball diamond (e) tennis court

(f) basketball court (g) ground track (h) harbor (i) bridge (j) vehicle

Fig. 3 Detection results of various targets

Acknowledgements. This work was supported in part by the Tianshan Young Talent Program, Xinjiang Uygur Autonomous Region under Grant 2018Q024, in part by the Natural Science Foundation of China under Grant 61771089 and Grant 61961040, and in part by the Regional Cooperative Innovation Program of Autonomous Region (Aid Program of Science and Technology to Xinjiang) under Grant 2019E0214.

References

1. Cao X, Wu C, Yan P et al (2011) Linear SVM classification using boosting HOG features for vehicle detection in low-altitude airborne videos[C]. In: 2011 18th IEEE International conference on image processing(ICIP). Brussels, IEEE, pp 2421–2424
2. Simonyan K, Zisserman A (2015) Very deep convolutional networks for Largc-Scalc image recognition[C]. In: The 3rd International conference on lcarning represcntation. San Dicgo, Canada, pp 1–5
3. Szegedy C, Liu W, Jia Y et al (2014) In: Going deeper with convolutions[J]
4. He K, Zhang X, Ren S, et al (2015) In: Deep residual learning for image recognition[J]
5. Girshick R, Donahue J, Darrell T et al (2014) In: IEEE 2014 IEEE conference on computer vision and pattern recognition (CVPR), Columbus, OH, USA (2014.6.23–2014.6.28)]. 2014 IEEE Conference on computer vision and pattern recognition—rich feature hierarchies for accurate object detection and semantic segmentation[J], pp 580–587
6. Girshick R (2016) Fast R-CNN[C]. In: 2015 IEEE international conference on computer vision (ICCV), IEEE
7. Dai J, Li Y, He K et al (2016) R-FCN: object detection via region-based fully convolutional networks[J]
8. Ren S, He K, Girshick R et al (2017) Faster R-CNN: towards real-time object detection with region proposal networks[J]. IEEE Trans Pattern Anal Mach Intell 39(6):1137–1149
9. Cheng G, Han J, Zhou P, Guo L (2014) Multi-class geospatial object detection and geographic image classification based on collection of part detectors. ISPRS J Photogram Remote Sens 98:119–132

Research on Airport Surface Simulation Method Based on Dynamic Path Planning

Shenghao Fu[1,2], Xiaowen Wang[1], Bin Dong[1](✉), and Yan Liu[1]

[1] State Key Laboratory of Air Traffic Management System and Technology, Nanjing 210007, China
fushenghao@126.com, nuaa_db@163.com

[2] The 28th, Research Institute of China Electronics Technology Group Corporation, Nanjing 210007, China

Abstract. With the development of the civil aviation industry, the scale and complexity of the airport are increasing. How to simulate airport surface operations to evaluate and improve airport capabilities is a hot issue. Firstly, the airport surface network model is constructed. Secondly, a dynamic path planning method based on improved A* algorithm is proposed to ensure the accuracy and efficiency of the planned path. Finally, the simulation algorithm based on pre-sequencing is introduced to realize airport operation simulation. Taking Pudong Airport as an example, the results show that the method can quickly and accurately plan the required path and provide effective support for the airport simulation and evaluation.

Keywords: Dynamic path planning · Pre-sequencing · Surface simulation

1 Introduction

In recent years, with the rapid development of the air traffic control industry, the density of air traffic has continued to increase [1]. Airports have become more and more busy, and aircraft conflicts have also become increasingly prominent. How to accurately and effectively deal with the simulation of airport surface has received extensive attention. The path planning of the airport surface is one of the key technologies of airport surface simulation, which is mainly composed of the airport surface network model and path planning algorithm.

At present, the main surface path models include the simplified directed graph model and the Petri net model [2]. Both of these models simplify the surface and cannot accurately describe the path of the target taxiing. In terms of path planning algorithms, there are precise algorithms represented by Dijkstra's algorithm [3] and A* algorithm [4] and intelligent algorithms represented by genetic algorithm [5], ant colony algorithm, and simulated annealing algorithm.

In this paper, a multi-attribute point-line topology model is built on the basis of the directed graph model, which not only solves the problem that the route cannot be accurately portrayed, but also solves the problem that the directed graph lacks support for

Q. Liang et al. (eds.), *Artificial Intelligence in China*, Lecture Notes
in Electrical Engineering 653, https://doi.org/10.1007/978-981-15-8599-9_45

the surface operation rules. On the basis of A* algorithm, the concepts of dynamic logical distance and redundant point removal strategy are introduced to make the algorithm more practical for airport operation and improve the operation efficiency.

2 Airport Surface Network Model

The airport surface activity area includes the area that provides aircraft takeoff, landing, taxiing, and parking. Constructing the airport surface network model is the process of converting the airport surface activity area into a mathematical model suitable for subsequent calculation and analysis. The original airport surface data generally exists in AutoCAD data format. Firstly, use GlobalMapper software for data preprocessing, convert AutoCAD data into shp format point and line data, and construct airport surface key points and taxi path metadata. In order to support the rules of surface operation, a multi-attribute model is built on surface metadata in combination with dbf format attribute files.

2.1 Construction of Surface Structure Model

In order to simplify and standardize the aircraft taxiing process, the key elements of the airport surface are abstractly modeled. The modeling principles include the following:

(1) Abstract the airport activity area into points and broken lines;
(2) Airport stands are identified by points, runways, and taxiways, rapid exit ways are identified by polyline segments, and one runway or taxiway is composed of multiple polyline segments;
(3) The intersection of runway centerline, taxiway, and exit ways centerline is abstracted as a key point;
(4) The polyline segment between two adjacent key points is defined as path metadata;

The model abstraction is shown in Fig. 1. A–F is the key point, and the broken line between AB, AC, BC, CE, DE, and EF is the path metadata.

The key point metadata object stores a set of key points associated with it and identifies the path metadata composed of any associated key points. The path metadata object stores the indexes of the two key points that constitute the path. Taking the key point C and the metapath CA as an example, the key point C data records the key points A, B, and E associated with it and the metapath CA, CB, and CE. In addition, key points C and A as the beginning and end of the metapath are recorded in the metapath CA. In this way, if a certain key point is known as the current key point, the adjacent related key point and related metapath can be obtained and even the relationship between the entire airport surface, as shown in Fig. 2.

2.2 Multi-attribute Metadata Model

The surface structure model is only a set of points and polylines, lacking the relevant physical meaning, and it is difficult to effectively support the airport surface operation rules in the subsequent algorithm implementation stage. Therefore, this article uses the GlobalMapper tool and the dbf format file to set multiple attributes for the metapath.

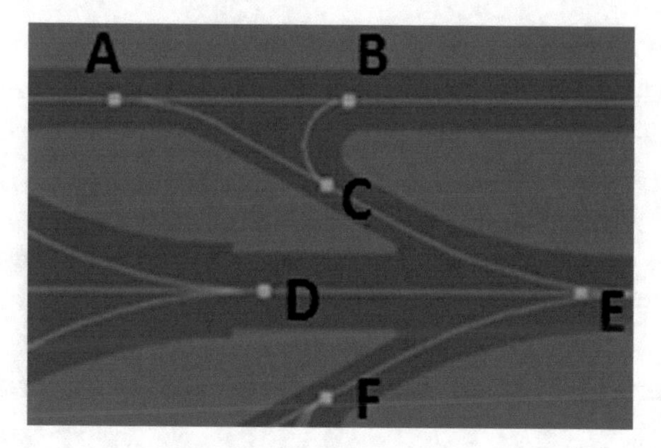

Fig. 1. Diagram of abstract model

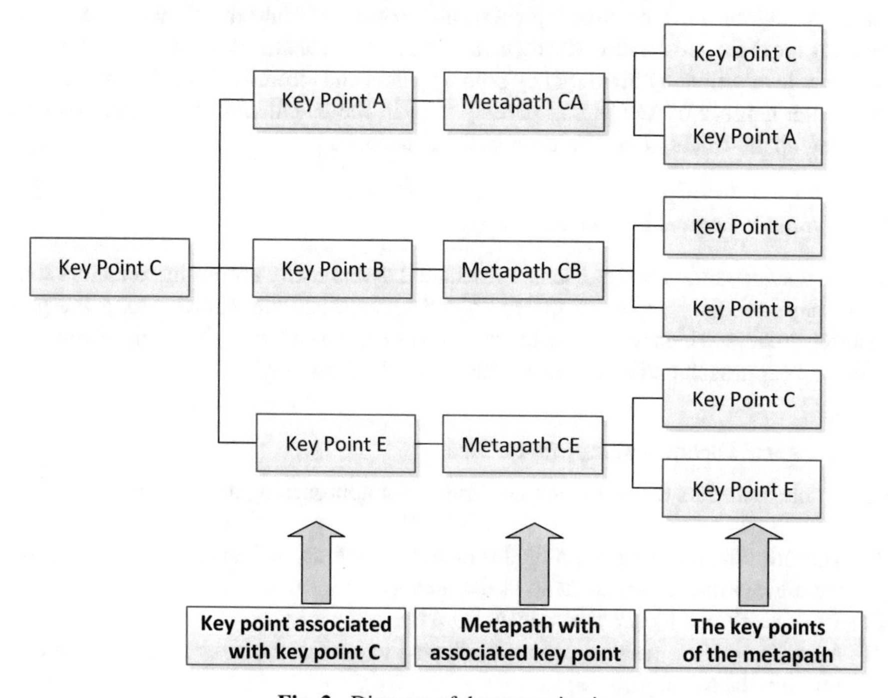

Fig. 2. Diagram of data organization

2.2.1 Runway Attributes

Runway attributes refer to the relevant constraints during runway allocation and use, including runway operation mode, runway operation direction, takeoff interval, landing interval, interception distance, etc.

2.2.2 Taxiway Attributes

Taxiway attributes refer to the relevant constraints during taxiway use, including taxiway direction attributes, available model attributes, speed limit attributes, etc.

2.2.3 Stand Attributes

Stand attributes refer to the constraints related to the allocation and use of stands, including airlines, aircraft type constraints, use priority, etc.

3 Path Planning Method Based on Improved A* Algorithm

A* algorithm is a typical shortest path algorithm, which can calculate the shortest path from one vertex to the remaining vertices in a directed graph. In this paper, based on the classic A* algorithm, a dynamic logical distance strategy is introduced. By dynamically updating the logical distance of the metapath, the surface rules are integrated into the path planning. In addition, by dividing key points by area and eliminating invalid calculation points, the number of key points participating in the calculation is reduced, thereby speeding up the calculation efficiency of the algorithm.

3.1 Dynamic Logical Distance Strategy

Logical distance refers to the distance calculated dynamically according to factors such as the physical distance of the metapath and operational constraints during the path planning process, which represents the reachability of the metapath, including the logical distance of approach taxi and logical distance of departure taxi.

3.1.1 Logical Distance of Approach Taxi

This article considers the following constraints for approaching aircraft taxiing:

(1) The aircraft has a certain sliding distance after landing, and it is not allowed to leave the runway directly at the head of the runway;
(2) The aircraft must leave the runway from the exit ways;
(3) After leaving the runway, it is not allowed to taxi through other runways before entering the parking space;
(4) After leaving the runway, try to reduce the behavior of crossing the runway;
(5) Avoid taxiing on taxiways with mismatched models.

Based on the above constraints, the non-exit ways associated with the target runway are set to unavailable state before the planning calculation, that is, the logical distance is infinite; the metapaths unrelated to the target runway are set to unavailable state; the exit way that is on the opposite side with the stand relative to the target runway is set to an unavailable state; set the taxiway metapath that is not suitable for the current model to the unavailable state.

3.1.2 Logical Distance of Departure Taxi

This article considers the following constraints for departure aircraft taxiing:

(1) It is not allowed to enter the runway from the exit ways;
(2) It is not allowed to taxi on any runway before entering the runway waiting point;
(3) Avoid taxiing on taxiways with mismatched models.

Based on the above constraints, the exit way is set to unavailable state before planning calculation; the metapath not corresponding to the target runway is set to unavailable status; set the taxiway metapath that is not suitable for the current model to the unavailable state.

3.2 Redundancy Removal Strategy

For large airports, the airport activity area is complex and contains a large number of key points. If traversing all key points for path planning, it will have a greater impact on the calculation efficiency and cannot meet the application requirements of real-time calculation. Therefore, the method of removing redundant points is adopted. Thus, the number of key points involved in the calculation is reduced, and the calculation complexity is reduced. There are two types of redundant points, isolated points, and independent area points.

3.2.1 Isolated Point Removal

On the basis of the dynamic logical distance strategy, the key points with infinite logical distance to the associated metapath are defined as isolated point, which is useless for path planning, so isolated points should be found and removed before planning.

3.2.2 Independent Area Point Removal

The airport surface is divided into areas, and the effective key point set is determined according to the area where the starting and ending points of the path are, so as to remove a large number of invalid key points which are not in the focus area. As shown in Fig. 3, taking the scene of runway 17R taxiing to stand 339 as an example, the dotted box range is the effective area, and the key points outside the effective area can be removed.

4 Simulation Algorithm Based on Pre-sequencing

Based on the path planning algorithm, the flight plan is generated according to the flights' distribution characteristics and pre-sequenced according to the conflict judgment algorithm. Finally, the flight plan is simulated by the aircraft dynamics model. The process is shown in Fig. 4.

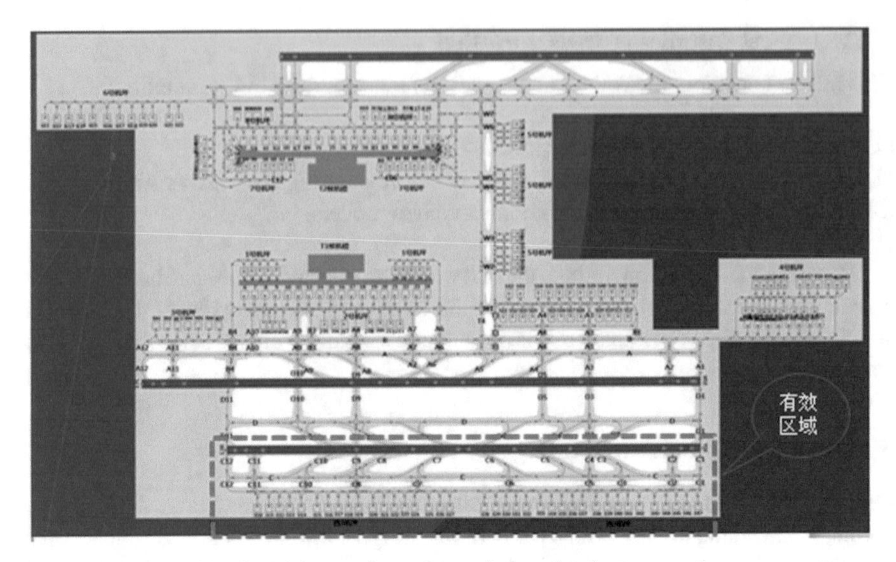

Fig. 3. Diagram of area key point's redundant removal

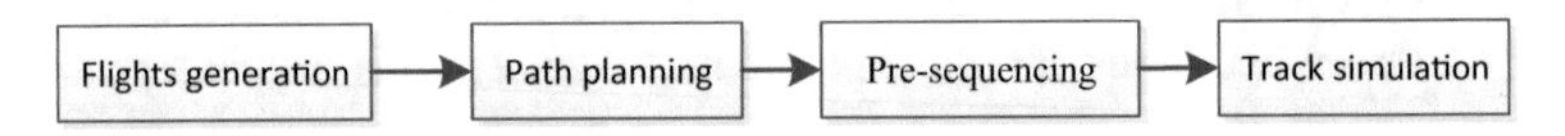

Fig. 4. Diagram of simulation process

4.1 Flights' Generation

Considering the distribution characteristics of flight takeoff and landing time, Poisson distribution is used to describe the number of flight takeoff and landing times per unit time. As shown in Formula (1), λ is the number of takeoff or landing aircraft per hour.

$$P(X = k) = \frac{\lambda^{k}}{k!} e^{-\lambda}, k = 0, 1, \ldots \quad (1)$$

Based on this setting, the flight takeoff and landing intervals follow a negative exponential distribution, as shown in Formula (2).

$$P(t) = 1 - e^{-\lambda t} \quad (2)$$

Combined with the statistics of the stand and runway, flight plan data which includes stand, runway, taxiway, and arrival time is formed for simulation.

4.2 Pre-sequencing

The conflict judgment is made for the flight plan after the path assignment, and the conflict is resolved by waiting at key point, so as to form pre-sequencing flight plans. Taking metapath and key point as resource object, the conflict of resource usage is calculated. The time range of a flight occupying the metapath is assumed to be T_{in} to

T_{out}, and then the use time of other flights for this resource must meet certain conditions, as shown in Formula (3).

$$T'_{\text{out}} < T_{\text{in}} || T'_{\text{in}} > T_{\text{out}} \tag{3}$$

where T'_{in} and T'_{out} are the start and end time of the other flights using the resource. The flight plan that does not meet the resource use conditions needs to wait.

4.3 Track Simulation

Track simulation is to calculate the arrival time of the target at each key point according to the kinematic model after the taxiing path and motion attributes are determined.

As shown in Fig. 5, a simulation path is shown, in which the solid point is the key point, including key points 0, 1,…, *n,* and so on and the hollow point is the component point of the metapath, such as the component point 0, 1, 2,…, *k* between the key points $n-1$ and n The attribute information of any key can be expressed as $(X_{n-1}, Y_{n-1}, V_{n-1}, T_{n-1}, T_{\text{stay}(n-1)})$, where $X_{n-1}, Y_{n-1}, V_{n-1}, T_{n-1}$ is the position, speed, and time when the aircraft reaches the point, $T_{\text{stay}(n-1)}$ is the waiting time at this point. The distance between any two adjacent composition points can be expressed as $\text{Dis}_{(n-1)(k-2)}$. The arrival time of point *n* can be calculated from the arrival time of point $n-1$, as shown in Formula (4).

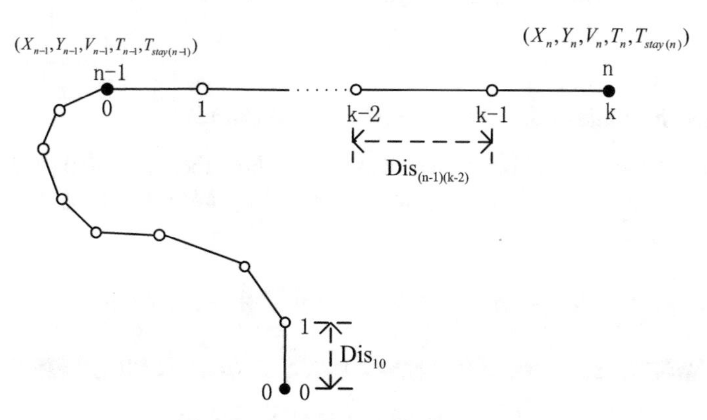

Fig. 5 Diagram of track simulation

$$T_n = T_{n-1} + \sum_{0}^{k} \text{Dis}_{(n-1)k}/(v_{n-1} + v_n) + T_{\text{stay}(n-1)} \tag{4}$$

5 Simulation Result

5.1 Application of Dynamic Logical Distance Strategy

Take the path of stand 339 taxiing to runway 17R and waiting for takeoff as an example, before the dynamic logical distance strategy is adopted, only the shortest physical path is planned, as shown in Fig. 6 before applying the strategy. In this case, the aircraft taxis directly into the runway from exit ways C6 and taxies on the runway that is not in accordance with the airport operation rules. After the dynamic logical distance strategy is adopted, the aircraft enters the runway head from taxiway C12, which complies with the airport operation rules, as shown in Fig. 6 after applying the strategy.

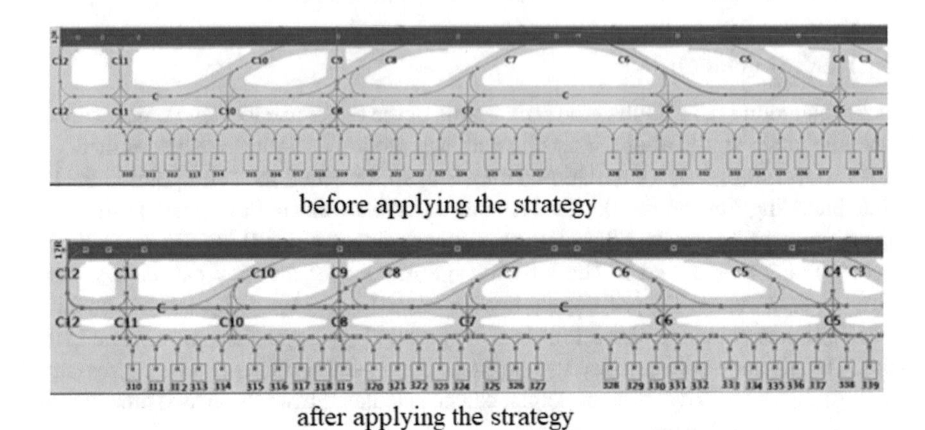

Fig. 6 Diagram of dynamic logical distance strategy

5.2 Target Simulation Based on Predetermined Path

With the aid of the target simulation processing module, the simulation track target can glide along the given path accurately according to the predetermined speed and course, as shown in Fig. 7.

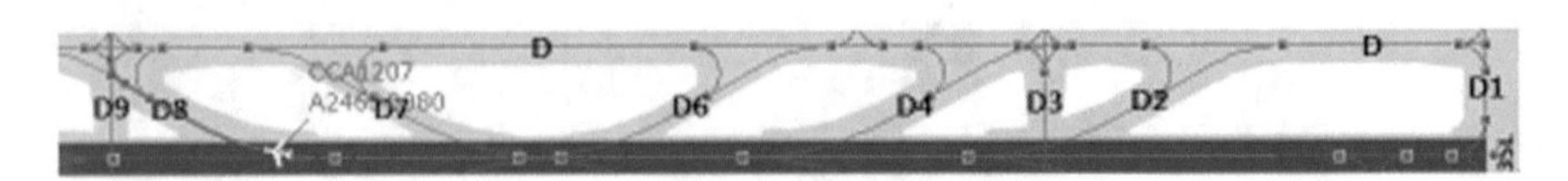

Fig. 7 Diagram of target simulation

5.3 Efficiency Comparison

The algorithm efficiency is greatly improved by using redundant point removal strategy, and the comparison of operation efficiency before and after strategy application is shown in Table 1.

6 Conclusions

Airport surface simulation is the basis for airport verification and evaluation. This article first constructs airport surface network model, based on the classic A* algorithm, we

Table 1. Performance comparison of redundant removal

	Scene 1	Scene 2	Scene 3	Scene 4
Number of calculated points (before/after)	1254/254	1254/305	1254/331	1254/350
Time/*S* (before/after)	0.18/0.01	0.43/0.04	0.49/0.05	0.78/0.06

propose a dynamic logic distance strategy and a redundant point removal strategy, and this method can effectively integrate airport surface operation rules into planning algorithms, thereby planning the simulation path quickly and accurately. Combined with the pre-sequencing simulation algorithm, we realize the airport surface simulation which was verified in the Pudong Airport simulation.

References

1. Chen Z (2016) Technological challenges of future air traffic control system development. Command Inf Syst Technol 7(6):1–5
2. Zhu X (2016) Petri-net-based aircraft taxiing route assignment for A-SMGCS. Inf Control 45(01):101–107
3. Dai Y (2016) Planning and design of conveyor route based on improved Dijkstra algorithm. Geospatial Inf 14(2):34–35
4. Li N (2012) Researchon taxing optimization for aircraft based on improved A* algorithm. Comput Simul 29(07):88–92
5. Liu Z (2008) Airport scheduling optimization algorithm based on genetic algorithm. J East China Univ Sci Technol 34(3):392–398

Triple-Channel Feature Mixed Sentiment Analysis Model Based on Attention Mechanism

DeGang Chen[1], Azragul[1](✉), and Bingcai Chen[1,2]

[1] School of Computer Science and Technology, Xinjiang Normal University, 102 Xinyi Road, 830054 Ürümqi, China
Azragul2010@126.com

[2] School of Computer Science and Technology, Dalian University of Technology, 116024 Dalian, China

Abstract. Sentiment analysis is an important branch of the field of natural language processing, and opinion sentiment analysis is the current research hotspot. With the development of artificial intelligence, it is very challenging to effectively extract important emotional information from a large amount of text information and analyzes the emotional tendency of views in response to the increasing number of emotional opinions with rich connotations. Today's methods mostly use shallow emotional factors, while ignoring deeper text semantic information, and cannot explore the semantic connection between words. To make up for this deficiency, this paper proposes a triple-channel feature-mixed sentiment analysis model Tri-BiGRU-Atten based on attention mechanism. This model combines different semantic feature mixed modeling to enable it to mine deeper emotional information in the context. Compared with the traditional attention mechanism, LSTM, Bi-LSTM, and other models, the emotion classification effect is more effective.

Keywords: Sentiment analysis · Feature mixed · Triple-channel · Attention

1 Introduction

The field of sentiment analysis, also known as opinion mining, mainly analyzes and analyzes the content with emotion in the text comment information generated by users through NLP-related technologies to determine the sentiment tendency.

Deep learning technology has gained new life in recent years. CNN uses the convolutional sliding window mechanism to continuously extract local effective features on input samples and further reduces the feature dimension through pooling operations. The RNN recurrent neural network can extract contextual content well for sequence text, but for too long sequence content, gradient explosion or gradient disappearance may occur. Therefore, in 2014, Cho et al. [1] first proposed the GRU model and used it. Applied to statistical machine translation, it has achieved good results. Both it and the LSTM [2] model are variants of RNN, and the RNN has been improved accordingly to solve the

Q. Liang et al. (eds.), *Artificial Intelligence in China*, Lecture Notes in Electrical Engineering 653, https://doi.org/10.1007/978-981-15-8599-9_46

problem that the traditional recurrent neural network may have semantic loss for longer information and the above mentioned problems. The attention mechanism proposed by Google [3] in "attention is all you need" has aroused widespread concern among people in various fields. Even without the use of networks such as RNN and CNN, a single attention mechanism can be achieved in the field of image and text engineering. Better income.

Based on the above processing methods, the contribution of this article mainly has three aspects:

(1) A Tri-BiGRU-Atten model is proposed, which uses a bidirectional gating unit and attention mechanism to form triple-channel parallel input under the condition of multi-feature fusion, which greatly improves the deep emotional semantic features of the model mining.
(2) Fully integrate the original text features, part-of-speech features, location features, and dependent syntactic features to establish connections, so that the emotional information of the input text can be better used.
(3) Experiments show that our proposed Tri-BiGRU-Atten model is superior to existing models in performance, verifying the effectiveness of our model in sentiment analysis tasks.

2 Related Work

The continuous development of deep learning technology has attracted more and more people's attention in recent years, and it is very popular in the NLP field.

In the field of emotion recognition using CNN, Kalchbrenner et al. [4] proposed a DCNN model in the paper. By modifying the structure of the pooling layer and retaining the k maximum information in the pooling stage, the global information is retained. Keunwoo et al. [5] combined RNN and CNN, used the front and back bidirectional RNN to connect the front and back words of the current word in the text with the word, and then, through the pooling operation, the context information of the word can be obtained more evenly. Bahdanau et al. [6] first proposed the attention mechanism in the paper, which breaks the limitation of the traditional encoder–decoder that relies on a fixed-length vector in the encoding stage. Attention this mechanism will perform better when the text predicts the emotional tendency and can better explain the impact of keywords in the model on the analysis of the text's emotional tendency. In the paper, Nikolaos et al. [7] combined RNN and attention mechanism. A HAN neural network structure is proposed, in which two levels of attention mechanisms are used at the encoding stage, word level and sentence level. The word encoder and sentence-level encoder both use bidirectional GRU for encoding, and the importance of each sentence and word in the classification task is well represented. Ji et al. [8] combined vocabulary features with location features, introduced bidirectional GRU and multi-head attention mechanisms, and achieved high results in sentiment analysis tasks.

The triple-channel attention mechanism sentiment analysis model Tri-BiGRU-Atten proposed in this paper fully exploits the deep implicit semantics of sentiment texts and does not rely on traditional sentiment dictionaries. It solves the problem that CNN fixed convolution kernel size has a relatively limited field of view; it is difficult to obtain longer sequence semantic effective information, which leads to the lack of expression of some semantic features of excessively long sequences. At the same time, the attention mechanism is introduced to solve the problem that RNN and CNN are not interpretable. The deep semantic mining of input sentiment text is brought into full play to obtain better classification results.

3 Model

In this section, we will describe the details of our model. The overall structure of Tri-BiGRU-Atten proposed in this paper is shown in Fig. 1, which is mainly composed of embedding layer, triple-channel layer, fusion layer, self-attention layer, DropOut layer, and emotion output layer.

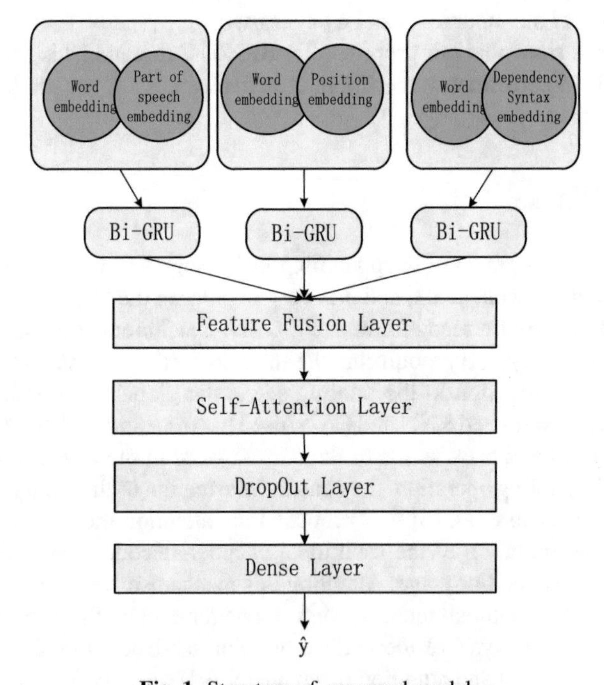

Fig. 1 Structure of proposed model

3.1 Word Embedding Layer

The word embedding layer is a triple-channel parallel input model formed by combining word vectors with part-of-speech vectors, position vectors, and dependent syntax vectors, respectively.

3.1.1 Word Vector

The traditional use of word vectors is represented by one-hot encoding. This method is relatively simple and straightforward, but its vector matrix density is high in latitude and sparse, which easily causes dimensional difficulties and is difficult to train.

In addition, the inherent algorithm idea of one-hot coding determines that the vector representations between words are orthogonal to each other. If it is used in the sentiment analysis task, unexpected important information will be missing, which will have a greater impact on the training of the underlying model. Therefore, in 2013, Mikolov [9] proposed the word2vec algorithm, which effectively solved the problem of missing one-hot encoding semantics. The paper proposed two models, CBOW and Skip-gram. The CBOW model and Skip-gram model are shown in the Formulas (1) and (2), respectively:

$$\text{Loss}_C = \sum \log p(w_{\text{i}}|\text{Context}(w_i)) \tag{1}$$

$$\text{Loss}_S = \sum \log p(\text{Context}(w_i)|w_{\text{i}}) \tag{2}$$

By comparing the two loss functions to be optimized, it can be seen that the a priori probabilities used by the two are exactly opposite. The CBOW model uses the information before and after the w word at the i-th position to predict the probability of the current word w_i, and the Skip-gram model uses the current word w_i to predict the before and after information. In this paper, we choose word2vec word embedding to obtain a vectorized representation of words. $t_i \in V^a$, where the part-of-speech vector of the i-th word is t_i, and the dimension of the vector is a. If you input a sentence s of length n, the vectorized matrix calculation is shown in Eq. (3):

$$V^a = t_1 \oplus t_2 \oplus t_3 \cdots \oplus t_n \tag{3}$$

3.1.2 Part-of-Speech Vector

Part of speech, that is, a single vocabulary as a unit, is divided into different categories according to the grammatical rules of the language it uses and the meaning of the word itself.

In this paper, HowNet, English sentiment sets are used to perform part of speech tagging on the original input text s, and then, the labeled vocabulary information is feature vectorized to form a multi-dimensional continuous-valued part-of-speech vector matrix.$\text{tag}_i \in V^b$, where the part-of-speech vector of the i-th word is a tag_i, and the dimension of the vector is b. If you input a sentence s of length n, the calculation is shown in Eq. (4):

$$V^b = \text{tag}_1 \oplus \text{tag}_2 \oplus \cdots \oplus \text{tag}_n \tag{4}$$

By vectorizing part-of-speech features into input, we can study the effect of feature input on sentiment analysis tasks after fusing part-of-speech features.

3.1.3 Position Vector

In the sentiment analysis task, the attention mechanism is mainly used to mine the similarity between the words in the input text and the lack of mining the time series information of the sentence.

Since the word order information of the sentence also has a great impact on the performance of the model, we also input the layer references the position information, encodes the position information of the original input text, and then combines it with the word vector to obtain the position vector. We adopt Vaswani et al. [3] to propose sinusoidal position encoding (Sinusoidal Position Encoding) in the paper. The calculation Formula is shown in (5) and (6):

$$\mathrm{POSE}_{(\mathrm{pos},2i)} = \sin(\frac{\mathrm{pos}}{10{,}000^{2i/d}}) \tag{5}$$

$$\mathrm{POSE}_{(\mathrm{pos},2i+1)} = \cos(\frac{\mathrm{pos}}{10{,}000^{2i/d}}) \tag{6}$$

where i represents the i-th word, the word at the pos position is mapped to a vector of dimension d in the Formula. Its dimension is the same as the word vector dimension of the original text, and the position vector of the fusion position feature can be obtained by matrix addition [10]. The position vector can be expressed as $\mathrm{pos}_i \in V^{\mathrm{c}}$, where the syntactic feature vector of the i-th word is pos_i, and the dimension of the vector is c. If you input a sentence s of length n, the calculation Formula is shown in Eq. (7):

$$V^{c} = \mathrm{pos}_1 \oplus \mathrm{pos}_2 \oplus \cdots \mathrm{pos}_n \tag{7}$$

3.1.4 Dependency Syntax Vector

Dependency syntactic analysis, by analyzing the input text sentences, mining the syntactic structure, hierarchical relationship and interdependence between words in the sentence.

The syntactic features analyzed for each sentence are mapped into a multi-dimensional continuous-valued vectorized representation. $\mathrm{gram}_i \in V^{d}$, where the syntactic feature vector of the i-th word is gram_i, and the dimension of the vector is d. If you input a sentence s of length n, the calculation Formula is shown in Eq. (8):

$$V^{\mathrm{d}} = \mathrm{gram}_1 \oplus \mathrm{gram}_2 \oplus \cdots \mathrm{gram}_n \tag{8}$$

By vectorizing the input of syntactic features, we can study the impact of the input of various types of features on the sentiment analysis task.

3.2 Triple-Channel Feature Fusion Bi-GRU Layer

3.2.1 Triple-Channel Feature Fusion

In order to prevent the model fed into the neural network from being too complicated, we combined and spliced the features separately to form the following Formulas (9), (10), and (11):

$$V_{\mathrm{Single}}^{a+b} = V^{a} \oplus V^{b} \tag{9}$$

$$V_{\text{Double}}^{a+c} = V^{a} \oplus V^{c} \tag{10}$$

$$V_{\text{Triple}}^{a+d} = V^{a} \oplus V^{d} \tag{11}$$

where V_{Single}^{a+b} represents the fusion of word vector and part-of-speech vector of the first channel, and a + b represents the vector dimension after the fusion of the two. V_{Double}^{a+c} and V_{Triple}^{a+d} are the same as above.

3.2.2 Bi-GRU

GRU is the gated recurrent unit. GRU sets the state of neurons entering the neural network through the gating mechanism. The internal structure of GRU is expanded as shown in Fig. 2:

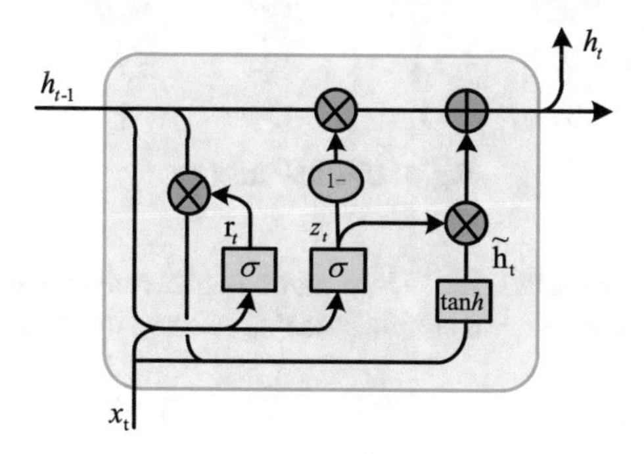

Fig. 2 Structure of proposed model

The calculation Formula of the core algorithm is shown in (12) to (15):

$$r_t = \sigma(W_r \cdot [h_{t-1}, x_t]) \tag{12}$$

$$z_t = \sigma(W_z \cdot [h_{t-1}, x_t]) \tag{13}$$

$$\tilde{h}_t = \tanh(W \cdot [r_t \otimes h_{t-1}, x_t]) \tag{14}$$

$$h_t = (1 - z_t) \otimes h_{t-1} + z_t \otimes \tilde{h}_t \tag{15}$$

where r_t is the reset gate, and z_t is the update gate.r_t and z_t combine the hidden state of the previous h_{t-1} with the current input x_t and combine the current hidden state information $\tilde{h}_t$ to obtain the current output h_t.

Although GRU can solve the problem of long-term dependence of inputting long text information, when the model trains sequence data, it will consider too much the information transmitted at the previous moment and ignore the incoming data information in the future. For long text sentiment analysis tasks, we need to obtain more sentiment semantic features. Using the bidirectional GRU model, we can fully consider the history and future information. Therefore, we use the Bi-GRU structure. The structure of Bi-GRU is shown in Fig. 3:

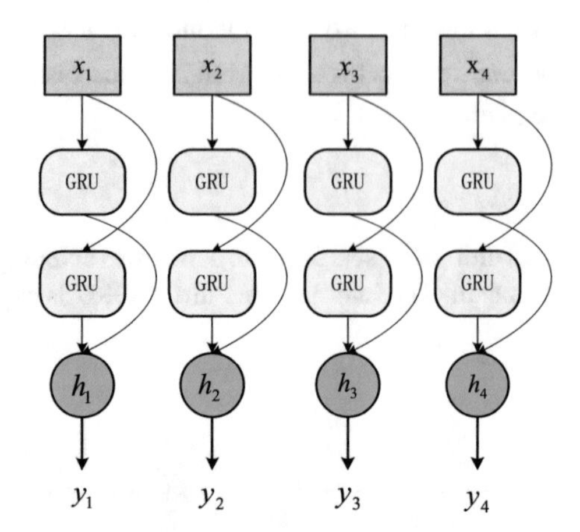

Fig. 3 Bi-GRU structure

After that, we will pass the hidden information processed by the triple-channel GRU to the fusion layer and also use a simple stitching method to connect the three kinds of fusion information.

3.3 Self-attention Layer

In the sentiment analysis task, by amplifying the local semantic information, the attention of the text content with less influence on the sentiment classification results is reduced.

Therefore, we introduce attention mechanism at the model level. The process is shown in Eq. (16):

$$\text{Attention}\,(Q, S) = \sum_{1}^{n} \text{softmax}\,(\text{similarity}\,(Q, K))V \tag{16}$$

In the NLP field, the key is often the value of the value, and the attention mechanism can be understood as: firstly calculating the similarity between query and values, secondly using the softmax function to normalize the weights, and finally weighting and summing the weights and values, get the final attention.

3.4 Fully Connected Layer

The DropOut layer randomly inactivates some neurons, reduces the excessive dependence of the neural network model on some neurons, enhances the robustness of the model, and then transfers the hidden information H to the fully connected layer, and the classification after nonlinear changes is performed through the sigmoid function; the result is output. The classification result is shown in Formula (17):

$$\hat{y} = \text{sigmoid}(wH + B) \tag{17}$$

where w is the weight system matrix, B is the corresponding offset, and $\hat{y}$ is the prediction result.

4 Experiment

4.1 Datasets and Settings

To test the performance of the Tri-BiGRU-Atten model, we used 50,000 movie review data sets from IMDB (www.imdb.com) to verify the validity of the model, including 25,000 positive reviews and 25,000 negative reviews.

We performed relevant preprocessing on the data set: cleaning out html tags and punctuation marks, converting the input text to lowercase, and removing related vocabulary that is meaningless to the sentiment analysis task. There are 88,273 words in the final datasets. We used the Stanford CoreNLP tool to perform word segmentation, part of speech tagging, and dependency syntax analysis on the IMDB data set. Throughout the experiment, we set the dimension of the word vector to 300, the dimensions of the remaining feature vectors are all 30, the size of Batch_Size in training is 32, and the parameter of DropOut is 0.5.

4.2 Result and Analysis

As shown in Table 1, we compare the constructed Tri-BiGRU-Atten model with other models on the IMDB data set. Compared with the common LSTM, Bi-LSTM, and MA-BGRU model proposed by Ji et al. [8], the performance of our model is higher than the above model in accuracy, precision, recall rate, and F1 value, which proves that our model is fully excavated. Deep semantic information of emotional text.

Table 1 Accuracy, precision, recall, and F1 of proposed model and other models

Model	Acc%	Pre%	Recall%	F1%
LSTM	87.6	86.2	89.5	87.8
Bi-LSTM	88.1	88.6	87.6	88.0
CNN-LSTM	88.0	86.7	89.9	88.2
MA-BGRU	90.0	88.7	90.5	89.6
Tri-BiGRU-Atten	**91.1**	**89.7**	**92.5**	**91.0**

Through the above experiments, we can verify the excellent performance of our Tri-BiGRU-Atten model. In the sentiment analysis task, the model we built combines the feature fusion triple-channel bidirectional gated attention mechanism to obtain the best results, proving the effectiveness of the model in the field of sentiment analysis.

5 Conclusion

This paper proposes a Tri-BiGRU-Atten model, which can play a better effect on the task of text sentiment analysis. The model combines vocabulary, part of speech, word position, and syntactic dependence, which are recognized by human beings and merges with different features for modeling. Considering that multi-feature input will increase the complexity of the model, we have constructed three parallel channels to fuse different features with each other. The bidirectional gated recurrent units Bi-GRU is used to extract the hidden representation of the context, and then, the extracted hidden information is fused to add an attention mechanism. Deeply dig into semantic relations and perform well on text sentiment analysis tasks. In the next step, we will further optimize the existing model and try to integrate other features for modeling to improve model performance.

Acknowledgements. This work was supported in part by the Natural Science Foundation of China under Grant 61662081, in part by the Natural Science Foundation of Xinjiang Uygur Autonomous Region under Grant 2017D01A58, in part by the Social Science Foundation of Xinjiang Uygur Autonomous Region under Grant 2016CYY067, in part by the Xinjiang Uygur Autonomous Region Youth Science and technology innovation talent training project under Grant QN2016BS0365, in part by the National Social Science Fund of China under Grant 14AZD11, in part by National language resources monitoring and Research Center minority language sub center project under Grant NMLR201602, and in part by Xinjiang Normal University Computer Application Key Discipline and Xinjiang Normal University Data Security Key Laboratory Funded Project under Grant XJNUSYS102017B01.

References

1. Cho K, Van Merrienboer B, Gulcehre C et al (2014) Learning phrase representations using RNN encoder-decoder for statistical machine translation. arXiv Preprint. arXiv:1406.1078
2. Hochreiter S, Schmidhuber J (1997) Long short-term memory [J]. Neural Comput 9(8):1735–1780
3. Vaswani A, Shazeer N, Parmar N et al (2017) Attention is all you need [J]. arXiv
4. Kalchbrenner N, Grefenstette E, Blunsom P (2014) A convolutional neural network for modelling sentences[J]. Eprint Arxiv, 1
5. Re Lai S, Xu L, Liu K, Zhao J. (2015) Recurrent convolutional neural net-works for text classification. vol 333. AAAI, pp 2267–2273
6. Bahdanau D, Cho K, Bengio Y (2014) Neural machine translation by jointly learning to align and translate[J]. Comput Sci
7. Yang Z, Yang D, Dyer C, et al (2016) Hierarchical attention networks for document classification[C]. In: Proceedings of the 2016 conference of the north american chapter of the association for computational linguistics, Human Language Technologies
8. Ji L, Gong P, Yao Z (2019) A text sentiment analysis model based on self-attention mechanism[A]. In: 2019 World symposium on artificial intelligence[C], Xi'an
9. Mikolov T (2013) Distributed representations of words and phrases and their compositionality[J]. Adv Neural Inf Process Syst 26:3111–3119
10. Gehring J, Auli M, Grangier D et al (2017) Convolutional sequence to sequence learning [J]

National Food Safety Standard Graph and Its Correlation Research

Li Qin[1,2](✉) and ZhiGang Hao[1](✉)

[1] College of Informatics, HuaZhong Agricultural University, Wuhan, China
qinli@mail.hzau.edu.cn, hzau111hzg@163.com

[2] Hubei Engineering Technology Research Center of Agricultural Big Data, Wuhan, China

Abstract. Food safety standards are nationally enforced food production criterion, and there are many kinds of national food safety standards, which involve a wide range of contents and are complex to refer to each other. For studying the content and structure of national food safety standards in a systematic way, this paper uses the knowledge graph technology to extract the content and the reference relationship in the national food safety standards as the knowledge triples to construct the food standard knowledge graph, and then use community discovery algorithm to analyze the graph structure in the knowledge graph and find out the correlation between the standards. Through this research, the high-impact national food safety standards in the graph can be discovered, and these will guide the formulation and updating of national food safety standards.

Keywords: National food safety standard · Knowledge graph · Community discovery algorithm · Correlation analysis

1 Introduction

Food safety is an important issue of people's livelihood. Among the many food standards, the most important standards are the food safety standards. They are the only mandatory food standards. Food safety standards are divided into two categories: national food safety standards and local food safety standards. Among them, national food safety standards include four standards that are general standards, product standards, production and operation specifications, inspection methods, and procedures, which mainly cover eight aspects, food, food additives, pathogenic microorganisms in food-related products, pesticide residues, veterinary drug residues, biological toxins, limits for pollutants such as heavy metals, and other substances that endanger human health. However, the food safety standards are complex, large in number, cover a wide range of contents, and have various forms of mutual quotation. It is difficult for ordinary people to sort out the relationship, and it is very difficult to read. So, there are few systematic analyses and structural studies on food safety standards. This paper aims at the relationship between the content of food safety standards and the reference between each other, using knowledge graph technology to perform content extraction and correlation analysis, to achieve the mapping of food safety standards, and then to use community discovery algorithm to find the correlation between the standards.

Q. Liang et al. (eds.), *Artificial Intelligence in China*, Lecture Notes
in Electrical Engineering 653, https://doi.org/10.1007/978-981-15-8599-9_47

2 Related Researches

The knowledge graph is essentially a knowledge base called the "Semantic network", this method can be applied in many fields and aims to describe the concepts, entities, events, and relationships among the objective world [1, 2]. At present, many knowledge graph resources can be found in Zhishi.me [3], DBpedia [4, 5], Yago [6, 7] and Wikidata [8], etc.; the open-source graph databases include gStore of Peking University [9], neo4j [10], jena [11], GraphDB [12], and so on. The application of knowledge graphs is quite extensive, but the research on establishing knowledge graphs for national food safety standards is rarely involved. This paper mainly studies how to apply knowledge graph technology to the knowledge mining of national food safety standards. We collected the national food safety standards currently in use in China,[1] classified those standards according to the text structure and grammatical characteristics, and designed different extraction strategies and algorithms for each type of standards to achieve their knowledge graph.

3 Food Safety Standards Knowledge Graph

This paper extracts the knowledge and relationship of national food safety standards, mainly including the reference relationship between food standards and the provisions and restrictions in food safety standards. When extracting triples, we found that machine learning and deep learning [13–15] are not accurate in analyzing such standard documents. The reason is that there are a large number of tables in the text, and the correlation between the text is not strong. In order to ensure the accuracy of knowledge, this paper mainly uses the method based on rules and keyword matching. The process of knowledge extraction and relationship mining in national food safety standards is shown in Fig. 1.

3.1 Data Collection

The food safety standards of China are divided into general standards, inspection standards, production management standards, and food product standards, and different strategies are used to extract triplet for different types of standards.

3.2 Triplet Extraction

3.2.1 Reference Relationship

The reference relationship is more common between food product standards and general standards, also between production and operation specification standards, so the reference triplets are mainly extracted from these standards. The extraction method is based on rules and regular expression matching. The two entities in the reference relationship triplet are the names of food safety standards, which are mostly in the form of "GB" or "SN" combined with a set of numbers, and regular expressions can be used to get

[1] http://down.foodmate.net/ziliao/sort/41/52236.html.

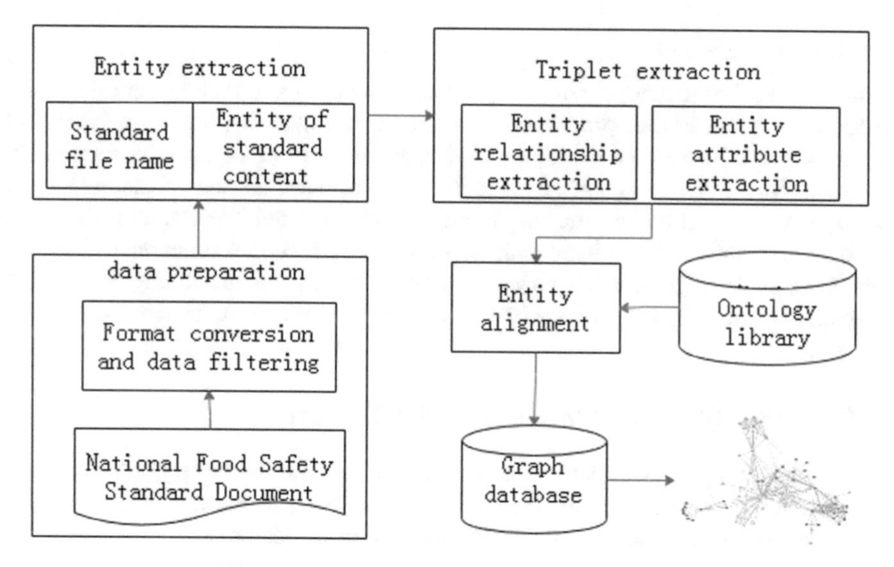

Fig. 1. Construction method of knowledge graph

the name of the food standard more accurately. The relationship between entities is the name of the reference item between these two standards. For example, the reference items between food product standards and general standards are mostly "contaminant limits", "mycotoxin limits", "food additives", "pesticide residues", etc. These common reference items can be constructed as a keyword thesaurus. The basis for extracting triplet is to determine whether there is a corresponding keyword between the two entities, and if so, extract them into triplets for output.

3.2.2 Checked Item and Its Value in Standards

The general standards in the food safety standards stipulate the specific limits of various foods in food additives, pesticide residues, etc., so this paper mainly conducts the knowledge extraction of checked item and its value of the food standards. Because the content of those parts are mostly displayed in the form of tables, and the current processing of PDF tables is not very well, so the extraction of those parts mainly depends on manual methods. Among them, the food type is regarded as the head entity of the triplet, the name of the checked item is regarded as the entity relationship, and limited value of checked item is regarded as the tail entity of the triplet.

3.2.3 Title of Standards

In addition to the knowledge mentioned above, this paper also extracts the triples from the title of food safety standards. For example, "GB 2713-2015 National Food Safety Standard-Starch Products.pdf", using regular expressions to extract "GB2713-2015" and "Standard-Starch Products" as the head entity and tail entity of the triplet, the entity relationship is "standard content ".

3.3 Entity Alignment

In order to realize the mapping of food production processes and food safety standards, this paper converts the food production ontology we have made in previous research [16] into triplets and merges them with the triplets extracted from the national food safety standards. However, due to some entities in the triplets have the same meaning but not exactly the same expression, such as "Potassium sorbate" and "Potassium sorbate and its Potassium salt", so the entity alignment is required. At present, common methods for large-scale entity alignment are word vector analysis [17] and embedding [18], and in this paper, we simplify the problem by establishing a thesaurus for the small number of synonyms.

4 Food Safety Standards Community Discovery

This paper digitally encodes the reference data of all national food safety standards, uses the Louvain algorithm [19] to divide the community, and try to analyze the central point of the reference relationship between these food safety standards.

4.1 Community Discovery Algorithm

The Louvain algorithm separates all initial nodes into a community independently and set the edge weight as zero. Then the algorithm consists of two steps: First step is to scan all nodes, traverse all neighbor nodes for each node, calculate the value of module gain in the community when the neighbor node is joined, select the neighbor node corresponding to the maximum value, add the node to the corresponding community, and repeat this process until the attribution of each node no longer changes; the second step is to consider each community as a node, calculate the weight of the edges between the new communities and the weight of the edges between all the nodes in the community, and then return to step one.

4.1.1 Relative Gain

In the first step, decide which community a node should join according to the value of relative gain $\triangle Q$. When $\triangle Q$ between a node and a certain community is the largest, the node joins the corresponding community. The value of $\triangle Q$ can be calculated according to Formula (1):

$$\Delta Q = \left[\frac{\sum \text{in} + k_{i,in}}{2m} - \left(\frac{\sum \text{tot} + k_i}{2m}\right)^2\right] - \left[\frac{\sum \text{in}}{2m} - \left(\frac{\sum \text{tot}}{2m}\right)^2 - \left(\frac{k_i}{2m}\right)^2\right] \tag{1}$$

In practical programming, the upper formula can be simplified as Formula (2) to reduce the computational complexity.

$$\Delta Q' = k_{i,in} - \frac{\sum \text{tot} \times k_i}{m} \tag{2}$$

Among them,$k_{i,in}$ represents the sum of the weights of node i pointed to community C, $\sum$ tot represents the total weight of community C pointed by node i, k_i represents the total weight of the node i, and m is the sum of the weights of all edges.

4.1.2 Modularity

The condition for the end of the iteration depends on the modularity Q, and its calculation formula is Formula (3)

$$Q = \frac{1}{2m}\sum_{i,j}\left[A_{ij} - \left(\frac{k_i k_j}{2m}\right)\right]\delta(c_i, c_j) \tag{3}$$

where A_{ij} represents the weight of the edge between node i and node j, k_i is the sum of the weights of all edges connected to node i, ci is the community number of node i, and the δ (c_i, c_j) function indicates that if nodes i and j are in the same community, the return value is 1; otherwise, it returns 0. If Q does not change after a round of iterations, the iteration is stopped.

4.2 Correlation Analysis

When assigning weights to edges, different node relationships obtain different weights. For the food product standard, the weights of the edges to the general standards are 3, to the detection standards are 2, and to the remaining standards are 1. The reason for this is that the relationship represented by the edge with the higher weight has a higher degree of manual participation in the extraction, so it has the higher credibility and the greater influence in the community division. After screening and processing, the reference graph has 972 nodes and 2654 edges; moreover, 23 communities are found from knowledge graph. The effect is shown in Fig. 2, where the nodes of the same community are marked with the same color, and the display size of the node label is adjusted according to the in-degree of the node. Nodes with larger in-degrees have larger labels and vice versa.

Through the observation of the food standards graph, it is found that the nodes of national food safety standards such as GB5009.12, GB14454.2, GB2762, and GB11540 are more prominent. Among them, standards such as GB5009 belong to the standard of physical and chemical testing methods, which stipulate the testing methods and standards for detecting chemical substances in food; standards such as GB14454 and GB11540 stipulate the determination method of spices; GB2762 is a general standard that specifies the limit of pollutants in food. The information indicates that among all food safety standards, the most frequently cited standards are testing standards and general standards that stipulate common unqualified items in food. At the same time, these standards are quoted from each other, which are the more influential food safety standards among all standards.

5 Summary

This paper constructed a knowledge graph of national food safety standards by extracting the content entities and reference relationship from the national food safety standards currently in use and analyzes the reference relationship in the food safety standard knowledge graph through the community discovery algorithm. Experimental results show that this method is feasible and can realize the discovery of "core" standards and will guide the formulation and updating of national food safety standards. However,

Fig. 2. Reference relationship graph of national food safety standards

there are still some works to be optimized in the future, one is to use machine learning to complete rapid knowledge extraction, and the other is to achieve knowledge fusion between multiple knowledge graphs through large-scale entity alignment.

References

1. Qi G, Gao H, Wu T (2017) The research advances of knowledge graph. Technol Intell Eng 3(1):004–025
2. Li J, Hou L (2017) Reviews on knowledge graph research. J Shanxi Univ 3:454–459
3. Niu X, Sun X, Wang H (2011) Zhishi.me—weaving chinese linking open data. In: International Semantic Web Conference. Springer, Berlin, Heidelberg, pp 205–220
4. Bizer C, Lehmann J, Kobilarov G (2009) DBpedia-a crystallization point for the web of data. Web Seman Sci Serv Agents World Wide Web 7(3):154–165
5. Auer S, Bizer C, Kobilarov G (2007) DBpedia: a nucleus for a web of open data. Seman Web 4825:722–735
6. Suchanek FM, Kasneci G, Weikum G (2007) Yago: a core of semantic knowledge. In: Proceedings of the 16th international conference on World Wide Web. ACM, pp 697–706
7. Suchanek FM, Kasneci G, Weikum G (2008) Yago: a large ontology from wikipedia and wordnet. Web Seman Sci Serv Agents World Wide Web 6(3):203–217
8. Vrande D, Tzsch M (2014) Wikidata: a free collaborative knowledgebase. Commun ACM 57(10):78–85
9. Zou L, Özsu MT, Chen L (2014) Gstore: a graph-based sparql query engine. VLDB J 23(4):565–590
10. Percuku A, Minkovska D, Stoyanova L (2017) Modeling and processing big data of power transmission grid substation using neo4j. Procedia Comput Sci 113:9–16

11. Wilkinson K (2006) Jena property table implementation. In: Smart PR (ed) Proceedings of the 2nd International workshop on scalable semantic web knowledge base systems. Athens, pp 35–46
12. Ontotext (2018) GraphDB. http://graphdb.ontotext.com/
13. Li F, Yu H (2019) An investigation of single-domain and multidomain medication and adverse drug event relation extraction from electronic health record notes using advanced deep learning models. J Am Med Inf Assoc JAMIA
14. Hai-hong E, Zhang W, Xiao S (2019) Survey of entity relationship extraction based on deep learning. J Softw 30(06):1793–1818
15. Fenia C, Thy TT, Kumar SS (2019) Adverse drug events and medication relation extraction in electronic health records with ensemble deep learning methods. J Am Med Inf Assoc: JAMIA
16. Li Q, Hao Z, Yang L-P (2020) Question answering system based on food spot-check knowledge graph. In: Proceedings of 2020 the 6th international conference on computing and data engineering, pp 168–172
17. Luo Y, Liu D, Yin K (2019) Weighted average Word2Vec entity alignment method. Comput Eng Des 40(7):1927–1933
18. Guan S, Jin X, Wang Y (2019) Self-learning and embedding based entity alignment. Knowl Inf Syst 361–386
19. Vincent B, Jean-Loup G, Renaud L (2008) Fast unfolding of communities in large networks. J Stat Mech Theory Exper

Research on the Hydraulic System of the Expandable Shelter Improved by the Servo Motor Pump

Fang Bai[1](✉) and Shucheng Wang[2]

[1] The 28th Research Institute of China Electronics Technology Group Corporation, Nanjing 210007, China
76964313@qq.com
[2] PLA 31649, Jiesheng Town, Shanwei, Guangdong, China

Abstract. In view of the shortcomings of the hydraulic system of expandable shelter with constant displacement pump and throttle valve, such as difficult to adjust the speed, poor stability, and large energy consumption, the servo motor plus constant displacement pump control mode is proposed and the circuit is upgraded. The motion parameters of top flap and bottom flap are analyzed by simulation and the results show that the improved hydraulic system can fully meet the needs of the main engine and has obvious technical advantages.

Keywords: Elevation-type expandable shelter · Servo motor pump · Hydraulic system · Energy saving

1 Preface

With the continuous increase of the loading equipment and personnel in the shelter and the enhancement of the support function, higher requirements are put forward for the loading space in the shelter. The expandable shelter has been widely used for its good maneuverability and large working space [1, 2].

There are many types of expandable shelter, such as side pull type, side flap folding, type and multi-level folding type, which can only realize the single direction expansion of length, width, and height. Considering that the height of bridge, tunnel, and transport aircraft is limited when entering and leaving the cabin door, as well as meeting the comfort requirements of personnel working in the cabin, an elevation-type expandable shelter emerges [3].

At present, the hydraulic drive is used as the control mode of the flap expansion shelter, mainly using three-phase asynchronous motor plus quantitative pump power component, overflow valve plus throttle valve as the control component for speed regulation [4]. However, in this way, the expansion and recovery speed of the top and bottom plates of the expansion cabin can not be adjusted and the working process is not smooth and continuous; the system has many energy losses and low operating efficiency. In this

Q. Liang et al. (eds.), *Artificial Intelligence in China*, Lecture Notes in Electrical Engineering 653, https://doi.org/10.1007/978-981-15-8599-9_48

paper, the servo motor is used to replace the original asynchronous motor, the internal gear pump is used to replace the original external gear pump, and the flow and pressure of the hydraulic system are controlled by the speed of the motor controlled by the servo driver.

2 Improvement of Hydraulic System

2.1 Operation Requirements of Elevation-Type Expandable Shelter

The elevation-type expansion shelter adopts a large plate assembly structure, which is mainly composed of the main cabin, the movable main cabin and the top plate, bottom plate, front plate, back plate and side plate of expansion cabin. The lifting of the top plate of the movable main cabin and the expansion cabin is driven by the lifting hydraulic cylinder in the expansion casing (expansion inner casing and expansion outer casing) at four corners of the expansion cabin, and the specification of the lifting cylinder is Φ50/28–470; the top plate of the expansion cabin is driven by four expansion top plate hydraulic cylinders, and the specification of the top plate cylinder is Φ50/28–410; the bottom plate of the expansion cabin is driven by four expansion bottom plate hydraulic cylinders, and the specification of the top plate cylinder is Φ63/32–480, and two groups are arranged at the front and rear ends of the expansion cabin.

2.2 Parameter Analysis of Hydraulic System

The power parameters of the expandable shelter hydraulic system are shown in Table 1. The pressure and flow required for each stage are very different. The pressure is from the minimum 0.5 MPa to the maximum 3.1Mpa, and the flow is from the minimum 13.3 to 35.9L/min. The power output required in each stage is also very different. When the original system is controlled by asynchronous motor and constant power variable pump, the output power of the hydraulic pump remains unchanged all the time. When the set flow and pressure exceed the process requirements, the load pressure and flow are regulated by the overflow valve and throttle valve at the outlet of the hydraulic pump, and according to statistics, the energy loss caused by overflow and throttle under this control is as high as 36–50% [5].

If the required load and speed of the actuator change, the speed of the servo motor can be adjusted to control the flow of the pump to adapt to the load speed, the throttling control of the hydraulic system can be cancelled, and the system overflow and throttling loss can be effectively reduced [6].

2.3 Working Principle of Hydraulic System of Servo Motor Pump in Expandable Shelter

In the driving part of the new hydraulic system, the original asynchronous motor is replaced by the servo motor, and the original vane pump and plunger pump are replaced by the internal meshing gear pump. In addition, the servo driver is added to control the pressure and flow required for production, respectively. The structure diagram is shown in Fig. 1.

Table 1. Power situation list of expandable shelter

S/N	Action	Pressure(MPa)	Flow(L/min)	Power(KW)
1	Main cabin rise	2.4	22.1	0.9
2	Top flap extended	3.1	19.3	1.0
3	Bottom flap extended	0.5	35.9	0.3
4	Bottom flap retraction	2.8	26.6	1.2
5	Top flap retraction	0.5	13.3	0.1
6	Main cabin descent	0.5	15.2	0.1

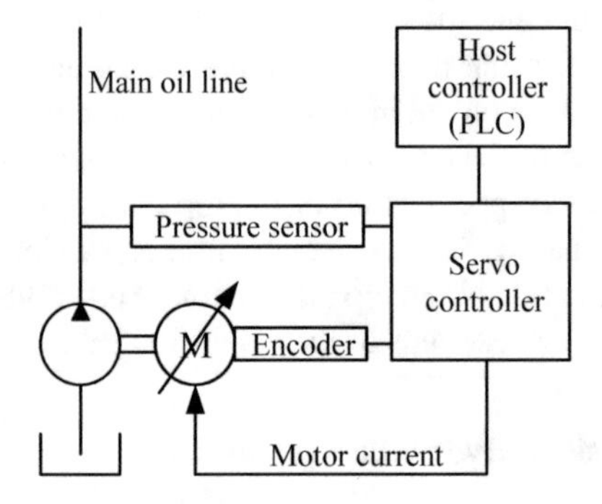

Fig. 1. Control diagram of servo motor pump source

Its working principle is the flow of the hydraulic pump is proportional to the speed of the motor, and the output pressure of the pump is also proportional to the output torque of the motor. When the oil pressure is not established, run the oil pump with the flow proportional to the speed. After the oil pressure is established, the deviation is used for speed control, and the oil pressure can be stable at a given value.

When the pressure does not reach the given value, the servo motor speed is controlled by the flow command; when the pressure reaches, the servo motor speed is controlled by the speed calculated by the pressure command and the pressure feedback difference. At the same time, the pump drive system obtains the corresponding system pressure by obtaining the control signal of the pressure valve, and compares it with the lowest reliable and stable running speed curve set by the motor under different pressures, so as to obtain the lowest frequency of the motor running under the current pressure, so as to avoid the occurrence of pressure pulsation.

2.4 Loop Improvement

As the system can control the motor speed through the servo controller, and then control the flow of the system, the schematic diagram of the redesigned hydraulic system is shown in Fig. 2, and the main features of the system are as follows:

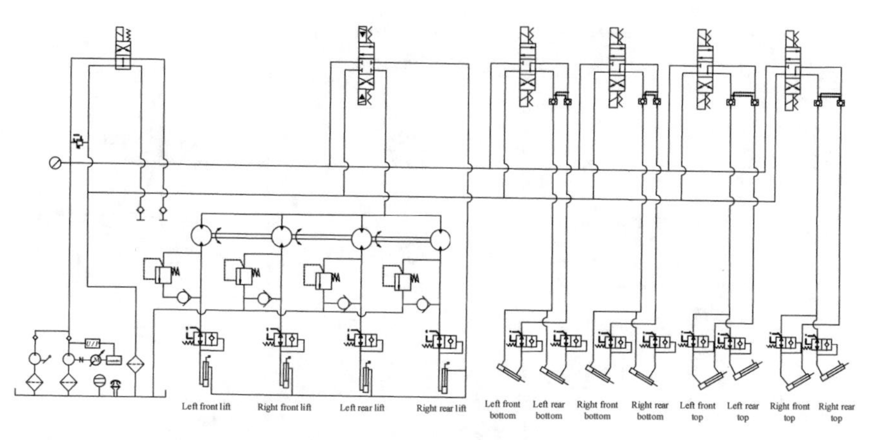

Fig. 2. Principle of hydraulic system of expandable shelter

1. Adopt the hydraulic drive automatic expansion mode, the expansion and recovery time can be adjusted automatically, and the required manpower is reduced; the power source of the hydraulic system adopts the control mode of servo motor quantitative pump group combined with manual hydraulic pump, in which the programmable controller in the servo motor quantitative pump group can realize the data exchange with the external hydraulic system. Through setting the speed signal of the programmable controller, it can control the movement speed of the main cabin lifting and expanding cabin retracting in real time. The four pairs of bottom and top hydraulic cylinders of the expanding cabin can realize stable and reliable retracting and locking.
2. The lifting part of the main cabin is composed of a synchronous motor. The purpose is to realize the synchronous control of the lifting of the main cabin. The lifting is stable without hydraulic impact. The four pairs of lifting hydraulic cylinders of the main cabin are, respectively, equipped with limit photoelectric switches, which are used to measure the displacement of the piston rod of the hydraulic cylinder quickly and output it to the coding circuit, so that the lifting cylinder of the main cabin can be limited in time.

3 Simulation of Hydraulic System

In order to analyze the influence of the change of the force on the hydraulic system and the setting of the servo motor, a simulation model is established to analyze the system.

AMESim software provides a modeling and simulation platform for a variety of disciplines to build complex systems, and different physical models have been strictly tested and tested. AMESim software is used for modeling and simulation. According to the hydraulic schematic diagram, the simulation model is obtained as shown in Fig. 3.

The whole model consists of four parts: pump source part, jacking part, top driving part and bottom driving part.

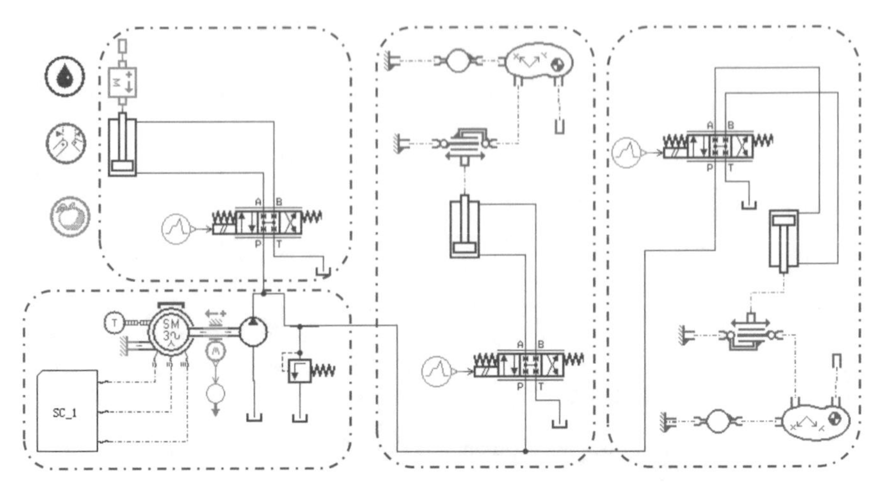

Fig. 3. Simulation model of hydraulic system of expandable shelter

3.1 Simulation Analysis of the Driving Part of the Top Flap

In the process of retraction and opening, the hydraulic cylinder of the top flap is affected by the heavy torque of the side cabin roof and the friction torque generated by the rotation of the hinge. The total torque direction makes the side cabin roof always close, and the side cabin roof is under the positive load during the process of opening. The schematic diagram of the retraction and extension position of the top plate is as follows (Fig. 4):

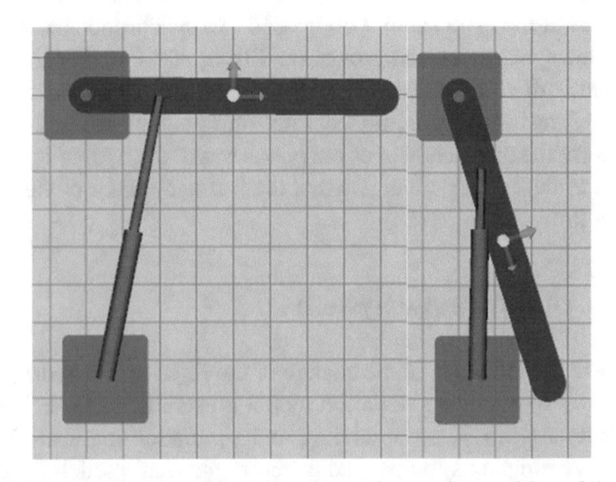

Fig. 4. Schematic diagram of the extension and retraction position of the top flap

3.1.1 When the Top Flap is Extended

Figure 5 is the load simulation curve of hydraulic cylinder when the top flap is extended. It can be seen that after 0.5 s starting, the load of the cylinder rises rapidly to 2.9KN, then starts to fall rapidly to 1.7KN, and then starts to rise slowly to 2.0 KN. During the process of cylinder extension, the output of the cylinder always changes.

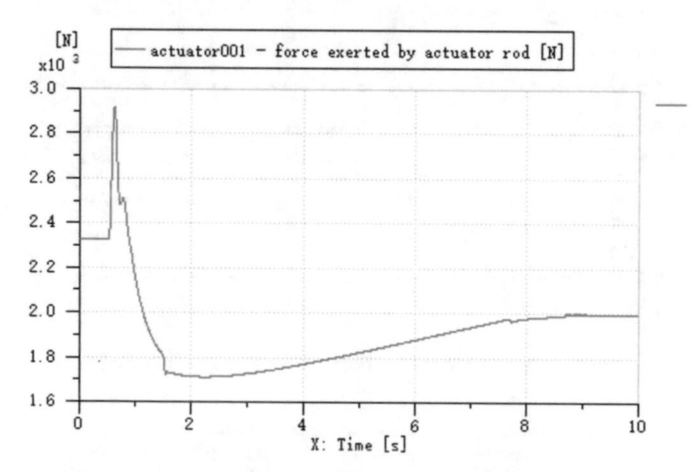

Fig. 5. Load curve of oil cylinder when the top flap is extended

Figures 6 and 7 are the angular displacement and angular velocity curves when the top flap is extended. From 0.5 s to 8.5 s, the top flap rotates from −80° to 0°, and the middle process is relatively smooth, but the slope changes. The speed of rotation increased greatly in the initial stage, and increased to 3.3rev/min in 1.5 s. After 4 s, it became stable slowly. This is because the set speed signal of the motor in the simulation is given linearly according to the extension displacement of the hydraulic cylinder (Fig. 8).

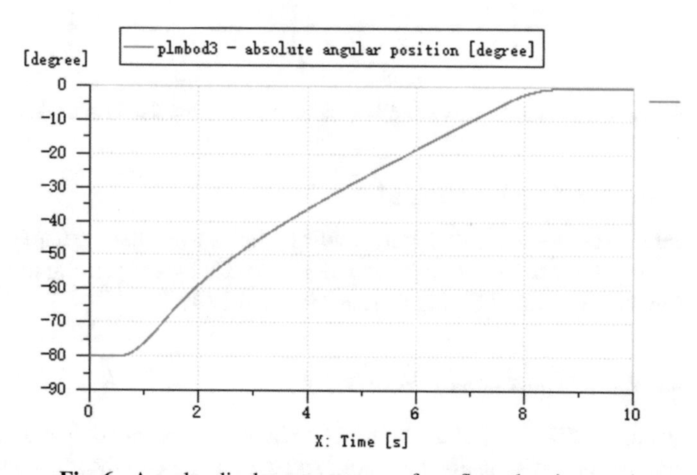

Fig. 6. Angular displacement curve of top flap when it extends

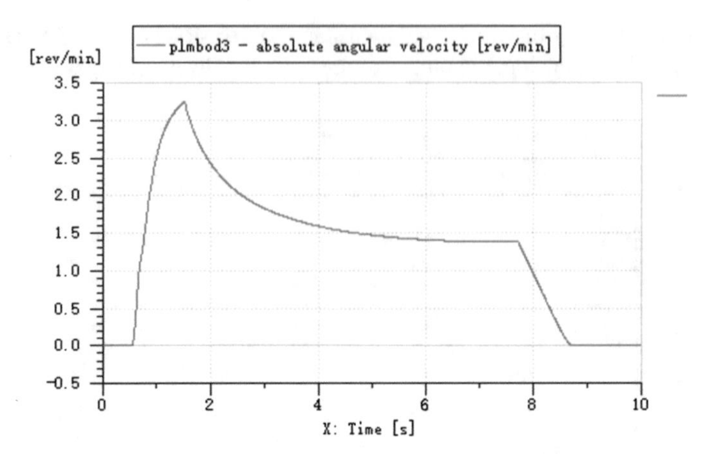

Fig. 7. angular velocity curve of top flap when it extends

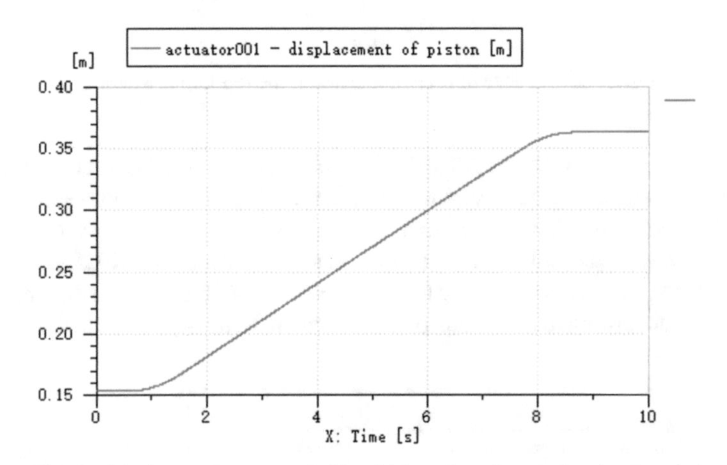

Fig. 8. Displacement curve of oil cylinder when the top flap is extended

3.1.2 When the Top Flap is Retracted

Figure 9 is the load simulation curve of hydraulic cylinder when the top flap is retracted. It can be seen that the load of hydraulic cylinder is 2.0KN before starting in 0.5 s, then it starts to slow down to 1.67KN, and then increases to 1.9KN.

3.2 Analysis of Part Load of Low Flap Drive

Due to the installation position of the bottom flap hydraulic cylinder, the bottom flap is retracted when the rod cavity is filled with oil, and the top plate of the low tank is under the positive load during the retraction process. The drawing below shows the retraction and extension position of the bottom plate (Fig. 10):

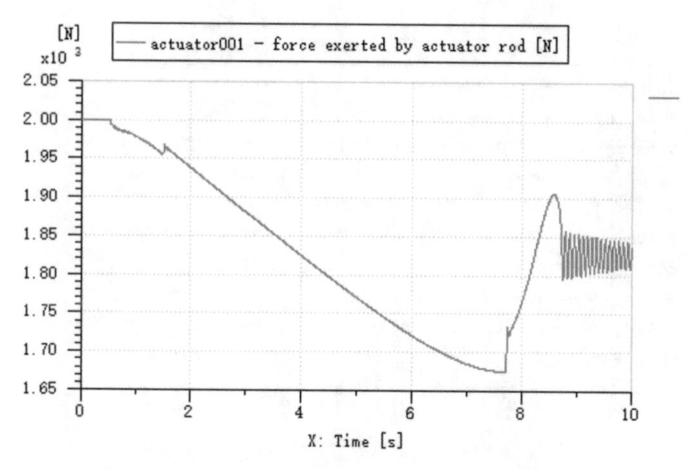

Fig. 9. Load curve of oil cylinder when the top flap is retracted

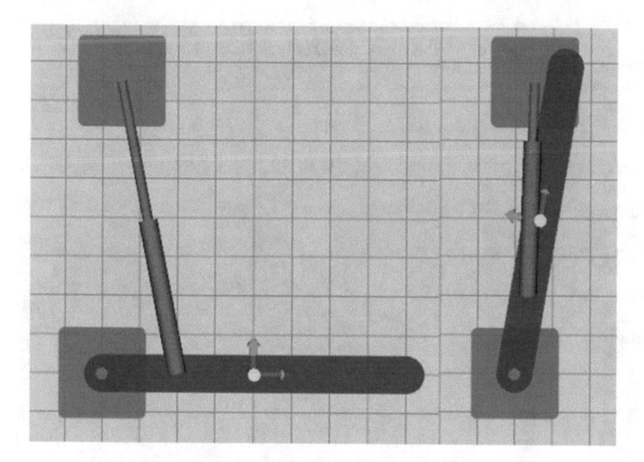

Fig. 10. Schematic diagram of extension and retraction position of bottom flap

3.2.1 When the Bottom Flap is Retracted

Figure 11 shows the load simulation curve of hydraulic cylinder when the low flap is retracted. It can be seen that the load of hydraulic cylinder is 3.0KN before the start-up in 0.5 s. After the start-up, the load slowly drops to 2.67 KN, and then slowly rises to 2.7KN. Because there is no limit in the end section of the simulation, light micro-oscillation occurs.

Figures 12 and 13, respectively, show the angular displacement and angular velocity curves of the bottom flap when it is retracted. From 0.5 s to 8.5 s, the top flap rotates from 0° to 75°. The rotation speed is relatively stable at the initial stage, and increases rapidly after 5S to 2.6rev/min when it reaches 7.7 s (Fig. 14).

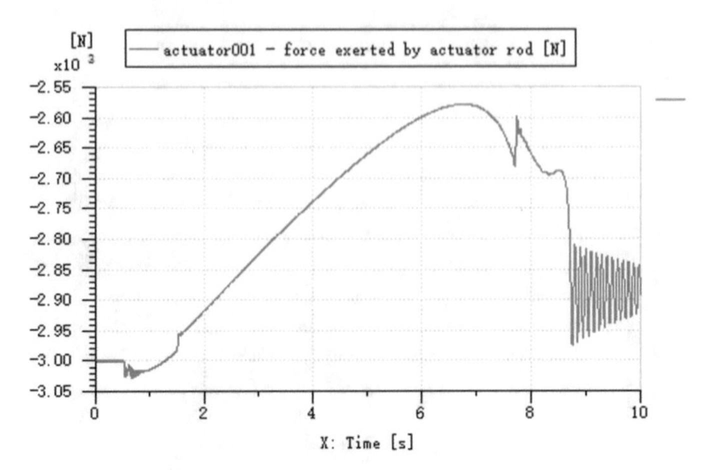

Fig. 11. Load curve of oil cylinder when bottom flap is retracted

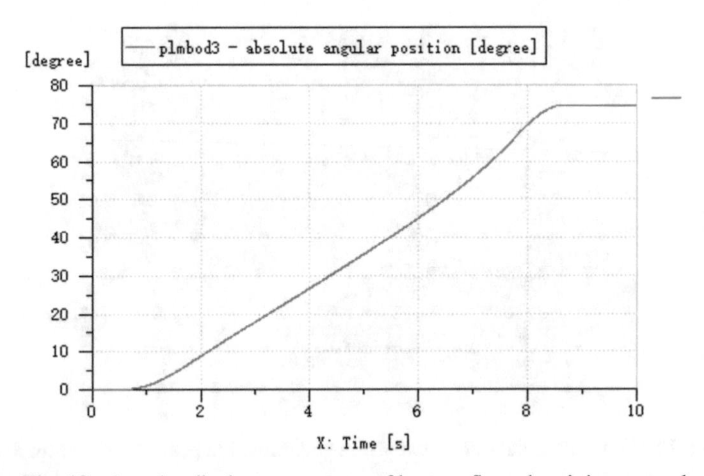

Fig. 12. Angular displacement curve of bottom flap when it is retracted

3.2.2 When the Bottom Flap is Extended

Figure 15 shows the load simulation curve of hydraulic cylinder when the bottom flap is extended. It can be seen that the load of hydraulic cylinder is 3.5KN before starting in 0.5 s. Then, it starts to slowly decrease to 2.6KN, and then increases to 3.0KN.

3.3 Given Signal and Power Curve of Motor

3.3.1 Given Curve of Motor Speed

Figure 16 shows the given curve of motor speed when the top flap is extended. It can be seen that the motor starts from 0.5 s, reaches the maximum value of 1480 r/min in 1.5 s, starts to decline in 6.5 s, and stops in 8.5 s.

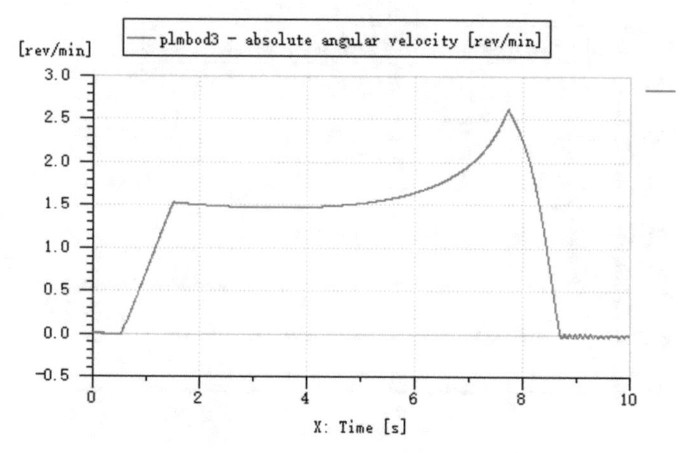

Fig. 13. Angular velocity curve of bottom flap when it is retracted

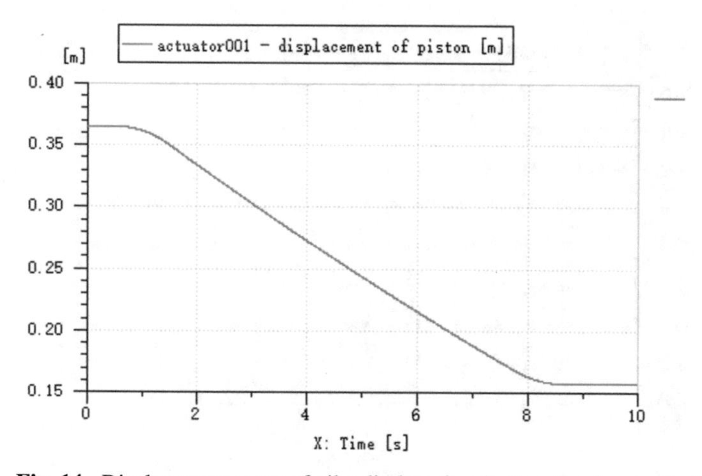

Fig. 14. Displacement curve of oil cylinder when bottom flap is retracted

3.3.2 Motor Power Curve

Figure 17 is the motor power simulation curve (four cylinders) when the top flap is extended. It can be seen that the motor power changes with the demand of the flap motion power. Compared with the constant power of the asynchronous motor and the quantitative pump system, the efficiency can be improved and the energy-saving effect can be achieved.

4 Conclusion

In view of the shortcomings of the extended crude oil hydraulic system, the servo motor and internal gear pump are used as new power components, and the rest parts are improved.

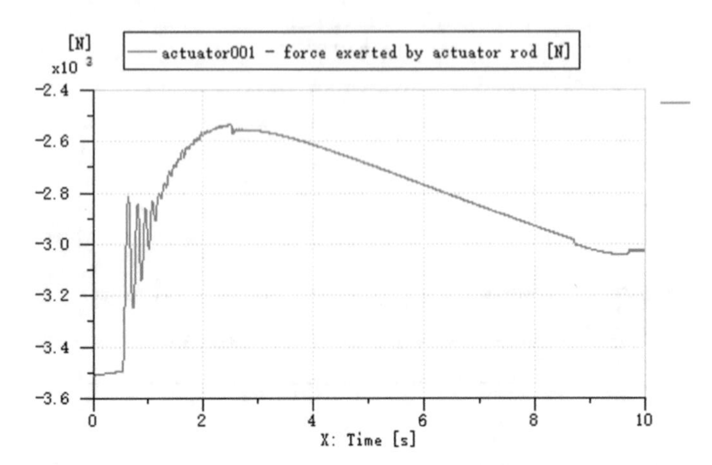

Fig. 15. Load curve of oil cylinder when the bottom flap is extended

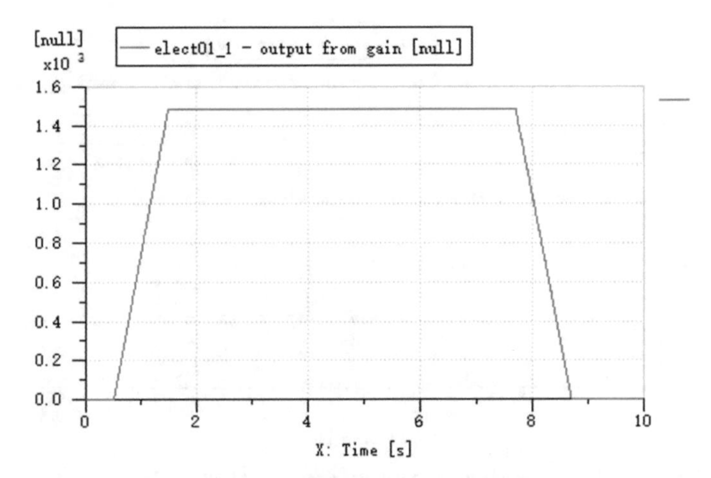

Fig. 16. Given curve of motor speed

Through the simulation analysis, it can be seen that the top flap and the bottom flap in the process of movement meet the flow demand by using servo motor pump due to the large change of rotation parameters. Compared with the common hydraulic system, the power consumption is smaller.

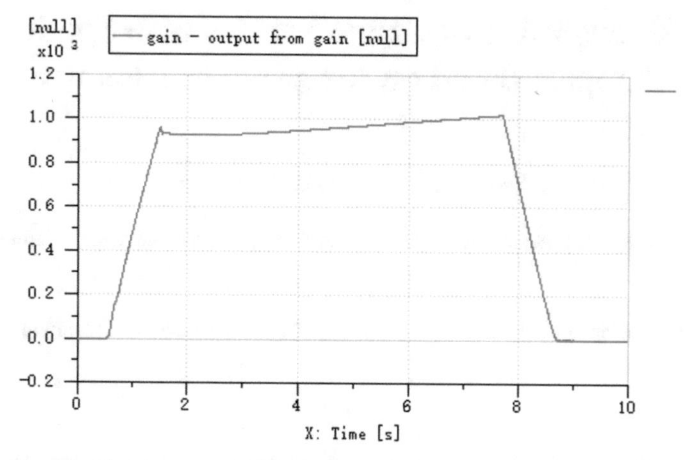

Fig. 17. Simulation curve of motor power when the top copy is extended

Acknowledgements. Foundation item: Advanced Research Project of the 13th Five-year Plan of the Equipment Development Department of the Central Military Commission.

References

1. Xiaowei T, Lihua Z (2017) Combined application of the extended shelter. Command Inf Syst Technol 8:94–98
2. Wang Z, Zhou X (2020) Construction idea of new mobile command and control system equipment. Command Inf Syst Technol 11:89–94
3. Chang J, Gong S, Liu S (2013) Design of hydraulic system of an elevation type expandable shelter. New Technol New Proc 6:70–73
4. Tong Y, Jing H, Song Y et al (2016) Hydraulic system design for a military shelter wall-plate display and closed. Machinery 1:60–64
5. Zhang T, Li B, Li Z (2010) Research on energy-saving technology of plastics injection molding machine driven by servo motor-driven hydraulic pump. Mech Electri Eng Technol 8:73–75
6. Ai T, Yao N, Zhang L et al (2016) Application research of HFST servo-hydraulic system for tire curing press. China rubber/plastics technology and equipment 42(19):60–65

Recognition of Grape Species with Small Samples Based on Attention Mechanism

Yanuo Lu[1] and Bingcai Chen[1,2(✉)]

[1] School of Computer Science and Technology, Xinjiang Normal University, Urumqi 830054, China
china@dlut.edu.cn

[2] School of Computer Science and Technology, Dalian University of Technology, Dalian, Liaoning 116024, China

Abstract. Take Turpan as an example, in recent years, the grape industry in Turpan has become one of the most important pillar industries for rural economic development and farmers' income increase in Turpan. However, there are still many problems in the development of grape industry in Turpan area. Due to many and complicated grape varieties, it is difficult to identify them, and current identification efficiency is far from enough. Moreover, there is no large-scale open data set in grape recognition at present, and each image itself taken for the grape has much noise effect, which leads to low recognition accuracy. In this paper, a small sample of grape variety recognition method based on attention mechanism is used to fine-tune the CNN model and process the dataset image differently, and is compared with the traditional method. The experiment results show that the method can recognize different grape image types accurately with an accuracy of 93.72% under the condition of small sample. By the method, we can not only improve the efficiency of intelligent recognition, but also reduce the manpower cost, and thus realize the intelligent recognition of grape types.

Keywords: Small sample · Attention mechanism · Convolutional neural network · Image recognition · Transfer learning

1 Introduction

Xinjiang grapes are the most famous in the world, especially the Turpan grapes [1]. As the production of grapes rose steadily, the sorting work became unusually mechanical and boring. The traditional manual sorting work is to observe the various characteristics of grapes by the naked eye for classification, and then this has a relatively large limitation. In recent years, with the rapid development in the field of computer vision, in order to improve the efficiency of the classification of grapes, people have begun to study the extraction of deeper features on the surface of grapes, and use the method of transfer learning to carry out automated classification and sorting.

Looking at home and abroad, many people have done a lot of research work on fruit classification, and have achieved many results. Amara et al. adopted deep learning

Q. Liang et al. (eds.), *Artificial Intelligence in China*, Lecture Notes
in Electrical Engineering 653, https://doi.org/10.1007/978-981-15-8599-9_49

methods to achieve automated banana disease leaf classification [2], Youwen et al. [3] used computer image processing technology and support vector machine recognition methods to study the recognition of grape leaf diseases. The test results show that support vector machine recognition The method achieves a better recognition effect than the neural network method. Xue et al. [4] used the data-enhanced convolutional neural network fire identification method to identify fires, which is meaningful in the use of convolutional neural networks for small samples. Jinyi et al. [5] proposed the multi-scale image data fusion classification model MS-EAlexNet, which is excellent in classification work.

Convolutional neural network (CNN) has a certain technical foundation in various image feature extraction and recognition [6], which has been widely used in various commercial sorting systems, greatly improving the economic value of fruits. There are many researches on fruit image recognition based on deep learning, but most of them are using a lot of training data, and acquisition requires a lot of manpower and cost, and there is no large public data set on the network, which makes acquiring a large number of sample images It becomes very difficult, so research on small sample data sets is necessary. In this study, on the basis of deep learning, by adding an attention module to the VGG-16 network that is easy to optimize, the three channels of RGB are weighted to find more important feature channels and perform experiments. By comparing and analyzing models, the combination of different pre-processing methods is used to improve the accuracy of small sample grape image classification.

2 Neural Network and Attention Mechanism

2.1 VGG-16 Network

VGG-16 is a network model proposed by Karen Simonyan and Andrew Zisserman at ILSVRC 2014. It performs better than googLeNet in multiple transfer learning tasks. Moreover, the VGG model is the preferred algorithm for extracting CNN features from images. In this competition, GoogleLeNet won the first place [7]. The VGG network uses a smaller 3*3 convolution kernels, and the deepest network of the year, the stack of two 3*3 convolution kernels is relative to the field of view of the 5*5 convolution kernels, and the three 3*3 convolution kernels. Stacking is equivalent to the field of view of a 7*7 convolution kernel. This reduces the number of parameters on the one hand, and has more nonlinear transformations on the other, which increases CNN's ability to learn features. Compared with the later ResNet network, the simple model, few layers and easy optimization make it still one of the more classic and used networks. The network model structure is shown in Fig. 1.

Taking an image with a size of 224*224 as an example, the model includes 13 convolutional layers, 5 pooling layers, 3 fully connected layers, and 1 classification layer. Among them, 1–2 layers of convolution layer are 64 convolution kernels, 3–4 layers are 128 convolution kernels, 5–7 layers are 256 convolution kernels, 8–13 layers are 512 convolutions In the kernel, the number of neurons in the first fully connected layer is 4096 and the number of neurons in the second is 5. The number of output categories of the classification layer is the final number of categories.

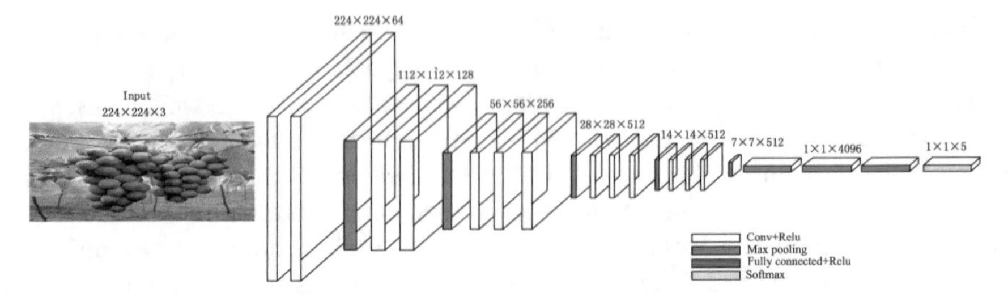

Fig. 1. VGG-16 model

2.2 Overview of Attention Mechanism

Attention plays a very important role in the human visual system. In recent years, the research work of combining deep learning with visual attention mechanism has mostly focused on using mask to form attention mechanism. The principle of the mask is to identify the key features in the picture data through another layer of new weights. Through learning and training, the deep neural network learns the areas that need attention in each new picture, which forms attention. With the transfer of ideas, two different types of attention are formed, one is soft attention and the other is hard attention.

Soft attention focuses more on channels and spaces, and soft attention is deterministic. It can calculate the gradient through the model for forward propagation and backward update to learn the model weight, and it can be directly generated by training.

The difference between hard attention and soft attention is that it is more inclined to random prediction, that is to say, hard attention pays more attention to discrete position information, and emphasizes the dynamic changes of the model. Therefore, it is difficult to form hard attention by end-to-end training, which is mostly done through reinforcement learning. In this study, we used soft attention [8] (Figs. 2, 3).

Fig. 2. Distribution of human visual attention

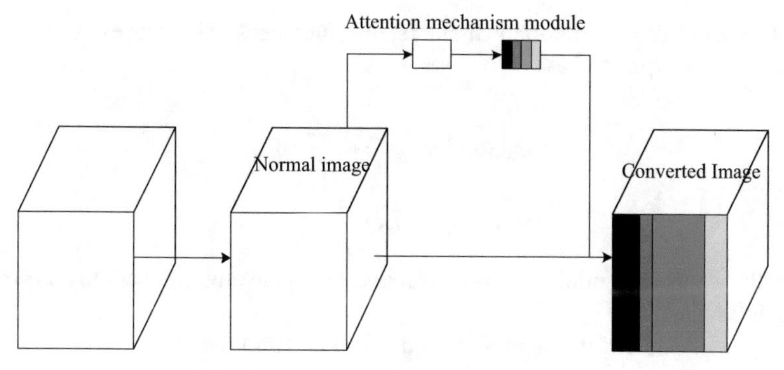

Fig. 3. Attention and mechanism model

2.2.1 Principle and Formula of Attention Mechanism

Represent attention input as $X_{1:N} = [x_1, \ldots, x_n], i = 1, \ldots, N$ (N represents the amount of input information), When calculating these characteristic information in the network model, the data information that is not related to the need is removed. Given a query vector q related to the current situational task, select the location of the information $z \in [1, N]$ to indicate. When $z = i$ indicates the i-th feature information. In this study, soft attention is used, and each input is determined by the calculated probability, with α_i representing the weight of the attention mechanism, the process of producing attention [9] is as follows:

$$\begin{aligned} \alpha_i &= p(z = i | X_{1:N:q}) \\ &= \operatorname{soft}\max_i(s(x_i, q)) \\ &= \frac{\exp(s(x_i, q))}{\sum_{j=1}^{N} \exp(s(s_i, q))} \end{aligned} \tag{1}$$

Among them $s(x_i, q)$ represents the scoring function, the choice of this function is determined by the current task and external factors, $s(x_i, q)$ can be directly provided by external information, or can be generated in model training. Before the model calculation, $s(x_i, q)$ can be represented by the following additive model:

$$s(x_i, q) = v^T \tanh(W_{xi} + U_q) \tag{2}$$

You can also use the dot product model:

$$s(x_i, q) = x_i^T W_q + B \tag{3}$$

Among them, the W, v, U parameters of the above formula are all learnable parameters, and B is the amount of paranoia.

After the weight of α_i the attention mechanism is obtained, the input feature needs to use the attention weight to perform a self-select operation. This operation is to encode the feature to filter out important input feature messages. The process based on soft attention can be expressed as:

$$\text{attention}(x_{1:N,q}) = \sum_{i=1}^{N} a_i x_i$$
$$= E_{z \sim p_{q(z|x_i)}}[X] \quad (4)$$

Finally, we use a sigmoid function to convert the convolution result to a normalized weight of 0 ~1.

The general form of the sigmoid function can be simplified to:

$$y = \varsigma(z) \quad (5)$$

where $z = ax + b$.

3 Small Sample Image Recognition

3.1 Materials and Methods

3.1.1 Data Set

This study takes five different kinds of grapes as the research object. Because the data in the agricultural field is not easy to obtain, and the noise of the pictures taken in the field is relatively large and related to many factors, part of the data set used in this study comes from the Internet and part from the field shooting. This can not only increase the generalization of model training and increase the reliability of the model. A total of 958 original images were collected in this experiment, including 201 green grapes, 168 red grapes, 224 summer black grapes, 189 jasmine grapes and 176 golden finger grapes. These 5 types of pictures are placed in different folders, and labeled with labels, label 0 represents green grapes, label 1 represents red grapes, label 2 represents summer black grapes, label 3 represents jasmine grapes, label 4 represents golden finger grapes. The transfer learning dataset uses the fruit pictures in the Imagenet dataset. The ImageNet dataset is the world's largest image dataset established by Fei-Fei Li et al. This dataset contains millions of color images and is divided into more than 1000 different categories. In practice, due to the lack of performance of the equipment and the limitations of the experimental conditions, the experiments in this paper did not select all the images, but selected 20 of them. The training set of each category took 400 tests. This experiment will collect good data. The size of the set picture is uniformly cropped to 224*224*3 pixels, and then the model training can be carried out. The data set is divided into training data set, verification set and test data set, the ratio is 8:2, 6:4, 7:3 three ratios to split the data set and test data set.

3.2 Image Enhancement

Under the premise of small samples, the lack of training samples in the data set has become the current primary problem. The most direct method is to increase the data training samples. In order to realize the image recognition work of small samples, this paper introduces image data enhancement, which not only expands the number of samples in the data set, but also improves the visual effect of the image, making the features of the sample image easier to extract. Common data enhancement methods include the following: grayscale transformation, grayscale equalization, pseudo color enhancement, smoothing, sharpening, filtering, etc.

In this experiment, the method of histogram equalization and white filled background fusion was used.

3.2.1 Histogram Equalization

Histogram equalization is one of the most commonly used methods in spatial image enhancement. It mainly uses image histogram to adjust the image contrast, especially when the contrast of the target area is quite close. Histogram equalization can make the brightness more uniform in the histogram Distribution, so that local contrast can be enhanced without affecting the overall contrast [10]. One of the main advantages of histogram equalization is that it is a fairly intuitive technique and is a reversible operation. If the transformation function is known, the original histogram can be restored, and the amount of calculation is small. One disadvantage of this method is that it does not choose the data to be processed, which may increase the contrast of the background noise and reduce the contrast of the useful signal, resulting in excessive enhancement of some areas and the generation of artificial traces.

For an image with picture pixels $N*N$, assuming that the pixel value range of the image is $\{0, 1, 2, 3... L-1\}$, the probability of the gray value r appearing in the image is the image The histogram of the calculation Formula is as follows:

$$P_r = \frac{n_r}{M \times N} \quad r = 0, 1, 2, \ldots, L-1 \tag{6}$$

In the Formula: $N*N$ is the total number of pixels in the image, which means the number of pixels in the image whose gray value is r. Then, through the transformation function, the pixel with the gray value of r in the original input image is mapped to the corresponding pixel with the gray value of S in the output image. The calculation formula of the transformation function is as follows:

$$S_r = \sum_{i=0}^{r} \frac{n_i}{M \times N} \tag{7}$$

Histogram equalization can reduce the contrast of high-contrast images, making their image features easier to extract.

The following diagram illustrates the effect of histogram equalization (Figs. 4, 5):

3.2.2 Gamma Transformation

Gamma transformation, also known as exponential transformation or power transformation, is another commonly used grayscale nonlinear transformation. The gamma transformation of the image gray level is generally expressed as the formula:

$$D_B = c \times D_A^{\gamma} \tag{8}$$

(1) Normal image (2) processed by Histogram equalization

Fig. 4. Histogram equalization processing effect

(1) processed by Histogram equalization (2) processed by Gamma Transform

Fig. 5. Gamma transform processing effect

When $\gamma > 1$, it will stretch the area with higher gray level in the image and compress the part with lower gray level.

When $\gamma < 1$, it will stretch the area with lower gray level in the image and compress the part with higher gray level.

When $\gamma = 1$, the grayscale transformation is linear, and the original image is changed in a linear manner at this time (Figs. 6, 7).

3.2.3 Image Enhancement Fusion

This article uses a combination of histogram equalization and image grayscale gamma transformation. The essence of histogram equalization is to redistribute the pixel values of the image, adding many local contrasts. The overall contrast has not changed much, of course there are The disadvantages are as follows: the gray level of the converted image is reduced, and some details are reduced; some images have high peaks, and the contrast is unnaturally enhanced after processing, so the single use of histogram equalization is

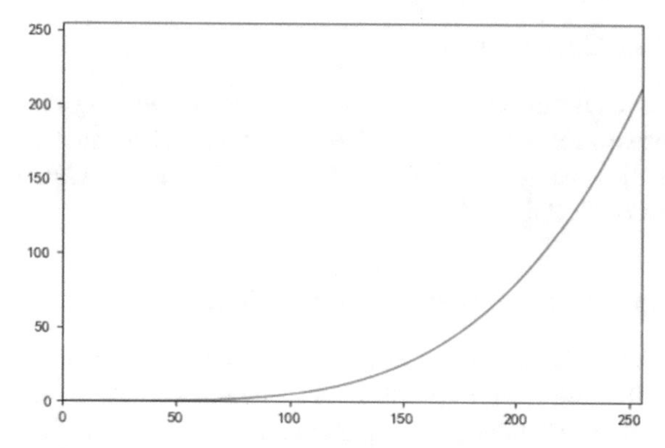

Fig. 6. Gamma transformation function when $\gamma > 1$

Fig. 7. Migrating sample instances

not good. In view of this, we introduce the combination of image gray gamma and it can solve the problem of image contrast enhancement and obtain features more easily. The transformation is as follows:

$$f' = n \times S + (1 - n) \times D_B \tag{9}$$

In the above Formula: S is the image after histogram equalization conversion, and is the image after grayscale gamma conversion, and the fusion degree is adjusted by the value of n $(0 \leq n \leq 1)$.

4 Experiment and Result Analysis

This study designed an experiment to compare the effect of different scale data set division and different processing of small sample data sets on the recognition effect.

4.1 Experimental Environment

The experimental environment is Windows 8 64-bit operating system, TensorFlow is used as the framework, Python is used as the programming language, the local computer memory is 32G, the processor is Intel (R) Core (TM) i7-4210 M CPU @2.60 GHz, and the graphics card is NVIDIA GTX 1060.

4.2 Model Preparation and Hyperparameter Setting

First download the VGG network model, add an attention mechanism to the fully connected layer, so that the features get different weights. As the pre-training model for this transfer learning, of course, some parameters need to be set before training, for example, the learning rate is set between 0.01–0.0015, and 0.001 is selected as the final learning rate after the experiment. As a very important hyperparameter, the learning rate represents the rate of model weight update. If the setting is too large, the cost function fluctuates too much, and the test results are not accurate enough. If the setting is too small, the network model converges slowly and the training time is extended. Another hyperparameter drop_out can be set to randomly inactivate 30–50% of neurons. After cross-validation, the 50% inactivation rate model has the best convergence effect. The number of softmax classifications is set to 5, which satisfies the classification tasks of the five grape varieties in this study. In order to view the model training process more conveniently, a checkpoint is set every 100 times during the training process to test the data in the testing machine, output the test accuracy rate, and finally save the model with the highest accuracy.

4.3 Model Migration Training

Given a source domain *Ds* and a learning task *Ts*, a target domain *Dt* and a learning task *Tt*, the purpose of transfer learning is to use the knowledge on *Ds* and *Ts* to help improve the learning of the prediction function $f()$ on the target domain *Dt*, Where $Ds \neq Dt$ or $Ts \neq Tt$ [11]. Convolutional neural network models often require a large number of training samples to ensure the effectiveness of the model, and the deeper the network, the greater the number of training samples required. The use of transfer learning [10] can effectively avoid the problem of small samples.

4.4 The Proportion of Experimental Samples

It is not only the amount of sample data that affects the overfitting of the model, but for a brand new data set, different proportions of the training test set will also have a great impact on the model training effect. So I divided the samples of the data set into three proportions: 80–20% (80% of the samples are used for training), 70–30% (40% of the samples used for training), 60–40% (40% of the samples are used for training) (Table 1). In theory, the sample ratio should be set as much as possible, so that the effect of the experimental deviation on the result will be smaller. You can choose a more accurate ratio of the training test set for training. In view of the limited experimental conditions, I only chose this Three representative ratios.

Table 1. Number of data sets with different ratios

Small sample grape dataset		
Training set: Test set	Training set	Test set
8:2	766	192
7:3	670	288
6:4	574	384

4.5 Experimental Comparison of Different Data Set Ratios

In this section, the VGG-16 network is used as the reference network. By adding attention to the network to fine-tune the model, the small sample data set (Table 2) is divided into 3 ratios, 8:2, 7:3, 6:4, the accuracy rate is calculated by comparing experiments to determine which training set test set ratio has the greatest effect on model training improvement.

Table 2. Comparison results of different scale data sets

Small sample grape dataset			
Model Fine-tuned VGG16-Net	Training set: Test set	Initial accuracy (%)	Average accuracy (%)
	8:2	48.4	94.3
	7:3	42.5	92.6
	6:4	38.9	91.2

From the above table, it can be seen that the recognition rate of 8:2 is more than other ratios. This is also demonstrated through experiments. When the sample data set is fixed, the more samples in the training set, the better the model can obtain its characteristics. In the follow-up experiments, in order to ensure that the model has excellent recognition ability, the training set test set ratio is 8:2.

4.6 Comparison of Data Set Experiments with Different Processing Methods

In order to verify the feasibility of this research method, in addition to the experiment on the self-built data set, we also took another three sets of pre-processing experiments compared with the method proposed in this article. Experiments show that the recognition accuracy after histogram equalization and gamma-transformed image enhancement fusion has reached XX% (see Table 3). Compared with other single enhancement and no processing methods, the recognition accuracy has been greatly improved. And achieved the ideal recognition effect. Generally speaking, in the experimental environment of this study, the accuracy of this method is higher in the image recognition task of small sample data sets, and it is more robust.

Table 3. Comparison results of different pretreatment methods

Pretreatment method	Single recognition accuracy					Average recognition accuracy (%)
	Green grapes (%)	Red grapes (%)	Black grapes (%)	Jasmine grapes (%)	Golden finger grapes (%)	
No treatment	90.7	89.5	85.6	91.6	88.1	89.10
Histogram equalization	91.2	92.7	87.2	92.5	91.5	91.02
Logarithmic transformation	92.3	91.9	85.9	90.1	92.8	90.60
Histogram equalization + Gama	95.6	94.5	90.1	92.2	96.2	93.72

5 Conclusion

In this paper, for the classification of small sample of grapes, a small sample learning method based on attention mechanism is proposed. First, we use the Imagenet dataset to train the feature extraction layer of the network. After the attention module, the features have their corresponding weights, and then the classification work is performed through the enhanced fusion operation of histogram equalization and gamma transformation. Through multiple sets of comparative experiments, it can be shown that the learning method has a good classification effect on grape recognition, and its recognition accuracy is higher than that of traditional methods, which is very significant for the use of convolutional neural networks for small samples in the future.

Acknowledgements. This work was supported in part by the Tianshan Young Talent Program, Xinjiang Uygur Autonomous Region under Grant 2018Q024, in part by the Natural Science Foundation of China under Grant 61771089 and Grant 61961040, and in part by the Regional Cooperative Innovation Program of Autonomous Region (Aid Program of Science and Technology to Xinjiang) under Grant 2020E0247 and Grant 2019E0214.

References

1. Man Baode (2017) Research on the development strategy of grape industry in Turpan, Xinjiang[D], Shihezi University
2. Amara J, Bouaziz B, Algergawy A et al (2017) A deep learn- ing-based approach for banana leaf diseases classification [C]. In: Mitschang B (ed) Lecture notes in informatics, Bonn, pp 79–88
3. Youwen T, Tianlai L, Chenghua L et al (2007) Grape disease image recognition method based on support vectormachine[J]. Trans Chinese Soc Agr Eng (Transaction of the CSAE) 23(6):175–180

4. Xue Wu, Xiaoru Song, Song Gao, Chaobo Chen (2020) Convolutional neural network fire recognition based on data enhancement [J]. Sci Technol Eng 20(03):1113–1117
5. Jinyi Q, Luo J, Xiu L, Wei J, Ni F, Feng H (2019) Multi-scale grape image recognition method based on convolutional neural network [J]. Comput Appl 39(10):2930–2936
6. Li Y, Hao Z, Lei H (2016) Summary of research on convolutional neural networks [J]. Comput Appl 36(9):2508–2515
7. Simonyan K, Zisserman A (2014) Very deep convolutional networks for large-scale image recognition. arXiv preprint arXiv:1409.1556
8. Ju M, Luo J, Wang Z, Luo H (2020) A multi-scale target detection algorithm fused with attention mechanism [J/OL]. Acta Opt 1–15
9. Pu X (2019) Research on plant disease and insect pest identification based on attention mechanism-CNN compression model [D]. Sichuan University
10. Li N, Wang Y, Xu S, Shi L (2019) Small sample surface floating object recognition based on AlexNet [J]. Comput Appl Softw 36(02):245–251
11. Han F, Yan L, Chen J, Teng Y, Chen S, Qi S, Qian W, Yang J, Moore W, Zhang S, Liang Z (2020) Predicting unnecessary nodule biopsies from a small, unbalanced, and pathologically proven dataset by transfer learning[J]. Springer International Publishing 33(10)

F-Measure Optimization of Forest Flame Salient Object Detection Based on Boundary Perception

Tiantian Tang[1,2] and Bingcai Chen[1,2](✉)

[1] School of Computer Science And Technology, Xinjiang Normal University, 10763 Xinjiang, China
cbc9@qq.com

[2] School of Computer Science And Technology, Dalian University of Technology, 116024 Dalian, China

Abstract. Relaxed F-measure is applied to the improved boundary perception model to optimize c. Shorten the model of training time, after one hundred iterations can study to the characteristics of forest fire; Highlight the significant areas quickly. With immediate proximity activation, Floss also maintains a large gradient; Make network model produce polarized activations, speeds up the model convergence. Solve forest fires due to wide dispersion of targets; A problem where foreground and background are not easily distinguishable.

Keywords: Relaxed F-measure · Polarized activations · Floss

1 Introduction

Forest fires spread rapidly, destroying the ecological environment and bringing economic losses and casualties. Therefore, rapid detection of accurate flame information can eliminate fire hazards in time. Past by target detection to detect fire information, such as the traditional fire detection method by the flame color [1], texture and static characteristics and movement information [2]. In the static characteristics, Kong to HSI color space, setting threshold segmentation method to extract the flame area, but individual space model of the flame identification efficiency is low; Sam et al. used texture features of HSVYCbCr color space and grayscale symbiosis matrix to identify flame, but this method has limitations for texture classification at pixel level. In moving object detection, there are background difference method, optical flow method, and frame difference method. Through the analysis of the dynamic characteristics of the image, each pixel of the image point of the velocity vector changes detected moving target. Movement information comparison process extended the detection time, and is not conducive to raising the speed of the fire.

The traditional method can not obtain the comprehensive flame information and the accuracy is low. Salient object detection [3] by simulating human attention mechanism to extract video and image of the most attractive areas significantly. Thus, effective information can be retained quickly and redundant information can be removed. The

Q. Liang et al. (eds.), *Artificial Intelligence in China*, Lecture Notes
in Electrical Engineering 653, https://doi.org/10.1007/978-981-15-8599-9_50

salient object detection [4] method is adopted to obtain the significance diagram to detect the flame, which can quickly determine the flame significance target. Improve the efficiency of flame detection. The convolution operation of SuperCNN and DNN-G methods is at the pixel block level [5], so the edge of salient target is blurred and the image definition is significantly reduced. Overlap exists when computing and storage redundancy. Both DHSNet and Liu [6] hope to reduce the complex influence of the object inside and background while preserving the details of the object. Liu and Han achieve the detection and localization of significant objects through GV-CNN [7] and HRCNN [8], but cannot save the details of the image. Wang and Borji proposed the circular network structure [9]. The crude saliency map and local saliency information of the input image are extracted circularly, and a clearer saliency map is generated by combining the pyramid pooling structure with the crude saliency map. InFCN networks used jump connection to help restore more image detail information in the process of deconvolution; DSS by bouncing connection combines HED structure in different scale lateral output, makes deep lateral output directly affect shallow lateral output, and raises significantly the accuracy of test results.

No matter it is a loop module or a side output, in a conventional CNN, the convolution operation will adopt the same processing method for each element in the feature diagram. However, the importance of different positions in the feature map is different, and the same processing of these information may have a certain impact on the final result. So attention mechanism is used to fine processing characteristic graph is put forward, in the process of convolution highlight significant regional characteristics of the picture, and suppress background related areas; Thus, the boundary information of the target is enhanced [10]. But the forest flame is widely distributed, the forest flame detected by the original model is incomplete [11], and the boundary is fuzzy. Slowed down the fire fighting. Moreover, the iteration times of the model are millions and the learning speed is slow. When approaching the activation target, the model does not converge easily. As a result, our perception model for boundary is improved, and F-measure Floss loss function optimization. To accelerate the training speed of the model, the forest flame area can be highlighted after 100 iterations. Make a clear distinction between foreground and background in a scattered scene.

2 Residuals Refinement Improvement

The boundary perception salient detection model Uses an encoder and a corresponding decoder to construct an attentional feedback mechanism [10], which refines the rough prediction scale and helps to capture the overall shape of the target. And by using the boundary of the boundary enhanced damage is fine, help target outline on the significant prediction study. The original boundary perception of significant target detection model in the forest of flame spreads wide and extracts only a part of the flame information; Narrow the scope of the flame, to assist people right to judge the severity of the fire. So we improved the model as shown in Fig. 1:

On the basis of the original RRM model, (b) Add a layer of 3×3 convolution kernel, batch normalization, residual error, and max pooling in the coarse-grained feature part; add the part of the fine-grained character corresponding to it. The increase in convolution

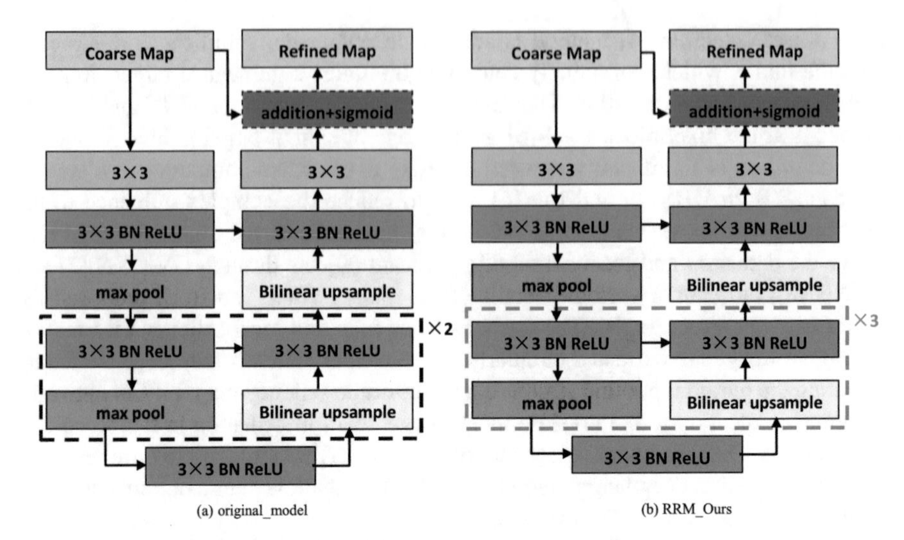

Fig. 1. Original model and croase-grained part improvements

layer can obtain more abundant flame characteristics, and pass by bilinear interpolation sampling will enlarge characteristics; make the edge of the flame target smooth and natural. Batch normalization is adopted to prevent the gradient disappearance of network model due to the increase of layers. By max pool retains the flame characteristics and at the same time reduces dimension model parameters.

Aiming at the problem of incomplete flame extraction and blurring edge of the original model in the flame dispersion scene. (d) part in fine-grained RRM model, increase a 3×3 layer of convolution kernels, batch of standardization, residual, Max pooling and dual linear differential sampling. Model to the original coarse granularity characteristics of part of the flame information again to study, and amplified; Prevent dispersing fire information loss. As shown in Fig. 2:

3 F-Measure Optimization

In the measure, real class, false positive class, and negative positive class are defined as the corresponding sample number:

$$
\begin{aligned}
\mathrm{TPY}^t, Y &= \sum_{\mathrm{i}} 1\left(y_i == 1 \text{ and } y_i^t == 1\right), \\
\mathrm{FPY}^t, Y &= \sum_{\mathrm{i}} 1\left(y_i == 1 \text{ and } y_i^t == 1\right), \\
\mathrm{FNY}^t, Y &= \sum_{\mathrm{i}} 1\left(y_i == 1 \text{ and } y_i^t == 0\right),
\end{aligned} \tag{1}
$$

Y is the gound value, Y^t is the binary prediction of the threshold t, Y is the ground value of the saliency map. $1(\cdot)$ is an indicator function, which is equal to 1 when the argument is true and 0 otherwise. In order to integrate F-measure into CNN and optimize

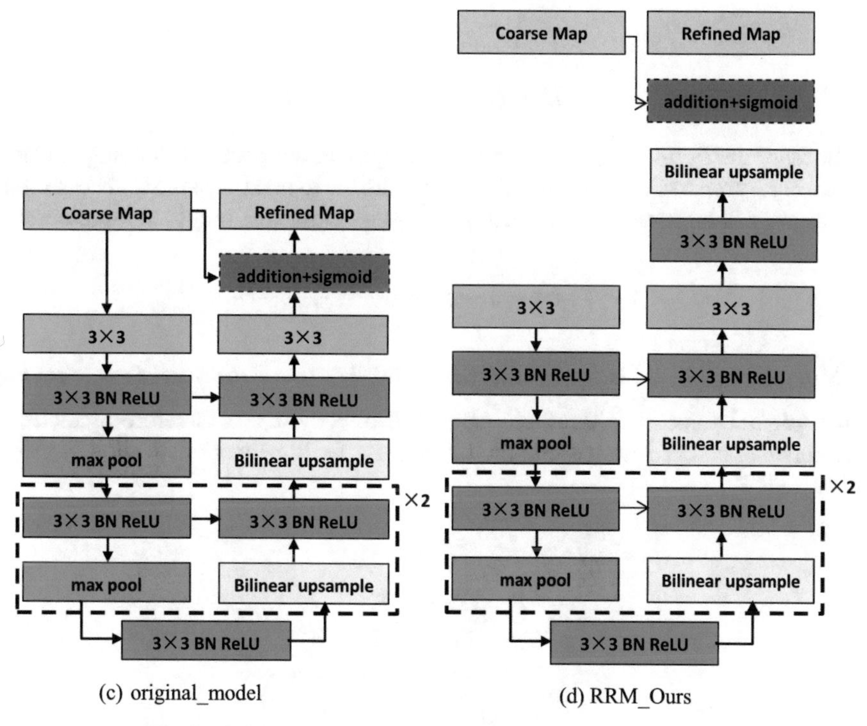

(c) original_model (d) RRM_Ours

Fig. 2. Original model and croase-grained part improvements

through the end-to-end method. Defined the decomposable F-measure; For posterior $\hat{Y}$ differentiable. True class, false positive class, and false negative class are redefined on the basis of continuous posterior Y^t :

$$
\begin{aligned}
\mathrm{TP}\hat{Y}, Y &= \sum_i 1\hat{y}_i \cdot y_i, \\
\mathrm{FP}\hat{Y}, Y &= \sum_i 1\hat{y}_i \cdot (1 - y_i), \\
\mathrm{FN}\hat{Y}, Y &= \sum_i \left(1 - \hat{y}_i\right) \cdot y_i,
\end{aligned} \tag{2}
$$

Based on the Formula (2), precision rate and recall rate are presented:

$$
P\hat{Y}, Y = \frac{\mathrm{TP}}{\mathrm{TP} + \mathrm{FP}}, \quad r\hat{Y}, Y = \frac{\mathrm{TP}}{\mathrm{TP} + \mathrm{FN}} \tag{3}
$$

The final relaxation F-measure can be written:

$$
\begin{aligned}
F\hat{Y}, Y &= \frac{(1 + \beta^2)p \cdot r}{\beta^2 P + r} \\
&= \frac{\left(1 + \beta^2\right)\mathrm{TP}}{\beta^2(\mathrm{TP} + \mathrm{FN}) + (\mathrm{TP} + \mathrm{FP})}
\end{aligned}
$$

$$
\begin{aligned}
&= \frac{(1+\beta^2)\mathrm{TP}}{H} \\
H &= (\mathrm{TP}+\mathrm{FN}) + (\mathrm{TP}+\mathrm{FP})
\end{aligned}
\tag{4}
$$

Because the Formulas (3) (4) can be decomposed, integrating it training structure and the back-prop to CNN. For in end-to-end approach to maximize the CNN relaxation F-measure, will be based on loss function FLoss we define the proposed F-measure:

$$
L_F\left(\hat{Y}, Y\right) = 1 - F = 1 - \frac{(1+\beta^2)\mathrm{TP}}{H} \tag{5}
$$

Minimize $L_F\left(\hat{Y}, Y\right)$ is equal to the maximum relaxation F-measure。L_F is according to the original forecast $\hat{Y}$ threshold calculated directly. L_F is therefore predicted $\hat{Y}$ differentiable, can be inserted into the CNN.I loss L_F to network activation in partial derivative is $\hat{Y}$:

$$
\begin{aligned}
\frac{\partial L_F}{\partial \hat{y}_i} &= -\frac{\partial F}{\partial \hat{y}_i} \\
&= -\left(\frac{\partial F}{\partial \hat{y}_i} \cdot \frac{\partial \mathrm{TP}}{\partial \hat{y}_i} + \frac{\partial F}{\partial H} \cdot \frac{\partial H}{\partial \hat{y}_i}\right) \\
&= -\left(\frac{(1+\beta^2)y_i}{H} - \frac{(1+\beta^2)\mathrm{TP}}{H^2}\right) \\
&= -\left(\frac{(1+\beta^2)\mathrm{TP}}{H^2} - \frac{(1+\beta^2)y_i}{H}\right)
\end{aligned}
\tag{6}
$$

Another form of Formula (6), can be F-measure logarithmic likelihood maximization:

$$
L_{\log}\left(\hat{Y}, Y\right) = -\log(F) \tag{7}
$$

The corresponding gradient is:

$$
\frac{\partial L_{\log}}{\partial \hat{y}_i} = \frac{1}{F}\left[\frac{(1+\beta^2)\mathrm{TP}}{H^2}\right] \tag{8}
$$

To demonstrate that FLoss is superior to other options, we compared the definitions, gradients, and surface maps of the three loss functions. CELoss is defined as:

$$
L_{\mathrm{CE}}\left(\hat{Y}, Y\right) = -\sum_{i}^{|Y|}\left(y_i \log \hat{y}_i + (1-y_i)\log\left(1-\hat{y}\right)\right) \tag{9}
$$

I for the input image space position, $|Y|$ for the pixel number of the input image; the gradient L_{CE} predicted by y_i is:

$$
\frac{\partial L_{\mathrm{CE}}}{\partial \hat{y}_i} = \frac{y_i}{\hat{y}_i} - \frac{1-y_i}{1-\hat{y}_i} \tag{10}
$$

As shown in Eqs. 7 and 9, the gradient of CELOSS depends only on the prediction or truth value of one pixel i. However, $\frac{\partial L_F}{\partial \hat{y}_i}$ is determined by the of all the pixel image prediction or true value. Relaxed F-measure overcomes the indifferentiability in the standard F-measure formula. The Floss can be decomposed and thus attached to CNN for supervision. FLoss, moreover, even in the saturation region also has a large gradient, and makes the prediction of polarization relative threshold is stable.

4 Conclusion

In order to solve the problem of forest fire to extract the incomplete fuzzy boundaries. In this paper, on the basis of modifying the residual refinement structure of the original model, F-measure was taken as the optimization target, and relaxed F-measure was adopted to overcome the indifferentiability in the standard F-measure formula. Maximize the F-measure. The decomposed Floss loss function is added after CNN, resulting in polarized activation of the network. Model F-measure of the ability to learn, to speed up the model convergence; Distinguish forest flame image foreground and background clearly and extract flame target completely. Thus, get the forest fire information timely and accurately, minimize the harmful effects of forest fire.

Acknowledgements. This work was supported in part by the Tianshan Young Talent Program, Xinjiang Uygur Autonomous Region under Grant 2018Q024, in part by the Natural Science Foundation of China under Grant 61771089 and Grant 61961040, and in part by the Regional Cooperative Innovation Program of Autonomous Region (Aid Program of Science and Technology to Xinjiang) under Grant 2020E0247 and Grant 2019E0214.

References

1. Horng WB, Peng JW, Chen CY (2005) A new image based real-time flame detection method using color analysis. In: Proceedings of IEEE international conference on networking, sensing and control, vol 10, pp 100–105
2. Liu ZG, Yang Y, Ji XH (2016) Flame detection algorithm based on a saliency detection technique and the uniform local binary pattern in the YCbCr color space. Signal Image Video Process 10(2):277–284
3. Shen S (2008) Research on flame flicker frequency identification based on image processing. Great science and technology in China, pp 223–235
4. Liu T, Yuan Z, Sun J et al (2007) Learning to detect a salient object. In: Proceedings of IEEE conference on computer vision and pattern recognition, vol 5, issue 6, pp 17–22
5. He S, Rynson WH, Lau et al (2015) A super pixel wise convolutional neural network for salient object detection. IJCV
6. Lee C-Y, Xie S, Gallagher P, Zhang Z, Tu Z (2015) Deeply supervised nets, AISTATS
7. Liu N, Han J (2016) DHSNet: deep hierarchical saliency network for salient object detection, CVPR, pp 22-35
8. Feng Y, Zhang Z, Zhao X (2018) Group-view convolutional neural networks for 3D shape recognition, CVPR
9. Chen S, Tan X, Wang B, Hu X (2018) Reverse attention for salient object detection. In: European conference on computer vision

10. Sandler M, Howard AG, Chen L-C (2018) Mobilenetv2: inverted residuals and linear bottlenecks. In: CVPR
11. Wang W, Shen J, Shao L (2019) An iterative and cooperative top-down and bottom-up inference network for salient object detection. CVPR, pp 5968–5977

Cluster Analysis of Student Scores Based on Global K-Means Algorithm

Jiashan Cui[1], Mei Nian[1(✉)], Jun Zhang[1,2], and Bingcai Chen[1]

[1] School of Computer Science and Technology, Xinjiang Normal University, No. 102, Xinyi Road, Urumqi 830054, China
2468830639@qq.com

[2] Xinjiang Institute of Physical and Chemical Technology, Chinese Academy of Sciences, Urumqi 830011, China

Abstract. Aiming at the problem that the initial clustering center in the K-means algorithm is easily affected by outliers, it is proposed to analyze the data based on the global k-means algorithm. The global k-means algorithm is used to improve the determination process of the initial clustering center and reduce the influence of the random initial clustering center on the clustering result. First, preprocess the data of the four courses of the first semester of the 2018–2019 academic year for undergraduates, and save them in csv format and then, through the experimental comparison of the real traditional k-means algorithm and the global k-means algorithm, we get the clustering indicators such as Jaccard coefficient, accuracy, F value, etc. The experimental results show that the global k-means algorithm is 7.9% more accurate than the original k-means clustering algorithm, and it is verified that the global k-means clustering algorithm is better than the traditional k-means algorithm.

Keywords: Data mining · Cluster analysis · Global k-means algorithm

1 Introduction

Educational data mining is a research hotspot in the context of big data, and has a wide range of application prospects. The cluster analysis method is one of the most promising achievement analysis methods at present. Its advantage is that its conclusion is concise and intuitive, and it is easy to find the hidden rules from it. The goal of the k-means algorithm is to divide M points in N dimensions into K clusters, so that the accuracy and recall rate reach the maximum value [1]. However, the initial cluster centers randomly selected in the analysis of the k-means algorithm are easily affected by outliers. When there are outlier objects that deviate from the data-intensive area, the calculation of the average value will be affected. In this paper, the global k-means algorithm is applied to the score analysis, which improves the process of determining the initial clustering center and reduces the negative impact of the screening of the initial clustering center on the k-means algorithm. The k-means clustering algorithm based on the global analysis of student results can get more accurate clustering results.

Q. Liang et al. (eds.), *Artificial Intelligence in China*, Lecture Notes in Electrical Engineering 653, https://doi.org/10.1007/978-981-15-8599-9_51

2 Clustering Algorithm

Clustering algorithm analysis is widely used in student score analysis. Common clustering algorithms are: K-means clustering, k-center clustering, CLARANS algorithm, DIANA algorithm, BIRCH algorithm, Chameleon algorithm, EM algorithm, OPTICS algorithm, DBSCAN algorithm, etc. As a typical unsupervised learning algorithm, clustering algorithms are mainly used to automatically classify similar samples into a category. Cluster analysis divides a large number of data objects into multiple clusters according to the differences in the nature of the data. The data in each cluster has a high degree of similarity, and there are certain differences between different clusters [2].

2.1 Global K-Means Clustering Algorithm

K-means clustering algorithm is one of the most commonly used clustering algorithms at present. Its advantage is that it is easy to implement and obtain a local optimal solution and the disadvantage is that in the initial clustering center using random selection, as the number of iterations increases, the result will produce obvious errors [3]. When using k-means algorithm for cluster analysis, first select K points randomly as the initial cluster center, then divide other data points into the cluster closest to the K initial cluster centers to complete an iterative process, each time iteratively calculate the mean value of each type of data point, update the cluster center, and finally, repeat the previous steps until the obtained cluster center point no longer changes, and then, complete the clustering process [4].

In this paper, a global k-means algorithm is used to analyze student achievements. Based on the k-means algorithm, the initial clustering center selection process is improved and the traditional k-means algorithm is optimized. The global k-means clustering algorithm simplifies the clustering process and divides a problem into multiple sub-problems.

2.2 The Basic Principle of the Global K-Means Algorithm

The basic principle of the global k-means algorithm: first calculate the centroid of the sample set, use it as the first optimal clustering center, and set it to $i = 1$, then when $i = i + 1$, $i > K$, the algorithm terminates [5]. Finally, take the remaining samples as the optimal clustering center, and then calculate the square error criterion function with the known i-1 optimal clustering centers. The sample point that minimizes this function is selected as the $i - 1$ th best clustering center, that is, the optimal clustering center of $K = i$, and so on.

2.3 Related Formulas

(1) Jaccard coefficient: Used to compare the similarity and difference between effective sample sets. Given two sets A, B,the Jaccard coefficient is defined as the ratio of the size of the intersection of A and B to the size of the union of A and B, which is defined as follows:

$$J(A,B) = \frac{|A \cap B|}{|A \cup B|} = \frac{|A \cap B|}{|A| + |B| - |A \cap B|} \tag{1}$$

(2) Accuracy: It is a very good and intuitive evaluation index.

$$\text{Accuracy} = \frac{\text{TP} + \text{TN}}{\text{TP} + \text{TN} + \text{FP} + \text{FN}} \tag{2}$$

(3) Precision: The ratio of the number of samples actually divided into positive examples by the classifier and the total number of positive examples.

$$\text{Precision} = \frac{\text{TP}}{\text{TP} + \text{FP}} \tag{3}$$

(4) Recall: The ratio of the number of samples that are actually positive and divided by the classifier into positive examples is the sum of the number of instances that are actually positive and divided by the classifier into positive examples and the number of instances that are actually positive but divided by the classifier into negative examples.

$$\text{Recall} = \frac{\text{TP}}{\text{TP} + \text{FN}} \tag{6}$$

(5) F_1-score: It is the arithmetic mean divided by the geometric mean.

$$F_1 = \frac{2\text{TP}}{2\text{TP} + \text{FP} + \text{FN}} \tag{7}$$

(6) Ave-Precision:

$$\text{AP} = \frac{\sum \text{Precision}}{\text{total\_image}} \tag{8}$$

3 Data Preprocessing

3.1 Data Selection

Data selection as the first step of data preprocessing directly affects the results of data mining. The research object of this article is the score data of 4 courses in the first semester of the 2018–2019 academic year for undergraduates, with a total of 430 person. The data information mainly includes the student number and the scores of various subjects, some of which are shown in Table 1.

Table 1. Final semester results of class 16-1 in the 2018–2019 school year

Student ID	Operating system principle	C# programming	Principles of computer organization	Computer network
20161601141007	85.3	91.5	85.9	94.2
20161601141002	87.4	86.7	91	85.2
20161601141005	86.8	83.6	92.5	83.8
20161601141012	81.7	92.2	90.7	83.2
20161601141013	82.9	78.7	79.6	70
20161601141014	83.2	89.4	76.3	81.2
…	…	…	…	…

3.2 Data Cleaning

The relevant original student data is generally noisy and incomplete. The noise or incompleteness of the data is not only affected by the lack of attribute values of the data itself but also restricted and affected by some inevitable objective factors, thereby reducing the quality of the original data. Therefore, an important part of data preprocessing is data cleaning, processing, or excluding some original data with missing values or noise [6]. The experiment mainly clears the data of related exchange students, clears blank data of semester students who have not completed credits, fills in missing values, and deletes duplicate values.

3.3 Data Integration

In student data mining, when encountering data from different sources, there are great differences in different data structure attributes, so the data must be integrated during data preprocessing, so that data from multiple sources are merged and stored in the same database [7]. However, in data integration, data redundancy often occurs, and sometimes different types of data encoding settings are different, resulting in data conflicts, so the main purpose of data integration is to unify the data source and structure and form to reduce data redundancy residuality and conflict. In the process of processing the data set, this paper integrates the documentation of the two semesters of each class into a complete excel file.

3.4 Data Conversion

After data cleansing and data integration, data structure transformation is required to ensure data consistency, which lays the foundation for data mining [8]. By stratifying different redundant data and generalizing the original data of the lowest layer with higher-level data, the data quality can be improved through data compression. In this paper, after data preprocessing, all the student data is converted into an array of strings, stored in a.csv file, and then the processed student score data set is imported into Python, using the traditional K-means algorithm and the global-based K-means algorithm performs cluster analysis on the data set.

4 Experimental Analysis

In order to verify the effectiveness of the global k-means algorithm, this paper compares the traditional ki-means algorithm with the global k-means algorithm. Experimental tool: Intel(R) Core(TM) i5-10210U CPU @ 1.60 GHz 2.10 GHz, 8G memory and experimental environment: 64-bit operating system win10, programming software: Python, experimental data set: students preprocessed with previous data in the 430 data sets, 301 data sets are used as the training set, and 129 data sets are used as the test set. In terms of validity verification, the k-means algorithm and the global k-means algorithm are compared using Jaccard coefficient, accuracy, precision, recall, and F_1-score.

Figure 1 shows a P–R curve graph comparing the traditional k-means algorithm and the global k-means algorithm. The abscissa represents the recall rate, and the ordinate represents the accuracy rate. It can be drawn from the P-R curve graph that in the same student score data set, the global-based k-means algorithm performs better than the traditional k-means algorithm.

Table 2 shows the comparative test conducted in the experiment. The experimental results on the AP. AP represents the average accuracy. From the table, it can be seen that the global-based k-means clustering algorithm is more accurate than the traditional k-means clustering algorithm. The average value of s is improved by 7.9%, indicating that the global-based k-means algorithm has high accuracy.

Regarding the evaluation of the results of the clustering algorithm, in addition to the commonly used indicators such as accuracy, the article also analyzes the clustering indicators such as Jaccard coefficient, accuracy rate, and F value, as shown in Table 3. The experimental results all prove the effectiveness of the global k-means algorithm.

From the above comparison of clustering evaluation indicators, it can be concluded that the global-based k-means algorithm reduces the impact of the initial clustering center on the clustering results in clustering, so that it shows more in the evaluation index of clustering results. Excellent clustering effect and accuracy.

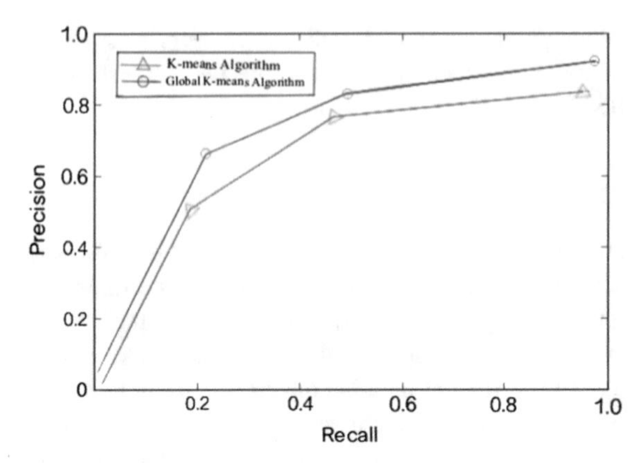

Fig. 1. P-R curve comparison between traditional K-means algorithm and global K-means algorithm

Table 2. Comparison of AP results

Method	AP/%
K-means	88.3
Global k-means	96.2

Table 3. Comparison of other clustering evaluation indicators

	Jaccard	Accuracy	Precision	Recall	F_1
K-means	0.494	0.838	0.845	0.886	0.928
Global k-means	0.743	0.958	0.954	0.962	0.959

5 Conclusion

This paper uses a clustering algorithm based on global k-means and a traditional k-means clustering algorithm to conduct a comparative experiment on student score analysis. The comparative experiment shows that the selection of clustering centers based on global k-means algorithm is reasonable and highly accuracy, but further research is needed: when the amount of data in the future is too large, how to quickly and efficiently implement cluster analysis.

Acknowledgements. This project was supported in part by the Open Research Fund of Key Laboratory of Data Security, Xinjiang Normal University, under Grant XJNUSY102017B04 and University Scientific Research Project, Xinjiang Autonomous Region under Grant XJEDU2017S032.

References

1. Gu X, Xu F, Yang Y et al (2019) Analysis of college student performance based on global K-means algorithm [J]. J Changchun Univ Sci Technol (Natural Science Edition) 5
2. Duan G, Liu S, Zou C (2020) Application of global center clustering algorithm in class programming[J]. Comput Digital Eng (3)
3. Zou C, Yang Y (2018) Based on maximum distance product and minimum distance and collaborative K clustering algorithm [J]. Comput Appl Softw 035(005):297–301,327
4. Liu Y (2014) Research and application of cluster analysis and association rule technology in score analysis [D]. Central China Normal University
5. Xiujuan Sun, Xiyu Liu (2008) Improved K-means algorithm based on new clustering effective function [J]. Comput Appl 28(12):3244–3247
6. Xingfei Ma, Yin Li (2016) Application of improved K-means algorithm in college student consumption data [J]. J Wuxi Commercial Vocat Coll 16(06):82–85
7. Yu Z, Qin H (2018) K-means algorithm based on improved bee colony algorithm [J]. Control Decis 033(001):181–185
8. Ruiyu Jia, Yugong Li (2018) K-means algorithm with self-determination of cluster number and initial center point [J]. Comput Eng Appl 054(007):152–158

Microblog Rumors Detection Based on Bert-GRU

Lianjin Han, Weimin Pan(✉), and Haijun Zhang

School of Computer Science and Technology, Xinjiang Normal University, No. 102, Xinyi Road, 830054 Urumqi, Xinjiang, China
379483304@qq.com

Abstract. Sina Weibo is an ideal place to spread rumors in China, and automatic debunking rumor is a crucial problem. To detect rumors, the rumors detection method based on deep learning adopts the static pretraining model for text representation. After word vector training, it will not change any more, and there are the problems that the vector cannot represent the polysemy of words in different contexts. In order to solve this problem, this paper presents a novel method that rumors detection based on Bert-GRU and proposes an implementation method using the Bert model for the microblog text representation, while using the gated recurrent unit (GRU) networks to learn features for rumors detection. Experimental results show that the proposed method has a good effect on detecting rumors in Weibo.

Keywords: Rumors detection · Pretraining · Bert · GRU

1 Introduction

Weibo is a leading social media platform that helps people create, disseminate, and discover content. Users can use the Internet or mobile phones and other media to publish content in real time and share information with more users. Due to its simplicity, it has attracted a large number of Internet users. However, since any user can freely publish and disseminate various Weibo messages, Weibo is inevitably injected with a lot of rumors. These rumors make it difficult for users to obtain useful information and may cause misunderstandings or negative emotions, which may cause public panic and anxiety, thereby disrupting economic and social order. Therefore, research on how to efficiently detect rumors in Weibo can help to purify the ecological environment of Weibo, help users to identify effective information, and make Weibo play an active role in the guidance of information dissemination.

Early rumor detection research generally used machine learning techniques, using labeled data sets for supervised learning. However, feature engineering is time-consuming and labor-intensive and requires certain professional background knowledge. With the development of deep learning and the use of natural language, the use of deep learning for rumors detection can automatically learn the features contained in the data

Q. Liang et al. (eds.), *Artificial Intelligence in China*, Lecture Notes
in Electrical Engineering 653, https://doi.org/10.1007/978-981-15-8599-9_52

set, to a certain extent, solve the problems of the methods based on machine learning. However, the text representation based on the deep learning method adopts the static pretraining models, and the word vector does not change after the training. There is the problem that the vector cannot represent the polysemies of words in different contexts. In order to solve this problem, a rumor detection method based on Bert-GRU is proposed, using the Bert model to represent the microblog text, and at the same time using the GRU network to learn features for rumors detection. Experimental results show that the proposed method can improve the accuracy of rumors detection.

2 Related Works

At present, researchers at Chinese and abroad have carried out extensive research work based on the rumors on the Twitter and Sina Weibo platforms, and built rumor detection models from different perspectives. In the early days, the methods based on machine learning were commonly used, and its core technologies include manual feature extraction and classifier training. Castillo et al. [1] extracted text features, user information features, topic features and message propagation features, and the accuracy of the J48 decision tree method reached 86%; Kwon et al. [2] proposed the time, structure, and language based on the spread of rumors characteristic. The rumors are classified according to the selected characteristics, and the recall rate is between 87% and 92%; Ma et al. [3] extended the model using dynamic time series based on Kwon et al. [2], Mao et al. [4] Considering deep semantic features such as sentiment tendencies, opinion leader influence, etc., the proposed method based on deep features and integrated classifier can effectively improve the performance of rumor detection; Wang et al. [5] proposed event popularity, ambiguity, and spread. There are three new features, and the new features have greatly improved the automatic detection of rumors. Generally speaking, the methods based on machine learning have achieved initial success, moving from proposing surface features to extracting potential deep features. However, this method relies on feature engineering, and the way of manually designing features requires a lot of manpower, material resources and time, and requires certain professional background knowledge.

In order to solve the problems of the methods based on machine learning, researchers began to explore methods based on deep learning to automatically learn the features contained in data. This method first uses a language model to map all text content into vectors and then uses deep neural networks to automatically learn effective features to detect rumors. Ma et al. [6] used deep learning models for the first to conduct rumor detection research, using tf-idf to calculate the microblog text vectors of each time period, and then separately trained the recurrent neural network (RNN) and its variants (LSTM and GRU) automatically The characteristic representation of learning text obtained 91% detection accuracy on the double-layer GRU network; Yu [7] used the doc2vec method to obtain the text vector of Weibo within the time period and used convolutional neural network (CNN) to automatically learn hidden layer feature detection rumors. The experimental results are superior to comparison algorithms such as support vector machines in terms of indicators; Ren et al. [8] used the Word2Vec model to train 50-dimensional distributed word vectors and used deep learning models such as LSTM and GRU to detect rumors. The detection accuracy rate reached 92.66%. Although the methods based on deep learning can automatically learn features, it has achieved better detection results. However, the research work uses a static pretraining model for text vectorization. There are problems that the polysemous word cannot be dealt with in the text representation [9].

3 Microblog Rumors Detection Model Based on Bert-GRU

This paper proposes a rumor detection model based on Bert-GRU as shown in Fig. 1. First use the Bert model to represent the microblog text, then use the GRU network to automatically learn features, and finally perform the Softmax operation to calculate the probabilities of the two categories of rumors and non-rumors.

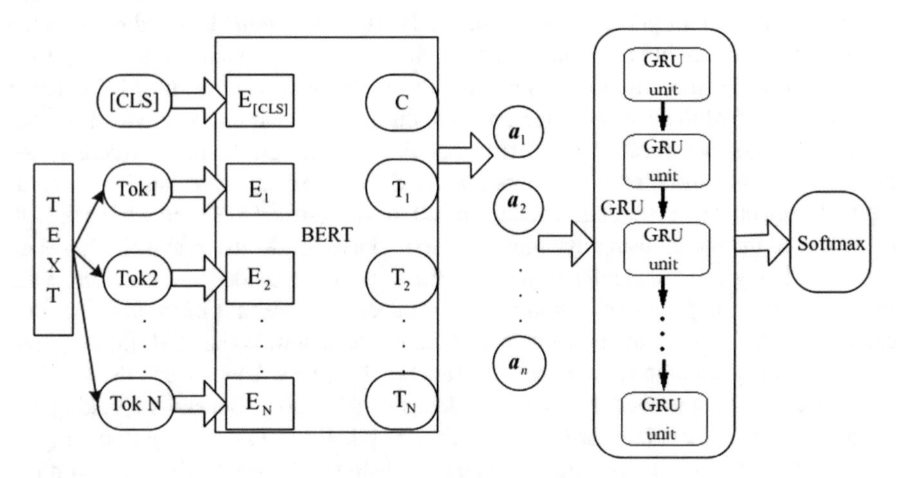

Fig. 1. The Microblog rumor detection model based on Bert-GRU

3.1 Bert Model

The Bert model was proposed by Devlin et al. [10] of the Google in 2018 and is suitable for a lot of natural language processing tasks. Using the Bert model can not only obtain rich grammatical and semantic features, but also solve the problem of ignoring polysemy.

The most important part of the Bert model is the transformer coding unit. The model structure is shown in Fig. 2. The transformer model contains two sublayers. One is the attention mechanism layer, which uses multi-head attention to make the model focus on different; the other is the fully connected forward neural network. In addition, transformer added residual network and layer normalization to improve the degradation problem. Transformer is the first model built using only the attention mechanism. It replaces the traditional encoder–decoder architecture and must combine the inherent model of CNN or RNN. Compared with the recurrent neural network, it can capture a longer range of information and achieve parallelization. The calculation speed has also been improved.

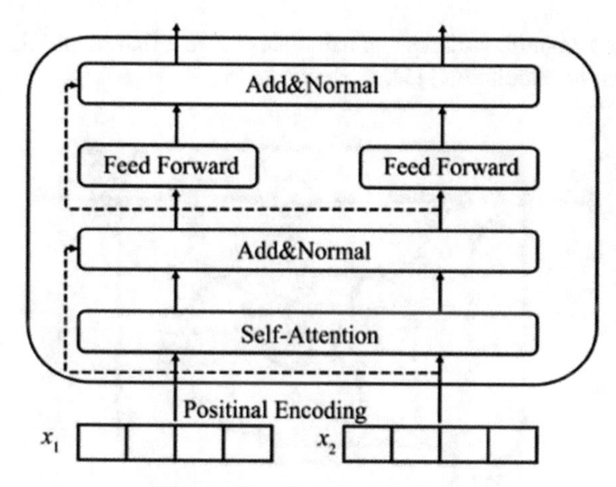

Fig. 2. Transformer coding unit

Bert's pretraining objective function uses Masked Language Model, which is to randomly replace words in a sentence according to rules, and then learn to represent the fusion of text in two different directions by predicting the replaced words. For the input of the Bert model, the expression of each word is generated by adding Token Embeddings, Segment Embeddings, and Positional Embeddings, as shown in Fig. 3. Among them, the first mark of each input sentence is always a special classification embedding ([CLS]), which corresponds to the output of the transformer, which can represent the entire sentence and can be used for downstream classification tasks. The mark [SEP] is used to separate two sentences.

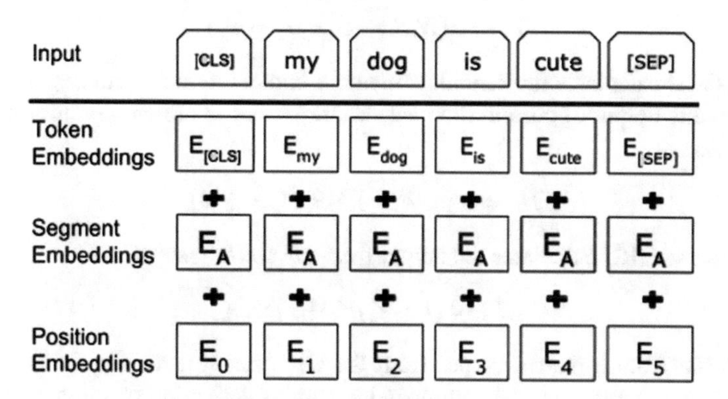

Fig. 3. Bert input representation

3.2 GRU Network

The microblog text vector representation generated by the Bert model will be used as input to the subsequent GRU network model. A gated recursive unit (GRU) can learn grammatical and semantic features and calculates vectors that output fixed dimensions, including four parts calculation [11], as shown in Fig. 4.

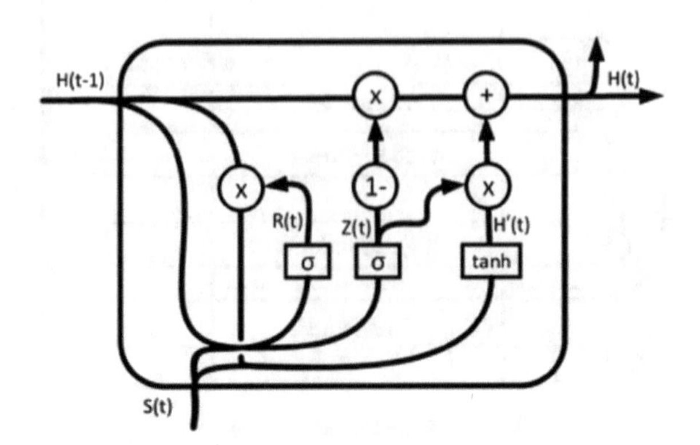

Fig. 4. GRU structure

First is the reset gate. GRU uses reset gate to choose which information to abandon at the previous moment, where W_r and U_r are weight information, and H_{t-1} is the input of the previous moment, B_r is the bias:

$$R_t = \sigma(W_r S_t + U_r H_{t-1} + B_r) \tag{1}$$

Next is the update gate. GRU selects and updates which information at the current moment through the update gate, where W_z and U_z are weight information, and H_{t-1} is the input of the previous moment, and B_z is the bias:

$$Z_t = \sigma(W_z S_t + U_z H_{t-1} + B_z) \tag{2}$$

Then GRU calculates the candidate memory content, which is an important step in calculating the output of current moment, where W and U are weight information and B is the bias:

$$\hat{H}_t = \tanh(\mathrm{W}S_t + \mathrm{UR}_t H_{t-1} + \mathrm{B}) \tag{3}$$

Finally, the GRU calculates the output from the above results:

$$H_t = (1 - Z_t)\hat{H}_t + Z_t H_{t-1} \tag{4}$$

Then, the feature matrix output from the GRU network is transferred to a fully connected layer composed of a neuron and a softmax activation function, and Softmax operations are performed to calculate the probabilities of the two categories of rumor and non-rumor.

4 Experiment and Analysis

In order to verify the text representation ability of the Bert model and the detection effect of rumors based on the Bert-GRU model, this paper uses the data set for rumor detection disclosed by Ma et al. [6]. This data set is a classic data set on the social media rumors detection. It contains 4664 events and their corresponding labels, including 2313 rumors and 2351 non-rumors. At the same time, the Bert Chinese pretraining model "Bert-Base, Chinese" released by Google was used in the experiment. The model uses a 12-layer transformer, which outputs a dimensional vector of size 768, and the multi-head attention parameter is 12.

4.1 Verify the Text Representation Ability of the Bert Model

In order to verify the validity of the text representation of the Bert model, this paper conducts a comparative experiment with the Word2Vec model. Firstly the input microblog using Bert model and Word2Vec model to train the word vector representation, then input it as a feature to RNN network and GRU network for learning, and finally classify the rumors. The experimental results are shown in Table 1.

Table 1. Different model performance verification results (R: rumor, N: non-rumor)

Model	Accuracy	Class	Precision	Recall	F_1
Word2Vec-RNN	0.788	R	0.792	0.769	0.780
		N	0.785	0.807	0.796
Bert-RNN	0.883	R	0.856	0.917	0.885
		N	0.913	0.849	0.880
Word2Vec-GRU	0.828	R	0.805	0.855	0.829
		N	0.853	0.802	0.827
Bert-GRU	0.927	R	0.905	0.952	0.928
		N	0.950	0.902	0.926

It can be seen from Table 1 that the experiment based on the Bert model has achieved a greater improvement in various indicators than the experiment based on the Word2Vec model. Both the RNN network and the GRU network have about 10% improvement in accuracy. The validity of the Bert model in text feature representation is verified, which can well represent the semantic relationship between words in the text and solve the problem of polysemy in different contexts; at the same time, it can also be seen from Table 1 that whether it is based on the Bert model or the experiment based on the Word2Vec model, the GRU network is about 4% higher than the RNN network in accuracy. The remaining indicators are also fully exceeded, proving the correctness of our choice of GRU network structure.

4.2 Verify the Bert-GRU-2 Model

In order to capture higher-level feature interactions, we added a second GRU layer and developed a multi-layer structure based on GRU. The model is named Bert-GRU-2. In order to verify the validity of the Bert-GRU-2 model, experiments will be carried out on the same data set as the selected benchmark method. The experimental results are shown in Table 2. The following four benchmark methods are selected in this paper:

Table 2. Different model performance verification results (R: rumor, N: non-rumor)

Model	Accuracy	Class	Precision	Recall	F_1
DTC	0.831	R	0.847	0.815	0.831
		N	0.815	0.847	0.830
SVM-TS	0.857	R	0.839	0.885	0.861
		N	0.878	0.830	0.857
Bert	0.889	R	0.896	0.878	0.887
		N	0.883	0.900	0.891
GRU-2	0.910	R	0.876	0.956	0.914
		N	0.952	0.864	0.906
Bert-GRU-2	0.930	R	0.924	0.934	0.929
		N	0.935	0.925	0.930

DTC model [1]. The model extracts features such as sentiment score, number of microblogs containing URL, number of user registration days, number of published microblogs, average number of followers, average number of followers, etc., and uses J48 decision tree for classification.

SVM-TS model [3]. The model uses dynamic time series to capture the changes of existing Weibo event features over time and connects the captured time-related features with the existing Weibo event features, and uses SVM classifier for classification.

Bert model. After pretraining the text features on the corpus using the Bert model, it is directly input into the Softmax classifier through a fully connected layer.

GRU-2 model [6]. The model firstly uses an algorithm to segment the Weibo events and then uses the tf-idf method to calculate the text representation of each time period. Finally, a two-layer GRU network is used to learn the hidden layer representation of each Weibo event, and further implementation classification of Weibo events.

It can be seen from Table 2 that directly using the Bert model to rumor detection and obtaining better performance than the machine learning method, illustrating the power of the Bert model. However, based on machine learning DTC and SVM-TS to detect rumors by manually extracting features, it is highly subjective and cannot learn deep potential features and their relationships, so the accuracy rate is low. At the same time, the accuracy rate of the method proposed in this paper reaches 93%, and the F1 values of rumors and non-rumors are also optimized in the benchmark method. Compared with the benchmark method, the method proposed in this paper can show a better effect on the detection rumors of Weibo, thus verifying that the method in this paper can improve the accuracy of the detection rumors of Weibo.

5 Conclusion and Future Works

This article proposes a microblog rumor detection based on Bert-GRU and introduces the Bert model, and a new method is proposed to solve the problem that the static pretraining model cannot handle polysemy in different contexts. Then input it to the GRU network to learn features to classify rumors. This model has achieved ideal results in experiments for microblog data sets, with an accuracy rate of 93%, providing a new and effective method for the rumors detection of Weibo. At present, a lot of news in Weibo is accompanied by multimedia information such as pictures and videos. How to integrate the feature information and text features of pictures and videos and detect rumors is the focus of future research.

Acknowledgements. This work is supported by Xinjiang Joint Fund of National Science Fund of China(U1703261) and the Xinjiang Normal University Graduate Research and Innovation Fund (XSY202002005).

References

1. Castillo C, Mendoza M, Poblete B (2011) Information credibility on twitter. In: Proceedings of the 20th international conference on world wide web, pp 675–684
2. Kwon S, Cha M, Jung K, et al (2013) Prominent features of rumor propagation in online social media. In: IEEE 13th international conference on data mining. IEEE, pp 1103–1108
3. Ma J, Gao W, Wei Z et al (2015) Detect rumors using time series of social context information on microblogging websites. In: Proceedings of the 24th ACM international conference on information and knowledge management, pp 1751–1754
4. Mao E, Chen G et al (2016) Research on detecting microblog rumors based on deep features and ensemble classifier. Appl Res Comput 33(11):3369–3373
5. Wang Z, Guo Y (2019) Automatic rumor event detection in Chinese microblogs. J Chin Inform Process 33(6):132–140
6. Ma J, Gao W, Mitra P et al (2016) Detecting rumors from microblogs with recurrent neural networks
7. Yu F, Liu Q, Wu S et al (2017) A convolutional approach for misinformation identification
8. Ren W, Qin B et al (2019) Rumor detection based on time series model. Intell Comput Appl 9(03):307–310
9. Chen Z, Ju T (2020) Research on tendency analysis of microblog comments based on BERT and BLSTM. Theory Appl, Information Studies
10. Devlin J, Chang MW, Lee K et al (2018) Bert: pre-training of deep bidirectional transformers for language understanding. arXiv preprint arXiv:1810.04805
11. Liu J, Yang Y, Lv S, Wang J, Chen H (2019) Attention-based bigru-cnn for chinese question classification. J Ambient Intell Humanized Comput

Intelligent Ocean Governance—Deep Learning-Based Ship Behavior Detection and Application

Peng Qin[1(✉)] and Yang Cao[2]

[1] China Electronics Technology Group Corporation, 100041 Beijing, China
125529995@qq.com
[2] Huazhong University of Science and Technology, 400074 Wuhan, China

Abstract. The development and utilization of marine resources by mankind has brought out a series of practical problems such as the destruction of marine ecology, the damage of seabed assets, and the disputes over marine sovereignty. How to use information technology tools to profile and monitor ships, accurately classify and identify ship behaviors through multi-source data fusion analysis, and timely alert and invert abnormal behaviors have become an important means of intelligent ocean governance. In response to the above needs, this paper classifies the ship's behavior, designs a new data structure ShipInfoSet that represents the ship's multi-source heterogeneous spatio-temporal information, and proposes a deep learning-based ship behavior-monitoring algorithm ML-Dabs. Accurate identification of the ship's behavior based on deep learning has realized the monitoring and warning of different types of ships' profiles and abnormal behaviors. This paper designs an intelligent ocean information port architecture, which can be implemented by deploying the algorithm.

Keywords: Deep learning · Data intelligence · Ocean information port · Ship behavior monitoring · Intelligent ocean governance

1 Introduction

Since the birth of the universe 15 billion years ago, the earth has appeared after a long period of more than 10 billion years. The earth's surface area is 510 million square kilometers, of which the ocean area is 367 million square kilometers, accounting for about 71% of the entire earth's surface area. The ocean is the cradle that nurtures life, provides water circulation for life, stores energy for the earth, and makes the earth suitable for life. The development of human civilization is also inseparable from the ocean. The ocean is a natural cornucopia, a key road for transportation, and a test field for modern technology research and development. It provides ample room for people to explore nature and promote economic transformation. Since the age of great navigation, the rise of Western powers such as Spain, Portugal, the Netherlands, the UK, and France

Q. Liang et al. (eds.), *Artificial Intelligence in China*, Lecture Notes
in Electrical Engineering 653, https://doi.org/10.1007/978-981-15-8599-9_53

has contributed to the direct control of the maritime traffic and the direct control of the ocean.

Developing the ocean, using the ocean, and managing the ocean have become effective ways for various countries to solve the problems of population expansion, environmental pollution, and resource shortage. The development and utilization of the ocean by mankind is inseparable from an important carrier-the ship. However, in the process of our voyage, overfishing caused the depletion of marine fishery resources, the anchoring of ships caused the destruction of high-value assets such as submarine optical cables and oil and gas pipelines, offshore mining and transportation caused the leakage of crude oil, and human advanced the development of resources in marine disputed areas. All of the above have brought out a series of practical problems such as the destruction of marine ecology, the damage of seabed assets, and the dispute of marine sovereignty. Therefore, the use of information technology tools for ship profiles and monitoring, accurate classification and identification of ship behaviors through multi-source data fusion analysis, and timely warning and inversion of abnormal behaviors have become an important means of intelligent ocean governance.

In response to the above practical needs, this paper first classifies the ship's behavior, designs a new data structure that characterizes the ship's multi-source heterogeneous spatio-temporal information, and proposes a deep learning-based ship behavior-monitoring algorithm ML-Dabs. By virtue of multi-source heterogeneous data cleaning and processing, we can achieve the fusion analysis of ship information and then conduct the training of behavior models using labeled data and the identification and warning of abnormal behavior based on deep learning, realizing real-time monitoring of different types of ships' profiles, and abnormal behaviors. At the same time, an intelligent ocean information port architecture is designed for algorithm deployment and implementation. The main innovations are as follows:

(1) This paper creatively classifies the behavior of ships, categorizes the typical behavior of fishery fishing into ten categories, categorizes the typical behavior of seabed asset warning into five categories, categorizes the typical behavior of marine shipping into eight categories, and proposes a new type of ship heterogeneous data structure ShipInfoSet;
(2) This paper originally proposes a deep learning-based ship abnormal behavior-monitoring algorithm ML-Dabs, which can be effectively used for the identification of ship behavior status and real-time detection and warning of abnormal behaviors;
(3) This paper designs a new ocean information port architecture for intelligent ocean governance. The architecture not only supports the operation of the ML-Dabs algorithm, but also can effectively deploy other marine applications and has very strong scalability.

2 Related Works

2.1 Domestic and Foreign Marine Development Strategies

With the development and utilization of marine resources, major countries have launched new marine development strategies. The USA has implemented a marine action plan and mapped out a road map for the development of marine science and technology in the next decade. Britain has formulated the "Ocean Plan 2025" to fully develop marine technology. Canada focused on the Arctic waters and implemented a marine action plan. Japan has launched the "Draft Marine Basic Law" to comprehensively advance the strategy of a strong maritime nation. Russia relies on science and technology to build a strong marine shipping country. Since the "Twelfth Five-Year Plan", China has also proposed a strategy to build a strong maritime country, mainly for the purpose of developing the ocean, using the ocean, and strategizing the ocean, in order to better solve the practical problems of resource shortage, environmental pollution, fishery fishing, and maritime law enforcement [1, 2].

2.2 Network and Intelligent Technology

2.2.1 AI and Machine Learning Technology

The "knowledge engineering" and "expert system" that emerged in the 1970s and reached their climax in the 1980s were the earliest forms of artificial intelligence. In the past 10 years, we most often heard the following terms related to artificial intelligence: artificial intelligence, machine learning, neural networks, and deep learning. So what is the relationship between these words? Artificial intelligence [3–8] can be divided into two parts: One part is called artificial learning, which is an expert system; the other part is called machine learning, which is that the machine learns by itself. There are five major factions in machine learning: symbolism, Bayesianism, analogy, connectionism, and evolutionism. The connectionism is currently the most popular neural network and deep learning. It occupies the dominant position among the five major factions. Neural networks include shallow learning and deep learning. In the past, when the chip integration was low, only a few neurons can be imitated. Now due to the improvement of chip integration, more neurons can be imitated. When many neurons are formed into a multi-layer network, we call it deep learning. The artificial intelligence we talk about today is actually the neural network and deep learning in machine learning. In general discussions, these concepts are often mixed. The ML-Dabs algorithm proposed in this paper refers to an AI algorithm based on deep learning (Fig. 1).

2.2.2 SDN Technology

The concept of software-defined networking (SDN) [9–13] was proposed by Nick McKeown et al., a professor at Stanford University in the USA. He first introduced the concept of SDN in 2008, that is, to separate the two functional modules of the data plane and control plane of traditional network equipment and manage and configure various devices through a centralized controller with standardized interfaces. Unlike traditional network technologies, SDN technology has three major features: separation of control forwarding, centralized logic control, and open network programming API, which brings programmable features to the network and provides more possibilities for resource management and use. Therefore, SDN technology is a major innovation and development and is listed as one of the ten key technologies in the IT field. This paper uses the global perspective of the OpenFlow controller in the intelligent ocean information network to

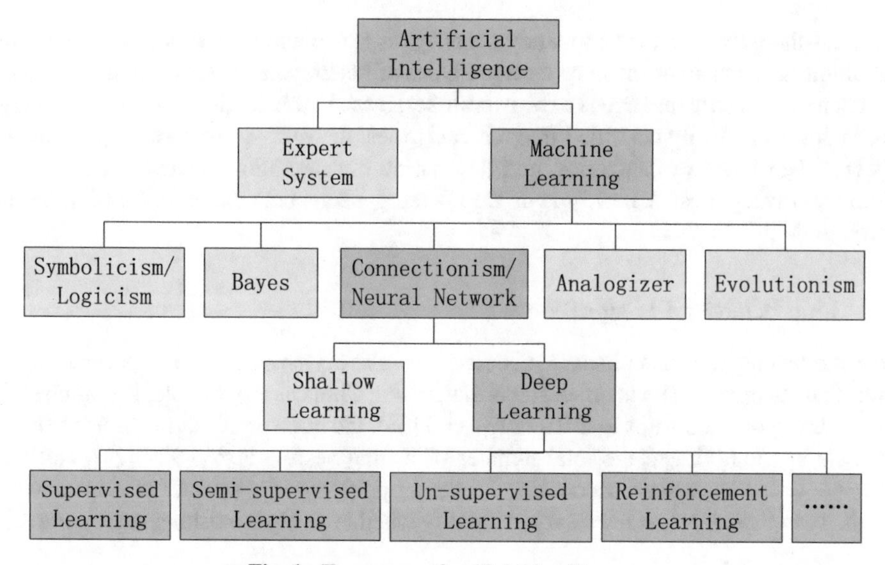

Fig. 1. Taxonomy of artificial intelligence

implement global task scheduling, unified resource allocation, dynamic performance optimization, and high-quality QoS guarantee.

2.2.3 Network Coding and Network Tomography Technology

As a new type of network technology, network coding was proposed around 2000, mainly to solve the problems of maximizing network throughput and optimizing network structure. Since the data packet needs to be encoded at the passing network node, the encoded data packet itself is accompanied by a series of information related to the network structure. The information can be used to monitor important indicators such as network topology, link packet loss rate, and transmission delay, so as to play an important role in optimizing the network structure and improving routing efficiency [14–18]. Compared with traditional network tomography methods [19–22], network tomography based on network coding not only adopts passive detection, but also has lower network overhead and higher detection accuracy. This technology is applied to the intelligent ocean information port network coding and communication service module and has natural advantages for the route optimization and network monitoring of the ocean information port.

3 Behavior Classification: Typical Ship Behavior Classification and Data Structure Design

Ocean governance and ocean safety involve both physical oceans (mainly including marine environment, marine minerals, marine fisheries, etc.) and human oceans (mainly including seabed assets, marine transportation, marine pastures, etc.). As an important carrier for the development and utilization of the ocean, classifying the behavior of ships is the only way for us to study ocean governance and endow ocean wisdom. In addition, ship behavior involves multiple types of heterogeneous spatio-temporal data, and a new data structure needs to be overall designed. In this section, the fishery fishing behavior is creatively classified into ten categories, the seabed asset warning behavior is classified into five categories, and the marine transportation behavior is classified into eight categories. A new type of ship heterogeneous data structure ShipInfoSet is proposed.

3.1 Ship Behavior Classification

For the fishing of marine fishery resources, we classify ships into fishing vessels and non-fishing vessels. The main reason is that, taking China as an example, the number of water transportation ships in 2019 is 131,600 [23], the number of marine motor fishing vessels is 156,000, and the total number of fishing vessels is 863,900 [24], and the number of fishing vessels accounting for the proportion of the number of ships is very high. Therefore, it is first necessary to identify whether a ship is a fishing vessel (Fig. 2).

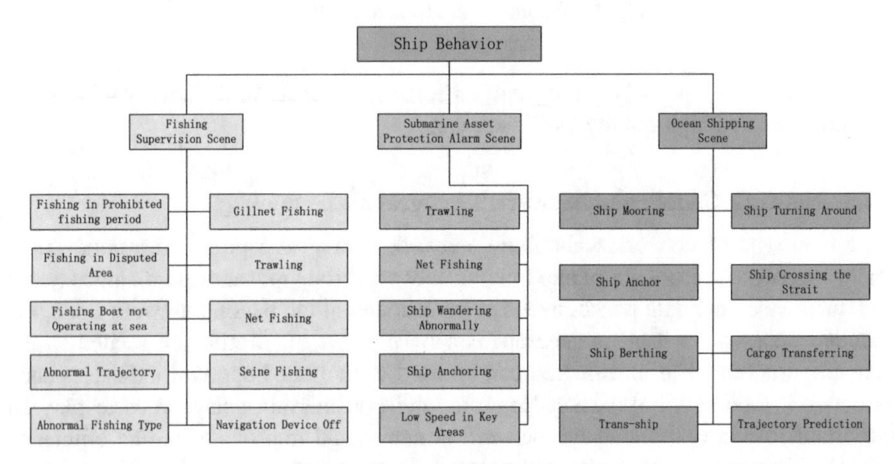

Fig. 2. Category of ship behavior

According to the operating characteristics of fishing vessels, fishing-related behaviors can be grouped into ten categories, namely fishing in prohibited fishing periods, fishing in disputed areas, fishing vessels not operating at sea, abnormal fishing trajectory, abnormal fishing type, gillnet fishing, trawl fishing, net fishing, purse seine fishing, and navigation device off, etc. Generally speaking, the six types of behaviors such as fishing in the prohibited fishing period and fishing in disputed areas are abnormal behaviors of fishing vessels.

When a ship is sailing in an offshore area, its behaviors such as anchoring, trawling, and low speed of ships in key areas are extremely susceptible to damage to submarine assets such as submarine optical cables, oil and gas pipelines, and further serious environmental pollution. Therefore, for the protection of submarine assets in key areas, we can classify ship behaviors into five categories: fishing trawling, net fishing, abnormal ship wandering, ship anchoring, and low speed of ships in key areas. The above behaviors are all abnormal behaviors that cause safety warnings for submarine assets.

For the marine shipping business, according to the business characteristics, the ship's behavior can be summarized into eight categories: mooring, anchor, berthing, tranship, turning around, crossing the strait, cargo transfer, and trajectory prediction. In the operation of the algorithm, it will determine which abnormal behaviors of the ship need to be alarmed according to the specific business.

3.2 Design of Ship Information Data Structure

The new ship information data structure is shown in ShipInfoSet, and the main content includes ship identification data ShipID, ship category data ShipType, Lloyd's ship file data LRF data LD, AIS data AD, GPS positioning data GD, shore-based radar data RD, fusion trajectory data TD, ship behavior status flag ship_behavior, ship behavior abnormal flag if_abnormal, and so on.

```
Typedef struct ShipInfoSet
{
    int ShipID;
    int ShipType;
    LRFData LD;
    AISData[] AD;
    GPSData[] GD;
    RadarData[] RD;
    TrajectoryData[] TD;
    int ship_behavior;
    int if_abnormal;
}ShipInfoSet;
```

4 ML-Dabs Algorithm: Machine Learning-Based Detection for Abnormal Behavior of Ships

This section originally proposes a deep learning-based ship anomaly detection algorithm ML-Dabs, which can be effectively used for the identification of ship behavior status and real-time detection of abnormal behaviors.

Algorithm 1

ML-Dabs：Machine Learning based Detection for Abnormal Behavior of Ships

---

Input: ShipInfoSet0 for ships under supervision
Step0: Start
Step1: Data cleaning and preprocessing for ShipInfoSet0, and obtain ShipInfoSet1.
Step2: Data fusion for ShipInfoSet1, and obtain ShipInfoSet including ship trajectory information and ship environmental information.
Step3: Machine training for ships' behavior using relevance pattern mining and machine leanrning according to deep neural networks based on marked data. Obtaining ship behavior recognition library, ShipRecLi1 for fishing ship and ShipRecLi2 for other types.
for (ships i from 1 to n)
Step4: Feature extration using decision tree with ShipInforSet, and obtain the ShipInforSet[i].ShipType for ship i.
If (ShipInforSet[i].ShipType == fishing ship)
Step5: Obtaining the ShipInforSet[i].ship_behavior and ShipInforSet[i].if_abnormal for fishing ship according to ShipRecLi1;
  If(ShipInforSet[i].if_abnormal==abnormal)
    Alarm and output ShipInforSet[i].ship_behavior;
  End if;
else
Step6: Obtaining the ShipInforSet[i].ship_behavior and ShipInforSet[i].if_abnormal according to ShipRecLi2.
  If(ShipInforSet[i].if_abnormal==abnormal)
    Alarm and output ShipInforSet[i].ship_behavior;
  End if;
Step7: Finish.

As shown in Algorithm 1, the input of the algorithm is ShipInfoSet0, the original ship data. Step1 cleans and preprocesses ShipInfoSet0 to get ShipInfoSet1. Step 2 performs massive heterogeneous data fusion operation on ShipInfoSet1 to obtain ShipInfoSet, including ship trajectory data and ship environment data. Step 3 uses the model of association pattern mining and deep neural network-based machine learning to train the typical behavior recognition library of ships, namely the fishing boat behavior recognition library ShipRecLi1 and the non-fishing boat behavior recognition library ShipRecLi2.

Step 4, Step 5, and Step 6 perform a loop operation on all monitored ship objects 1 to n. Among them, Step 4 first extracts ShipInforSet[i].ShipType of ship i based on ShipInfoSet data set and then judges whether it belongs to a fishing boat. If it is a fishing vessel, execute Step 5, extract the behavior information of vessel i and compare it with the fishing vessel behavior recognition library ShipRecLi1, and immediately output and alarm when an abnormality is found. If it is not a fishing vessel, Step 6 is executed to extract the behavior information of vessel i and compare it with ShipRecLi2, a non-fishing vessel behavior recognition library, and output and alarm immediately when an abnormality is found.

5 Application Deployment: Intelligent Ocean Information Port

In this section, a new ocean information port architecture is designed for intelligent ocean governance. This architecture not only supports the operation of the ML-Dabs algorithm, but also effectively deploys other marine applications. It has very powerful scalability. As shown in Fig. 3, the intelligent ocean information port is divided into three layers, such as shared infrastructure, common service platform, and typical application system, and a vertical operation of three layers of security operation and maintenance and standard specification system.

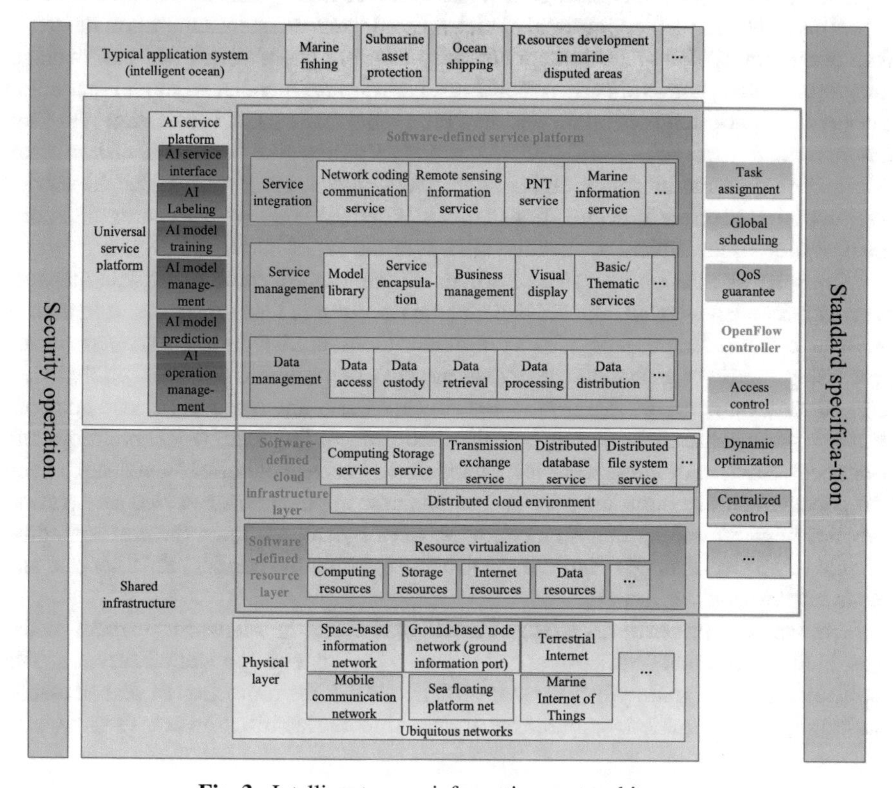

Fig. 3. Intelligent ocean information port architecture

The common infrastructure is composed of three parts: physical layer, resource layer, and cloud infrastructure layer. Among them, **the physical layer** mainly includes physical entities such as space-based information network, ground-based node network (ground information port), terrestrial Internet, mobile communication network, floating network on the sea surface, and Internet of Things. **The software-defined resource layer** uses software definition and virtualization technology to form a virtualized resource pool to provide basic support for cloud infrastructure on-demand calls. **The software-defined cloud infrastructure layer** builds a distributed cloud environment and calls down resource pool resources to provide unified computing, storage, transmission switching, and distribution upwards various services such as database and file system. By virtue of SDN technology, the software-defined cloud infrastructure layer can realize the global scheduling of virtual resources in the information port.

The general service platform includes AI service platform and software-defined service platform. **The AI service platform** includes functions such as AI labeling, AI model training, AI model management, AI model prediction, AI operation management, and AI service interface. Among them, the AI labeling can automatically label space-time data such as ship information and marine environment, greatly reduce the labeling cost, and construct positive and negative samples for model data training. AI model training designs algorithms through a visual interface and optimizes parameters of the model parameters to achieve optimal model. AI model management supports mainstream data programming languages, integrates deep learning frameworks, and can manage, integrate, publish, and offline own models and third-party models. AI model prediction can predict spatio-temporal data and has the ability to request verification and flow monitoring. AI operation management can identify high-risk behaviors of the ocean information port and promptly alert to effectively ensure data security. The AI service interface is based on the abovementioned modules of the AI service platform and supports the operation of the typical application system of the upper intelligent ocean.

The software-defined service platform is composed of three parts: data management, service management, and service integration. Among them, **data management** is oriented to spatio-temporal data from different sources, unified data access, storage, and processing, using the global perspective of the SDN controller to provide a unified view to the outside, support the integration and efficient query and retrieval of data, and conduct safe and efficient data sharing and distribution according to network conditions and user needs. **Service management** is a standardized package of common and supporting components that are often used in various business applications. **Service integration** provides network coding communication services by calling down the corresponding module of service management, remote sensing information service, PNT service and ocean information service, etc.

The typical application system of intelligent ocean is oriented to vertical industries, individuals, and specific users and provides customized application services. The application services mainly include marine fisheries fishing, protection of seabed assets, marine ship transportation, resource development in marine disputed areas, etc.

6 Analysis and Discussion

By designing an AI service platform, not only the automatic labeling of ship's spatio-temporal data is realized, the labeling cost is greatly reduced, and the model is optimized through deep learning and training of the ship's behavior. It can also predict spatio-temporal data and ship behavior. It has the capability of request verification and flow monitoring, which effectively supports the efficient operation of the upper-layer intelligent ocean application system. At the same time, by virtue of the SDN technology, the OpenFlow controller will control the software-defined service platform and the software-defined resources and software-defined cloud infrastructure layer of the common infrastructure in a unified manner. Using the global perspective of the OpenFlow controller, the ocean information port will achieve the global task scheduling, unified resource allocation, and dynamic performance optimization, and high-quality QoS guarantees will eventually realize the intelligent, green, and efficient operation of the ocean information port.

7 Conclusion

In summary, this paper classified the ship behavior in typical scenarios for the problem of intelligent ocean governance related to ship behavior and designed a new data structure ShipInfoSet that characterizes the ship's multi-source heterogeneous spatio-temporal information. This paper proposed a deep learning-based ship behavior detection algorithm, namely ML-Dabs, and proposed an intelligent ocean information port architecture to support the implementation of the algorithm. The ML-Dabs algorithm and construction of ocean information port will strongly support intelligent ocean governance and protection.

Acknowledgements. This work was supported by National 863 Program of China (No. 2015AA015701) and NCFC (No. 91338201).

References

1. Erickson A, Goldstein LJ (2010) China, The United States and 21st century sea power: defining a maritime security partnership. Naval Institute Press, Annapolis
2. President Xi Jinping talks about building a maritime power. Available: www.politics.people.com.cn/GB/n1/2018/0813/c10
3. Domingos P (2018) The master algorithm: how the quest for the ultimate learning machine will remake our world. Basic Books, New York
4. Kotsiantis SB (2007) Supervised machine learning: a review of classication techniques. Informatica 31(3):249–268
5. Alom MZ, Taha TM, Yakopcic C, Westberg S, Sidike P, Nasrin MS, Asari VK (2019) A state-of-the-art survey on deep learning theory and architectures. Electronics 8(292), 1–66
6. Kaelbling LP, Littman ML, Moore AW (1996) Reinforcement learning: a survey. J Artif Intell Res 4:237–285
7. Fadlullah ZMd, Tang F, Mao B, Kato N, Akashi O, Inoue T, Mizutani K (2017) State-of-the-Art deep learning: evolving machine intelligence toward tomorrow's intelligent network traffic control systems. IEEE Commun Surv Tutorials 19(4):2432–2455
8. Bkassiny M, Li Y, Jayaweera SK (2013) A survey on machine-learning techniques in cognitive radios. IEEE Commun Surv Tutorials 15(3):1136–1159
9. Qin P, Dai B, Huang B, Xu G (2017) Bandwidth-aware scheduling with SDN in hadoop: a new trend for big data. IEEE Syst J 11(4):2337–2344
10. Mckeown N et al (2008) Openflow: enabling innovation in campus networks. ACM SIGCOMM Comput Commun Rev 38(2):69–74
11. Qin P, Li J, Xue X, Jiang C, Wang Y (2019) A green and high efficient architecture for ground information port with SDN. In: CSPS 2019 International Conference
12. Amin R, Reisslein M, Shah N (2018) Hybrid SDN networks: a survey of existing approaches. IEEE Commun Surv Tutorials 20(4):3259–3306
13. Qin P, Liu H, Zhao X, Gao Y, Lu Z, Zhou B (2018) Software defined space-based integration network architecture. In: CSPS 2018 international conference
14. Sattari P, Fragouli C, Markopoulou A (2013) Active topology inference using network coding. Phys Commun 6:142–163
15. Qin P, Dai B, Huang B, Xu G, Wu K (2014) A survey on network tomography with network coding. IEEE Commun Surv Tutorials 16(4):1981–1995

16. Yao H, Jaggi S, Chen M (2010) Network coding tomography for network failures. In: Proceedings of IEEE INFOCOM, San Diego, CA, USA, pp 1–5, Mar 2010
17. Fragouli C, Markopoulou A, Srinivasan R, Diggavi S (2007) Network monitoring: it depends on your points of view. In: Proceedings of ITA Workshop, San Diego, CA, USA, pp 1–10, Jan 2007
18. Sattari P, Markopoulou A, Fragouli C (2011) Maximum likelihood estimation for multiple-source loss tomography with network coding. In: Proceedings of international symposium on NetCod, Beijing, China, pp 1–7, July 2011
19. Qin P, Dai B, Huang B, Xu G, Wu K (2016) Taking a free ride for routing topology inference in peer-to-peer networks. Peer-to-Peer Network Appl 9(6):1047–1059
20. Castro R, Coates M, Liang G, Nowak R, Yu B (2004) Network tomography: recent developments. J. Stat. Sci. 19(3):499–517
21. Qin P, Dai B, Huang B, Xu G (2015) DCE: a nover delay correlation measurement for tomography with passive realization. Comput Sci
22. Zhang X, Phillips C (2012) A survey on selective routing topology inference through active probing. IEEE Commun Surv Tutorials 14(4):1129–1141
23. 2019 Statistical Bulletin of Transportation Industry Development. Available: www.gov.cn
24. Ministry of Agriculture of China. Available: http://m.chyxx.com/view/787015.html

The Research on Disruptive Technology Identification Based on Scientific and Technological Information Mining and Expert Consultation: A Case Study on the Energy Field

Lucheng Lyu[1,2], Xuezhao Wang[1,2](✉), Wei Chen[2,3,4](✉), Xin Zhang[1,2], Xiaoli Chen[1,2], and Xiwen Liu[1,2]

[1] National Science Library, Chinese Academy of Sciences, 100190 Beijing, People's Republic of China
wangxz@mail.las.ac.cn
[2] Department of Library, Information and Archives Management, School of Economics and Management, University of Chinese Academic of Sciences, 100190 Beijing, People's Republic of China
chenw@whlib.ac.cn
[3] Wuhan Library Chinese Academy of Sciences, 430071 Wuhan, People's Republic of China
[4] Hubei Key Laboratory of Big Data in Science and Technology, 430071 Wuhan, People's Republic of China

Abstract. Disruptive technology identification is of great significance to the development of the countries and enterprises. In this paper, both quantitative analysis and qualitative analysis are combined to propose a disruptive technology identification method based on scientific and technological information mining and expert consultation. In the part of the quantitative analysis, three kinds of scientific and technological information data are involved including papers, patents, and projects, and the word embedding model and clustering technology are applied to dimension reduction and aggregation the scientific and technological information which results in finding "seeds" or "sprouts" of disruptive technologies from the mass scientific and technological information. In the part of the expert consultation, the ranking of potential disruptive technologies is evaluated by an expert questionnaire based on a "three-dimensional index system of technology disruptive potential evaluation." The result of the energy field research is that 45 potential disruptive technologies are identified and their disruptive ranking is determined.

Keywords: Disruptive technology identification · Multi-resource data · Text mining · Cluster analysis · Word2vec model · Energy field

1 Introduction

The concept of disruptive technology is put forward first by Harvard Business School professor C M. Christensen in "the innovator's dilemma: when new technologies cause great

Q. Liang et al. (eds.), *Artificial Intelligence in China*, Lecture Notes in Electrical Engineering 653, https://doi.org/10.1007/978-981-15-8599-9_54

enterprise failure" [1]. The technologies are generally able to replace the existing traditional mainstream ones and change the original technical performance trajectory with the characteristics of simple operation, easy carry, and low cost. After C M. Christensen, many scholars have studied disruptive technologies from the perspectives of "technology driven" [2, 3] and "market driven" [4, 5]. From the perspective of technology promotion, the disruptive effect can be produced by both the technological breakthrough and the new field application of an original technology. From the perspective of user demand, the user demands for products in the market lead to the new market disruption.

For a country, disruptive technologies can promote scientific and technological innovation, stimulate market vitality, and be helpful to find the technological breakthroughs by using the existing low-end consumption and new market consumption in undeveloped markets, so as to improve national economic strength. For an enterprise, disruptive technologies often bring great impact on the present mainstream enterprises or market resulting in rapid changes in the economic benefits. How to identify disruptive technologies before they are widely used in the market is of great significance.

This paper proposes a disruptive technology identification method based on scientific and technological information mining and expert consultation, which combines quantitative analysis and qualitative analysis method, expands the quantitative data source and introduces word2vec word vector model and other technical means. Finally, the empirical research in the field of energy is carried out.

2 Related Work

As shown in Fig. 1, the methods of disruptive technology identification include subjective and objective ones.

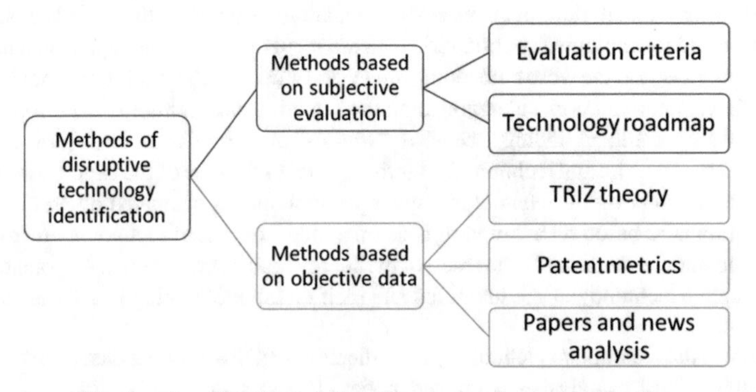

Fig. 1. Classification of disruptive technology identification

2.1 Methods Based on Subjective Evaluation

In the early stage, evaluation criteria or technology road map were usually made by expert consultation. The evaluation criteria are constructed based on the concept and mechanism of disruptive technology which involves technology, product, industry, and external environment.

Collins et al. [6] established the assessment criteria from the usefulness of the technology, the quality of the technology's output, the technology's utility, the technology's compatibility, and the technology participants' willingness. Sun et al. [7] constructed an evaluation criterion from technical breakthrough, alternative products, broad market, and industry revolutionary. Furthermore, Guo et al. [8] constructed the metrics of disruptive innovation including technical characteristics (integration degree, leadership, maturity, diffusivity, simplified), market dynamics (emerging markets, value network, cost reduction), and the external environment (policy and macro economy). The assessment factors of the above research were relatively comprehensive, and there was no unified evaluation standard, which caused the subjective and limited evaluation results.

Additionally, the technology road map can be used to identify disruptive technologies. Kostoff et al. [9] found the technical alternatives and experts then made the technical route planning by the literature searching. Vojak et al. [10] proposed that the changes of industry standards, architecture, relationship among various elements in the super system, the integration and decomposition of elements in various forms, and the substitution within the subsystem should be considered in the process of developing technology road map. Drawing technology road map requires multi-party cooperation, which is conducive to information and knowledge exchange among relevant personnel and potentially disruptive technologies identification. However, a large amount of time, money, and brainpower are needed, which is more suitable for operation at the national level.

2.2 Methods Based on Objective Data

Gradually, objective data were used to identify disruptive technology. The technology maturity and system function analysis were used to identify the environment of disruptive technology and the possible innovation direction in TRIZ theory. Xu et al. [11] used the indicators of technology level, patent numbers, technical performance, and technology profits to analyze the maturity of technology and found that disruptive innovation should occur in the mature stage of original technology. Sun.et al. [12] also used TRIZ to obtain the lagging technology which could be improved to make innovations.

Patents were often used to identify disruptive technologies which include their external characteristics text information and citation network. Buchanan et al. [13] used the indicators of patent applications, a large number of new applicants, patents per capita to identify potential disruptive innovations. Su [14] considered the effectiveness of technology to reveal the evolution track of disruptive technology from patent quantity, citations, and citation rate. Dotsika et al. [15] established a keywords co-occurrence network to determine potential disruptive keywords. J. Kim et al. [16] explained the keywords strength by calculating the visual growth rate and diffusion growth rate of keywords. L C. Huang et al. [17] calculated the difference between the attribute sets before and after the appearance of a new technology to measure the disruptive technology. Besides, they [18] analyzed the disruptive potential from the recombination of the IPC classification and the different reference structure of the IPC classification. Momeni [19] constructed patent citation network to extract the main path, identified the underlying theme by k-core analysis and topic model, and verified the disruptive technology by the performance of the identified topic in the scientific paper and the function of the technology.

Disruptive technology has been also identified by the relationship between patent and scientific knowledge. Zhang [20] measured technological change by using the keywords, research topics, and subject classification to calculate the word frequency changes of new and repeated keywords. Bai et al. [21] studied the mutation of scientific knowledge topics by calculating the semantic relevance of technical topics on the timeline, including the emergence of new topics, the convergence of original topics, and the fusion of multiple topics. Zhao [22] analyzed the changes in the proportion of topics by calculating the number differences in the literature, patent, and web news, and then, he obtained the key technical topics, and used the three-year patent citation rate, average independent patent claims, and the literature growth rate to identify disruptive technology topics.

Based on the above analysis, the methods of subjective evaluation and objective data analysis have their own advantages, but both of them need to be improved. The research results based on the subjective evaluation are authoritative and reliable but limited by the expert knowledge and difficult to verify the recognition effect. The research on disruptive technology identification based on objective data analysis is mostly about the validation of disruptive technology, where the specific methods are used to analyze and validate the specific data. The calculation process is objectively traceable, but it is not possible to predict potential disruptive technologies from unknown technical domains.

To avoid these problems, our research combines the advantages of subjective evaluation and objective data analysis and proposes the method based on science and technology information mining and expert consultation. By using data mining technology, potential disruptive technologies are obtained from high-quality scientific and technological information, and by scoring by experts, the ranking of score for disruptive potential of these technologies is calculated.

3 Methodology

The method of disruptive technology identification proposed in this paper is divided into two parts: the generation of the list of potential disruptive technologies based on scientific and technological information mining and the evaluation of technology disruptive potential based on expert consultation (as shown in Fig. 2).

Fig. 2. Basic framework of method

In the first step, using the high-quality papers, high-value patents, important scientific research projects and other scientific and technological information data in a certain technology field, we apply the word embedding model and data mining technology to do dimension reduction and aggregate the scientific and technological information. As a result, the candidates of potential disruptive technologies are obtained, on the basis of which the domain information analysts synthesize (if needed) and interpret the potential disruptive technologies. Finding "seeds" or "sprouts" of disruptive technologies from the mass scientific and technological information are accomplished.

In the second part, a "three-dimensional index system of technology disruptive potential evaluation" including technology dimension, industrial economy dimension, and external environment dimension is designed, based on which an expert questionnaire is formed. Through expert consultation, we evaluate the technology disruptive potential in the list of potential disruptive technologies, then calculate the ranking of technology disruptive potential, and finally get the list of disruptive technologies in the field.

3.1 The Generation of the List of Potential Disruptive Technologies Based on Scientific and Technological Information Mining

Figure 3 shows the process framework of a potential disruptive technology list generation method based on scientific and technological information mining, which is divided into five steps: data preparation, text preprocessing, text vectorization, cluster analysis, and expert interpretation.

(1) **Data preparation**. Combined with the characteristics of a certain field, we determine the data sources of scientific and technological information including high-quality papers, high-value patents, and important scientific research projects in this field and extract data from source papers, patents, and project databases.
(2) **Text preprocessing**. The extracted paper, patent, and project data are preprocessed by term extraction, stop words filtering, word stemming, and other text preprocessing, so as to prepare for the next data mining. Specifically, term extraction can adopt the method of term recognition based on dictionary matching; the filtering of stop words can be realized by domain stop word list filtering; word stemming can be realized by software tools such as Porter Stemmer and Snowball Stemmer. In addition, it also needs to filter invalid characters such as simple numbers, punctuation marks, blank characters.
(3) **Text vectorization**. This study adopts word2vec model to vectorize the text data of papers, patents and projects. The word2vec model used in this method needs to be trained based on large-scale corpus in the target field. The calculation method of text vectorization is as follows:

    (a) Feature selection. Because of the inconsistency of text length, this study applies the method of feature engineering to extract key features for text representation, such as adopting TF-IDF to extract the top n keywords of each text to represent text.
    (b) The word vectors of each key word are obtained from word2vec model and combined to form a two-dimensional array.
    (c) The elements of the two-dimensional array are summed one by one to form a one-dimensional array consistent with the length of the word2vec vector.

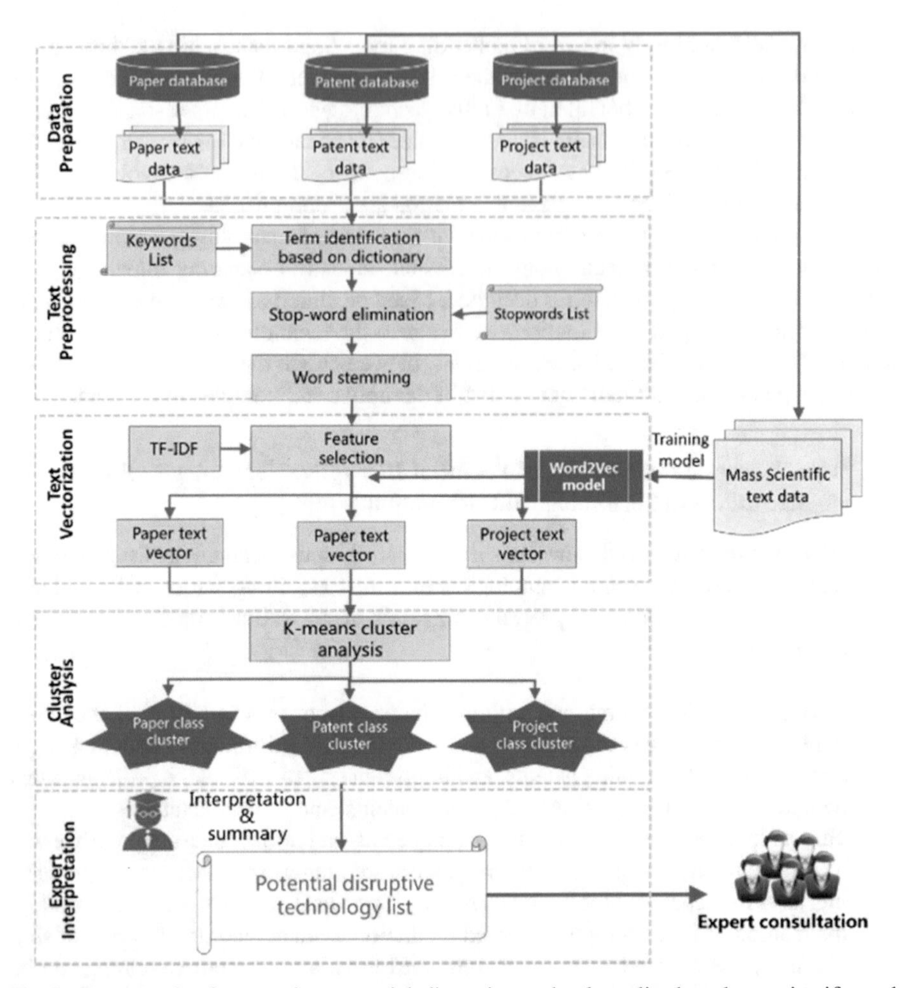

Fig. 3. Framework of generating potential disruptive technology list based on scientific and technological information mining

(d) Normalize the one-dimensional array by the vector module length corresponding to the one-dimensional array. The normalized formula is as follows.

$$\text{stArray} = \text{Array/sqrt}((\text{Array} * \text{Array}).\text{sum}()) \tag{1}$$

(e) The calculation result stArray is the sentence vector of target text.

(4) **Cluster analysis**. Using the patent text vector, paper text vector, and project text vector, k-means clustering method is used to cluster, respectively, and the optimal number of clusters is selected according to the Silhouette Coefficient [23] to form high-quality paper cluster, high-value patent cluster, and important scientific research project cluster.

(5) **Expert interpretation**. Domain information analysts will rely on domain knowledge to interpret, merge, refine, and summarize clusters one by one, so as to form a list of potential disruptive technologies in a certain field.

3.2 The Evaluation of Technology Disruptive Potential Based on Expert Consultation

The technology disruptive potential evaluation based on expert consultation aims to evaluate the disruptive potential of the target technology on the list of potential disruptive technologies through expert consultation. This research designs a "three-dimensional index system of technology disruptive potential evaluation" including the dimensions of science and technology, industrial economy, and external environment, as shown in Table 1. Based on the index system, technical experts can score the technologies on the list of potential disruptive technologies one by one.

Because some technologies on the potential disruptive technology list may not be disruptive technology, experts need to select the disruptive technologies they think from the potential disruptive technology list before scoring the disruptive potential.

After gathering the results of expert evaluation, we integrated the number of experts who recognized technology as disruptive technology and the evaluation score of experts for technology disruptive potential. Through normalization, we calculated the disruptive potential value of each technology. The disruptive potential score for each technology is calculated as follows.

$$\text{Score} = \text{Score 1} + \text{Score 2} \tag{2}$$

$$\text{Score 1} = {}^{m}\!/_{\max_1} \tag{3}$$

$$\text{Score 2} = \sum\nolimits_{i=1}^{n} \sum\nolimits_{j=1}^{k} \text{score}_{ij}/n * \max_2 \tag{4}$$

In formula (3), m is the number of experts who recognize the technology as disruptive technology, and $\max_1$ is the maximum number of experts who are recognized as disruptive technology.

In formula (4), i is the number of experts, n is the number of experts to score the technology, j is the number of indicators, k is the number of scoring indicators, score_{ij} is the score of the ith expert on the jth indicator of the technology, and $\max_2$ is the maximum value of the average score of experts on the technology disruptive potential.

4 Empirical Research

Energy is an important material basis for human survival and development. At present, major countries and regions in the world regard energy technology as the breakthrough point of a new round of scientific and technological revolution and industrial revolution. Therefore, this paper takes the energy field as an example to conduct the empirical research on disruptive technology identification method. The following is a detailed introduction of empirical research.

Table 1. Three-dimensional index system of technology disruptive potential evaluation

First-grade index	Second-grade index	Evaluation criteria (score range: 1–5)
Technology dimension	Contribution of basic theory	The breakthrough degree of the basic theory of the technology
	Maturity of technology development	The level of development and maturity of the technology
	Wide application of technology	Application scope of the technology, such as electric power, chemical industry, manufacturing industry, transportation industry, or military
	Technical safety	The security performance of the technology, such as external dependence of resources or internal security of technology
	Technical performance improvement	The improvement degree of the technology compared with the original technology in the field in terms of key performance indicators
Industrial economy dimension	Industrial market prospect	The industrial market prospect of the technology, such as the scale of creating new industries or new formats or new services or the possibility of subverting the original industry
	Size of beneficiary groups	Scale of stakeholders involved in the technology, such as producers, consumers, or service providers
	Impact on people's lives	The degree or influence of the technology on the change of people's production and life style, such as reduction of living cost, improvement of convenience, or change of behavior mode
External environment dimension	Macro policy	The extent to which the state supports the development of the technology, such as whether to issue policy planning, implement scientific research projects, or establish specialized agencies

(*continued*)

Table 1. (*continued*)

First-grade index	Second-grade index	Evaluation criteria (score range: 1–5)
	Infrastructure and supporting system	Whether the infrastructure and supporting system related to the technology are perfect or not

4.1 Dataset

In the empirical research, the data of high-quality papers, high-value patents, and important scientific research projects are, respectively, from highly cited papers and hot papers in the energy field of the Web of Science (WOS) paper database, the quintuple patent families in the past decade in the energy field of the Derwent Innovation (DI) patent database, and the energy field projects published on the official website of Advanced Research Projects Agency Energy (ARPA-E) of the United States.[1] The terms are explained as follows.

- **Highly Cited Papers** are selected from the most recent 10 years of data and reflect the top 1% of papers by field and publication year.
- **Hot Papers** are selected by virtue of being cited among the top one-tenth of one percent (0.1%) in a current bimonthly period.
- **Quintuple Patent Families** in the past decade refer to patents applied in China, the United States, Europe, Japan, and South Korea in the most recent 10 years.

The retrieval strategy and other descriptive information of data are shown in Table 2.

4.2 Text Preprocessing

The keywords dictionary used for term recognition is built by the keywords that are not single words in the energy field from the WOS paper database, with a total of 601,178. The list of stop words is constructed manually gathered by analyzing the text, including 927 stop words. The word stemming tool is Snowball Stemmer integrated in Natural Language Toolkit (NLTK) natural language processing package. It also filters invalid characters including simple numbers, punctuation marks, blank characters.

4.3 Text Vectorization

This research trains different word2vec models for the vectorization of different data types. The energy paper word2vec model is trained by title and abstract of 965,126 WOS papers in the field of energy. The energy patent word2vec model is trained by title

[1] https://arpa-e.energy.gov/?q=projectlisting&field_program_tid=All&field_project_state_value=All&field_project_status_value=All&term_node_tid_depth=All&sort_by=field_organization_value&sort_order=ASC.

Table 2. Data source and description

Field	Type		
	Paper	Patent	Project
Data source	WOS paper database	DIpatent database	ARPA-E website of the United States
Data retrieval strategy	WC = Nuclear Science Technology OR WC = Energy Fuels,SCI-EXPANDED	CPC = ((YO4) OR (Y02E))	Grab all data
The amount of data	96,5126	702,499	845
The amount of data for cluster analysis	6338(highly cited papers and hot papers)	7760 (quintuple patent families in the past decade)	809 (excluding the projects with no published details)
Data language	English	English	English
Retrieval time	Aug 19, 2019	Sep 3,2019	Dec 19,2019

and abstract of 702,499 DI patents in the field of energy. Because the amount of project data is not enough to train the word2vec model and the writing of project text is similar to the patent text, and the energy patent word2vec model is used to vectorize project data.

The major parameters used in word2vec model training include that word vector dimension is 300, minimum word frequency is 5, window (maximum distance between the current and predicted word within a sentence) is 10, and training algorithm selects Skip-Gram algorithm.

Considering the inconsistency of text length, TF-IDF is used to extract the top 30 keywords of each text to be clustered, and the text representation vector is calculated based on these keywords.

4.4 Cluster Analysis

K-means clustering algorithm is used to cluster the data of 6338 high-quality papers, 7760 high-value patents, and 809 important scientific research projects. In particular, the number of clusters is determined by the domain information analysts first and then by the calculation of Silhouette Coefficient. According to the amount of data for each type of data and technology distribution in the field of energy, domain information analysts, respectively, determined the number of paper clustering interval between 140 and 170, the number of patent clustering interval between 140 and 170, and the number of project clustering interval between 20 and 50.

Through the calculation, the distribution of the Silhouette Coefficient of clustering analysis of papers, patents, and projects is, respectively, shown in Fig. 4. By selecting the largest Silhouette Coefficient as the optimal clustering number, 152 paper clusters, 162 patent clusters, and 31 project clusters were obtained.

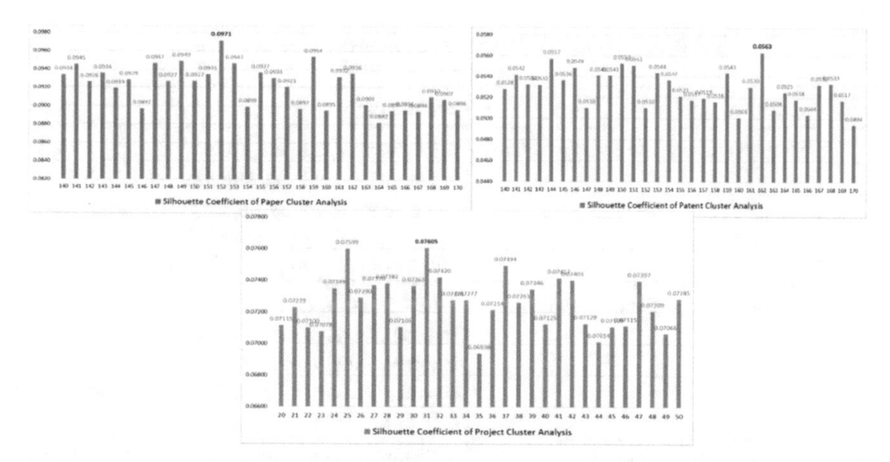

Fig. 4. Silhouette Coefficient distribution of paper text cluster analysis、 patent text cluster analysis、 project text cluster analysis in the field of energy

4.5 Expert Interpretation

In this research, 152 paper clusters, 162 patent clusters, and 31 project clusters are submitted to energy field information analysts for interpretation, merging, refining, and summary, and forty-five potential disruptive technologies in the energy field are finally obtained, which are classified into five categories according to the field of technology.

4.6 Evaluation of Technology Disruptive Potential

Based on above three-dimensional index system, this research designs the "Expert Questionnaire on Technology Disruptive Potential in the Field of Energy" to conduct expert consultation. The targets of consultation include project leader of clean energy project, academic committee of Clean Energy Innovation Research Institute and users of Energy Express, and a total of 63 questions were recovered.

Through the calculation of expert scores, the evaluation results of disruptive potential of 45 technologies are obtained. Table 3 shows the top five energy technologies with disruptive potential, among which the top three are green hydrogen production technologies (including renewable energy electrolytic water hydrogen production, photocatalytic hydrogen production, biomass hydrogen production, thermochemical cycle hydrogen production), advanced fuel cell technologies (proton exchange membrane fuel cell, anion exchange membrane fuel cell, proton ceramic fuel cell, solid oxide fuel electricity Pool) and multi-energy complementary integrated energy system technology.

5 Conclusion

This paper proposes a disruptive technology identification method based on scientific and technological information mining and expert consultation then conducts an empirical study. The advantages and innovation of the method are as follows.

Table 3. Top five disruptive technologies in the energy field

Rank	Technical field	Technology	Score
1	New and renewable energy sources	3.10 Green hydrogen production technology (renewable energy electrolysis water hydrogen production, photocatalytic hydrogen production, biomass hydrogen production, etc.)	1.84
2	New and renewable energy sources	3.12 Advanced fuel cell technology (proton exchange membrane fuel cell, anion exchange membrane fuel cell, proton ceramic fuel cell, solid oxide fuel cell, etc.)	1.70
3	Energy system integration	5.5 Multi-energy complementary integrated energy system technology	1.57
4	New and renewable energy sources	3.3 Photocatalytic fuel production or high-value chemical (artificial light synthesis) technology	1.44
5	Advanced energy storage	4.6 All-solid lithium battery technology	1.40

(1) Combining the quantitative analysis method with the qualitative analysis method. Firstly, text mining technology is used to find the potential disruptive technology from the scientific and technological data of target technical field, so as to ensure the validity and objectivity of the source of the technology list. Then, expert consultation method is used to evaluate and select the subversive technology from the potential subversive technology list, so as to make up for the problem that the influencing factors are not fully considered when the quantitative analysis method is used to identify subversive technology.
(2) Expanding the quantitative data source of disruptive technology identification. In the generation of potential disruptive technology list based on scientific and technological information mining, this study comprehensively considers the data of papers, patents, and projects and selects high-quality papers, high-value patents, and important scientific research projects from them for analysis and mining. Compared with the existing disruptive technology based on one or two kinds of the literature data, the results of technology identification are more comprehensive and reliable.
(3) Applying a new technology to carry out disruptive technology identification. In the process of technology information mining, word2vec model is used for text vectorization, which makes up for the problem that synonyms cannot be distinguished when general methods are used to vectorize text, and optimizes the effect of text clustering.

Acknowledgements. This work is financially supported by the projects of Strategy research (GHJ-ZLZX-2018-41, GHJ-ZLZX-2019-31-2) from Bureau of Planning and Strategy, Chinese Academy of Sciences.

References

1. Christensen CM (1997) The innovator's dilemma: when new technologies cause great firms to fail. Harvard Business School Press, Boston
2. Nagy D, Schuessler J, Dubinsky A (2016) Defining and identifying disruptive innovations. Ind Mark Manage 57:119–126
3. Zheng L, Chunping L, Hui L (2016) A brief analysis of the connotation and cultivation of subversive technology—paying attention to the basic scientific research behind the subversive technology. Global Sci Technol Econ Outlook 31(10):53–61
4. Katz R, Paap J (2004) Anticipating disruptive innovation. Eng. Manag. Rev. IEEE 32(4):74–85
5. Ganguly A, Nilchiani R, Farr JV (2010) Defining a set of metrics to evaluate the potential disruptiveness of a technology. Eng. Manage. J. 22(1):34–44
6. Collins R, Hevner A, Linger R (2011) Evaluating a disruptive innovation: function extraction technology in software development. In: Hawaii International conference on systems science
7. Sun YF, Wang LH (2017) Research on the connotation and selection of disruptive technology that causes industrial transformation. Chin Eng Sci 19(5):9–16
8. Guo JF, Pan JF, Guo JX (2019) Measurement framework for assessing disruptive innovations. Technol Forecast Soc Chang 139:250–265
9. Kostoff RN, Boylan R, Simons GR (2004) Disruptive technology roadmaps. Technol Forecast Soc Chang 71(1):141–159
10. Vojak BA, Chambers FA (2004) Roadmapping disruptive technical threats and opportunities in complex, technology-based subsystems: the SAILS methodology. Technol Forecast Soc Chang 71(1):121–139
11. Xu ZH, Zhang GY (2016) Research on disruptive technology selection environment based on TRIZ theory. Ind Eng 19(4):43–47
12. Sun J, Gao J, Yang B (2008) Achieving disruptive innovation forecasting potential technologies based upon technical system evolution by TRIZ. In: IEEE International conference on management of innovation & technology
13. Buchanan B, Corken R (2010) A toolkit for the systematic analysis of patent data to assess a potentially disruptive technology. Intellectual Property Office, United Kingdom (2010)
14. Su JQ, Liu JH (2016) Evolutionary trajectories and early identification of disruptive technologies. Sci Res Manage 37(3):13–20
15. Dotsika F, Watkins A (2017) Identifying potentially disruptive trends by means of keyword network analysis. Technol Forecast Soc Chang 119:114–127
16. Kim J, Park Y, Lee Y (2016) A visual scanning of potential disruptive signals for technology roadmapping: investigating keyword cluster, intensity, and relationship in futuristic data. Technol Anal Strategic Manage 1–22
17. Huang LC, Cheng Y, Wu FF (2015) Exploration of disruptive technology identification framework. Sci Res 33(5):654–664
18. Huang LC, Jiang LB, Wu FF (2019) Research on disruptive technology identification in the embryonic stage. Sci Technol Progress Countermeasures 1:10–17
19. Abdolreza M, Rost, K (2016) Identification and monitoring of possible disruptive technologies by patent-development paths and topic modeling. Technol Forecast Social Change 104:16–29
20. Zhang JZ, Zhang XL (2016) A review of research on the identification of breakthrough innovations based on patent science citation. J Intell 35(9):955–962

21. Bai GZ, Zheng YR, Wu XN (2017) Research and demonstration of disruptive technology prediction method based on literature knowledge association. Intell Mag 9:42–48
22. Zhao G (2017) Disruptive technology identification based on multivariate heterogeneous data. Huazhong University of Science and Ttechnology, Wuhan
23. Rousseeuw PJ (1987) Silhouettes: a graphical aid to the interpretation and validation of cluster analysis. Comput Appl Math 20:53–65

Research on Transfer Learning Technology in Natural Language Processing

Ruilin Shen and Weimin Pan(✉)

School of Computer Science and Technology, Xinjiang Normal University, 830054 Urumqi, China
panweiminss@163.com

Abstract. In the field of natural language processing, traditional machine learning methods rely on a large amount of labeled data to obtain better results and require training data and test data to satisfy the assumption of independent and identically distributed. Application of transfer learning technology provides a good solution to the problems above and promotes the research progress of natural language processing. This paper studied the application of transfer learning in natural language processing, made the definition of transfer learning, classified the transfer learning in natural language processing, and prospected the development of transfer learning in natural language processing.

Keywords: Transfer learning · Natural language processing · Machine learning · Deep learning

1 Introduction

With the penetration of language into all aspects of human life, natural language processing has become indispensable for computers and developing models that can understand human language has become a key research direction in the field of artificial intelligence. At present, there are two methods commonly used in NLP, which are based on traditional machine learning methods and methods based on deep neural networks. Based on traditional machine learning methods, mathematical models are used to automatically learn rules from data without the need for manual rule construction, but manual feature labeling is required, and manual feature engineering still requires a lot of manpower, material resources, and time and to complete specific tasks, get the eigenvector of is not robust. Deep neural network-based methods can even automatically extract features from massive amounts of data and learn deeper knowledge representations [1]. However, the development of deep learning has derived demands for data volume and data quality.

In many areas of NLP, there are only a few data resources, such as research on small languages, and some research areas have fewer labeled data resources, such as the Chinese rumor detection. The high cost of labeling data makes it difficult to obtain high-quality labeled data. In this regard, researchers began to apply transfer learning to NLP.

Q. Liang et al. (eds.), *Artificial Intelligence in China*, Lecture Notes in Electrical Engineering 653, https://doi.org/10.1007/978-981-15-8599-9_55

This paper studies the transfer learning technology of NLP. The first part introduced what transfer learning is and gave the definition of transfer learning; the second part introduced the classification of transfer learning in NLP and also introduced the definition and application scenarios of each category; the third part made an outlook on the future development of transfer learning.

2 The Concept of Transfer Learning

The identification-based transfer (DBT) algorithm [2] formulated by Patter et al. in 1993 is considered to be the earliest cited work on transfer learning. Transfer learning is a new machine learning method that uses existing knowledge to solve problems in different but related fields [3]. For example, the knowledge used to identify bicycles can also be used to improve the ability to identify motorcycles. Transfer learning breaks the assumption that the training data and test data in traditional machine learning must satisfy independent and identical distribution and solves the problem of insufficient labeled data.

Transfer learning is in the stage of continuous exploration in the field of natural language processing, and now it has been applied in many fields, such as sentiment analysis, named entity recognition, text summary, patent vocabulary recognition, and machine translation. The future development of transfer learning is what every researcher pays attention to. Enda Wu once said in the tutorial at NIPS 2016 “After supervised learning, transfer learning will lead the commercialization of the next generation of machine learning technology.”

3 Transfer Learning’s Classification

In recent years, a large number of researchers have conducted extensive research on transfer learning. Among them, [4, 5] and others have made a good summary of the work of transfer learning in the past few years and introduced the development process, classification, and related technologies of transfer learning. This article divides transfer learning into four categories according to whether the source task and the target task are the same and the amount of labeled data in the source and target data: multi-task learning, sequential transfer learning, domain adaptation, and cross-language learning.

3.1 Multi-task Learning

Multi-task learning (MTL) belongs to the category of inductive transfer learning. MTL was first proposed in 1997 to put multiple related tasks together and learn in parallel, complement each other through hidden layer sharing information, promote each other, and improve the performance of each model. By sharing the knowledge of multiple tasks, the problem of less labeled data in a single task and easy overfitting can be solved [6–9]. The process of single-task learning and multi-task learning is shown in Fig. 1.

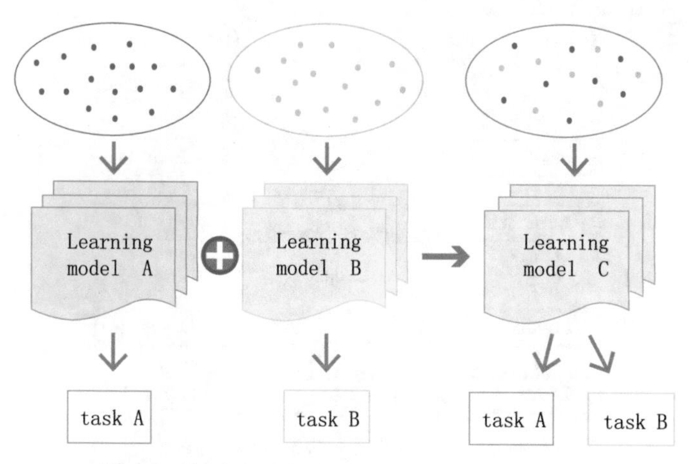

Fig. 1. Single-task learning and multi-task learning

3.2 Sequential Transfer Learning

Sequential transfer learning (STL) belongs to the category of inductive transfer learning. Sequential transfer learning first uses the source data to train the neural network model and then applies the trained parameters and network structure in the model to the target task to improve the performance of the target model. Sequential transfer learning is currently the most widely used transfer learning method and the assumption that needs to be satisfied: The tasks of transfer learning must be related [10–12]. The process of sequential transfer learning is shown in Fig. 2.

The process of sequential transfer learning can generally be divided into two phases: the pre-training phase (Fig. 2a) and the adaptation phase (Fig. 2b).

3.3 Domain Adaption

Domain adaptation belongs to the category of transductive transfer learning. The purpose of domain adaptation is to use a large amount of labeled data in the source domain to assist the target domain in learning by mapping the source domain and the target domain to a similar feature space, as shown in Fig. 3.

In natural language processing, the main methods used in current domain adaptation can be roughly divided into three categories: feature-based domain adaptation, instance-based domain adaptation, and self-labeled domain adaptation. Plank et al. [13] used the probability generated by the domain classifier as the weight of the instance. The self-labeling method belongs to the category of semi-supervised domain adaptation. It learns labeled instances in the data and then assigns pseudo-labels to unlabeled instances [14].

3.4 Cross-Lingual Learning

Cross-lingual learning belongs to the category of direct push transfer learning. Cross-language learning aligns the source language data and the target language data in the

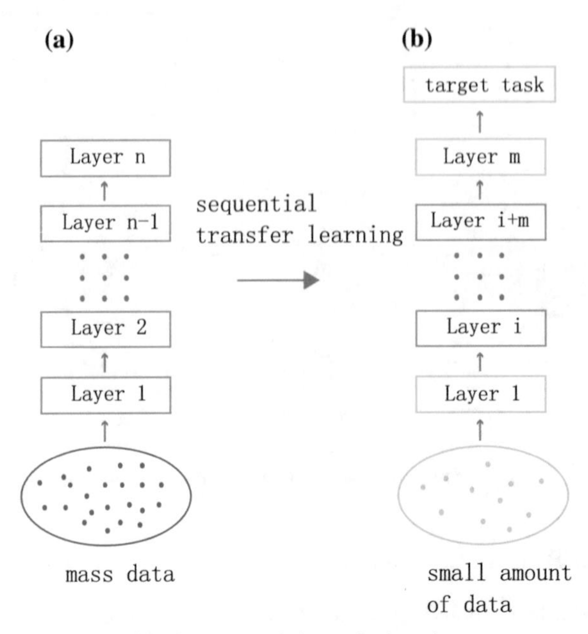

Fig. 2. Sequential transfer learning

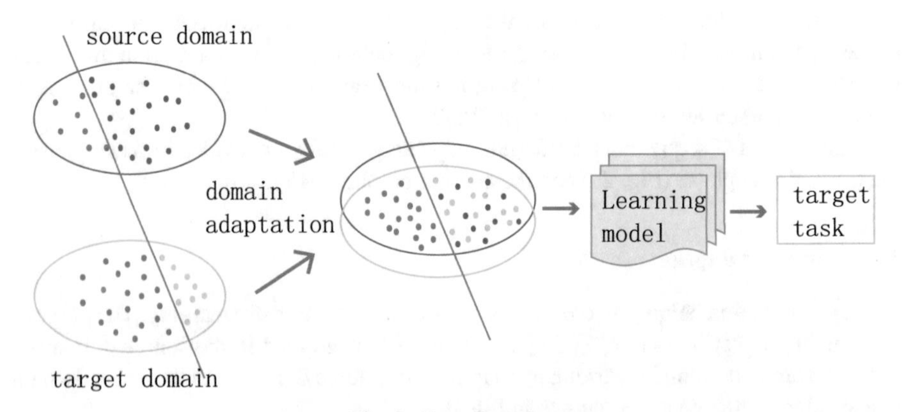

Fig. 3. Pre-adaptation

same representation space, so that the data of different languages with the same meaning have the same vector representation, thereby transferring the knowledge in the source language data to the target language, as shown in Fig. 4.

Cross-language transfer learning can be divided into parallel data alignment and indirect alignment according to the alignment method. Parallel data alignment refers to the one-to-one correspondence between source language data and target language data through a bilingual dictionary or translation mechanism. Indirect alignment uses images, Wikipedia, etc. as the transition to align the two languages [15]. Many pictures on the Internet have corresponding descriptions, so that images and text are related, and then

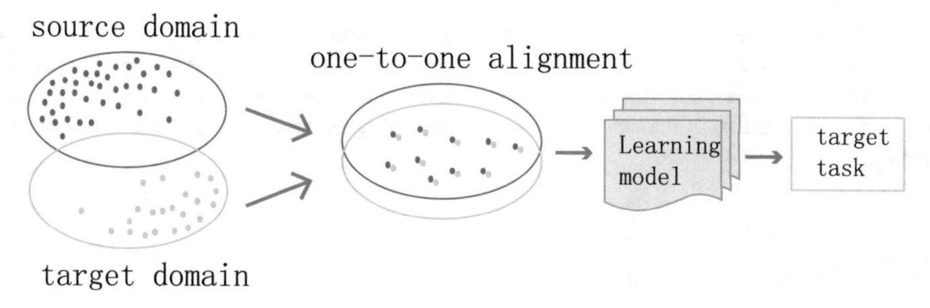

Fig. 4. Cross-language learning

different languages are aligned one by one according to different language descriptions of similar pictures [3]. In addition, some researchers have introduced hearing as an intermediate basis.

4 Conclusion

By studying the technology of transfer learning in natural language processing, this paper summarizes the characteristics of different transfer learning: Multi-task learning not only solves the problem of less label data, but also realizes an efficient working method of multiple tasks processing simultaneously; sequential transfer learning is the most popular transfer learning method currently; domain adaptation will provide powerful help for unsupervised learning; cross-language learning breaks the barriers between languages.

Transfer learning breaks through the limitations of traditional machine learning, provides more possibilities for machine learning, and promotes the development of artificial intelligence. In the research and application of transfer learning, we should pay special attention to the following aspects:

(1) Avoid negative migration. Negative transfer is often encountered in transfer learning. Analyzing the causes of negative transfer and avoiding the problem is that transfer learning should be further studied.
(2) One of the purposes of transfer learning is to improve the generalization ability of the model. Although current transfer learning can achieve cross-language and cross-domain learning, but to meet the prerequisites of similar tasks or similar domains. Therefore, it is very meaningful to study how to improve the generalization ability of the transfer learning model. It is expected that in the future, transfer learning can be improved toward a more generalized ability.
(3) Special pre-training model. It is hoped that the deep pre-training representation will become a common tool for NLP practitioners, and its practicality is similar to the pre-trained word embedding in the past few years.
(4) Using one or more transfer learning methods to efficiently deal with various problems will replace single method transfer learning.

Fund Project. NSFC-Xinjiang Joint Fund, a key support project of the National Natural Science Foundation of China, "Network Rumor Detection and Public Opinion Guidance Algorithm Research" (U1703261); Xinjiang Normal University Key Laboratory Project (XJNUSYS2019B13); Xinjiang Uygur Autonomous Region graduate education reform and innovation program in 2020.

References

1. Ian G, Yoshu B, Aaron C (2016) Deep learning. MIT Press, Cambridge, pp 1–29
2. Pratt LY, Pratt LY, Hanson SJ (1993) Discriminability-based transfer between neural networks. In: NIPS conference: advances in neural information processing systems. Morgan Kaufmann Publishers, San Francisco, pp 204–211
3. Kiela D, Vulić I, Clark S. Visual bilingual lexicon induction with transferred Conv Net features. In: Proceedings of conference on empirical methods in natural language processing. ACL, Stroudsburg, pp 148–158
4. Pan JL, Yang Q (2010) A survey on transfer learning. IEEE Trans Knowl Data Eng 22(10):1345–1359
5. Ruder S, Breslin JG, Ghaffari P (2019) Neural transfer learning for natural language processing. National University of Ireland, Galway, pp 42–129
6. Liu PF, Qiu XP, Huang XJ (2016) Recurrent neural network for text classification with multi-task learning. In: International joint conference on artificial intelligence. Morgan Kaufmann, San Francisco, pp 2873–2879
7. Misra I, Shrivastava A, Gupta A et al (2019) Cross-stitch networks for multi-task learning. In: IEEE conference on computer vision and pattern recognition. IEEE, pp 3994–4003
8. Xiao LQ, Zhang HL, Chen WQ et al (2018) Learning What to Share: Leaky Multi-Task Network for Text Classification. In: Proceedings of the 27th international conference on computational linguistics. ACM, New York, pp 2055–2065
9. Zheng RJ, Chen JK, Qiu XP (2018) Same representation, different attentions: shareable sentence representation learning from multiple tasks. In: International joint conference on artificial intelligence. Morgan Kaufmann, San Francisco, pp 4616–4622
10. Kirkpatrick J, Pascanu R, Rabinowitz N et al (2017) Overcoming catastrophic forgetting in neural networks. PNAS 114(13):3521–3526
11. Thrum S (1996) Is learning the n-th thing any easier than learning the first? In: Annual conference on neural information processing systems. MIT Press, Cambridge, pp 641–646
12. Wiese G, Weissenborn D, Neves M (2017) Neural domain adaptation for biomedical question answering. In: Conference on natural language learning. CoNLL, Belgium, pp 281–289
13. Plank B, Johannsen A, Søgaard A (2014) Importance weighting and unsupervised domain adaptation of POS taggers: a negative result. In: Proceedings of the 2014 conference on empirical methods in natural language processing. ACL, Stroudsburg, pp 968–973
14. Clark K, Luong T, Le QV (2018) Semi-supervised sequence modeling with cross-view training. In: Proceedings of conference on empirical methods in natural language processing, ACL, Stroudsburg, pp 1–15
15. Duong L, Kanayama H, Ma T (2016) Learning crosslingual word embeddings without bilingual corpora. In: Proceedings of the 2016 conference on empirical methods in natural language processing. ACL, Stroudsburg, pp 1285–1295

Identification of Key Audience Groups Based on Maximizing Influence

Jie Zhou(✉), Weimin Pan, and Haijun Zhang

School of Computer Science and Technology, Xinjiang Normal University, 10763 Xinjiang, China
522671725@qq.com

Abstract. As the development of social networks has promoted scientific research in various fields, the Sina Weibo platform has become popular with users because of its real-time information and has become a new communication method and research tool for public opinion topics. The key audience groups of Weibo topics have an important influence on public opinion, so the research on key audience groups of Weibo hot topics has become one of the main contents of public opinion analysis. The research focuses on the characteristics of Weibo social platforms and the relationship between participants and forwarders in popular topics on Weibo. On the one hand, the main improvement's overreliance on algorithms makes it impossible to achieve the greatest impact on scalability. By combining data driving with machine learning, some practical improvements are made. On the other hand, to improve the quality of the seed set for maximizing the impact of propagation impact prediction, a multi-dimensional Weibo audience user influence maximization measurement algorithm maximize your influence (MYI) is proposed. The experimental results show that the MYI algorithm can more fully identify the main audience groups in information dissemination and play an extremely important role in guiding public opinion.

1 Introduction

With the rapid development of the Internet, Weibo has become a representative platform. The hot topic under Weibo has become the focus of public life and an indispensable force of public opinion. Users under the hot topic of Weibo have different functions. Some users have become the so-called central points under a hot topic. They are called the key audience under the topic and play the role of media under the hot topic. The audience group has formed a secondary communication. With the Internet celebrity effect and star effect in recent years, the spread of the formation of key audience groups has a significant impact on the public opinion orientation of society [1]. In the information dissemination of the Weibo network, the key audience groups under the hot topic of Weibo will accumulate a large number of reposts, comments, and likes. As the influence increases, the conference will change from a general audience to an opinion leader and become a leader of network information. The characteristics of media attributes are revealed. Under all kinds of topics, there are different key audiences from different

Q. Liang et al. (eds.), *Artificial Intelligence in China*, Lecture Notes in Electrical Engineering 653, https://doi.org/10.1007/978-981-15-8599-9_56

fields and industries. However, some big stars V may not show their own influence under various topics, and some blog posts published by the audience of "passers-by" can more attract the public's topic participation and affect the participation, behavior, and opinions of other passers-by. These will lead to a certain degree of public opinion guidance.

Lin and Yang [2], etc., carried out secondary emotional feature extraction based on the machine learning Naive Bayes algorithm for popular blog posts and hotel reviews and experimented with different combinations and preprocessing of special symbols. Liu and Liu [3] and others adopted three feature selection algorithms and used Naive Bayes, IG, and SVM to compare different topic comments under different movie weights for different feature weights and found that the tendency of emotion depends on different topics comment style. Lee [4] uses the number of Weibo text repositories and the number of subscribers of text users to calculate the influence of Weibo users and identify key audiences. This method finds the top ten key audiences among the popular topics crawled crowd, and the result shows that the key audience identified by this method has an important impact on its general audience, but the number of microblog main fans who post the article under this hot topic may not be very large. Xu et al. [5, 6] improved the leader rank algorithm to make the calculation more accurate and have a higher degree of recklessness and added emotional factors to Weibo users, taking into account their activity and reduction. In view of the influence of malicious registered users, the top 20 key audience groups were extracted, and the results showed that the coverage of affected users was more extensive.

In summary, the identification methods of key audience groups are mainly from the improvement of the influence algorithm of the communication network and the basic attributes of the individual, and these two points are considered. This article summarizes the shortcomings and deficiencies of the existing methods of identifying key audience groups in China, and there are two problems at present: (1) Only the characteristics of the user's basic attributes are considered to evaluate the influence of the key audience groups, and they are spreading. In the process of network information, personal emotional factors are ignored. (2) Several factors should be considered for the calculation of the influence index, and the potential influence of the audience users cannot be ignored. It is the static influence of the audience users.

In view of the above two problems in order to reasonably measure the influence of Weibo audience users. This article considers the three dimensions of user basic attributes, user interaction behavior, and sentiment of blog content, focusing on Baidu open-source deep learning platform PaddlePaddle [7], design blog sentiment analysis [8, 9] neural network LSTM, combined with improved MYI algorithm, a microblog public opinion event key audience prediction model based on maximizing influence.

2 Maximize the Impact of Key Audience Identification Model

This article studies the audience users and content under different themes of Weibo. Firstly, based on the diffusion of the emotional factors of user blog posts, an emotional tendency classification model based on LSTM is established. Secondly, an improved MYI algorithm is proposed for the attribute characteristics of Chinese microblogs to calculate the influence of key audience groups.

2.1 Emotional Tendency Analysis of Weibo Blog Posts

The sentiment classification modeling of LSTM in this paper is based on PaddlePaddle core framework. Compared with recurrent neural network (RNN), there are problems such as gradient explosion and disappearance. LSTM can control the gradient convergence and maintain the advantage of long-term memory during training. Compared with SVM, it overcomes the shortcomings of calculating the word vector of the sentence and losing the sequence information of the word sentence due to averaging, leaving semantic information between words and words. The hidden emotion information in the word vector can be extracted better by using complex nonlinear computation. Compared with LSTM and SVM, the accuracy of its method has improved a lot, and its effect has performed well on emotion classification.

The training model of sentiment analysis in this paper is based on the data collected by 230 subject keywords related to "new coronary pneumonia". Among them, 100,000 pieces of data are manually marked, which are 1 (positive), 0 (neutral), and −1 (negative). First, preprocess the content of 100,000 labeled blog posts with Jieba word segmentation and Harbin Institute of Technology stop word list. Then define a long-term and short-term memory network to feed feature data for training to get an emotion classification model. The preprocessed blog post generates an emotion dictionary, and the train and test data are divided. Generate a data dictionary and data list, and convert the feature vector Weibo Vector of the blog post into the required data format. Import paddle.fluid to build the network of LSTM, and define the data layer and get the classifier. Define the loss function and accuracy function of the evaluation standard and the optimization method. Train the model and use evaluation criteria for evaluation to ensure that the accuracy meets the standards. Use the trained model LstmModle for data prediction, and divide it into three categories: positive, neutral, and negative.

Compared with Naive Bayes classification and SVM algorithm, this classification algorithm has great advantages in capturing sequence features of text, long-distance dependent learning, and context-dependence, with higher accuracy and higher classifier performance. The classification algorithm in this paper is more suitable for emotion classification.

2.2 Calculation of Initial Influence of Real-Time Blog Posts

In the calculation of the MYI algorithm in this article, it is necessary to input the influence value of the initial audience user to iterate. Before the calculation, the factor analysis of real-time microblogs is required. Due to the large span of the collected feature data, such as some celebrity V users have many fans. However, the influence of the audience as a hot topic is not necessarily higher than that of ordinary users. In order to reduce the overall results of the audience's individual indicators that are too prominent, it is necessary to use the coefficient of variation method to calculate the weight ratio of each basic attribute. The coefficient of variation of each index is as shown in Eq. (1).

$$N_t = \frac{O_t}{y_t}(t = 1, 2, 3, 4, \ldots n) \tag{1}$$

Where N_t is the coefficient of variation of the t-th index, O is the standard deviation of the tth index, and Y_t is the average of the t-th index. The weight of each characteristic attribute index is formula (2), the ranking is calculated by calculating the comprehensive score as formula (3) and normalized after the total score is obtained, and the method adopts max-min normalization as formula (4).

$$P_t = \frac{N_t}{\sum_{t=1}^{n} m} \tag{2}$$

$$U = \sum_{t=1}^{n} H_t y_t \tag{3}$$

$$F = \frac{F - low}{high - low} \tag{4}$$

This article calculates the initial impact of real-time blog posts, and its characteristic attributes take into account the number of reposts, comments, and likes. Definition uses formula (5) to calculate the value of user u's own influence.

$$I = O_1 A_1 + O_2 A_2 + O_3 A_3 \tag{5}$$

where I is the initial influence value of user u, A_1, A_2, A_3 are the number of likes, comments, and retweets, and O_1, O_2, O_3 are the above weight coefficients. The three characteristics of this paper are calculated and assigned by the above algorithm (Table 1).

Table 1. Audience user interaction attribute weights

Interactive feature	Forwarding volume	Comment volume	Likes
Weights	0.22	0.46	0.32

2.3 MYI Algorithm

This paper improves the discovery algorithm of key audience groups based on the PageRank algorithm, referred to as the MYI algorithm, as shown in Eqs. (7) and (8). In the end, the iteration result of the audience user is IO(u). The algorithm sets the resistance value d to 0.7 through the interaction characteristics between users. HR_u represents a collection of users who have been reposted, liked, and commented on. $L(u, v)$ represents the proportion of user u in the interaction set of user v. HE_v represents the set of people that user v interacts with. Set the basic influence of user u to ZI(u) and the propagation probability to FITE(u). Compared with traditional algorithms, this paper considers the calculation of the initial influence value ZI(u) and the interaction behavior HE_v between users. The MYI algorithm is considered more comprehensive, and the results obtained are more objective.

$$\mathrm{IO}(u) = (1 - d) + d \sum_{v \in \mathrm{HR}_u} L(u, v) \times \mathrm{IO}(v) \tag{6}$$

$$L(u, v) = \frac{\mathrm{ZI}(u) \times \mathrm{FITE}(u)}{\sum_{k \in \mathrm{HE}_v} \mathrm{ZI}(k) \times \mathrm{FITE}(k)} \tag{7}$$

Suppose that the number of users who post blog posts on the Weibo network is N, where M is the user pointing to N, and ZI is the initial influence of the current user. There are two conditions for the algorithm to terminate the convergence, one is the number of iterations is 100 and the other is a value which is 0.01, indicating the difference of the influence value before and after each individual, that is, the current $\mathrm{IO}(u)$ value and the last iteration result $\mathrm{IO}(u)$ difference threshold of old value. After the code satisfies the condition of the end of the iteration, the final $\mathrm{IO}(u)$ value is obtained, and the maximum set is returned for the reverse order of the $\mathrm{IO}(u)$ value.

The algorithm needs to preprocess the input value in actual operation. Its damping factor, number of iterations, and other conditions are not unique. It can be debugged. The parameter set is determined by comparing the F_1 value of the result. The second step is to construct a directed graph model. The value of M can be set as an $n+$ sequence starting from zero according to actual needs. In summary, the improved MYI algorithm in this paper is fully suitable for the influence calculation of Weibo audience users and can be converged.

3 Experiment and Analysis

3.1 Data Set

This article uses Sina Weibo as the data source during the epidemic and then compares it on the new data set, so as to get the actual effect of the algorithm in this article. Data set of this paper was collected based on five subject keywords related to "car accident scene", and a total of 43,641 Weibo data was captured from March 10, 2020, to April 20, 2020, in order to reduce unnecessary calculations, from the crawled data, followers, and historical posts and delete less than 15 followers. 1780 microblog samples to be evaluated were extracted from the above data set, the number of comments plus likes and blog posts less than 5 were removed, and 436 microblogs were studied, which was more in line with the number of microblog real-time hot topics. Then using the sentiment tendency classification model in this paper, we get 181 sets of positive attitude K1, 98 sets of negative attitude K2, and 157 sets of neutral attitude K3.

3.2 Experimental Results and Analysis

In this paper, the MYI algorithm selects UIRank [10], PageRank, and fan ranking algorithm that have recently achieved excellent results for comparison. It is only inaccurate to determine the key audience groups by the number of fans and the number of reposts, so refer to the algorithm F_1 value defined in [11] to evaluate the effectiveness of each algorithm.

$$A_1 = \left(A_{\text{MYI}} \cap A_{\text{PageRank}}\right) \cup \left(A_{\text{MYI}} \cap A_{\text{UIRank}}\right) \cup \left(A_{\text{MYI}} \cap A_{\text{Fans}}\right) \cup \left(A_{\text{PageRank}} \cap A_{\text{UIRank}}\right) \cup \left(A_{\text{PageRank}} \cap A_{\text{Fans}}\right) \cup \left(A_{\text{UIRank}} \cap A_{\text{Fans}}\right) \tag{8}$$

In formula (9), A1 represents the total blogger ranking set of each algorithm. Use AMYI, A_{UIRank}, A_{PageRank}, and A_{Fans} to represent: this article, UIRank, PageRank, and the collection of key audience bloggers under the fan ranking. The calculation formula of the accuracy rate, recall rate, and F_1 value of the algorithm evaluation is shown in formula (9), (10), and (11). Among them, F_1 is the comprehensive evaluation of the accuracy rate and the recall rate. The larger the F_1 value, the better the effect of the algorithm.

$$P_{\text{MYI}} = \frac{A_{\text{MYI}} \cap A_1}{A_{\text{MYI}}} \tag{9}$$

$$R_{\text{MYI}} = \frac{A_{\text{MYI}} \cap A_1}{A_1} \tag{10}$$

$$F_1 = \frac{2 \times P_{\text{MYI}} \times R_{\text{MYI}}}{P_{\text{MYI}} + R_{\text{MYI}}} \tag{11}$$

In contrast to the F_1 value in Fig. 1, the user MYI algorithm in this article has achieved good results overall. Because the user's interaction in a certain field and topic is low and the activity is not high, the actual influence of the FANS algorithm is not high.

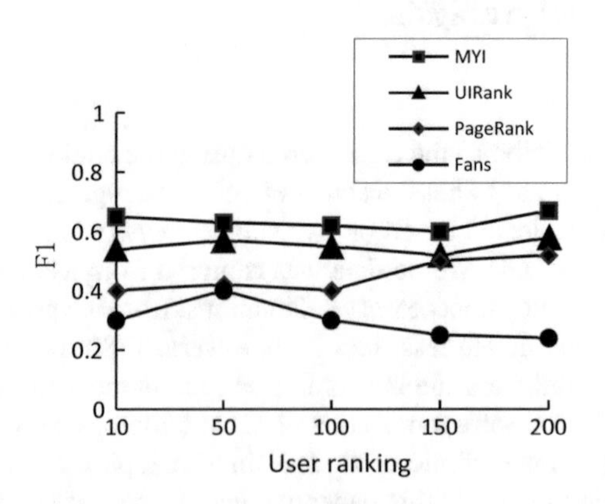

Fig. 1. F1

In order to show the breadth of the algorithm's influence in this paper, the coverage rate is introduced as an evaluation indicator to measure the users of key audience groups, which directly or indirectly affect the coverage ratio of other users, as shown in Eq. (12).

$$H(i) = \frac{\sum_{i=1}^{M} P(i)}{M} \tag{12}$$

The coverage of top i users is represented as $H(i)$, all audience users in the data set are M, and the nodes affected by key audience groups are represented as $P(i)$.

In Fig. 2, the MYI algorithm of this paper achieves the highest coverage rate of 0.7% in the Weibo interactive network composed of 23,446 audience users. It can be seen from the results that the key audience groups in Weibo are related to the emotional tendencies of the blog content. Although there are some negative blog posts whose forwarding and likes are far inferior to some normal Weibo content, there are more controversies in their comments and their impact is greater. Therefore, we should not only consider the users' fans, the general attributes of blog posts, but also the content nature of blog posts and the potential influence of blog posts.

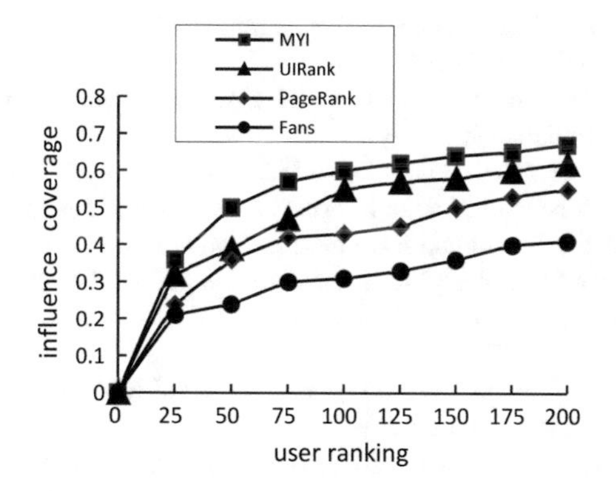

Fig. 2. Comparison of coverage of different algorithms

4 Conclusion

On the basis of considering the potential influence of Weibo users, the emotional tendency factor is added to the recognition model of Weibo key audience groups, and an improved MYI algorithm is proposed. The algorithm considers a wider range of microblog feature attributes and the weight distribution between attributes. Compared with other algorithms, F_1 value and coverage are better. Nowadays, Weibo often has many "water soldiers". If the interference factors of the "Water Army" can be eliminated and the time factor is taken into consideration, then the periodicity of the Weibo theme will be increased to find out the trend of time changes posted to the blog, and the final result will be more objective.

Acknowledgements. This work was partially supported by the National Natural Science Foundation of China NSFC-Xinjiang Joint Fund Key Support Project "Network Rumor Detection and Public Opinion Guidance Algorithm Research" (U1703261).

References

1. Pei W (2009) China' public opinion leaders in virtual communities influence public opinion research. Northeast Normal University, Changchun
2. Lin J, Yang A, Zhou Y, Chen J, Cai Z (2012) A Microblog emotion classification based on Naive Bayes. Comput Eng Sci 34(09):160–165
3. Liu Z, Lu L (2012) Empirical research on chinese weibo sentiment classification based on machine learning. Compu Eng Appl 48(01):1–4
4. Lee C, Kwak H, Park H et al (2010) Finding influentials based on temporal order of information adoption in Twitter. In: WWW 2010, Raleigh, North Carolina, April 2010
5. Xu J, Zhu F, Liu S et al (2015) Identifying opinion leaders by improved algorithm based on LeaderRank. Comput Eng Appl 1:23
6. Xu J, Zhu F, Liu S, Zhu B (2014) Opinion leader mining for improving LeaderRank algorithm. Comput Eng Appl 51(01):110–114 + 166
7. Guo H (2018) Research and implementation of advertising picture distribution strategy based on PaddlePaddle. Xinjiang University
8. Jun L, Chai Y, Yuan H, Gao M, Zan H (2015) Sentiment analysis based on polarity transfer and LSTM recurrent network. J Chin Inform Process 29(05):152–159
9. Zhou J, Xu W (2015) End-to-end learning of semantic role labeling using recurrent neural networks. In: Proceedings of the 53rd Annual meeting of the Association for Computational Linguistics and the 7th international joint conference on natural language processing, vol 1: Long Papers, pp 1127–1137
10. Jianqiang Z, Xiaolin G, Feng T (2017) A new method of identifying influential users in the micro-blog networks. IEEE Access 5:3008–3015
11. Huang X, Yang A, Liu X, Liu G (2019) An improved Weibo user influence evaluation algorithm. Comput Eng 45(12):294–299

Technical Theme Analysis of WeChat Graphic Based on Domain Science and Technology Information

Min Zhang[1,2], Rui Yang[1,2](✉), Wei Chen[1,2,3], Jun Chen[1,2], Jinglin Xu[1,2], and Yanli Zhou[1,2]

[1] Wuhan Library of Chinese Academy of Sciences, 430071 Wuhan, China
yangr@mail.whlib.ac.cn
[2] Hubei Key Laboratory of Big Data in Science and Technology (Wuhan Library of Chinese Academy of Sciences), 430071 Wuhan, China
[3] School of Economics and Management, University of Chinese Academy of Sciences, 100190 Beijing, China

Abstract. In order to enhance the WeChat communication influence and effect in the field of scientific and technological information, this paper explores the relationship between the technical theme of WeChat public account and the WeChat communication in the field of scientific and technological information. Taking the "Advanced Energy Science and Technology Strategic Information Research Center" WeChat public account as an example, this paper uses the singular value decomposition algorithm to cluster the WeChat technical theme, and calculates the WeChat communication index of each technical theme. Based on the calculation results, this paper analyzes the overall of technical themes, the correlation of technical themes and the typical graphics of technical themes. This work has a certain guiding and reference significance to the operation of the WeChat public account in the field of scientific and technological information.

Keywords: Scientific and technological information · WeChat public account · Technical theme · Clustering analysis · Communication influence

1 Introduction

With the development of self-media, the information fragmentation trend has become unstoppable. WeChat as a new way of information dissemination and services, more and more organizations had realized the importance of this platform [1]. In this environment, how to give full play to its own advantages, rapidly and accurately pushing the information hot spots in the field is something that scientific and technical information research institutions need to think about [2, 3]. The WeChat public account of "Advanced Energy Science and Technology Strategic Information Research Center" is a WeChat public account of "Wuhan Library of Chinese Academy of Sciences," it focuses on tracking the latest progress of domestic and foreign strategic policies, scientific and technological development, and industrial layout in the field of energy, and it provides the latest

Q. Liang et al. (eds.), *Artificial Intelligence in China*, Lecture Notes in Electrical Engineering 653, https://doi.org/10.1007/978-981-15-8599-9_57

information services for the strategic decision-making and scientific and technological innovation activities of the decision makers. This paper explores the case of "Advanced Energy Science and Technology Strategic Information Research Center" WeChat public account, using the singular value decomposition algorithm to cluster the technical themes of the graphical information of WeChat, and analyzes different technical themes in combination with the WeChat public account communication index, to obtain the audience's content preferences and propose optimization suggestions.

2 Research Design

The graphic information of the "Advanced Energy Science and Technology Strategic Information Research Center" WeChat public account mainly comes from the domestic and international energy strategy and technology news compiled by the team, as well as independently written review-type articles. From the perspective of basic indicators such as reading volume and like volume, the target group is relatively fixed, and the continuous and steady growth of users is concerned. The selection of this WeChat public account as a research object has certain representativeness and practicality. The overall research idea is shown in Fig. 1.

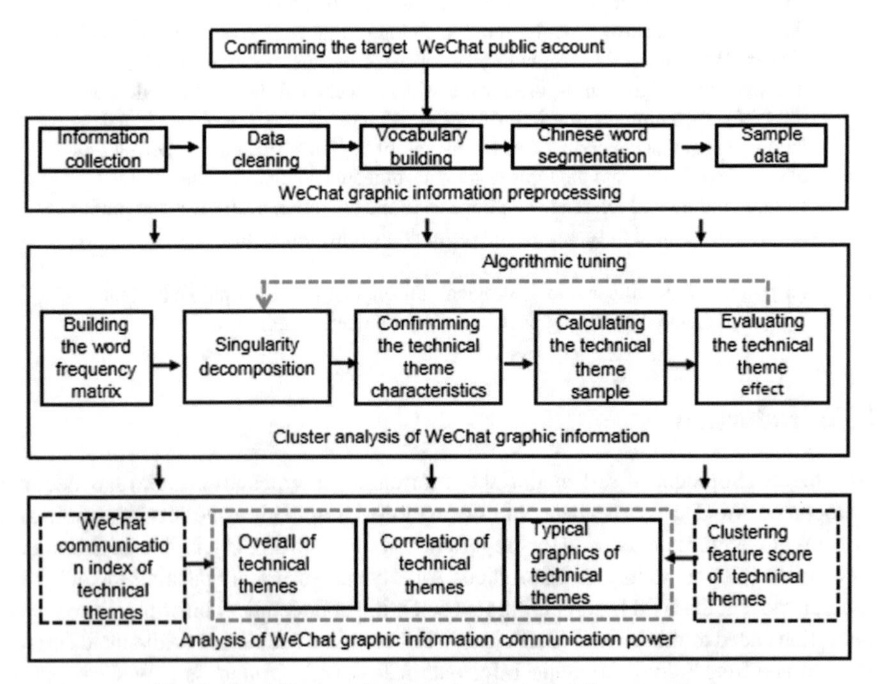

Fig. 1. Overall research approach of technical themes analysis

2.1 Sample Data Collection

This article uses manual collection and automatic collection methods to collect the graphic information of the WeChat public account of the "Advanced Energy Science and Technology Strategic Information Research Center" from the establishment on September 1, 2016 to January 31, 2019. Including the graphic title, graphic content, graphic readings, likes and other related content, the information was collected on February 1, 2019. Then filtered and rechecked the collected data, removed conferences, notifications, and other graphic information that are not highly relevant to the technical theme to form the original material. Cleaned the original material content, and removed various image information, promotional information, and copyright information that are not related to the analysis content. A total of 546 WeChat graphics were collected as sample data.

2.2 Chinese Word Segmentation

This article selects the HanLP natural language processing tools, a toolkit consisting of a series of models and algorithms [4], to complete word segmentation processing. Among them, the CRF conditional random field model is mainly used to mark and segment serial data, which has good effect on new word extraction. Select this model for word segmentation processing, and at the same time, combine professional terminology and vocabulary which were accumulated in the energy field of intelligence research for a long time. Finally, 5824 words were extracted for word segmentation after building a custom vocabulary through screening, as shown in Table 1.

Table 1. Technical terms and vocabulary in the field of energy

Vocabulary	Words
Fuel cell	Anode, carbon monoxide, catalyst, catalytic combustion, cathode, cold combustion, converter, electrolysis, electrolyte, energy carrier…
Nuclear power	Absorption dose, absorption rods, absorption coefficient, absorption cross-section, absorption ratio, acceptance guidelines…
Solar	Radiation measurement, radiation meter, total radiation meter, total heliograph, field of view angle, heliograph, sunshade…
Oil and gas	Oilfield, liquefied petroleum gas, liquefied natural gas, fractionation, cracking, separation, wells, mineral samples, crude oil, octane…
New energy	Energy-efficient technologies, sustainability, ecological balance, biofuels, green power, ecosystems, wind power, geothermal power…
Other vocabulary	Energy chemistry glossary, energy synthesis glossary, domain agency glossary, domain expert glossary, deactivation glossary…

2.3 Technical Theme Clustering

The traditional vector space model (VSM) cannot deal with the problem of polysemy [5], in order to analyze the technical theme features of WeChat graphic information, the article introduced the SVD (Singular Value Decomposition) model to perform clustering analysis of technical themes on WeChat graphic information, and obtain the names and feature scores of various technical themes at the same time [6, 7]. The principle of SVD algorithm is shown in Fig. 2.

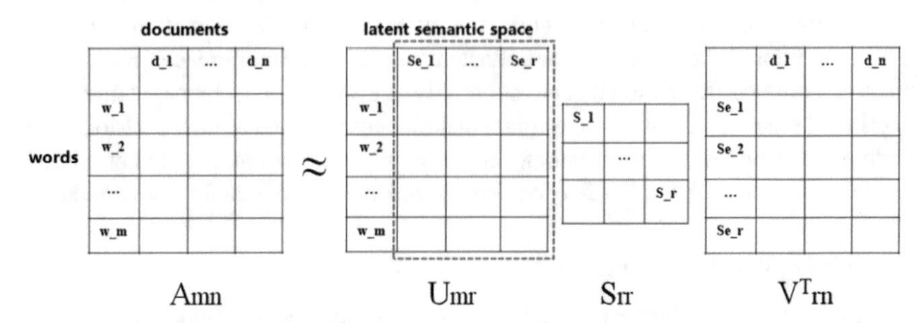

Fig. 2. Principle of singular value decomposition algorithm

According to the principle of singular value decomposition algorithm [8], the matrix A_{mn} is decomposed into n eigenvalues, the larger r values are selected and sorted, and the matrix A_{mn} is approximated by USV^T. Each column of the matrix U represents a latent semantic, which is composed of m words with different weights, r latent semantics form a semantic space, each singular value in S represents the importance of latent semantics, and each column in V^T represents a text in the text collection, the collection of all texts is mapped to the latent semantic space [9]. After repeated experiments, it was determined that the clustering effect was relatively ideal when $r = 17$. Through the features of the singular value decomposition algorithm and the tuning of various parameters including cluster calculation, theme generation, matrix model, etc., the clustering analysis results of technical themes of the "Advanced Energy Science and Technology Strategic Information Research Center" WeChat public account are shown in Table 2.

Table 2. Results of technical themes clustering analysis

No.	Technical themes	Num	Score	No.	Technical themes	Num	Score
1	Power system	176	71.27	10	Biofuels	82	120.84
2	Battery technology	165	18.48	11	Fuel cell	75	109.76
3	Renewable energy	133	35.69	12	Energy consumption	75	101.63
4	Wind power	121	112.10	13	Catalysts	70	35.28
5	Lithium ion battery	111	98.99	14	Reactors	31	26.53
6	Natural gas	109	23.83	15	Additive	20	12.15
7	Clean energy	99	80.61	16	Gas turbine	19	23.01
8	Perovskite	88	108.41	17	Transformers	5	14.58
9	Electric car	87	98.66				

2.4 Technical Theme Communication Index

The push of the "Advanced Energy Science and Technology Strategic Information Research Center" WeChat public account is mainly characterized by single graphic information, based on the Qingbo WeChat communication index WCI (V13.0) calculation formula [10], this paper combined and optimized WCI's overall communication index and headline communication index, and designed the formula for calculating the communication index based on technical themes [11], as shown in Table 3.

Table 3. Communication index table of technical themes

First-level indicators	Secondary indicators	Secondary weight	Standardized score
Overall communication (60%)	Daily readings (R/d) Daily likes (Z/d)	85% 15%	$85\% * \ln\left(\frac{R}{d}+1\right)$ $15\% * \ln\left(10 * \frac{Z}{d}+1\right)$
Average communication (30%)	Average readings (R/n) Average likes (Z/n)	85% 15%	$85\% * \ln\left(\frac{R}{n}+1\right)$ $15\% * \ln\left(10 * \frac{Z}{n}+1\right)$
Peak communication (10%)	Max readings (R_{max}) Max likes(Z_{max})	85% 15%	$85\% * \ln(R_{Max}+1)$ $15\% * \ln(10 * Z_{Max}+1)$

R is the total number of readings of all articles in the evaluation period; Z is the total number of likes of all articles in the evaluation period; d is the number of days in the evaluation period; n is the number of articles pushed by the WeChat public account during the evaluation period; R_{Max} and Z_{Max} are the highest number of readings and likes of articles pushed by the WeChat public account during the evaluation period.

$$\begin{aligned} \mathrm{WCI} = {} & 60\% * \left[85\% * \ln\left(\frac{R}{d}+1\right) + 15\% * \ln\left(10 * \frac{Z}{d}+1\right)\right] \\ & + 30\% * \left[85\% * \ln(R_{Max}+1) + 15\% * \ln\left(10 * \frac{Z}{n}+1\right)\right] \\ & + 10\% * [85\% * \ln(R_{Max}+1) + 15\% * \ln(10 * Z_{Max}+1)] \end{aligned}$$

The number of graphic readings, likes and WCI index of different technical themes were calculated, as shown in Table 4.

3 Empirical Analysis

From the clustering results of technical themes, the graphic information pushed by the WeChat public account of the "Advanced Energy Science and Technology Strategic

Table 4. Calculated results of technical themes communication index

No.	Technical themes	R	Z	d	n	R_{max}	$Zmax$	WCI
1	Power system	36,518	165	853	176	4208	15	183
2	Battery technology	30,152	124	853	165	4208	15	170
3	Renewable energy	25,794	129	853	133	1631	6	158
4	Wind power	21,821	101	853	121	1631	6	149
5	Lithium ion battery	15,988	51	853	111	4208	15	137
6	Natural gas	20505	106	853	109	1631	6	148
7	Clean energy	22,866	95	853	99	4208	15	163
8	Perovskite	14,847	57	853	88	4208	15	138
9	Electric car	13,542	66	853	87	1631	6	127
10	Biofuels	18,790	86	853	82	4208	15	156
11	Fuel cell	19229	79	853	75	4208	15	158
12	Energy consumption	12,231	77	853	75	1631	6	126
13	Catalysts	16,426	57	853	70	4208	15	149
14	Reactors	8628	30	853	31	4208	15	128
15	Additive	1772	2	853	20	408	1	53
16	Gas turbine	3856	20	853	19	763	6	89
17	Transformers	1471	6	853	5	822	5	74

Information Research Center" basically focused on the major changes in the international energy landscape and the general environment of increasingly fierce international competition for energy technology and industrial transformation, highlighted the idea that the energy system has shifted from the absolute dominance of fossil energy to the integration of low-carbon and multi-energy, and new industries and new formats are growing stronger. Use the Carrot [2] software [12] to draw the clustering and co-occurrence map of the technical themes, as shown in Fig. 3.

3.1 Overall Analysis of Technical Themes

The comparative analysis of the communication index and clustering feature score of various technical themes of the WeChat public account of the "Advanced Energy Science and Technology Strategic Information Research Center" is shown in Fig. 4.

The differences in the technical themes of the WeChat public account of the "Advanced Energy Science and Technology Strategic Information Research Center" are not very large, and the readers' attention on each technical theme is more balanced. This is consistent with the positioning of the WeChat public account for domestic and foreign strategic policies, technological development, and industrial layout in the energy field, which can present more comprehensive scientific and technological information in the field to readers. At the same time, through detailed analysis of the technical theme

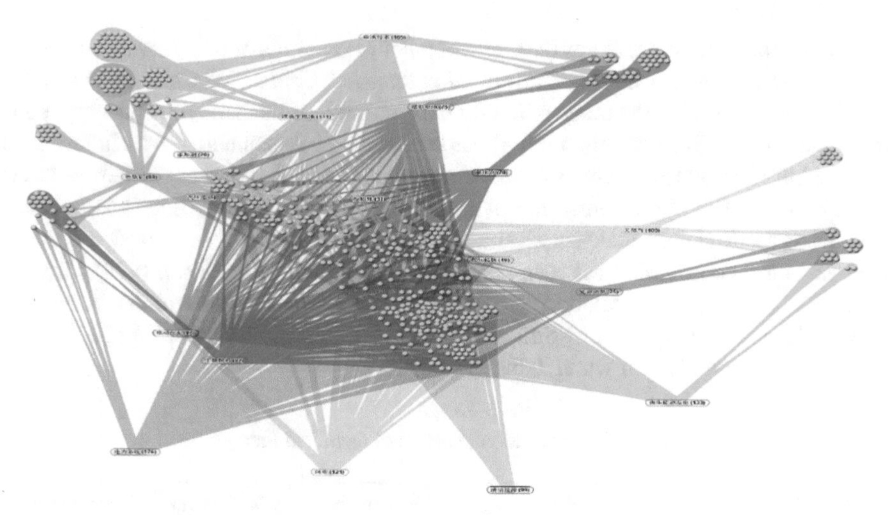

Fig. 3. Clustering co-occurrence map of WeChat public account technical themes

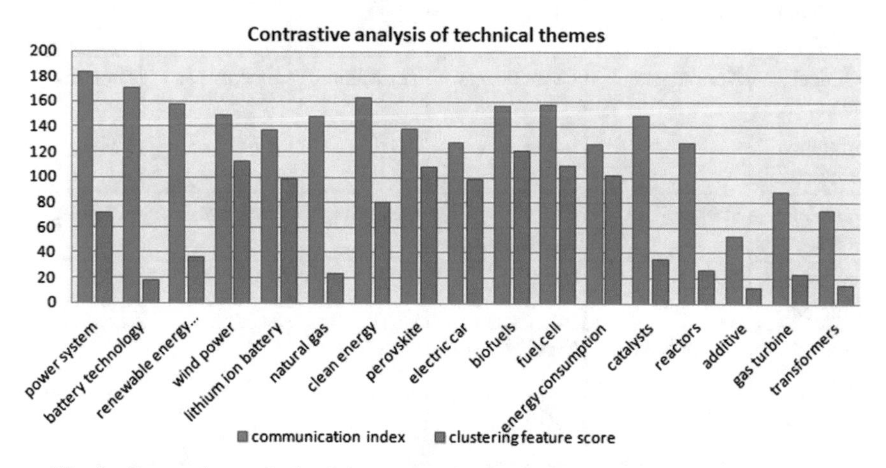

Fig. 4. Contrastive analysis of the communication index and clustering feature score

WeChat communication index and clustering feature scores, we found that the WeChat public account graphic information that can attract more readers' attention is mainly concentrated in the aspects of planning, projects, funding, reports and comments of energy-related field. Only by selecting hot themes that readers are really interested in can it attract more readers' attention and achieve the effect of disseminating scientific and technological information in the field.

3.2 Correlation Analysis of Technical Themes

The relationship between attention and acceptance was further analyzed by combining the basic indicators of graphic information of each technical theme, including number of readings (R), the number of likes (Z), the number of readings of headlines (Rt), and the number of likes of headlines (Zt). The number of readings of the "Advanced Energy Science and Technology Strategic Information Research Center" WeChat public account mainly concentrated in the range of 0–2000, the number of likes mainly concentrated in the range of 0–15. The readings number was used as the basic standard for the influence of graphic information to measure the reader's attention, and the number of likes was used as the basic standard for the influence of graphic information on the reader's acceptance, conducted a correlation analysis on the number of readings and likes to determine if there is a direct correlation between the two. This paper used the SPSS (v24) software to perform the Pearson correlation calculation on the technical themes of the "Advanced Energy Science and Technology Strategic Information Research Center" WeChat public account, the results are shown in Table 5.

Table 5. Correlation analysis of technical themes

Technical themes	Power system	Battery technology	Renewable energy	Wind power	Lithium ion battery
Correlation coefficient	0.815	0.802	0.655	0.734	0.932
Technical themes	Natural gas	Clean energy	Perovskite	Electric car	Biofuels
Correlation coefficient	0.670	0.853	0.870	0.745	0.900
Technical themes	Fuel cell	Energy consumption	Catalysts	Reactors	Additive
Correlation coefficient	0.855	0.757	0.924	0.969	0.897
Technical themes	Gas turbine	Transformers			
Correlation coefficient	0.897	0.958			

The correlation coefficients of technical themes such as "power system," "battery technology," "lithium ion battery," "clean energy," "perovskite," "biofuels," "fuel cell," "catalyst," "reactor," "additive," " "gas turbine" and "transformer" are between 0.8 and 1.0, and the correlation between "attention" and "acceptance" is very significant and belongs to a extremely strong correlation. The number of readings and likes of various technical themes are highly consistent in the overall trend, the more readings, the more likes.

3.3 Analysis of Typical Graphics

According to the analysis of the typical graphics with higher readings and likes under each technical theme, the features of graphic information that readers are mostly interested in and accepted mainly include three aspects.

The first is timeliness. Readers need to understand the "new events," "new progress," "new policies," "new plans" of scientific and technological information in the field. With the WeChat public account, readers can browse information and transfer information anytime and anywhere, fully utilizing their readers' fragmented time. If the WeChat public account can provide time-sensitive domain scientific and technological information [13, 14], allowing readers to receive first-hand domain scientific and technological information will naturally increase readers' attention to the WeChat public account. From the timeliness of the typical graphics of the WeChat public account of "Advanced Energy Science and Technology Strategic Information Research Center," it basically reflects the latest developments in related technical themes in the energy field. For example, in the "perovskite" technology theme that was pushed out in February 2017, the "Organic guanidine cations help perovskite solar cells achieve stable operation over a thousand hours" graphic information timely pushed the latest research results of the Swiss Federal Institute of Technology Lausanne. The relevant research paper was published in the second issue of "Nature Energy" in 2017, and the WeChat public account graphic information was basically consistent with it in time.

The second is the authority. The authority of the source of graphic information is also very important, it is directly related to the quality of graphic information [15]. Unreliable sources of graphic information are easy to cause readers to question and lose their attention to the WeChat public account. When pushing graphic information, a professional intelligence team needs to filter out reliable information from a variety of complex information sources and push it to readers. At the same time, careful checks should be carried out from content to editing, to strictly prevent false scientific and technological information. So this requires the intelligence team to have a certain background in the subject area and be familiar with key authoritative institutions in this area. From the authoritativeness of typical graphics of the WeChat public account technology theme of the "Advanced Energy Science and Technology Strategic Information Research Center," the source of the graphic information released basically comes from important international organizations, government agencies, industry associations, well-known universities, large enterprises and authoritative journals, including "International Energy Agency," "US Department of Energy," "EU Joint Research Center," "International Renewable Energy Agency," "European Commission," "BP," "World Energy Council," "Science Journal," "Nature Energy Journal," etc., the authenticity and authoritativeness of the graphic information is high, which is easy to arouse readers' attention.

Then there is the originality. If facts are simply stated or described in the graphic information, and there is no in-depth analysis, then the acceptance of the WeChat public account will be affected to some extent [16]. With the diversification of sources of scientific and technological information, the lack of personalized analysis and commentary will cause readers to naturally turn to other media for information, and gradually lose their attention. This requires the intelligence team to carry out comprehensive processing of information from a professional perspective, to grasp the needs of readers, and extract hierarchical and in-depth information to be recognized by readers. From the originality of the typical graphics of the WeChat public account technology theme of the "Advanced Energy Science and Technology Strategic Information Research Center," there is room for further improvement. The "likes" index of typical graphics is generally not high, it shows that readers' acceptance of graphic information needs to be strengthened, but there is no lack of good graphic information. For example, the graphic information of "Brief analysis of Trump's energy policy and its impact on China " in the theme of "clean energy" technology pushed in November 2016, it better analyzed the "energy independence" mentioned by Trump for many times during the campaign and the impact of restructuring US energy policies on the global energy landscape, at the same time, it analyzed the impact of Trump's energy policy on our country from various angles such as "improving commodity trade costs" "reducing production costs and slowing down the pressure of economic decline," "increasing the downward pressure on China's economy" and "not conducive to the implementation of renewable energy strategies."

4 Conclusions

This paper applied the computational theory approach of latent semantic analysis, representing information through a high-dimensional vector space model and mapping it to a low-dimensional latent semantic space, extracting and amplifying the typical features to realize the technical themes clustering of the WeChat public account in the field of science and technology intelligence. At the same time, for each technical theme, calculated the weighted communication influence from three dimensions of the overall communication power, the average communication power and the peak communication power, and analyzed from three aspects of the overall of technical themes, the correlation of technical themes and the typical graphics of technical themes. This has a certain guiding and reference significance to the operation of the domain scientific and technological information WeChat public account.

Acknowledgements. The work was supported by Hubei Key Laboratory of Big Data in Science and Technology (Wuhan Library of Chinese Academy of Science) (No. 20KF011011) and the Special Project of the Chinese Academy of Sciences for the Construction of Literature Information Capability "Documentation Information 'Data Lake' and Open Big Data Framework Construction" (Y1852).

References

1. Li K, Huang X, He X (2018) Research on the construction of the attractiveness of universities WeChat public accounts—taking communist youth league public platform of Guangxi University as an example. In: 2018 international joint conference on information, media and engineering (ICIME), Osaka, pp 183–186
2. Zhang L, Zhang X, Li Z, Lu H (2018) Research on the evaluation of information communication power of WeChat public platform of think tank based on BP neural network. Inform Stud Theory Appl 41(10):93–99
3. Wang D, Song M, Wei R (2018) Case study on information service model of WeChat public platform for academic research. Rese Library science 9:31–36
4. HanLP, [online]. Available: https://github.com/hankcs/HanLP

5. Liu X, Xiong H, Shen N (2017) A hybrid model of VSM and LDA for text clusteing. In: 2017 2nd IEEE international conference on computational intelligence and applications (ICCIA), Beijing, pp 230–233
6. Abidin TF, Yusuf B, Umran M (2010) Singular value decomposition for dimensionality reduction in unsupervised text learning problems. In: 2010 2nd international conference on education technology and computer, Shanghai, pp 422–426
7. Akter N, Hoque AHMS, Mustafa R, Chowdhury MS (2016) Accuracy analysis of recommendation system using singular value decomposition. In: 2016 19th international conference on computer and information technology (ICCIT), Dhaka, pp 405–408
8. Wang B, Tian Z, Zhang Y (2010) Optimization of singular vector decomposition algorithm. Acta Electronica Sinica 38(10):2234–2239
9. Xu J, Chang Z, Zhang X (2017) Image super-resolution based on alternate K-singular value decomposition. Geomatics Inform Sci Wuhan Univ 42(8):1137–1143
10. WeChat Communication Index WCI (V13.0), [online] Available: http://www.gsdata.cn/site/usage/
11. Wu F, Tong Y, L. Huang, H. Miao and X. Li (2018) The application prospect analysis of technology based on WeChat official accounts. In: 2018 Portland international conference on management of engineering and technology (PICMET), Honolulu, HI, pp 1–9
12. Carrot2, [online]. Available: http://project.carrot2.org
13. Fang J, Lu W (2016) a study on influential factors of WeChat public accounts information transmission hotness. J Intell 35(2):157–162
14. Yang, Zhao H (2019) Analysis of factors affecting attraction of medical teaching content based on WeChat platform. In: 2019 10th international conference on information technology in medicine and education (ITME), Qingdao, pp 279–283
15. Tian M, Xu G (2017) Exploring the determinants of users' satisfaction of WeChat official accounts. In: 2017 3rd international conference on information management (ICIM), Chengdu, pp 362–366
16. Xiong Y, Jiang S, Xing R (2018) Mechanism of transmission and guidance of WeChat public opinion. Inform Sci 11:54–60

Coronavirus Disease (COVID-19) X-Ray Film Classification Based on Convolutional Neural Network

Shixiang Yan[1] and Bingcai Chen[1,2](✉)

[1] School of Computer Science and Technology, Xinjiang Normal University, 102 Xinyi Road, Urumqi, Xinjiang 830054, China
china@dlut.edu.cn

[2] School of Computer Science and Technology, Dalian University of Technology, 2Linggong Road, Ganjingzi District, Dalian, Liaoning 116024, China

Abstract. The main incidence of pneumonia in Wuhan is that Coronavirus Disease (COVID-19) infects the lungs and causes respiratory failure [1]. Based on the technical research of lung infection detection, the medical image X-ray film can be detected by artificial intelligence technology to effectively improve the diagnosis accuracy and efficiency. We propose an end-to-end solution based on a deep learning method. This method uses a variety of convolutional neural networks to generate feature maps, and then stitch the feature maps by channel to obtain good performance on our data set. In our experiments, the accuracy rate obtained by experimenting with only one feature map is best at 0.89, and then the feature maps are stitched using four convolutional neural networks to obtain 0.91.

Keywords: CNN · COVID-19 · X-Ray

1 Introduction

At present, deep learning has numerous backbone networks and skills to improve network models for image classification. If these network models can be flexibly applied to the classification of new coronaty pneumonia to rationally allocate patients to the corresponding wards, this artificial intelligence method can effectively reduce the workload of doctors. Normally, the edge of the lungs of normal people is clearly visible, and as the virus becomes more serious, there will be blurring, even white lungs. But for some whitish X-ray, it may also be due to other conditions such as wearing heavy clothes or poor coloring of the device. This problem can be solved by obeying the normalization of the Gaussian distribution. The main classification features of this project are lung spots, lung contours and other salient features. The main classification features of this project

Q. Liang et al. (eds.), *Artificial Intelligence in China*, Lecture Notes in Electrical Engineering 653, https://doi.org/10.1007/978-981-15-8599-9_58

are lung spots, lung contours and other salient features. The main process of its internal parameter adjustment operation is that the convolutional neutral network and the optimization function reverse transfer are automatically adjusted to obtain features, and then the features extracted by the convolutional layer are classified through full connection or global normalizaton. The overall framework is as shown in Fig. 1.[1]

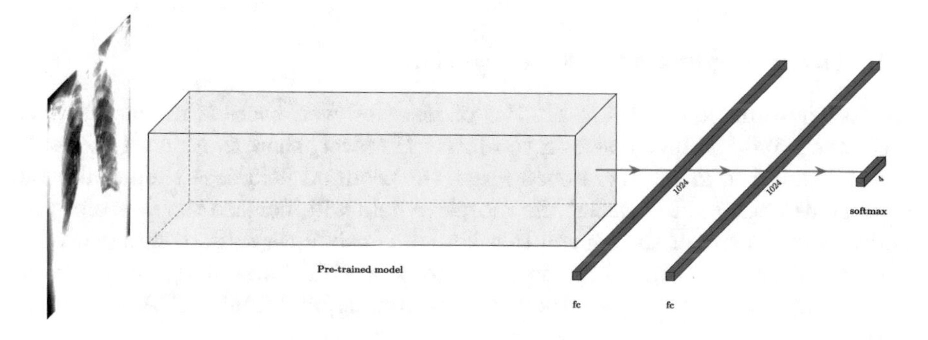

Fig. 1. System framework diagram

2 Related Work

Firstly, we summarize the current mainstream neural networks, and focus on the analysis of various related convolutional neural networks. Secondly, we deeply understand the reasons for the excellent performance of each mainstream convolutional neural network. Then, we combine the advantages of each network model and the skills to improve models to build a model suitable for this project. At the stage of comparing the main models, since the pre-trained models have been open sourced on some websites at home and abroad and a large amount of data has been given for feature extraction, what needs to be done is to load the training model into this project for X-ray four-category classification. The pre-trained models used are: VGG16, VGG19, ResNet50, InceptionV3, InceptionResNetV2, Xception, MobileNet, MobileNetV2, DenseNet121, DenseNet169, Dense201, ResNet101, ResNet152, ResNet101V2, ResNet152V2 [2–6]. Only part of these network models for feature extraction are adopted, and, and then the extracted features are classified via full connection or full convolution. After obtaining the best-performing network model through comparison, based on the model, the current mainstream techniques for improving the network model are used to construct the final model.

[1] The COVID-19 pandemic, also known as the coronavirus pandemic, is an ongoing pandemic of coronavirus disease 2019 (COVID-19), caused by severe acute respiratory syndrome coronavirus 2 (SARS-CoV-2). The outbreak was first identified in Wuhan, China, in December 2019.

3 Data

Different medical institutions use different equipment, which results in sharpness of the X image and increases the difficulty of extracting features[7]. And the original image is too large, which will affect the speed of subsequent network model training and the accuracy of the model. This project will adopt two preprocessing schemes, which are downsampling scheme and data enhancement.

3.1 Downsampling and Normalization

The original image are $224 \times 224 \times 3$ pixels. However, there is no need to use 3 channels and the image size in too large. However, there is no need to use 3 channels and the image size in too large, the required field area should also be as large as possible. Increasing the receptive field will increase the depth of the network or the size of the convolution kernel, which is very disadvantageous for training. First, the original image needs to be scaled down by a certain ratio, followed by feature averaging, then PCA whitening, and finally ZCA whitening [8,9].

3.2 Data Augmentation

In fact, there are not 10000 pictures in the data set. Therefore, a lot of data enhancement work needs to be done. There are many methods for data enhancement, such as adding some light to the image randomly, using adaptive histogram equalization to change the contrast, sharpen, bump, blur, jitter, random hue, brightness, transform, etc., which make the data more reliable [10].

4 Architecture

4.1 Convolutional Neural Network

Image classification refers to distinguishing images of different categories according to the semantic information of the images. Before 2012, machine learning methods were popular in image classification projects, that is, features were extracted from images for encoding, and machine learning methods were used for classification [11].

At present, the X-ray film detection based on deep learning uses CNN as the feature extractor on the front end, and uses a traditional classifier such as SVM on the back end [4]. This method is more troublesome to train and the detection speed is slower. If it is only for classification, an end-to-end network is needed to directly classify the input medical image of the new coronary pneumonia disease in order to obtain a double improvement in speed and accuracy. There is a big difference between medical images and general images. Applying the existing detection directly cannot achieve good results, so how to improve the existing detection network model to make it suitable for the field of medical image analysis is a key. Secondly, in the medical field, missed detections are often more fatal than false detections, but generally the two are a relative relationship. If the missed detection rate is low, the false detection rate is high, and vice versa.Therefore, how to make the missed and false detections reach an acceptable range is also a key.

4.2 Training

The backbone network is very numerous,and each backbone network has its own advantages. So this project intends to use multiple neural networks in parallel to broaden the structure of the network and make the accuracy of feature extraction higher. Then it is spliced from a variety of modules, and set up 4 networks to get the feature map as h_1, h_2, h_3, h_4. The input of the connection layer is a splicing vector generated by several network models and has the following form (Fig. 2).

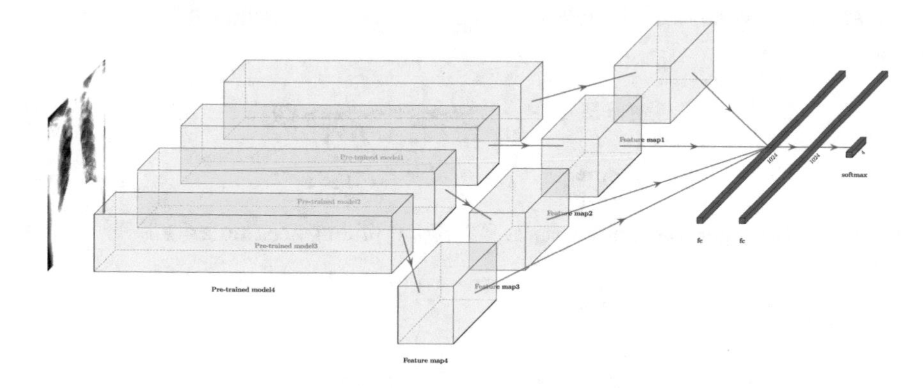

Fig. 2. System framework diagram

$$x_l = H_l(h_1, h_2, h_3, h_4) \quad (1)$$

where x_l is the output feature layer, $h_l(\cdot)$ is to splice the feature vectors generated by the four network models.

In order to prevent the wrong labeling of the image from affecting the model parameters, the use of label smoothing for model regularization can effectively solve the problem. The traditional cross-entropy loss function is calculated as follows.

$$l = -\sum_{k=1}^{K} log(p(k)q(k)) \quad (2)$$

For a certain category, its label $k \in \{0, 1\}, p(k)$ is the model to label k calculated probability, $q(k)$ is the true distribution of label samples.

Label smoothing is based on the true distribution of the label to make its input is no longer pure 1 or 0, but a number close to 1, such as 0.9. The negative label is a number close to 0 to achieve label smoothing. Suppose there is a real label y make $q(y) = 1$, while label k $q(k) = 0$.For this expression, the Dirac function can be used, $\delta_{k,y}$ to indicate: when k=y, $\delta_{k,y} = 1$,otherwise, $\delta_{k,y} = 0$.Set an error rate ϵ, indicating the probability of a certain label error, $u(k)$ represents a uniform distribution, which is calculated using the following formula $q(x)$.

$$\hat{q}(k) = (1 - \epsilon)\delta_{k,y} + \epsilon\frac{1}{k} \tag{3}$$

If the label is true, then $(1 - \epsilon)\delta_{k,y} = 1 - \epsilon$. Because ϵ it is usually very small, so $\epsilon\frac{1}{k} \rightarrow 0$ eventually $\hat{q}(k) \rightarrow 1$. When the label is false, it can also be pushed $\hat{q}(k) \rightarrow 0$.Finally, use $\hat{q}(k)$ instead to $q(k)$ get the final loss.

However, the learning rate should not be set in stone. Each parameter should have a suitable learning rate, so that the dimensions of the parameters that have little change in gradient direction are updated quickly. Therefore, the RMSPorp algorithm adds a cumulative term, so that the historical gradient can also participate in the change of the learning rate with a certain weight [8]. The formula is as follows:

$$\theta_{t-1} = \theta_t - \frac{1}{\sqrt{\gamma \cdot V_t + (1 - \gamma) \cdot g_t^2}} \cdot g_t \tag{4}$$

$$v_t = \gamma \cdot V_{t-1} + (1 - \gamma) \cdot g_{t-1}^2 \tag{5}$$

where γ is the hyperparameter, v_t is 0 to t-1 gradient accumulation.

5 Result

In our experiment, the feature extraction part uses a variety of networks to obtain pre-trained models in the ImageNet dataset. First, use an end-to-end network that uses only one pre-trained model on our self-made data set. These models have also achieved good results? Especially the Xception pre-training model. The gradient descent loss graph of Xception is as follows, Fig. 3.

Due to the small amount of data, Xcepiong has basically converged after 10 epochs and has achieved good performance.

5.1 Single Network Structure

Our experiments were conducted in different pre-trained network models: VGG16, VGG19, ResNet50, InceptionV3, InceptionResNetV2, Xception, MobileNet, MobileNetV2, DenseNet121, DenseNet169, Dense201, ResNet101, ResNet152, ResNet101V2, ResNet152V2 [12]. Our experimental results are shown in Table 1. It can be seen from the table that the VGG network cannot extract features well for this data, and the loss of the DenseNet network is very low due to overfitting, but the recall rate performance is very poor and is not applicable to this model. Finally, the Xception network is most suitable for image datasets among the methods using only one network model.

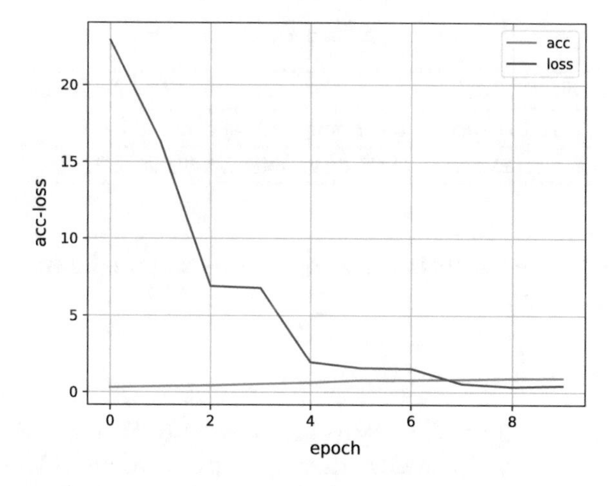

Fig. 3. The gradient descent loss graph of Xception

Table 1. Single network structure

Model	Loss	Accuracy	Recall rate
VGG16	2.678	0.621	0.237
VGG19	2.032	0.673	0.365
ResNet50	1.765	0.731	0.570
InceptionV3	0.598	0.800	0.671
InceptionResNetV2	1.175	0.792	0.701
Xception	0.098	0.937	0.754
DenseNet121	0.057	0.850	0.433
DenseNet169	0.009	0.897	0.278
ResNet101	1.008	0.790	0.751
ResNet152V2	0.056	0.801	0.698

5.2 Multi-network Structure

Our team also conducted corresponding experiments on the multi-network parallel method and obtained better results than the single network structure (Table 2).

6 Conclusion

Nowadays, the Coronavirus Disease (COVID-19) has broken out in various countries, and the lack of doctor strength has led us to urgently need to use artificial intelligence methods to share the X-Ray disease classification. In our framework, we use multiple well-known convolutional neural networks. We first extract feature maps, then stitch feature maps, and finally classify them. The framework

Table 2. Multi-network structure

Model	Accuracy
VGG16+Res50+Inceptionv3+Xception	0.907
Dense121+Res152v2+Inception+Xception	0.914

has obtained effective results in actual operations and can be applied to modern medical image recognition.

7 Future Work

In future work, we will use Generative Adversarial Networks (GAN) to expand the data set in order to obtain more data. At the same time, GAN can make the similarity between categories greater to improve the accuracy of the model, and we will use adaptive allocation of weights during the network splicing phase to splice the feature map of each pre-trained model.

Acknowledgments. This work was supported in part by the Tianshan Young Talent Program, Xinjiang Uygur Autonomous Region under Grant 2018Q024, in part by the Natural Science Foundation of China under Grant 61771089 and Grant 61961040, and in part by the Regional Cooperative Innovation Program of Autonomous Region (Aid Program of Science and Technology to Xinjiang) under Grant 2020E0247 and Grant 2019E0214.

References

1. Bengio Y, Courville A, Vincent P (2013) Representation learning: a review and new perspectives. IEEE Trans Pattern Anal Mach Intell 35(8):1798–1828
2. He K, Zhang X, Ren S, Sun J (2016) Deep residual learning for image recognition. In: Proceedings of the IEEE conference on computer vision and pattern recognition, pp 770–778
3. Hu J, Shen L, Sun G (2018) Squeeze-and-excitation networks. In: Proceedings of the IEEE conference on computer vision and pattern recognition, pp 7132–7141
4. Jayasundara V, Jayasekara S, Jayasekara H, Rajasegaran J, Seneviratne S, Rodrigo R (2019) Textcaps: handwritten character recognition with very small datasets. In: 2019 IEEE winter conference on applications of computer vision (WACV). IEEE, New York, pp 254–262
5. Lopes UK, Valiati JF (2017) Pre-trained convolutional neural networks as feature extractors for tuberculosis detection. Comput Biol Med 89:135–143
6. Mahajan D, Girshick R, Ramanathan V, He K, Paluri M, Li Y, Bharambe A, van der Maaten L (2018) Exploring the limits of weakly supervised pretraining. In: Proceedings of the European conference on computer vision (ECCV), pp 181–196
7. Novel et al (2019) Coronavirus Pneumonia Emergency Response Epidemiology. The epidemiological characteristics of an outbreak of 2019 novel coronavirus diseases (covid-19) in China. Zhonghua liu xing bing xue za zhi= Zhonghua liuxingbingxue zazhi 41(2):145 (2020)

8. De Sousa Ribeiro F, Leontidis G, Kollias S (2019) Capsule routing via variational Bayes. arXiv preprint arXiv:1905.11455
9. Simonyan K, Zisserman A (2014) Very deep convolutional networks for large-scale image recognition. arXiv preprint arXiv:1409.1556
10. Szegedy C, Liu W, Jia Y, Sermanet P, Reed S, Anguelov D, Erhan D, Vanhoucke V, Rabinovich A (2015) Going deeper with convolutions, pp 1–9
11. Touvron H, Vedaldi A, Douze M, Jégou H (2020) Fixing the train-test resolution discrepancy: Fixefficientnet. arXiv preprint arXiv:2003.08237
12. Fan W, Zhao S, Bin Yu, Chen Y-M, Wang W, Song Z-G, Yi H, Tao Z-W, Tian J-H, Pei Y-Y et al (2020) A new coronavirus associated with human respiratory disease in china. Nature 579(7798):265–269
13. Guan W, Ni Z, Hu Y, Liang W, Ou C, He J, Liu L, Shan H, Lei C, Hui DSC, et al (2020) Clinical characteristics of coronavirus disease 2019 in China. New England J Med 382(18):1708–1720
14. Tan M, Le QV (2019) Efficientnet: Rethinking model scaling for convolutional neural networks. arXiv preprint arXiv:1905.11946
15. Wei L, Xiao A, Xie L, Chen X, Zhang X, Tian Q (2020) Circumventing outliers of auto augment with knowledge distillation. arXiv preprint arXiv:2003.11342
16. Wong C, Houlsby N, Lu Y, Gesmundo A (2018) Transfer learning with neural AutoML. In: Advances in neural information processing systems, pp 8356–8365
17. Xie Q, Hovy E, Luong M-T, Le QV (2019) Self-training with noisy student improves imagenet classification. arXiv preprint arXiv:1911.04252
18. Zeiler MD, Fergus R (2014) Visualizing and understanding convolutional networks, pp 818–833

The Shortest Path Network Rumor Source Identification Method Based on SIR Model

Zhongyue Zhou, Hai-Jun Zhang(✉), Weimin Pan, Bingcai Chen, and Yanjun Li

School of Computer Science and Technology, Xinjiang Normal University, Urumqi 830054, China
zhj1p@163.com

Abstract. The rapid and effective identification of rumor sources are of great significance to reduce the harm of false information spreading. However, it is known to all that the identification of rumor sources is very difficult, because information is dynamic and rapid in the process of dissemination, and it is explosive and diffusible. Based on the SIR infectious disease model, this paper uses the betweenness center score to identify the rumor source. The feasibility of the proposed method is verified by a real-network simulation experiment. And the results show that the proposed method can achieve good results in rumor source identification.

Keywords: Rumor source · SIR model · Rumor centrality · MLE

1 Introduction

In recent years, with the rapid development of modern communication technology, the influence of social media on people has become more and more important. Social media systems provide people with convenient tools to analyze information, communicate, and express their personal opinions. Online social networking platforms have a large number of users and are widely used in people's life and study. However, the rapid spread of information also brings potential harm, such as false information, rumors, and so on, will also spread quickly and widely.

Since information spreads quickly and widely, it is of great significance to identify information sources effectively for the dissemination of social networks. The spread of Internet rumors is similar to the spread of infectious diseases in the real world, both of which spread based on a certain point or multiple points. Therefore, researches and methods on rumor sources are mostly based on three basic models of spreading disease, such as SI [1], SIS [2], and SIR [3]. In recent years, researchers have done a lot of research on the spread of rumor information, and the problem of rumor tracing and identification has attracted researchers in different fields. Based on the complex network center index, this paper proposes a shortest path approach to identify the rumor source.

The rest of this paper is arranged as follows: the second part mainly introduces the research work related to network rumor source identification and some mainstream

Q. Liang et al. (eds.), *Artificial Intelligence in China*, Lecture Notes
in Electrical Engineering 653, https://doi.org/10.1007/978-981-15-8599-9_59

methods at present; the third part describes the algorithm model and its principle of information source location; the fourth part carries on the simulation experiment to the proposed method and carries on the analysis to its result. Finally, the research work is summarized, and the future work is discussed.

2 Related Works

The study of rumor sources is of great significance to the dissemination of information. The identification of rumor sources can timely and effectively block the spread of unconfirmed information and reduce the harm to the society and people's life. So far, there are mainly two methods for rumor source identification [4]. One is to identify rumor source based on graph theory snapshot, and the other is to predict rumor source by deploying observation points.

Original source localization work is based on the network graph of detection methods, mainly the rumors centricity theory proposed by Shah and Zaman et al. [5] is the most influential. Under this study, assuming that the probability of each node in the network as the rumor source is the same, according to the spread of infected nodes subgraph computing node path probability of infection, so as to establish the likelihood function and compute nodes maximum likelihood function. The node with the highest likelihood function value is considered to be the information source point in the network. As shown in Eq. (1):

$$R(u, G) = \frac{|V|!}{\prod_{w \int V} T_w^u} \tag{1}$$

where T_w^u represents the number of nodes susceptible to infection in subspecies.

Since graph theory method needs to observe the whole network structure, which is difficult to realize in the actual network. Therefore, some researchers put forward the method of deployment for rumor source detection. Fang et al. [6] proposed a source location method consisting of an estimator based on correlation coefficient and a matrix that approximates the diffusion delay between nodes, aiming at the source location problem of asynchronous diffusion process in online social networks. The improved breadth first algorithm, the weighted length algorithm of minimum hops path, was used and verified by numerical simulation. Hu [7] proposed an optimized greedy algorithm for rumor source detection, which reduced the number of observation points.

Kumar et al. [8] predicted the rumor source based on rumor snapshot and proposed a possible estimator to infer the rumor source based on Markov chain tree theorem. The method combined Aldous algorithm and carried out simulation experiments using SI model. In this study, we adopt a shortest path algorithm based on intermedium centrality and use likelihood function to identify rumor sources.

3 Source Location Algorithm

This paper mainly studies the information source location in social networks. To be specific, we first focus on the nodes higher in the center, then calculate the shortest infection path, optimize the deployment of observation points, and finally calculate the likelihood function value to estimate the rumor source.

3.1 Problem Formulation

We consider a network represented by a finite undirected graph $G = (V, E)$. Where, node $v \in V$ represents the user in the network, node $e \in E$ represents the relationship between users, and information propagation is carried out in G. We assume that an unknown rumor source node $v* \in V$ randomly selected starts to diffuse at an unknown time $t*$. We can observe the diffusion process through a set of nodes called observable nodes. The task is to locate the source by monitoring a limited number of observable nodes. At the same time, it is reasonable to assume that the rumor reaches the sensor node via. the shortest path from the source $v*$. The rumor propagation model is SIR [3].

Assuming the rumor has been spread in the network, we build a new gateway of node G_I (V, E) expressed as the subgraph of $G(V, E)$. $\hat{v}$ represents the estimated source of rumors, and $G_I \in G$, $\hat{v} \in v^*$. Meanwhile, a set of sensor nodes $S = \{s_1, s_2, \ldots s_k\}$ is defined; the sensor is selected from node V. As we lack the prior knowledge of the source of infection, we assume that each node is a rumor source with equal probability, so maximum-likelihood estimator (MLE) algorithm is used to estimate the rumor source. This algorithm calculates the likelihood estimate based on the rumor center score.

$$\hat{v} = \arg\ \max\ \rho\left(\mathrm{G}_I | v^* = V\right) \tag{2}$$

where ρ $(G_I|v)$ is the probability of G_I observed under the propagation model SIR. Assuming that $v \in v^*$, therefore, we only need to find all of the $v \in G_I$ for the likelihood function value of $\rho(G_I|v)$ and then select the highest value.

3.2 Rumor Centrality

Rumor centrality was first proposed in the literature [5]. Rumor centrality is widely used in the research of rumor source prediction. Due to different propagation modes in real network, node v has multiple propagation paths in network G_I, and the number of paths is proportional to $\rho(G_I|v)$. Its formulation is as follows:

$$R(u, G_I) = \frac{|v|!}{\prod_{w \in V} T_w^{v_j}} \tag{3}$$

where T_w^u represents the number of nodes susceptible to infection in the network neutron tree. Suppose that the source point v^* has n neighbor nodes $\left(v_1^*, v_2^*, v_3^*, \ldots, v_n^*\right)$. Nodes in the subtree judge the propagation of a node based on whether its neighbor node receives the rumor information. Slots that allow one location in a node arrangement, the first location being the source by default. Where the permutation number of G_I tree is reduced to the following formula:

$$\begin{aligned} R(u, G_1) &= \frac{|V|!}{\prod\limits_{w \varepsilon V} T_w^u} = (I-1)! \prod_{i=1}^{n} \frac{R\left(v_i, T_{v_i}^v\right)}{T_{v_i}^v!} \\ &= (I-1)! \prod_{i=1}^{n} \frac{\left(T_{v_i}^v - 1\right)}{T_{v_i}^v} \prod_{v_{ij} \in T_{v_i}^n}^{n} \frac{R\left(v_{ij} - T_{v_i}^v\right)}{T_{v_i}^v} \end{aligned}$$

$$= (I-1)!\prod_{i=1}^{n}\frac{R(T_{v_i}^{v}-1)}{T_{v_i}^{v}}\prod_{v_{ij}\in T_{v_i}^{n}}\frac{R\left(v_{ij},T_{v_{ij}}^{v}\right)}{T_{v_i}^{v}!}$$

$$= (I-1)!\prod_{i=1}^{n}\frac{1}{T_{v_i}^{v}!}\prod_{v_{ij}\in T_{v_i}^{n}}\frac{R\left(v_{ij},T_{v_{ij}}^{v}\right)}{T_{v_i}^{v}!}$$

$$= I!\prod_{u\in G_i}\frac{1}{T_u^{v}}$$

For a better understanding, as shown in Fig. 1, let the network node be 1, then the number of nodes in the network $|V| = 5$, $T_1^1 = 7$, $T_2^1 = 5$, $T_3^1 = T_4^1 = T_6^1 = T_7^1 = 1$. The rumor centrality of node 1 can be obtained from the above formula: $R(1, G_I) = \frac{7!}{7\times5\times3\times1} = 48$.

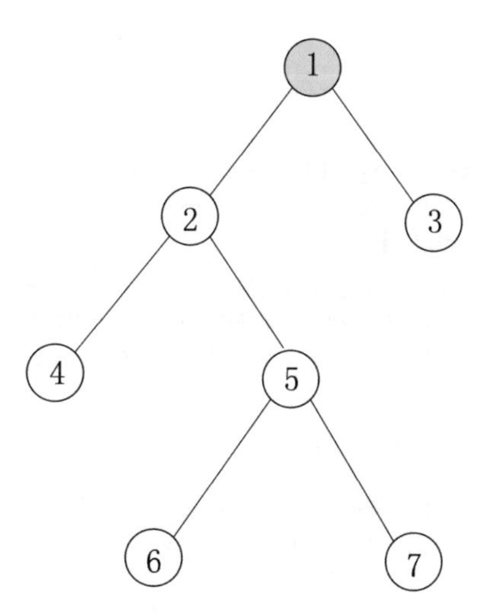

Fig. 1. Example diagram

3.3 Methods for Locating Rumor Sources

Based on the above, rumor centrality is an important index to measure the strength of network relationship. Different propagation paths of the network are composed of nodes in the network. Therefore, we assume that the shortest path enters into the network in the process of information transmission, the most appropriate sensor node should be the node with betweenness centrality (BC), shown in the formula (4).

$$C_b(V) = \sum_{s\neq v\neq t\in V}\frac{g_{st}^{x}(v)}{g_{st}} \tag{4}$$

where g_{st} represents the shortest path from node v_s to node v_t in the network, that is the number of shortest paths of $s \to t$. g_{st}^i represents the number of g_{st} shortest paths through a node v_x in a shortest path from node v_s to node v_t. The higher the value of a node in the network, the more important and influential it is. Meanwhile, we considered the order of intermediate number centrality, sorted the node according to its BC value as betweenness centrality rank (BCR), marked the highest score as 1, and optimized the deployment of the watch point scheme combining with the ordering of BCR. Finally, the likelihood function value of the above MLE algorithm was used to estimate the rumor source in the network..

4 Experiment and Analysis

4.1 Experimental Configuration

All of our experiments were based on python 3.6, and the editing tool used the PyCharm platform.

4.2 The Data Set

Next, we conducted simulation experiments on three large real-world networks. In order to ensure the reliability of the experiment, we ran the simulation experiment independently for 50 times. The total number of the cyclic process is denoted as y, where the number of the correct rumor source prediction is denoted as s, the probability of successful rumor prediction can be expressed as θ, and the formula can be expressed as $\theta = \frac{y}{s}$. The detailed information of the data set used in the experiment is shown in Table 1

4.3 Experimental Results and Analysis

In order to analyze the effectiveness of the algorithm, in this paper, the back propagation algorithm mentioned in method [9] is compared with the shortest path algorithm based

Table 1. Description of network data set

Network	|V|	|E|	E/V	Description
Power grid [10]	4941	6594	1.33	The network representing the topology of the Western States power grid of the United States
Wikipedia vote network [11]	7115	103689	14.57	The network contains all the Wikipedia voting data from the inception of Wikipedia till January 2008
[a]Autonomous systems	10670	22002	2.06	(AS) graphs of autonomous systems (AS) peering information inferred from Oregon route-views between March 31, 2001 and May 26, 2001

Data sets: http://snap.stanford.edu/data/index.html

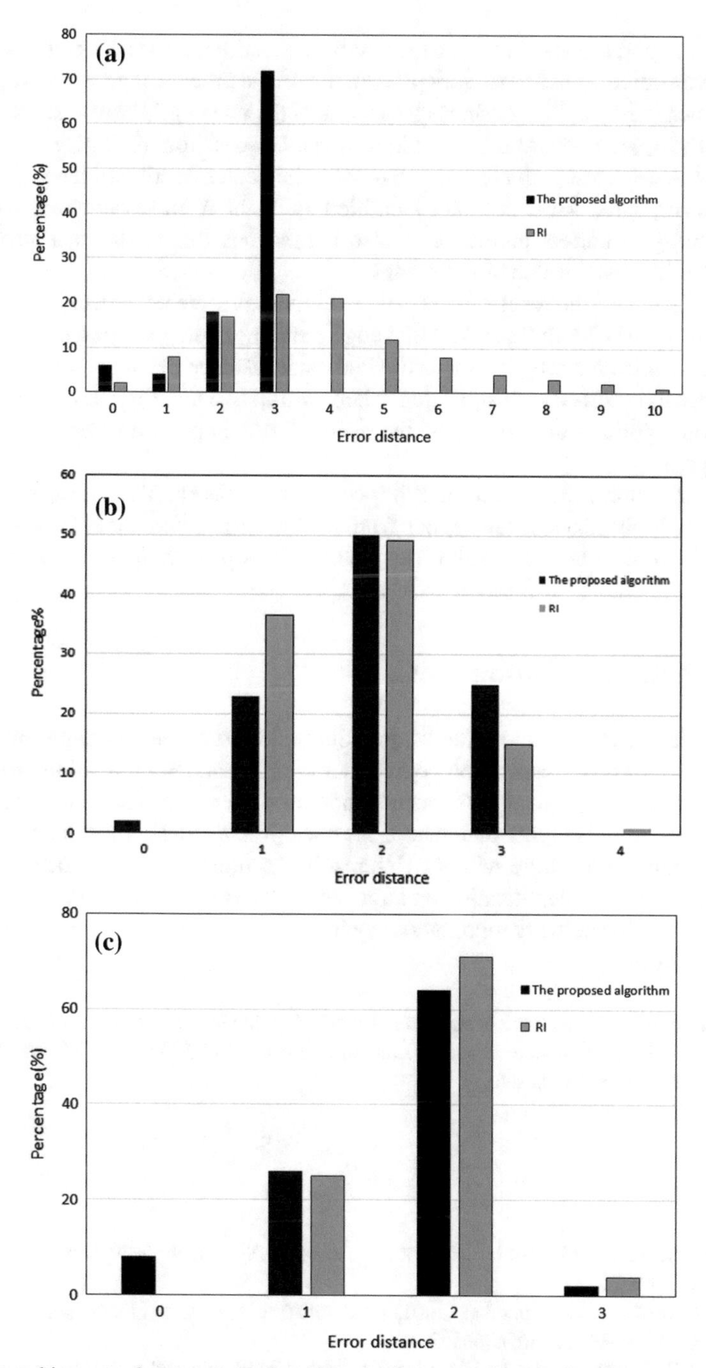

Fig. 2. Error histogram between two algorithms in three actual networks, **a** power grid network, **b** internet autonomous systems network, **c** Wikipedia vote network. The detailed introduction of the data set is shown in Table 1

on central fraction proposed in this paper. At the same time, a node is randomly selected as the rumor source point for network propagation in the process of network propagation.

As shown in Fig. 2, Fig. 2a shows the results of the power grid network. The network has a total of 4,941 nodes and 6,594 edges. It can be seen from the figure that the error distance of the shortest path algorithm based on the center fraction of the grid is within three hops compared with the rumor identified by the back propagation algorithm, and the probability of correct prediction is also higher than that of the back propagation algorithm, with a peak value of three hops.

Figure 2b shows the results of Internet autonomous systems network. The network is composed of 10,670 nodes and 22,002 edges, and the average degree of nodes is low. As can be seen from the figure, the estimated error distance of rumor source identified by our proposed shortest path algorithm is kept within two hops by 80%, while the back propagation algorithm does not show the result of zero hops, and its peak error is kept at two hops.

Figure 2c shows the results of Wikipedia vote network. The network has 7,115 nodes and 103,689 edges. It can be seen from the figure that the error of the shortest path algorithm is kept above 90% within two hops, and its peak value is two hops, among which the prediction probability is nearly 10%.

5 Concludes and Future Work

In this paper, we mainly adopt the rumor source detection method based on the central score optimization observation point under the SIR model. Simulation experiments through three large real networks, results show that the rumor recognition rate of the improved algorithm is higher than that of the back propagation algorithm, and the error distance remains above 80% within two hops. In the future work, our focus will be on the research of rumor identification on large networks. How to consider the relationship between users in the network more rationally is of great significance for the identification of rumor sources.

Acknowledgements. This work is supported by National Science Fund of China-Xinjiang Joint Fund (U1703261) and Graduate Research Innovation Project (XJ2019G231). We thank Stanford University for their open data sets.

References

1. Hurley M, Jacobs G, Gilbert M (2006) The basic SI model. New Direct Teach Learn 2006(106):11–22
2. Pastor-Satorras R, Vespignani A (2001) Epidemic dynamics and endemic states in complex networks. Phys Rev E 63(6):066117
3. Moreno Y, Pastor-Satorras R, Vespignani A (2002) Epidemic outbreaks in complex heterogeneous networks. Eur Phys J B-Condens Matter Complex Syst 26(4):521–529
4. Chen Y, Li Z, Liang X, Qi J (2018) Summary of online social network rumor detection. Chin J Comput 41(07):1648–1677

5. Shah D, Tauhid Z (2012) Rumor centrality: a universal source detector. In: Proceedings of the 12th ACM SIGMETRICS/PERFORMANCE joint international conference on measurement and modeling of computer systems
6. Fang M et al (2018) Locating the source of asynchronous diffusion process in online social networks. IEEE Access 6:17699–17710
7. Hu ZL, Wang L, Tang CB (2019) Locating the source node of diffusion process in cyber-physical networks via minimum observers. Chaos Interdiscipl J Nonlin Sci 29(6):063117
8. Kumar A, Borkar A, Karamchandani N (2017) Temporally agnostic rumor-source detection. IEEE Trans Sign Inf Process Netw 3(2):316–329
9. Zhu K, Ying L (2014) Information source detection in the SIR model: a sample-path-based approach. IEEE/ACM Trans Netw 24(1):408–421
10. Watts DJ, Strogatz SH (1998) Collective dynamics of 'small-world' networks. Nature 393(6684):440–442
11. Leskovec J, Huttenlocher D, Kleinberg J (2010) Predicting positive and negative links in online social networks. In: Proceedings of the 19th international conference on World wide web, pp. 641–650

Microblog Rumor Detection Based on Bert-DPCNN

Yan-Jun Li, Hai-Jun Zhang(✉), Wei-Min Pan, Ru-Jia Feng, and Zhong-Yue Zhou

School of Computer Science and Technology, Xinjiang Normal University, Urumqi 830054, China
zhjlp@163.com

Abstract. At present, most of the rumor detection methods take the content of Weibo text as the main target of rumor detection. This study uses user information and Weibo text as the target to detect Weibo rumors, and the focus is on user information. A rumor detection model based on Bert [1] combined with DPCNN [2] method is proposed, which can process Chinese data more conveniently, extract the characteristics of user information more accurately, and introduce the evaluation standard as the final evaluation index. Finally, a microblog rumor detection system based on user information is constructed to make the rumor detection more accurate.

Keywords: User information · BERT · DPCNN · Rumor detection

1 Introduction

Sina Weibo, as a hot network communication tool, has the characteristics of a series of communication tools such as forums, blogs, and social networks. This is why Sina Weibo has always been popular among the public. This is also due to the development of the Internet. As of March 2020, the number of Internet users in China reached 854 million. The age of Internet users is not limited, and there are various educations and occupations, so that some people cannot distinguish the rumors well when the rumors are generated, which brings an excellent opportunity for the spread of rumors. Therefore, the majority of Weibo users will be disturbed by the spread of rumors while enjoying the open and shared information of Weibo.

Although the current information science community has not clearly defined and declared rumors, most studies believe that rumors are false information that is passed between users in the cyberspace, social circles, and other information environments and attracts public attention [3]. There will always be some fluky people on the Internet spreading rumors on Weibo, causing great damage to the network ecological environment. Therefore, the fast reacting Weibo rumors recognition model and system are still one of the research hotspots. Therefore, this paper studies the construction of rumor detection model based on user information and proposes the BERT-DPCNN model. The second part of this article introduces the domestic and foreign related work of rumor detection. The third part introduces the rumor detection method of BERT-DPCNN. The fourth part analyzes the experiment. The final part is the conclusion.

Q. Liang et al. (eds.), *Artificial Intelligence in China*, Lecture Notes in Electrical Engineering 653, https://doi.org/10.1007/978-981-15-8599-9_60

2 Related Works

The rumor detection model generally adopts the technology of machine learning. The general extracted features include three aspects: extracting rumor detection features from text content, user attributes, and time attributes. This article focuses on user information. User information on Weibo can be divided into editing personal description, uploading user avatar, adding personal homepage, user Weibo name, user profile, Weibo authentication of user, user gender, user city, user registration Weibo, the initial time of the blog, etc. In addition, the user profile also has attention, activity, etc. At present, the rumor detection at home and abroad is mainly carried out from traditional machine learning methods and deep learning methods.

In machine learning, it mainly uses Bayes, decision tree, support vector machine, hidden Markov model, and so on. Dayani et al. [4] used K-Nearest Neighbors (KNN) classifier and NB classifier to detect rumors in Twitter by extracting user features and content features. Al-Khalifa [5] et al., Gupta et al. [6] detect false pictures spread on Twitter, but the detection method is still to extract the text features and user features of the message separately and then use the decision tree classification to perform. Compared with the test, the research result is that the test is effective. Ma et al. [7] proposed social context features based on the time series of the rumor life cycle, including microblog content features, user features, and propagation features and used linear SVM classifiers on the Twitter data set and DT, Random Forest (Random Forest, RF) and SVM-RBF method for comparison. The experimental results show that the accuracy of the proposed method takes less time. Xie Bolin et al. [8] proposed a method for early detection of false information on Weibo based on the behavior of gatekeepers. This method uses the hidden semi-Markov model whose model state duration probability is Gamma distribution to characterize the checkup behavior of information forwarders and commenters on popular real information and to identify the false information popular on Weibo as early as possible.

In terms of deep learning, it mainly uses methods such as recurrent neural network, convolutional neural network, long-term and short-term memory network, and attention. Chen et al. [9] used the features extracted from user comments and introduced the attention mechanism on the recurrent neural network to improve the accuracy of rumor detection. Ruchansky et al. [10] proposed a CSI (capture, core, integrate) model, which includes three modules: The first module uses RNN to capture the immediate mode of user behavior, and the second module captures the long-term mode of user behavior., Integrate these two modules into a third module to classify the messages falsely. Li [11] and others used a shared long-term short-term memory (LSTM) layer to learn a set of common functions related to two tasks, and each task can also learn its task-specific functions through its specific layer. The user credibility information is integrated into the rumor detection layer, and the attention mechanism is applied during the rumor detection process, which can get a good classification effect.

Although the above-mentioned user-based rumor detection can all achieve good results, for the growing demand for rumor detection, the existing user information rumor detection algorithms still have some shortcomings, mainly manifested in the following aspects: In the predecessor's rumor detection method, the user's characteristics are described as very important clues. When the predecessors used the user's characteristics to detect the rumor, the processed data did not pay attention to the relationship between them, and the processing efficiency can also be improved again. Therefore, considering the rumor detection from the above perspective, this paper mainly studies the Weibo rumor detection from the characteristics of user description, user profile, and so on.

3 Rumor Detection Model

In order to improve the correct rate of rumor detection, by analyzing user information in Weibo, referring to the relevant rumor detection methods at home and abroad, a fusion model of BERT-DPCNN is proposed.

3.1 BERT-DPCNN

BERT-DPCNN model uses BERT pre-trained model to process text data. One advantage of BERT is that it can directly process Chinese data. The advantage of DPCNN is that it can extract long-distance text dependencies, use DPCNN to extract user information and text features, and finally link the fully connected layer to output the classification results. The model structure diagram is shown in Fig. 1.

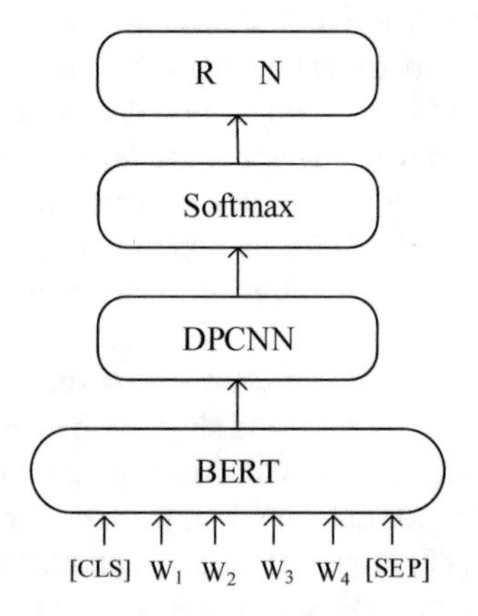

Fig. 1. BERT + DPCNN model structure diagram

3.2 User Information Extraction

According to the data set in Ma and the information in the crawled data set, user information is extracted by dictionary method, as shown in Table 1.

Table 1. User information

User information	Feature	Description
	Nickname	User's Weibo name
	Profile	User's profile
	Authentication	Weibo's authentication of users
	Gender	User's gender
	Location	User's city
	Registration time	The initial time for users to register on Weibo

3.3 User Information Into the Model

The frame diagram of the model is shown in Fig. 2.

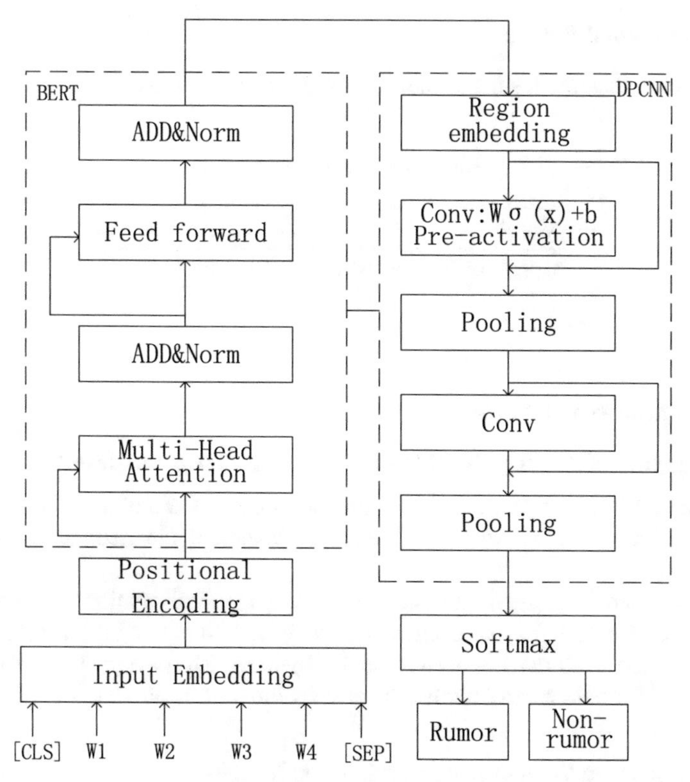

Fig. 2. BERT-DPCNN frame diagram

The workflow of integrating user information into the BERT-DPCNN model:

(1) Process the data set and divide the data set into non-rumor Weibo and rumor Weibo.

(2) Clean the data, remove useless content such as Weibo, and stop words.
(3) Extract user information
(4) Construct a data set containing user information
(5) Train the BERT-DPCNN model
(6) Verify the model.

4 Experimental Results and Analysis

4.1 Dataset Processing

Using Ma's data set, directly extract the user's nickname, profile, description, location and other user information, and text content, remove stop words, special characters, and divide the data set into training according to the ratio of 7:2:1 set, test set, and validation set. Among them, there are 2313 rumors and 2351 non-rumors. The total number of Weibo is 3805656.

4.2 Experimental Results

Verify the classification results of the rumors, as shown in Table 2.

Table 2. Experimental results

Model	Recall	F1-score	Accuracy
BERT-DPCNN	89.91%	88.62%	87.45%

4.3 Comparative and Analysis

4.3.1 Influence of Adding User Information on Experiment Results

Taking user information as the main feature, adding it to the model and not adding the model to do a comparative experiment, the experimental comparison results are shown in Fig. 3.

It can be seen from the figure that after adding user information to the model, the accuracy of rumor detection has been increased by 2.2%, the F1 value has been increased by 4.8%, and the recall rate has been increased by 3.7%. The data in the table shows that adding user information can improve the effectiveness of the model.

4.3.2 Comparison With Other Experimental Models

Comparing this method with other methods, the experimental results are shown in Table 3.

It can be seen from the table that the results obtained by the model in this paper are the best. Through these comparative experiments, the effectiveness of BERT-DPCNN can be proved.

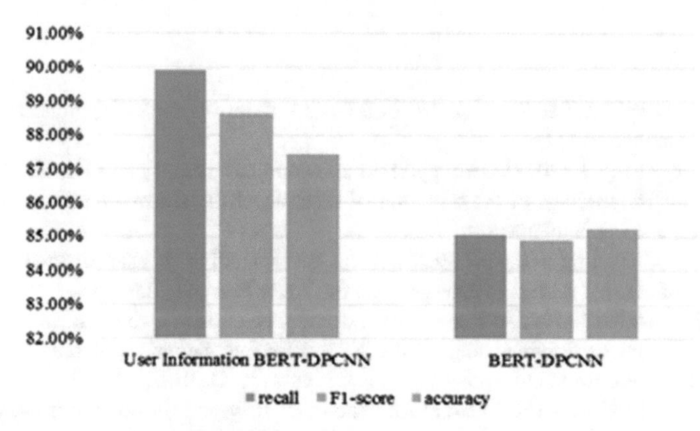

Fig. 3. Influence of user information on the model

Table 3. Comparison of experimental results of different models

Model	Recall (%)	F1-score (%)	Accuracy (%)
SVM	88.5	86.1%	83.9
RNN	96.4	88.4	81.6
LSTM	96.8	91.3	84.6
BERT-DPCNN	89.91	88.62	87.45

5 Conclusion and Future Works

This paper uses the combination of BERT and DPCNN models. Combining the advantages of the two models, BERT can handle the context more efficiently. Not only did BERT solve the problem of long-distance dependencies, but also DPCNN set up the model of the long-distance dependencies in the text, so the combination of the two models processed data and extracted user information more accurately. The experimental results obtained show that the accuracy of the model reaches 87.45%, which is better than other models and the model is effective. However, the model still has limitations, and the accuracy can be further improved. I hope to find a better way to improve the accuracy of the model again.

Acknowledgements. This work is supported by National Nature Science Foundation of China-Xinjiang Joint Fund (U1703261) and Graduate Research Innovation Project (XJ2019-G231). Special thanks to Ms. Ma Jing, the Chinese University of Hong Kong for publishing the rumor detection data set.

References

1. Devlin J, Chang MW, Lee K et al (2018) BERT: pre-training of deep bidirectional transformers for language understanding
2. Johnson R, Zhang T (2017) Deep pyramid convolutional neural networks for text categorization. In: Proceedings of the 55th annual meeting of the association for computational linguistics, vol 1: Long Papers
3. Yuliang Z, Chuanling J (2018) Research on the spreading mechanism and governance strategies of internet rumors in emergencies. Inf Theor Pract 41(5):91–96
4. Dayani R, Chhabra N, Kadian T et al (2015) Rumor detection in Twitter: an analysis in retrospect. In: ANTS 2015: proceedings of the 2015 IEEE international conference on advanced networks and tele communications systems. Piscataway, NJ: IEEE, pp 1–3
5. Al-Khalifa HS, Al-Eidan RM (2011) An experimental system for measuring the credibility of news content in Twitter. Int J Web Inf Syst 7(2):130–151
6. Gupta A, Kumaraguru P et al (2012) Credibility ranking of tweets during high impact events. In: Proceedings of the 1st workshop on privacy and security in online social media, Lyon, France, pp 2–8
7. Ma J, Gao W, Wei Z et al (2015) Detect rumors using time series of social context information on microblogging websites. In: Proceedings of the 24th ACM international on conference on information and knowledge management. Melbourne, Australia: ACM, pp 1751–1754
8. Xie B, Jiang A, Zhou Y et al (2016) Early detection method of false information on Weibo based on gatekeeper behavior. J Comput 39(4):730–744
9. Chen, Weiling, Zhang et al (2018) Unsupervised rumor detection based on users' behaviors using neural networks. Patt Recogn Lett
10. Ruchansky N, Seo S, Liu Y (2017) CSI: a hybrid deep model for fake news detection. In: Proceedings of ACM on conference on information and knowledge management, Singapore, pp 797–806; Liang G, He W, Xu C et al (2015) Rumor identification in microblogging systems based on users' behavior. IEEE Trans Comput Soc Syst 2(3): 99–108
11. Li Q, Zhang Q, Si L (2019) Rumor detection by exploiting user credibility information, attention and multi-task learning. ACL

A New Method of Microblog Rumor Detection Based on Transformer Model

Ru-Jia Feng, Hai-Jun Zhang(✉), Wei-Min Pan, Zhong-Yue Zhou, and Yan-Jun Li

School of Computer Science and Technology, Xinjiang Normal University, Urumqi 830054, China
zhjlp@163.com

Abstract. Traditional rumor detection methods rely on artificial features, which is inefficient and weak in generalization. Recurrent neural network has obvious advantages in processing sequential data, but gradient disappearance is difficult to solve. Aiming at the above problems, this paper proposes a microblog rumor detection method based on Transformer model. This method adopts the word embedding method of XLNet, extracts deep semantic features from microblog books through the encoder of Transformer, and then inputs the learned deep semantic features into Softmax layer to get the final classification result, and then realizes microblog rumor detection.

Keywords: Rumor detection · Transformer · XLNet

1 Introduction

Nowadays, we are living in an era of booming Internet. While providing convenience to people, all kinds of social networks also lead to the spread of rumors on social media, which seriously hinder people's access to the authenticity and reliability of information and even cause significant economic losses or public panic in emergencies. Therefore, it is of great urgency to construct an automatic detection model to identify rumors on social media as soon as possible in view of the possible adverse effects of rumors.

This paper proposes a rumor detection method based on Transformer model by studying the semantic information of Weibo data. The rest chapters of this paper are distributed as follows: The second part introduces the research status of rumor detection. The third part introduces the rumor detection method based on Transformer model. The fourth part analyzes the results of the experiment. Finally, the conclusion and future work are presented.

2 Related Works

In the early stages of rumor detection, a supervised learning approach was adopted to extract rumor characteristics from text content, user behavior, and propagation structure and establish a classifier. Based on Twitter data, Castillo et al. [1] screened 15

Q. Liang et al. (eds.), *Artificial Intelligence in China*, Lecture Notes
in Electrical Engineering 653, https://doi.org/10.1007/978-981-15-8599-9_61

most discriminative features from four aspects of content, user, dissemination, and topic and constructed the J48 decision tree classification model to achieve 86% accuracy in rumor recognition. Yang et al. [2] built a SVM classification model based on Sina Weibo data, combining content features and user features, and the model accuracy finally reached 78.7%. Takahashi et al. [3] based on the rumor events triggered by the Japanese tsunami on the Twitter platform extracted the rumor outbreak point, forwarding rate, and other propagation characteristics and built a rumor detection system based on the above characteristics, and achieved good results. However, the early rumor detection methods relied on manual extraction of features, which was time-consuming and labor-intensive. Moreover, the features of manual design generally had certain limitations and poor generalization performance.

To solve the above problems, the researchers propose to use deep neural networks to automatically learn high-level features to detect rumors. Ma et al. [4] used recurrent neural network to model reposts and capture hidden features from text content. Experiments show that the deep learning model can mine the hidden contents in the process of time change and has good classification performance. Yu et al. [5] used convolutional neural network to capture semantic features of text. Experiments show that the model can not only identify rumors effectively, but also perform well in the early detection of rumors. However, the problem of gradient disappearance and not easy parallelization exists in recurrent neural network. Although LSTM and GRU can alleviate the problem of gradient disappearance, they cannot solve it completely. The convolutional neural network is good at feature detection, but it has no memory function.

Aiming at the above problems, this paper proposes a microblog rumor detection method based on Transformer model. This method extracts deep semantic features from microblog text through Transformer encoder and then inputs the learned deep semantic features into Softmax layer to obtain the final classification result. This method can capture the deep semantic information of microblog, which is helpful to improve the performance of rumor detection model. In this paper, the Sina Weibo data set published by Ma et al. [4] in 2016 was used. Through the analysis of experimental results, the accuracy of this method reached 93.9%, which verified the effectiveness of the model.

3 Microblog Rumor Detection Model

3.1 XLNet-Based Pre-training Method

In this paper, a generalized autoregressive language model, XLNet, is used. First, the method maximizes the expected logarithmic likelihood of all possible factorization sequences to achieve bidirectional context learning. Secondly, XLNet does not rely on incomplete data. Therefore, XLNet will not have BERT [6] pre-training-fine-tuning difference. Finally, the goal of autoregression is to use the product rule to factor the predicted joint probability of token, which eliminates the independence hypothesis in BERT.

For example, given a rumor text $w = \{w_1, w_2, w_3, w_4\}$, to predict w_3. We use a sequence language to model the target, and the text sequence should be arranged in four! But in order to reduce the calculation amount, we can only randomly arrange the sampling parts, such as: $2 \to 3 \to 4 \to 1$, $3 \to 2 \to 4 \to 4 \to 4 \to 2 \to 2 \to 1 \to 3$. The following diagram shows the decomposition method as $1 \to 4 \to 3 \to 2$ (Fig. 1):

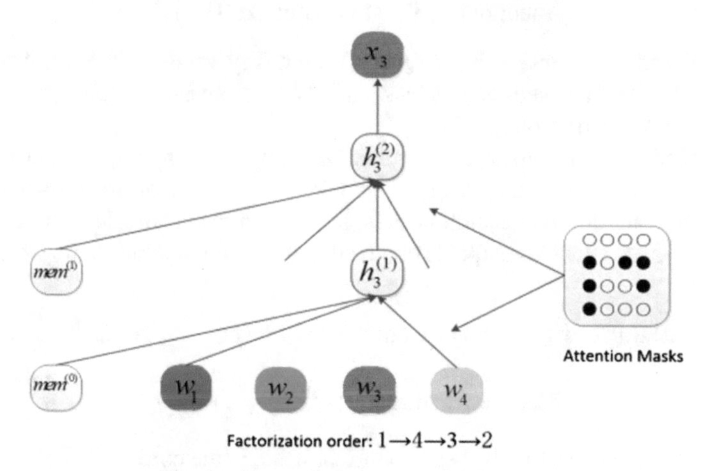

Fig. 1. Arrange language model diagrams

We can attend to w_1 and w_4 when predicting w_3, but we can't attend to w_2. When we have traversed through the above four permutations, we will have all the context information. In addition, it should be noted that our input is still in the order of the original sentence, while the arrangement is achieved through the attention mask. We can code w_3 so that it can attend to w_1, w_4, and mask w_2.

The final optimization goal is shown in formula (1):

$$\max_{\theta} E_{z \in Z_T}\left[\sum_{t=1}^{T} \log p_\theta(x_{zt}|X_{z \leq t-1})\right] \tag{1}$$

where Z_T represents the set of all permutations of a sequence with length T, and $z \in Z_T$ is a permutation method. Let's call it zt for the Tth element in the arrangement, and $z < t$ is the first to the $t-1$th element of z.

3.2 Transformer Encoder

The content of microblog contains clues to distinguish between rumor and non-rumor. There are certain differences between rumor and non-rumor in language expression and wording. Therefore, we use semantic information of microblog to identify rumors. The multi-head attention of Transformer model can solve this problem well, learning the long-distance dependency while parallel computing, and fully capturing the global semantic information of the input text.

We represent the microblog text of length n as $w = \{w_1, w_2, \ldots, w_n\}$, where w_i is the ith word in the microblog text w. Through preprocessing, the word vector is $x = \{x_1, x_2, \ldots, x_n\}$ x, which is the ith word vector in the x of the microblog. Learn the text characteristics of tweets through a Transformer encoder. In this paper, we use an encoder of the same structure as that in reference [7].

$$\text{Attention}(Q, K, V) = \text{softmax}(QK^T)V \tag{2}$$

where Q, K, and V are, respectively, Query Vector, Key Vector, and Value Vector. The attention is the weighted sum of V values, and Q and K are used to calculate the weight coefficients of the corresponding V.

Multi-head attention carries out H sublinear mapping of input Q, K, and V, and the parameters between the heads are not shared. The parameters of linear transformation of Q, K, and V are different each time. The h attention results obtained are connected, and the multi-head attention output is obtained after linear mapping again, which can be expressed as:

$$\text{MultiHead}(Q, K, V) = \text{Concat}(\text{head}_1, \text{head}_2, \ldots, \text{head}_h)W^O \tag{3}$$

$$\text{head}_i = \text{Attention}(QW_i^Q, KW_i^K, VW_i^V) \tag{4}$$

where H is the number of multi-head attention, head_i is the output of i heads, and W_i^Q W_i^K, and W_i^V are the parameter matrices of Q, K, and V, respectively.

The second layer of the Transformer encoder sub-layer is the fully connected feed forward neural network layer. If the output of multiple attention layers is expressed as Z, then the feed forward neural network layer can be expressed as:

$$H_z = \max(0, ZW_1 + b_1)W_2 + b_2 \tag{5}$$

where W_1 and W_2 are the weight matrix of the feed forward neural network, and b_1 and b_2 are the bias terms of the feed forward neural network. Transformer produces the final text representation by continually overlapping this level of attention mechanism with the normal nonlinear level of input text.

3.3 Output

Finally, the results of the Transformer H_Z using Softmax function carries on the corresponding calculation and then identify whether the Weibo for rumors, computation formula is as follows:

$$\hat{y} = \text{softmax}(WH_z + B) \tag{6}$$

where $\hat{y}$ is the output prediction category, W is the weight coefficient matrix, and B is the bias term corresponding to it.

4 Experimental Results and Analysis

4.1 Experimental Data

In this experiment, Sina Weibo data in the social media rumor detection data set published by Ma et al. [7] in 2016 was used. This data set contains a total of 4664 Weibo events, each event has a corresponding label, each event contains several Weibo, and the specific content has been fully provided. As shown in Table 1.

Table 1. Data set statistics table

Sina Weibo	Statistical
Event	4664
Rumor	2313
Non-rumor	2351
Weibo number	3,805,656

4.2 Experimental Parameter Setting

The ratio of training set and test set of the data set in this paper is 3:1. The experiment is implemented based on TensorFlow framework, using Adam optimizer, and the learning rate is 0.001. The specific parameter settings of the model are shown in Table 2:

Table 2. Experimental parameter setting

Parameter	Number
Transformer encoders	1
Multi-head attention	6
Vector dimension of words	100
Dropout	0.4
Epoch	20

4.3 Experimental Results and Analysis

This paper selects the following methods to compare with the model proposed in this paper.

(1) DTC model [1]: Castillo et al. manually extracted the emotional score, user characteristics, microblog number containing URL, and other characteristics of microblog book and constructed J48 decision tree classifier.
(2) SVM-RBF [2]: Yang et al. manually extracted the content characteristics, user characteristics, transmission characteristics, and subject characteristics of microblog and adopted the SVM model based on RBF kernel function for classification.
(3) GRU-2 model [4]: This model carries out vector representation of text time series and then uses the two-layer GRU network to capture the changing characteristics of the context information of relevant posts over time, so as to realize the classification of rumors.
(4) CAMI model [5]: This model divides microblog time periods into equal lengths, uses the Doc2VEC method to vectorize the text of each time period, and finally

uses the convolutional neural network to automatically obtain the key features of microblog events and then carries out the identification of rumor events.

The experimental results of each method are shown in Table 3, from which it can be seen that the performance order of the rumor recognition method is as follows: Transformer, CAMI, GRU-2, DTC, SVM-RBF. As can be seen from Table 3, for the method based on traditional machine learning, the accuracy of DTC and SVM-RBF models on Sina Weibo data set is 83.1% and 81.8%, respectively. In the method based on neural network, CAMI and GRU-2 models achieved 93.3% and 91.0% accuracy, respectively, on Sina Weibo data set. Compared with the method based on deep neural network, the performance of the method based on traditional machine learning is relatively poor. This may be due to the poor generalization ability of manual features or regular methods in complex social media scenarios. The method based on deep neural network can learn the high-level interaction between deep potential features, which makes the model closer to the real scene. However, the accuracy rate of the method proposed in this paper on Sina Weibo data set reaches 93.9%, and the performance of rumor recognition is slightly higher than CAMI and GRU-2 models. This may be because Transformer model can extract richer features of implicit semantic information in the text, making the performance of the model better.

Table 3. Comparison of experimental results

Method	Accuracy	Recall	F1
DTC	0.831	0.847	0.830
SVM-RBF	0.818	0.824	0.819
GRU-2	0.910	0.864	0.906
CAMI	0.933	0.921	0.932
Transformer	0.939	0.943	0.939

5 Conclusion and Future Works

This paper proposes a microblog rumor detection method based on Transformer model, which adopts XLNet word embedding method and utilizes Transformer model to learn the deep semantic features in microblog content. At last, Softmax function is used to calculate the feature vector to identify whether the microblog is a rumor. The validity of the proposed method is verified by experiments on real data sets. In future work, we will consider adding the emotional characteristics of Weibo comments to the rumor detection model, so that it can achieve better classification effect.

Acknowledgements. This work is supported by National Nature Science Foundation of China-Xinjiang Joint

Fund (U1703261) and Graduate Research Innovation Project(XJ2019G231).

References

1. Castillo C, Mendoza M, Poblete B (2011) Information credibility on twitter, pp 675-684
2. Yang F, Yu X, Liu Y (2012) Automatic detection of rumor on Sina Weibo. ACM
3. Takahashi T, Igata N (2012) Rumor detection on twitter. In: 2012 joint 6th international conference on Soft computing and intelligent systems (SCIS) and 13th international symposium on advanced intelligent systems (ISIS). IEEE
4. Ma J, Gao W, Mitra P et al (2016) Detecting Rumors from microblogs with recurrent neural networks. The 25th international joint conference on artificial intelligence (IJCAI 2016)
5. Yu F, Liu Q, Wu S, Wang L, Tan T (August 2017) A convolutional approach for misinformation identification. In: Proceedings of the 26th international joint conference on artificial intelligence, AAI Press, pp 3901–3907
6. Devlin J, Chang M W, Lee K et al (2018) BERT: pre-training of deep bidirectional transformers for language understanding
7. Vaswania, Shazeern, Parmarn etal (2017) Attention is all you need. Adv Neur Inf Process Syst, pp 5998–6008

Research on Image Classification Method Based on Improved Xception Model

Shuping Chen[1] and Bingcai Chen[1,2](✉)

[1] School of Computer Science and Technology, Xinjiang Normal University, 830054 Urumqi, Xinjiang, China
china@dlut.edu.cn

[2] School of Computer Science and Technology, Dalian University of Technology, 116024 Dalian, Liaoning, China

Abstract. Image classification is an important basic problem in computer vision research, and it is also the basis of other high-level vision tasks such as image segmentation, object tracking, and behavior analysis. Since the features extracted by the neural network are not necessarily all useful features, this paper is optimized on the basis of Xception, and a Convolutional Block Attention Module (CBAM) is introduced to learn channel attention and spatial attention information separately to enhance the discrimination of features. Experiments were conducted on the monkey breed classification dataset. The results show that the improved Xception model based on the CBAM module proposed in this paper can classify these images with an accuracy of 90.05%. Compared with related algorithms, the training accuracy of this model increased by 0.6–1.2%, which further proves that the CBAM module can improve the accuracy of Xception, so as to improve the reliability and stability of image classification.

Keywords: Image classification · CBAM · Deep learning · Xception · Depth-wise separable convolution

1 Introduction

Image classification is a very active research direction in the fields of computer vision, pattern recognition, and machine learning. It currently has a wide range of applications, such as content-based image retrieval in the Internet field, automatic album classification, and so on. Image classification has been applied to all aspects of people's daily lives. Computer automatic classification technology has also reduced the burden on people to a certain extent and changed the way of life of human beings.

In 2014, GoogLeNet proposed by Google Research Institute won the championship in ILSVRC image classification. GoogLeNet modifies the traditional convolutional layer in the network and proposes an inception structure that uses convolution kernels of different sizes in the same layer. Subsequent proposals for improved versions of Inception V2 [1], Inception V3 [2], and Inception V4 [3] have been proposed. Inception V2 uses

Q. Liang et al. (eds.), *Artificial Intelligence in China*, Lecture Notes in Electrical Engineering 653, https://doi.org/10.1007/978-981-15-8599-9_62

3 × 3 convolution kernels instead of 5 × 5 large convolution kernels on the one hand to build more complex nonlinear transformations while reducing parameters, and on the other hand uses batch normalization to reduce the training difficulty of neural networks. The formula is as follows:

$$y^{(k)} = \gamma^{(k)}\hat{x}^{(k)} + \beta^{(k)}, \quad (1)$$

γ and β are the normalization terms of the parameter $\hat{x}$ for a batch.

Xception is also an improvement based on Inception V3. It uses deep separable convolution to completely separate the convolution operation on the channel and the space. First, perform convolution on the channel to obtain feature maps and then use a 1 × 1 convolution kernel to fuse these feature maps. The residual connection mechanism similar to ResNet added by Xception significantly accelerates the convergence of Xception and obtains a higher accuracy rate.

The simplified Xception module is shown in Fig. 1.

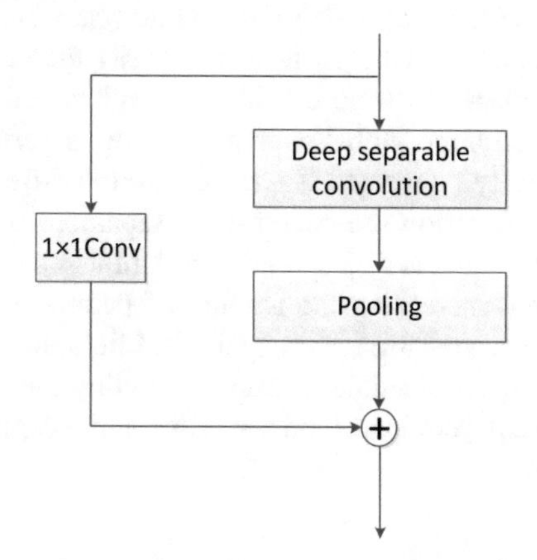

Fig. 1. Simplified Xception module

In this paper, CNN-based deep learning algorithms are used, and the convolutional block attention module is used to improve the existing network structure, and the final classification accuracy is 90.05%. This paper demonstrates the improvement of the recognition performance of the convolutional block attention module through detailed experiments and finally analyzes the effect of the attention mechanism through experiments.

Based on the above processing methods, the contribution of this paper mainly has three aspects:

(1) An improved Xception model is proposed, which uses the Convolutional Block Attention Module (CBAM) to improve the original Xception model, adding an attention mechanism, learning channel attention and space separately attention information enhances the discrimination of features.
(2) Image classification experiments were carried out on a monkey dataset using data enhancement preprocessing based on image geometric transformation. The results

show that the improved Xception model proposed in this paper can classify these images with an accuracy of 90.05%, further proving. The model proposed in this paper can meet the requirements of reliability and stability of image classification.

(3) The improved Xception model proposed in this paper provides a new reference idea for solving image classification problems in the future.

2 Related Work

2.1 Xception Module

Xception [4], the whole network is divided into three parts: Entry, Middle, and Exit. Entry flow contains eight convolutions, middle flow contains $3 * 8 = 24$ convolutions, exit flow contains four convolutions, so Xception has a total of 36 layers. As shown in Fig. 2, Xception is based on Inception V3 and combines depth-wise convolution. The advantage of this is to improve the network efficiency, and under the same parameter amount, the effect is better than Inception V3 on large-scale dataset. This also provides another "lightweight" idea: Given the hardware resources, increasing network efficiency and performance as much as possible can also be understood as making full use of hardware resources. Finally, the global average pooling is performed, and the results are output through the fully connected layer. Xception is different from conventional convolution operations. Xception uses depth-wise separable convolutions to consider the correlation between channels and spatial correlation separately and has achieved very good results without increasing the amount of parameters. Due to its excellent performance, Xception has been used in many fields. Literature [5] improved the U-Net architecture with the Xception module, thereby extracting the buildings in the remote sensing image. Literature [6] compares the capabilities of Xception and other networks for smoke detection.

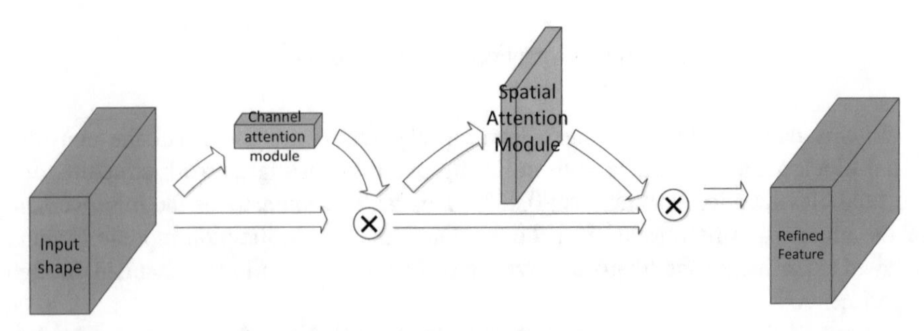

Fig. 2. Channel attention module

2.2 Convolutional Block Attention Module

In order to pay attention to important local detail areas and filter unimportant local information, this paper uses CBAM, which has the same starting point as other attention mechanisms: focus on important feature information and filter unimportant feature information [7]. However, the models proposed in previous studies are all applying residual attention networks in all dimensions [8] or just using the attention mechanism in a specific dimension. In contrast, CBAM uses channel attention module and spatial attention module to obtain feature maps in sequence. Therefore, for image recognition and classification problems, we can pay more attention to the more subtle local differences between images and magnify the representativeness of local features. The structure of the convolutional attention module is divided into a channel attention module as shown in Fig. 2 and a spatial attention module as shown in Fig. 3, and the mathematical forms of which are shown in formulas (2) and (3), respectively.

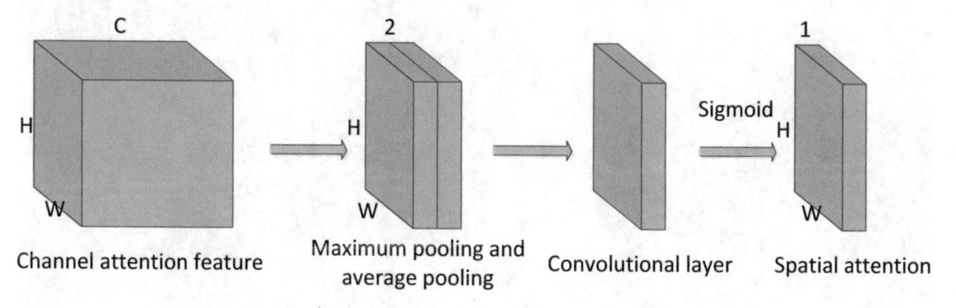

Fig. 3. Spatial attention module

$$\begin{aligned} \mathrm{M}_c(F) &= \sigma(MLP(\mathrm{AvgPool}(F)) + MLP(\mathrm{MaxPool}(F))) \\ &= \sigma\Big(W_1\Big(W_0\Big(F^c_{\mathrm{avg}}\Big)\Big) + W_1\big(W_0\big(F^c_{\mathrm{max}}\big)\big)\Big), W_0 \in^{\mathbb{R}^{C/r\times C}}, W_1 \in^{\mathbb{R}^{C\times C/r}} \end{aligned} \tag{2}$$

$$\begin{aligned} M_C(F) &= \sigma(MLP(\mathrm{AvgPool}(F)) + MLP(\mathrm{MaxPool}(F))) \\ &= \sigma\Big(W_1\Big(W_0\Big(F^c_{\mathrm{avg}}\Big)\Big) + W_1\big(W_0\big(F^c_{\mathrm{max}}\big)\big)\Big), W_0 \in^{\mathbb{R}^{C/r\times C}}, W_1 \in^{\mathbb{R}^{C\times C/r}} \end{aligned} \tag{3}$$

By using channel attention and spatial attention separately, you can focus on important features from the two dimensions of channel and space and filter out unimportant features. The CBAM module learns the channel attention mechanism and the spatial attention mechanism separately. Since the pooling layer does not introduce learnable parameters, it is implemented through a shared fully connected layer. Therefore, CBAM is a lightweight module, which greatly reduces the amount of parameters required to use the attention mechanism, making the model training process more efficient.

3 Experimental Evaluation

3.1 Monkey Dataset

The verification experiment was conducted on the monkey dataset. This dataset contains ten species, nearly 1400 monkey images, each image size is 400 × 300 pixels or larger and is in JPEG format. Among them, label N0 represents maned roar monkey, label

N1 represents red monkey, label N2 represents white bald monkey, label N3 represents Japanese macaque, label N4 represents dwarf marmoset, label N5 represents capuchin monkey, label N6 represents silver monkey, label N7 represents squirrel monkey, label N8 represents night monkey, and label N9 represents black langur. Figure 4 is some pictures of monkey dataset. 80% of all the data in each of the above datasets is divided into the training set, and 20% is divided into the test set.

Fig. 4. Some pictures of monkey dataset

3.2 Data Preprocessing

In the data preprocessing stage, considering as much as possible to keep the monkey pictures with rich color information and spatial information characteristics, this paper uses a data enhanced preprocessing method based on image geometric transformation. Then, in order to adapt to the input data size required by the model, the training dataset is zero averaged.

In order to enrich the deep learning samples, the generalization ability of the model is stronger. The data enhancement preprocessing method based on image geometric transformation is used, so that the rich color information and spatial information characteristics of the monkey picture can be retained as much as possible. At present, the use of reserved tags to generate data enhancement sets is a more effective data expansion method. In this paper, the method of left and right mirror image is used to enhance the data, so that the sample image of the monkey with a specific label is expanded by two times.

In order to adapt to the size of the input data required by the network, quadratic linear interpolation is used to uniformly shrink the original picture size to 299×299. Then perform zero-average processing, that is, subtract the mean of all pictures in the same dataset, so that the processed mean is zero, which can accelerate the convergence of the model during the back propagation process and then divide by the standard deviation of all pictures, so that feature standardization.

3.3 Experimental Environment

The experimental environment is Windows 7 64-bit operating system, TensorFlow is used as the framework, Python is the programming language, the experimental platform is Keras, the memory of the hardware environment is 32G, and the CPU is Intel(R) Core(TM) i7-7700 CPU @3.60 GHz 3.60 GHz, and GPU is NVIDIA GTX 1080ti.

3.4 Experimental Procedure

3.4.1 Image Format Conversion

Converting the original JPEG format dataset to TFRecords format (TFRecords is a standard binary data format of TensorFlow) can make the image reading more efficient. Use shuffle batch to randomly disrupt batch reading, read the image data into the network, go through a series of layers, and finally output the sparse representation of the classification results.

3.4.2 Loss Function

Loss function using Softmax cross-entropy loss:

$$L = -\frac{1}{N}\sum_{1}^{N}\log\left(\frac{e^{z_j}}{\sum_{l}^{k} e^{z_k}}\right) j = 0, 1, 2, \ldots, k - 1 \tag{4}$$

In formula (4), k is the number of predicted categories, z_j is the predicted output of the j th category, and N is the size of a training batch.

3.4.3 Optimizer

The optimizer is Adam, which can make the model converge faster while consuming less resources. Using exponential decay method, the learning rate update formula is as follows:

$$\mathrm{lr}_t = \mathrm{lr} \cdot \frac{\sqrt{1 - \beta_2^t}}{1 - \beta_1^t} \tag{5}$$

In Eq. (5), lr is the initial learning rate, β_1 is the attenuation coefficient of the first-order moment estimation of the gradient in the Adam optimization algorithm, β_2 is the attenuation coefficient of the second-order moment estimation, and t is the number of steps.

3.4.4 Improved Xception Model Training

In order to reduce parameter growth, the CBAM module is only embedded in the Exit flow part of Xception. The embedding method is shown in Fig. 5. Each input training image undergoes several depth separable convolutions to obtain a feature map. After adding the residual, the channel attention and spatial attention information are separately learned by the CBAM, which further enhances the feature discrimination.

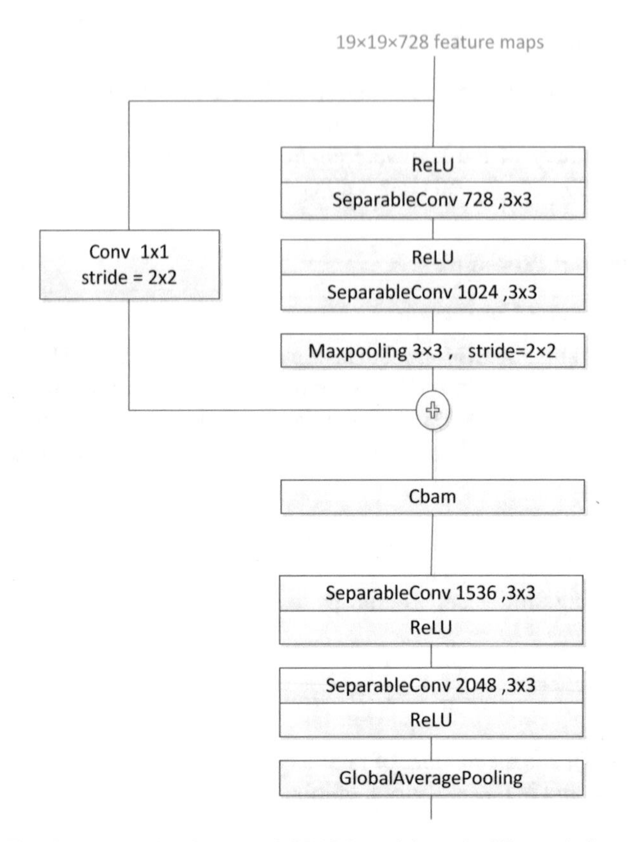

Fig. 5. Schematic diagram of CBAM module embedding exit flow

3.5 Results

Under the above conditions, the accuracy curves of the training set (blue) and test set (orange) of Xception and the improved Xception model are shown in Figs. 6 and 7, respectively.

The loss function curves of the training set (blue) and test set (orange) of Xception and the improved Xception model are shown in Figs. 8 and 9, respectively.

It can be seen from the experimental results that as the number of training steps increases, the model adjusts the weights and deviations through autonomous learning, thereby reducing the training error and gradually increasing the accuracy.

Compared with the original Xception model, the loss of the improved Xception model after training has been reduced to a certain extent. The accuracy curve has been improved by 0.6–1.2% after the improvement. The overall curve is much smoother, and the stability is better.

Use the trained model to test on the dataset. The accuracy of the test set is shown in Table 1.

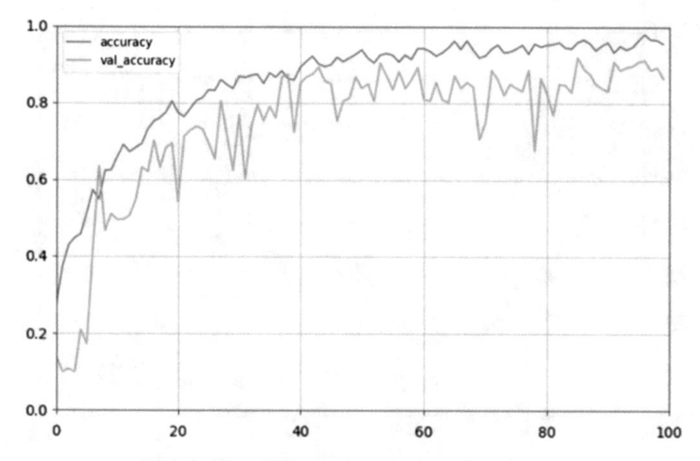

Fig. 6. Xception accuracy on monkey dataset

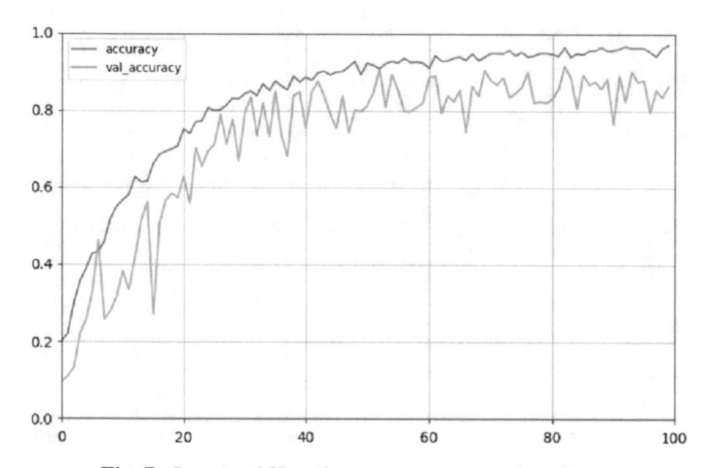

Fig. 7. Improved Xception accuracy on monkey dataset

From Table 1, the accuracy of the improved Xception on the monkey test set is higher than the original Xception. Therefore, the improved Xception can further improve the accuracy of Xception.

This experiment proves that the improved Xception has higher accuracy and can be better applied to image recognition problems by conducting experiments on the TensorFlow platform. In the recognition task of the monkey dataset, firstly, the picture data of ten kinds of monkeys are made into TFRecords format. After that, the Xception network was used for training. When inputting unprocessed raw data, the accuracy of the test set was only 88.79%. When testing with the improved Xception model, the results show that the accuracy of the test set is improved to 90.05%. Thus, an improved Xception model can be obtained, which can meet the requirements of image classification reliability and stability.

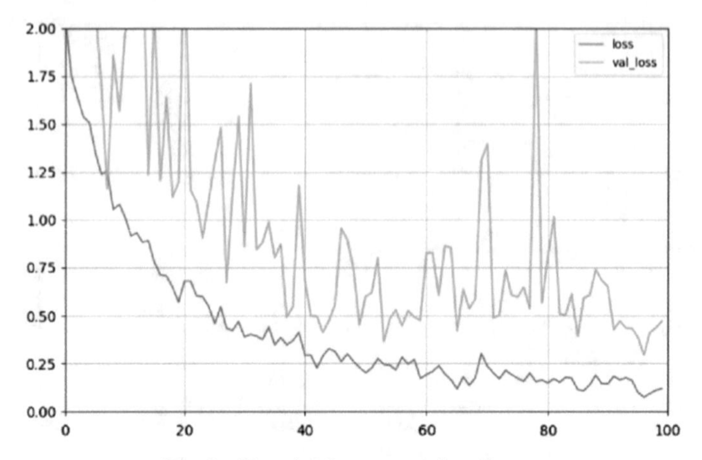

Fig. 8. Xception loss on monkey dataset

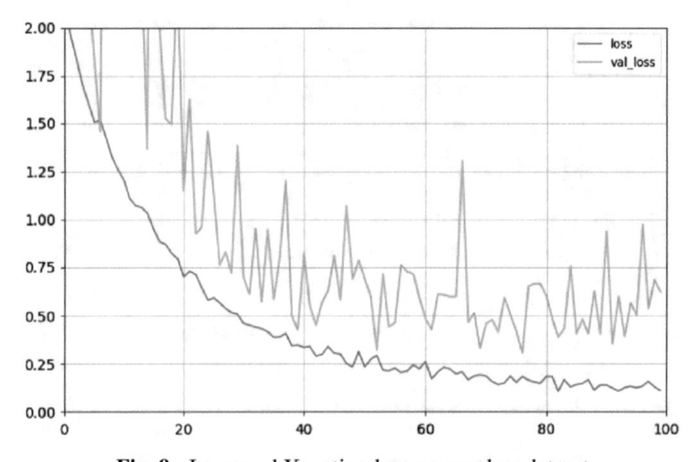

Fig. 9. Improved Xception loss on monkey dataset

Table 1. Comparison of the accuracy between Xception and Improved Xception

	Training set (%)	Test set (%)
Xception	87.52	88.79
Improved Xception	89.82	90.05

4 Conclusions and Future Directions

From the perspective of deep learning, this paper uses the advantages of convolutional neural networks in computer vision to propose an improved Xception model monkey image classification algorithm. Through the data enhancement preprocessing method, based on the geometric transformation of the image, the monkey image training data is data enhanced; then, the CBAM module is embedded in the exit flow layer of the Xception model. Experimental results show that the improved Xception model has stronger stability, higher recognition accuracy, and robustness.

Due to the small number of images in the dataset used in this paper, more images need to be added for training and verification in the future. On the one hand, how to obtain higher recognition accuracy under strong background interference; on the other hand, use the model for other animals image classification and recognition, such as birds, fish.

Acknowledgements. This work was supported in part by the Tianshan Young Talent Program, Xinjiang Uygur Autonomous Region under Grant 2018Q024, in part by the Natural Science Foundation of China under Grant 61771089 and Grant 61961040, and in part by the Regional Cooperative Innovation Program of Autonomous Region (Aid Program of Science and Technology to Xinjiang) under Grant 2020E0247 and Grant 2019E0214.

References

1. LeCun Y, Bengio Y, Hinton G (2015) Deep learning. Nature 521:436–444
2. Szegedy C, Ioffe S, Vanhoucke V, Alemi A (2016) Inception-v4, inception-resnet and the impact of residual connections on learning
3. Ioffe S, Szegedy C (2015) Batch normalization: accelerating deep network training by reducing internal covariate shift
4. Chollet F (2017) Xception: deep learning with depthwise separable convolutions. In: 2017 IEEE conference on computer vision and pattern recognition (CVPR). IEEE
5. Hui J, Du M, Ye X, Qin Q, Sui J (2019) Effective building extraction from high-resolution remote sensing images with multitask driven deep neural network. IEEE Geoence Rem Sens Lett 16(5):786–790
6. Filonenko A, Kurnianggoro L, Jo KH (2017) Comparative study of modern convolutional neural networks for smoke detection on image data. In: International conference on human system interactions. IEEE
7. Woo S, Park J, Lee JY, Kweon IS (2018) CBAM: convolutional block attention module
8. Wang F, Jiang M, Qian C, Yang S, Tang X (2017) Residual attention network for image classification. In: 2017 IEEE conference on computer vision and pattern recognition (CVPR). IEEE

Research on Real-Time Expression Recognition of Complex Environment Based on Attention Mechanism

Shunping Li[1], Cheng Peng[1](✉), and Bingcai Chen[1,2]

[1] School of Computer Science and Technology, Xinjiang Normal University, 102 Xinyi Road, Urumqi, Xinjiang 830054, China
371560906@qq.com

[2] School of Computer Science and Technology, Dalian University of Technology, Dalian 116024, China

Abstract. With the development of deep learning, facial expression recognition has already begun to show results, and facial recognition has a wide range of applications. Whether it is in the field of criminal investigation, education, or medical treatment, it has a bright application prospect. However, in complex situations, due to the influence of factors such as face posture, occlusion, and lighting, facial expression recognition still faces great challenges. In view of the current low accuracy of facial expression recognition in complex situations and the poor real-time performance caused by the diversity and complexity of network structures, this paper proposes a real-time facial expression recognition system based on attention mechanism, which includes separable CNN, residual network, and computer vision attention mechanism. Through the combination of separable CNN and residual network, the number of parameters is greatly reduced, and its real-time requirements are guaranteed. The attention mechanism is used to focus on the detection target and improves the recognition accuracy. Experiments on the face expression dataset of complex scenes in fer-2013 show that the attention mechanism can significantly improve the recognition rate of expressions, and the network also maintains a good real-time effect.

Keywords: Deep learning · Deep separable CNN · Real-time expression recognition · Attention mechanism

1 Introduction

Facial Expression Recognition (FER) has a wide range of application prospects in the prevention of public dangerous events, polygraph detection, driver fatigue driving, and medical services. Whether it is a controlled environment (no occlusion, posture standards, and light balance) or a complex environment, FER is a long-term research focus in the field of computer vision. Compared with FER under controlled conditions, FER in a complex environment has more practical application value. However, due to factors

Q. Liang et al. (eds.), *Artificial Intelligence in China*, Lecture Notes
in Electrical Engineering 653, https://doi.org/10.1007/978-981-15-8599-9_63

such as face posture, obstructions, and unbalanced lighting, it is more challenging to achieve real-time accurate FER in complex environments, and it is also a major problem in the current FER field.

The FER program is generally divided into three steps: facial image preprocessing, facial expression feature extraction, and facial expression determination. Among them, face alignment is a common preprocessing method for face images. By extracting the coordinates of face feature points and using affine transformation to rotate the coordinates of both eyes to the same horizontal line, the face is corrected. However, there are two drawbacks to face alignment processing: First, the extraction of face feature points itself consumes a lot of time, and in complex environments, due to factors such as occlusion, posture tilt, and unbalanced lighting, feature point extraction may fail; further, through double correcting the face with the mesh coordinates may lose the face pixels related to facial expression recognition. Kim et al. [1] proposed the Alignment–Mapping Network (AMN) in order to extract the failure of face feature points in natural environment and learned the mapping from non-aligned face state to aligned face state; Yu et al. [2] improved the accuracy of feature point extraction by combining three face feature point extraction methods (DCNN [2], JDA [3], MoT [4]). However, at present, there are few achievements in the research to solve the problem of the loss of facial pixels caused by the correction of feature points, which lead to a lower accuracy and poor robustness of complex facial expression recognition.

Traditional expression recognition methods generally use manual methods to extract facial expression features and use traditional machine learning methods to determine facial expressions. Reference [5] uses Local Binary Pattern (LBP) to extract facial expression features to reduce the influence of facial light imbalance on facial expression recognition and uses Support Vector Machine (SVM) to determine facial expressions. Zhao et al. [6] further enhanced the robustness of expression feature extraction by fusing LBP (LBP on three orthogonal planes) features, reduced the influence of factors such as facial posture and lighting, and combined K-Nearest Neighbors (KNN) and Hidden Markov Models (HMM) to determine the expression. Zhi et al. [7] proposed a circular preserving Sparse Non-negative Matrix Factorization (GSNMF) method to highlight the face based on the theory of Non-negative Matrix Factorization (NMF). Features improve the recognition rate of facial features, and the final facial discrimination is also implemented using KNN. In the research of traditional expression recognition, there is a lack of consideration of the efficiency of expression recognition. At the same time, the accuracy of expression recognition is low due to the limitations of traditional machine learning classification methods.

As the Convolutional Neural Network (CNN) exhibits excellent performance in the field of computer vision in image segmentation and image classification, more and more studies tend to use convolutional neural networks, to implement expression feature extraction, to improve expression, and to determine the robustness. Kim et al. [8] performed an exponentially weighted decision fusion by merging the results of multiple CNNs to determine the type of static expression; Li et al. [9] proposed a new CNN method for maintaining deep locality, aiming to maintain local tightness. At the same time, maximize the gap between categories to enhance the discrimination between expression categories. Kample et al. [10] analyzed the algorithm differences and performance impacts in multiple literatures and improved the accuracy of expression recognition by constructing a cascade CNN. Although the above research has improved the recognition accuracy to a certain extent, the complex network structure and connection methods make the training process quite cumbersome and difficult to achieve real-time recognition. In order to improve the recognition efficiency, Arriaga et al. [11] combined the residual module and the deep separable convolution layer to greatly simplify the network structure, so that the constructed CNN can realize the real-time recognition effect, but the model can only achieve the benchmark accuracy. In summary, the existing expression recognition research cannot take into account both recognition accuracy and recognition efficiency.

This paper proposes a CNN that integrates an attention mechanism, a residual module and a deep separable convolutional layer. Among them, the residual module is combined with the depth-separated convolutional layer and plays a role in extracting features. The convolutional layer is processed with an attention mechanism before the final classification, which deepens the network's accuracy of recognition and has a small amount of training parameters, and the advantages of taking up less memory and the innovations of this article are as follows:

1. Optimized the neural network (Xception) composed of the original residual module and the deep separable convolutional layer, replacing the last fully connected layer with a convolutional layer, reducing parameters and improving efficiency.
2. Adding the attention mechanism to the final classification makes the accuracy that was originally lost due to the reduction of parameters has been improved to a certain extent.

2 Related Work

The expression recognition framework of the convolutional network based on the attention mechanism constructed in this paper is shown in Fig. 1. The overall framework includes two parts: feature extraction and feature optimization. The convolutional neural network composed of the residual network and the separable convolutional layer is responsible for the features. Extraction and then optimize the extracted feature map through the attention module. In addition, unlike the general classification using a fully connected layer, the classification in this paper uses a convolution layer plus a global average pooling layer. This structure also greatly reduces the parameters compared to the fully connected layer. In the end, the complex face image is first extracted by the separable convolutional layer and then optimized by the attention module, and finally, the expression is determined by the combination of the convolutional layer and the fully connected layer.

2.1 Selection and Construction of Data Sets

All experiments in this paper are based on a classic dataset for facial expression recognition in a complex environment: FER-2013. FER-2013 [12] dataset, which was introduced

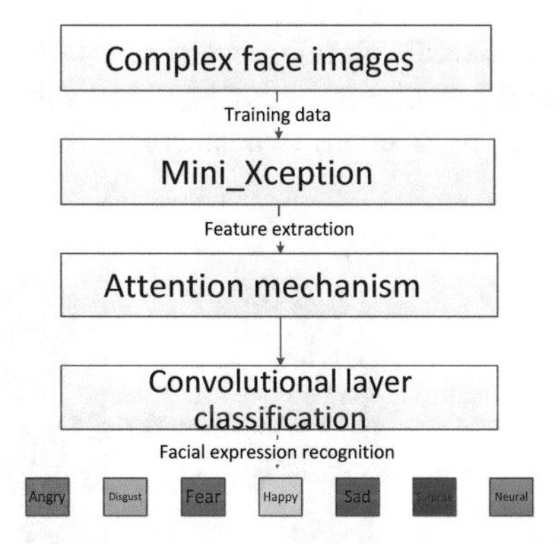

Fig. 1. Expression recognition framework based on SE module's attention mechanism

during the ICML2013 expression learning challenge and automatically collected by the Google image search API. And it is an unconstrained database. Among them, there are 28,708 test images (Training), 3589 public verification images (Public Test), and private verification images (Private Test). Each image is composed of a grayscale image with a fixed size of 48 × 48. There are seven expressions, respectively, corresponding to the digital labels 0 –6, and the labels and English corresponding to the specific expressions are as follows: 0 angry; 1 disgust; 2 fear; 3 happy; 4 sad; 5 surprised; 6 neutral.

2.2 Feature Extraction Network

Generally speaking, for the CNN network applied to facial expression recognition, the increase of network depth will make it have more accurate feature extraction effect. This view has also been confirmed in the study of Kample et al. [10]. The author proves by comparing and analyzing the expression recognition effect produced by CNNs of different depths; the deeper network imposes a stronger prior on the structure of the learning decision base, effectively overcomes the problem of overfitting, and makes the system robust enhance. On the other hand, with the continuous deepening of the network structure at this stage, more and more studies [13–15] also show that deeper networks do not necessarily have more parameters, and the depth of the network does not only depend on the number of layers of the network. What is more important is the ability to learn at a deeper level. This paper uses a convolutional neural network that combines a residual network and a deep separable convolutional layer, which has the advantages of few training parameters and relatively high-training accuracy.

Deep separable convolution is the key structure of many efficient neural network frameworks. The most prominent is the Xception structure proposed by Google in 2017 [14]. Its basic idea is that traditional convolution operations are divided into deep convolution and point-by-point convolution. In the two parts, the former performs lightweight filtering by applying a single convolution filter to each input channel, and the latter forms a 1 × 1 standard cross-channel in advance. Assuming that the size of the convolution kernel used on the $H \times W \times d_i$ feature map is $k \times k$ and the depth is dj, the standard convolution operation requires the parameter quantity to be calculated as $H \times W \times d_i \times$

$k \times k \times d_j$. The depth separable convolution can produce the same effect, and the parameter quantity of the standard convolution is the depth separable convolution $k^2 d_i/(k^2 + d_j)$. The structural difference between the two convolutions is shown in Fig. 2.

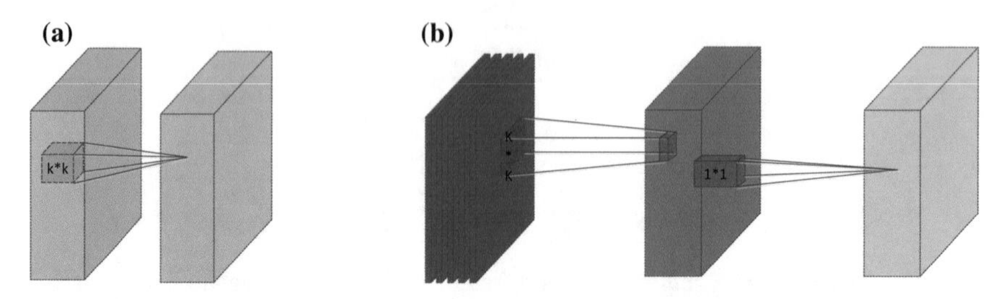

Fig. 2. Standard convolution layer (**a**) and depth separable convolution layer (**b**)

The idea of the residual network is to first perform a 1×1 convolution to increase the channel to a higher dimension and performs deep convolution to reduce the feature channel to the original dimension, which will make the shortcut link established in the number of channels between fewer feature presentation layers, and the effect is to greatly alleviate the problem of feature degradation caused by ReLU, increases feature learning expression and network capacity, and reduces running time and memory usage. The structure of the entire convolutional neural network is shown in Fig. 3.

2.3 SE Module

Attention mechanism was first applied to natural language processing, and now, more and more people apply it to images. In the case of no attention, the neural network treats the features of the picture equivalently. Although the neural network sees the features of the picture to classify it, the neural network does not pay attention to a specific area. The basic idea of the attention mechanism in computer vision is to let the system learn to ignore irrelevant information and focus on the key information. The attention mechanism used in this article is the Squeeze and Excitation Networks module (SE) [20] which can be divided into the following processes:

The first is the squeeze, which is a pooling operation in fact. We compress the spatial dimension so that each feature map becomes a real number. This real number actually has a global receptive field.

Secondly, the excitation is inspired by the mid-gate mechanism in the RNN network, and a weight is generated for each channel through the parameters, where the parameters are learned to explicitly model the feature channel to see the correlation.

Finally, there is the reshape operation, which considers the weight of the output of the excitation operation as the importance of each feature channel after feature selection, and then weights the previous features on a channel-by-channel basis through multiplication. The specific process is shown in Fig. 4.

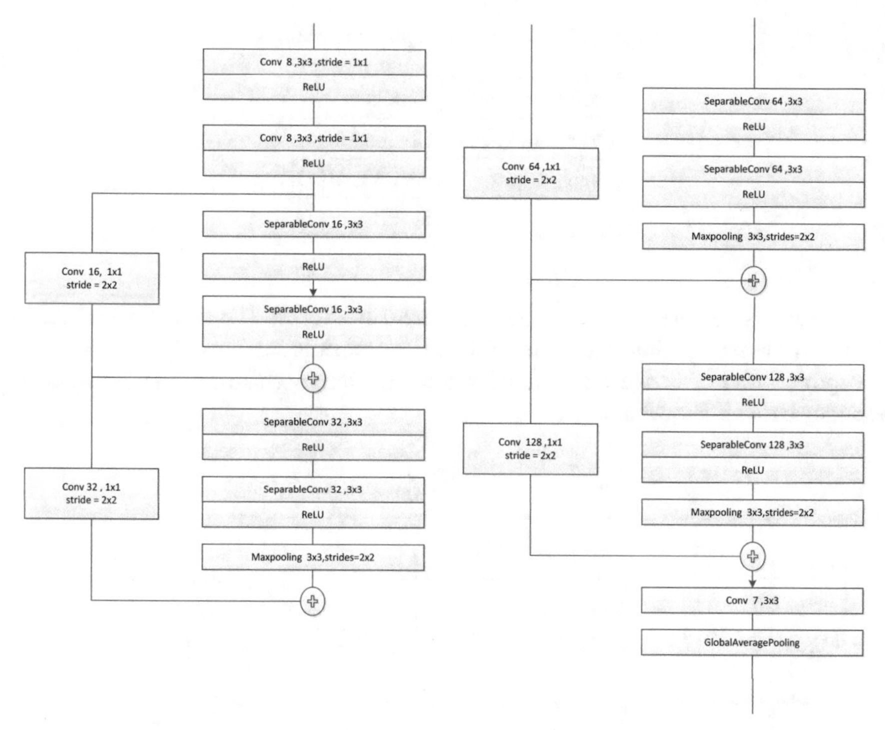

Fig. 3. Process of mini_Xception

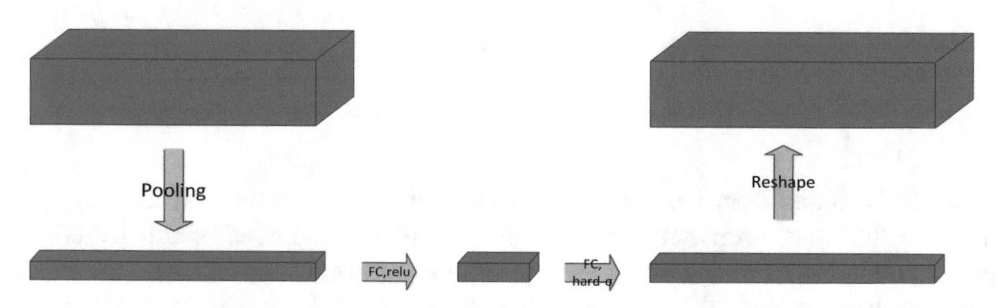

Fig. 4. Process of attention module

3 Experiment

3.1 Experimental Platform and Parameter Settings

All the experiments in this article are based on the Keras framework of the Python language. The rest of the configuration information is shown in Table 1:

3.2 Experimental Data Analysis

The image input size is the original 48 × 48 pixels of FER-2013. Then, it is compared with mini-Xception [11], which has the same characteristics of less parameters and

Table 1. Experimental platform settings

Lab environment	Details
Operating system	Ubuntu 16.04
Graphics card	NVIDIA 1080Ti
RAM	32 GB

small memory requirements, and other different networks. The experimental results show that the recognition rate of the FER-2013 dataset is significantly increased after the optimization of image features using the attention mechanism. The experimental accuracy curve is shown in Fig. 5.

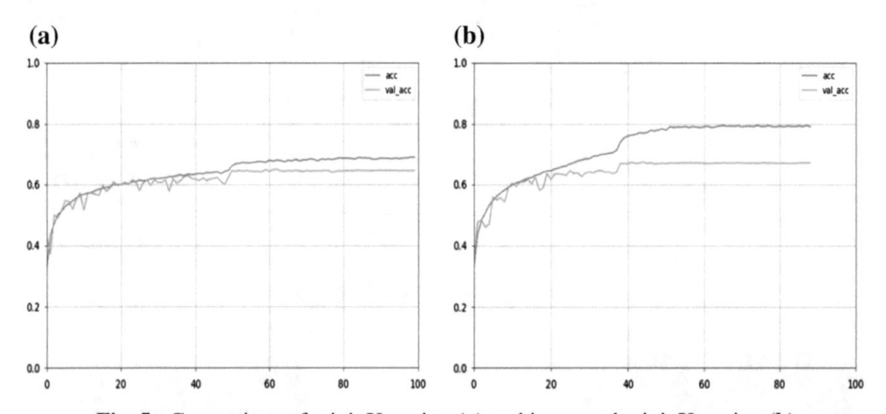

Fig. 5. Comparison of mini_Xception (**a**) and improved mini_Xception (**b**)

It can be found from the experimental curve that as the number of training steps increases, the model learns autonomously and adjusts the weights, and gradually reduces the loss function, obtaining an ascending accuracy curve. Compared with the model before the improvement, the improvement on the test set is 5.1–6.3%, and the improvement on the training set is more obvious, reaching more than 11%, and the curve is stable later, which also proves the stability of the model.

Table 2 shows the performance of different literature research methods on the FER-2013 dataset. The recognition rate of the proposed method on the FER dataset has reached 71%, which is 6% higher than the benchmark recognition rate of 65%. It is also significantly better 65% real-time accuracy in the mini_Xception model. Table 3 shows the accuracy of Kaggle's challenge to the FER-2013 recognition rate ranking model. This method is superior to all currently submitted network models.

In addition, through the observation of the data, it can be found that the three expressions of "happy," "surprised," and "angry" have a higher degree of recognition, while the three expressions of "sadness," "disappointment," and "fear" are extremely easy to cause confusion.

Table 2. Performance comparison of different methods on the FER-2013 dataset

Model name	Parameter/10^3	Feature	Accuracy/%	Real-time
SVM [16]	12,000	SVM	70.12	N
MCDNN [8]	4800	Weighted CNN	70.58	N
CNNs [17]	6200	Cascade CNN	68.75	N
DCN [18]	7300	Singleton CNN	69.30	N
DCNs [19]	21,300	Cascade CNN	72.62	N
DCNs + AMNs [1]	2400	CNN framework	71.86	N
mini_Xception [11]	57	Singleton CNN	65.76	Y
Mini_Xception + SE	196	CNN framework	71.33	Y

Table 3. Kaggle challenge FER-2013 recognition rate ranking

Model name	Accuracy/%
Visual attention + CNN	71.33
RBM	71.16
Unsupervised	69.27
Maxin Milakov	68.82
Radu + Marius + Cristi	67.48
Lor.Voldy	65.25
Ryank	65.09

4 Conclusion

This paper studies the expression recognition system in a complex environment based on the attention mechanism. It combines the residual network with the separable convolution layer for image feature extraction and replaces the last fully connected layer with a convolution layer and a pooling layer. Reduced the number of parameters and increased the detection rate. The SE module is added as an attention mechanism to optimize the feature map, so that the network can learn and recognize the key parts of the image, which improves the detection accuracy. And achieved good results on the complex environment expression recognition dataset FER-2013. The experimental results show that, compared with the existing technical methods, the proposed method exhibits better ideas and recognition performance, while ensuring the real-time effect while taking into account the accuracy of recognition. At the same time, the method proposed in this paper has an accuracy rate of 80% on the training data. The author believes that the FER-2013 dataset has some images with no faces or no pictures, which leads to differences in the accuracy of the training set and the test set. The next step will be to use a more complete complex environment expression dataset for research, and further modify the model to improve the recognition effect, especially for "sadness," "disappointment," and "fear" three kinds of expressions that are easily confused. Eventually, the purpose of transplanting the entire framework to the mobile terminal or the embedded device is achieved.

References

1. Kim BK, Dong SY, Roh J et al (2016) Fusing aligned and non-aligned face information for automatic affect recognition in the wild: a deep learning approach. IEEE conference on computer vision and pattern recognition workshops. IEEE, pp 48–57
2. Yu Z (2015) Image based static facial expression recognition with multiple deep network learning. ACM on international conference on multimodal interaction. ACM
3. Chen D, Ren S, Wei Y et al (2014) Joint cascade face detection and alignment. In: European conference on computer vision. Springer International Publishing
4. Zhu X, Ramanan D (2012) Face detection, pose estimation, and landmark localization in the wild. In: 2012 IEEE conference on comput vision pattern recognition (CVPR). IEEE
5. Shan C, Gong S, Mcowan PW (2009) Facial expression recognition based on local binary patterns: a comprehensive study. Image Vis Comput 27(6):803–816
6. Zhao G, Pietikainen M (2007) Dynamic texture recognition using local binary patterns with an application to facial expressions. IEEE Trans Patt Anal Mach Intell 29:915–928
7. Zhi R, Flierl M, Ruan Q et al (2011) Graph-preserving sparse nonnegative matrix factorization with application to facial expression recognition. IEEE Trans Syst Man Cybern Part B 41(1):38–52
8. Kim B-K, Lee H, Roh J (2015) Hierarchical committee of deep CNNs with exponentially-weighted decision fusion for static facial expression recognition
9. Li S, Deng W, Du JP (2017) Reliable crowdsourcing and deep locality-preserving learning for expression recognition in the wild. In: IEEE conference on computer vision and pattern recognition. IEEE
10. Pramerdorfer C, Kample M (2016) Facial expression recognition using convolutional neural networks: state of the art
11. Arriaga O, Valdenegro-Toro M, Paul P (2017) Real-time convolutional neural networks for emotion and gender classification
12. Goodfellow IJ, Erhan D, Carrier PL et al (2013) Challenges in representation learning: a report on three machine learning contests. Neur Netw 64:59–63
13. Adrian B, Tzimiropoulos G (2017) How far are we from solving the 2D and 3D face alignment problem? (and a dataset of 230,000 3D facial landmarks)
14. Ronneberger O, Fischer P, Brox T (2015) U-net: convolutional networks for biomedical image segmentation
15. Jonathan L, Shelhamer E, Darrell T (2015) Fully convolutional networks for semantic segmentation. IEEE Trans Patt Anal Mach Intell 39(4):640–651
16. Tang Y (2013) Deep learning using support vector machines. arXiv:1306.0239
17. Yu Z, Zhang C (2015) Image based static facial expression recognition with multiple deep network learning. The 2015 ACM. ACM
18. Mollahosseini A, Chan D, Mahoor MH (2015) Going deeper in facial expression recognition using deep neural networks
19. Zhang Z, Luo P, Loy CC et al (2015) Learning social relation traits from face images
20. Hu J, Shen L, Albanie S et al (2017) Squeeze-and-excitation networks. IEEE Trans Patt Anal Mach Intell

Identification Model of Crop Diseases and Insect Pests Based on Convolutional Neural Network

Yong Ai[1,2(✉)], Chong Sun[1,2], Anran Liu[1,2], Feng Ding[1,2], and Jun Tie[1,2]

[1] College of Computer Science, South-Central University for Nationalities, Wuhan, China
aiy_scuec@qq.com

[2] Hubei Provincial Engineering Research Center for Intelligent Management of Manufacturing Enterprises, Hubei, China

Abstract. Agricultural diseases and insect pests are one of the most important factors that seriously threaten agricultural production. Early detection and identification of pests can effectively reduce the economic losses caused by pests. In this paper, convolutional neural network is used to automatically identify crop diseases. The data set comes from the public data set of the AI Challenger Competition in 2018, with 27 disease images of ten crops. In this paper, Inception-ResNet-v2 model is used for training. The experimental results show that the overall recognition accuracy is 86.1%. The experimental results verify the effectiveness of the proposed model.

Keywords: Identification of pests and diseases · Deep learning · Convolutional neural network

1 Introduction

As a big country with more than 20% of the world's total population, China has been facing the problem of insufficient arable land resources. With the development of science and technology, agricultural production is progressing. However, due to various natural factors and non-natural factors, the yield of crops has not been greatly improved. Among all kinds of factors, the largest proportion is crop pests and diseases. According to statistics, the area of crops affected in China is as high as 7 billion acres every year, and the direct output loss is at least 25 billion kilogram. In recent years, this problem is on the rise and seriously threatens the development of planting industry.

With the rapid development of deep learning [1], especially in image recognition [2], speech analysis, natural language processing and other fields, it shows people the uniqueness and efficiency of deep learning. In the field of agricultural production, compared with traditional methods, deep learning is more efficient in the diagnosis of crop diseases. It can monitor, diagnose and prevent the growth status of crops in time. Image recognition of crop diseases and insect pests can reduce the dependence on plant protection technicians in agricultural production, so that farmers can solve problems at the first time. Compared with artificial recognition, the speed of intelligent network recognition

Q. Liang et al. (eds.), *Artificial Intelligence in China*, Lecture Notes
in Electrical Engineering 653, https://doi.org/10.1007/978-981-15-8599-9_64

is much faster than that of artificial detection. And the recognition accuracy is getting higher and higher in the continuous development. Establish a sound agricultural network, through the Internet and agricultural industry. Not only can solve the crop yield and other related issues, but also conducive to the development of agricultural information.

2 Related Works

Deep neural network is designed by imitating the structure of biological neural network, an artificial neural network to imitate the brain, using learnable parameters to replace the links between neurons [3]. Convolutional neural network, as a branch of feedforward neural network, is one of the most widely used deep neural network structures [4]. The appearance of the deeper AlexNet network [5] in 2012 is the beginning of the modern convolutional neural network. The success of AlexNet network model also confirms the importance of convolutional neural network model. Since then, convolutional neural networks have developed vigorously and have been widely used in financial supervision, text and speech recognition, smart home, medical diagnosis and other fields.

Convolutional neural networks are generally composed of three parts. Convolution layer is used for feature extraction. The convergence layer, also known as the pooling layer, is mainly used for feature selection. By reducing the number of features, the number of parameters is reduced. The full connection layer carries out the summary and output of the characteristics. A convolution layer consists of a convolution process and a nonlinear activation function ReLU [6].

At present, the typical convolutional neural networks widely used are as follows.

(1) LeNet-5 [7, 8]: Although proposed very early, LeNet-5 is a complete and successful neural network, especially in handwritten numeral recognition system applications. The LeNet-5 network has seven layers, including two convolution layers, two convergence layers (also called pooling layers) and three full connection layers. The input image size is 32 * 32, and the output corresponds to ten categories.
(2) AlexNet [9]: AlexNet consists of five convolution layers, three convergence layers and three full connection layers. AlexNet absorbs the idea and principle of LeNet-5 network and also makes many innovations. These include using the ReLU function instead of the sigmoid function to solve the gradient dispersion problem. Dropout is used at the fully connected level to avoid overfitting.
(3) Inception network [10]: Inception is different from the general convolutional neural network in that it contains multiple convolution kernels of different sizes in its convolution layer, and the output of Inception is the depth stitching of the feature map. GoogLeNet, the winner of the 2014 ImageNet Image Classification Competition, is the earliest version of Inception v1 used.
(4) Residual network [11]: The core idea of residual network is to make a nonlinear element composed of neural networks infinitely approximate the original objective function or residual function by using the general approximation theorem. Many nonlinear elements form a very deep network, which is called residual network.

3 Crop Disease Recognition Model

3.1 The Structure of Crop Disease Recognition Model

In this paper, Inception-ResNet-v2 network is used as the basic model of crop disease recognition. This hybrid network not only has the depth advantage of residual network, but also retains the unique characteristics of multi-convolution core of Inception network. After adding residual unit in Inception network, although the accuracy rate did not improve significantly, it effectively solved the problem of gradient disappearance and gradient explosion, accelerated the convergence speed of the model, improved the training efficiency and improved the performance slightly [12]. The structure of Inception-ResNet-v2 is shown in Fig. 1.

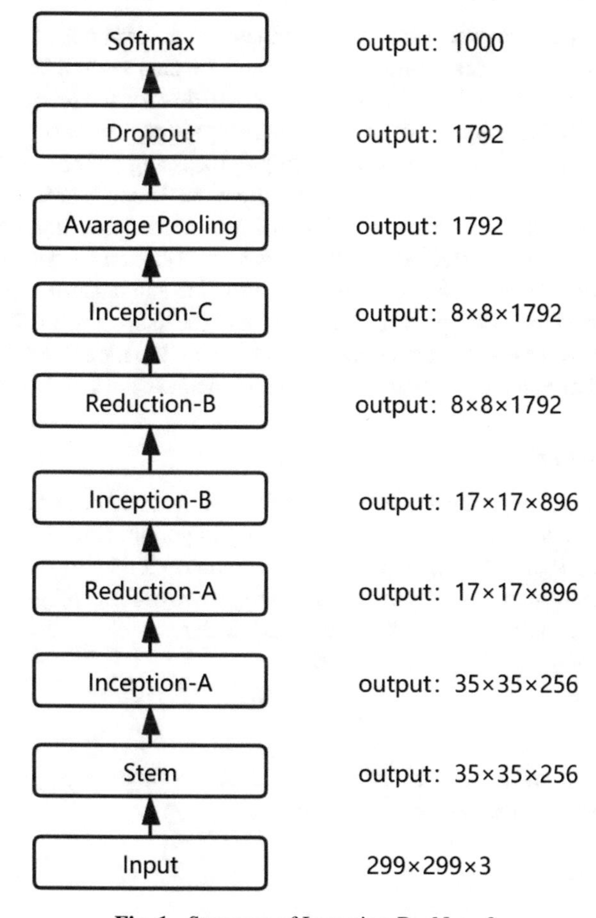

Fig. 1. Structure of Inception-ResNet-v2

The original Inception module takes a parallel structure to extract features and then stack them. In this model, the residual network unit is added to the cross-layer directly connected edge and multi-way convolution layer. Aft that end of the combined convolution operation, activation is performed by connecting the last entry ReLU function.

3.2 Data Set

The data set used in this paper is from the data set used in the Crop Disease Recognition Competition of the 2018 Artificial Intelligence Challenger Competition. The data set includes 47,363 images of 27 diseases related to ten crops (mainly tomatoes, potatoes, corn, etc.). The data set is divided into three parts: 70% training set, 10% validation set and 20% test set. Each picture contains only the leaves of a single crop.

3.3 Image Preprocessing

The purpose of image preprocessing is to eliminate the interference of useless information in data set to model recognition and to expand the data set to a certain extent. The neural network can achieve better training effect. In this way, the recognizability of the image can be effectively improved, so that the recognition accuracy of the model can be improved. At present, the commonly used preprocessing methods include geometric space transformation and pixel color transformation. The former includes flip, crop, rotate, zoom and so on. The latter includes changing contrast, adding Gaussian noise, color dithering and so on. Because of the uneven distribution of data sets, so in this paper, we mainly take the method of light transformation and random clipping. Enhance the feature information of the picture and the scale of the data set itself. The influence of the background factor and the data quantity problem on the model is weakened. It can make the model produce better learning effect and increase the stability of the model.

3.4 Normalized Processing

Aft that above steps are complete, the picture of the data set will be normalized. Normalization can be said to be an indispensable and very important part of the convolutional neural network. It scales the characteristics of each dimension to the same range. On the one hand, it is convenient to calculate data and improve the efficiency of operation. On the other hand, the association between different features is eliminated. Therefore, the ideal model training result can be obtained.

$$x' = \frac{x - \mu}{\sigma} \tag{1}$$

In Formula 1, x and x' are the data before and after normalization. And μ means the average value while σ means the covariance.

4 Experiment

4.1 Experimental Environment

The operating system of this experiment is Windows. The programming language is Python, and framework is TensorFlow deep learning framework. The specific equipment configuration is shown in Table 1.

Table 1. Experimental environment

Configure	Param
CPU	Intel(R)Core(TM) i7-6200u
Anaconda	Anaconda 3.6
TensorFlow	1.2.1
Operating system	Windows 10
Hard disk	512GSSD
RAM	8G

4.2 Training Strategy

In this paper, we use the Inception-ResNet-v2 model for migration. The network weight parameters trained by a large number of data sets are transferred to their own network for training, and the network is fine-tuned. The method comprises the following steps.

(1) The pre-training model is loaded first. We keep the parameters of the convolution layer and the pooling layer in the original model as the initial parameters and freeze the last fully connected layer. Set up a new full connection layer to achieve the classification problem of the target task.
(2) Set the parameters. First set the learning rate to 0.001 and the batch_size to 32. The workout count is set to five epochs, and the Dropout is set to 0.5.
(3) The loss function of the loss layer uses a cross-entropy loss function. The optimizer chooses to update the weights and biases using the Adam optimization algorithm.
(4) And that image in the preprocessed train set and the preprocessed verification set is randomly sent into an image with a batch size for training.
(5) After the model training, the recognition and classification are completed on the test set. A summary of the performance metrics analyzed for the data set.

4.3 Results and Analysis

The evaluation index used in this paper is the commonly used Top1 accuracy in classification problems. It refers to the accuracy rate ACC of the class with the largest recognition probability of the model and the actual class. The formula is shown as Formula 2, where N is the number of samples and R is the number of correct predictions.

$$\text{acc} = \frac{R}{N} \tag{2}$$

The images in the data set are preprocessed and then trained. After each epoch iteration, a verification is performed. The image convergence process is shown in Fig. 2.

It can be seen from the graph that the curve of the convolutional neural network training model used in this paper keeps stable after three epochs are trained, and its accuracy and loss keep a relatively stable state. The final accuracy is 86.1%, and the recognition effect reaches the expectation.

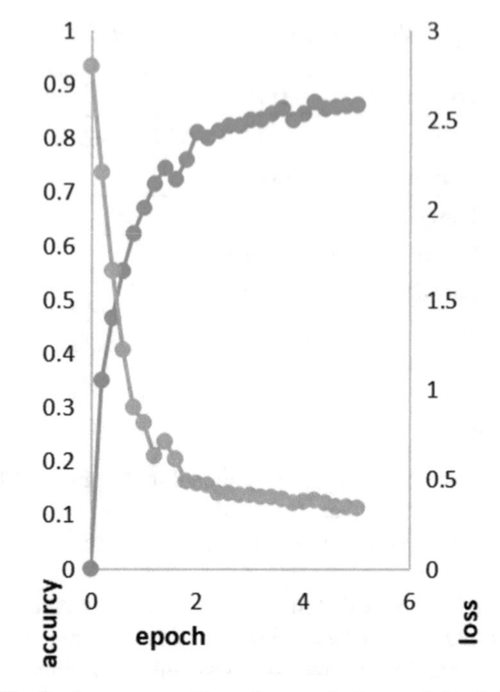

Fig. 2. Accuracy and loss changes during model training

5 Conclusion

In this paper, 27 kinds of disease recognition of ten kinds of crops were studied. The Inception-ResNet-v2 model is constructed by using deep learning theory and convolutional neural network technology. Experiments show that the model can effectively identify the data set, and the overall recognition accuracy is as high as 86. 1%. The results show that the recognition accuracy of this hybrid network model is relatively higher than the traditional model, and it can be effectively applied to the identification and detection of plant diseases and insect pests.

Acknowledgements. This work is supported by the Major Projects of Technological Innovation in Hubei Province (No. 2019ABA101) and the Fundamental Research Funds for the Central Universities, South-Central University for Nationalities (No. CZY19014).

References

1. Schmidhuber J (2015) Deep learning in neural networks: an overview. Neur Netw 61:85–117
2. Mohamed A, Dahl G, Hinton G (2012) Acoustic modelling using deep belief network. Audio Speech Lang Process 20(1):14–22
3. Bengio Y, Delalleau O (2011) On the expressive power of deep architectures. In: Proceeding of the 14th international conference on discovery science, vol 1, pp 18–36

4. Deng L, Abdel-Hamid O, Yu D (2013) A deep convolutional neural network using heterogeneous pooling for trading acoustic invariance with phonetic confusion. Acoustics, Speech and Signal Processing (ICASSP), pp 6669–6673
5. Krizhevsky A, Sutskever I, Hinton GE (2012) Image net classification with deep convolutional neural networks. In: International conference on neural information processing systems. Curran Associates Inc. pp 1097–1105
6. He K, Zhang X, Ren S et al (2016) Deep residual learning for image recognition
7. Lecun Y, Bottou L, Bengio Y, Haffner P (1998) Gradient-based learning applied to document recognition. IEEE, Nov 86(11):288–319
8. Belongie S, Malik J, Puzicha J (2010) Shape matching and object recognition using shape contexts. In: IEEE international conference on computer science and information technology. IEEE, pp 483–507
9. Alom Z, Taha TM, Yakopcic C et al (2018) The history began from alexnet: a comprehensive survey on deep learning approaches. Comput Vision Patt Recogn
10. Szegedy C et al (2014) Going deeper with convolutions. Comput Vis Patt Recogn 1409–1414
11. He K, Zhang X et al (2016) Deep residual learning for image recognition. IEEE Comput Vis Patt Recogn 770–779
12. Szegedy C, Vanhoucke V, Ioffe S et al (2016) Rethinking the inception architecture for computer vision. In: Proceedings of the IEEE conference on computer vision and pattern recognition, pp 2818–2826

UBHIC: Top-Down Semi-supervised Hierarchical Image Classification Algorithm

Jiang Qing Wang[1,2(✉)], Jian Quan Bi[1,2], Lei Zhang[1,2], Chong Sun[1,2], and Jun Tie[1,2]

[1] College of Computer Science, South-Central University for Nationalities, Wuhan 430074, China
wjqing2000@mail.scuec.edu.cn

[2] Hubei Provincial Engineering Research Center for Intelligent Management of Manufacturing Enterprises, Wuhan 430074, China

Abstract. At present, most of hierarchical image classification methods are built under the premise that label of images are known. For unlabel image classification, the accuracy and efficiency can be improved. To solve this problem, this chapter proposes a top-down semi-supervised hierarchical image classification algorithm (UBHIC) with efficiency as the optimization objective. Firstly, the AP clustering is used for the batch of unlabel images, then the similarity between cluster centers and nodes is compared from top-down. Finally, the best classification path is determined by evaluating multiple classification paths. Experiments show that the UBHIC has better accuracy and efficiency than the classical methods.

Keywords: Hierarchical image classification · Batch unlabel images · AP clustering · Top-down

1 Introduction

With the rapid development of the Internet, the number of images is expanding, and image classification has become a research hotspot of scholars at home and abroad. Meanwhile, hierarchical image classification algorithm can effectively index and classify image dataset. At present, the research of hierarchical image classification in academic circle is mainly divided into two kinds: bottom-up [1] and top-down [2], which are widely used in hierarchical image classification, but most of them only consider hierarchical classification of label images, while

Corresponding: Wang Jiang Qing(1964-), female, professor, Research interest: Intelligent algorithm.

Q. Liang et al. (eds.), *Artificial Intelligence in China*, Lecture Notes in Electrical Engineering 653, https://doi.org/10.1007/978-981-15-8599-9_65

research on hierarchical classification of unlabel images is rare. To solve this problem, BUHIC algorithm [3] has a high accuracy for the classification of unlabel images, but it does not consider the case of batch unlabel images, and there is still a large space to improve the efficiency of the model.

Therefore, a top-down semi-supervised hierarchical image classification algorithm (UBHIC) is proposed in this paper, which is suitable for the scene of batch unlabel images. Firstly, the AP clustering is used for the batch of unlabel images, and then similarity comparison is used between cluster centers and nodes from top-down. Finally, the best classification path is determined by evaluating multiple classification paths. Experiments show that the UBHIC has better accuracy and efficiency than the classical methods.

The other parts of this chapter are as follows, the Sect. 2 is related work, the Sect. 3 is the description of the specific implementation of UBHIC algorithm, the Sect. 4 is the experimental verification of UBHIC algorithm, and the Sect. 5 is the summary of this chapter.

2 Related Work

Hierarchical structure is one of the most popular data structure used in real-world classification tasks. Especially, in text classification [4] and image classification [5] tasks, hierarchical classification has become a research hotspot for researchers.

The key of hierarchical classification is to build hierarchical structure according to image categories. According to the difference of the basement methods, the construction algorithma can be roughly divided into four types: clustering-based [6], flat classifier-based [7], local classifier-based [8] and global classifier-based [9]. Among them, based on the idea of top-down hierarchical classification structure in literature [8], a classifier is trained on the parent node to distinguish its child nodes to the maximum extent, so that the image category prediction can be tracked by the top-down method.

According to the similarity of image categories, building hierarchical structure can effectively classify the image, but at present, most of the hierarchical image classification algorithms are aimed at the label image, and only a few algorithms are aimed at the classification of the unlabel images in the hierarchical structure. For example, Zhang et al. [10] combined with a constrained hierarchical document theme to generate a category hierarchy built by the model and automatically discover new categories. Fan et al. [11] developed a method to accurately detect new class features through cost sensitive learning, and used incremental learning algorithm to effectively train new class classifiers. These algorithms use node classifiers to classify. When the new image does not exist in hierarchy, it needs to relearn the classifiers, which is low performance. Therefore, this chapter uses the top-down hierarchical image classification method based on the AP clustering to classify the unlabel images.

3 The Proposed Methods

This chapter presents a top-down hierarchical image classification (UBHIC) algorithm, which is divided into three steps: (1) top-down classification, which compares the cluster center with the hierarchical nodes from top-down, designs a propagation error correction algorithm, which solves the possible classification propagation errors one layer more than one layer down each time, and prunes the different joint points. (2) path evaluation, when the classification continues but there are no categorizable nodes, the score of all current paths is calculated to obtain the best path. (3) new node generation, after determining the last classification node of the best path, judging whether a new node needs to be generated.

3.1 Top-Down Classification

In this chapter, the same hierarchy pre-construction method in the BUHIC algorithm is used. Given the cluster center, at the beginning of the classification, the cluster center is classified into the root node by default, with path length of 1, and then the layer $l(1 < l \leq H)$ is classified from top-down. The classification process is shown in Fig. 1. After the first cluster center classification is finished, the next cluster center is operated the same until the task of cluster center classification in all different batches is finished (Algorithm 1).

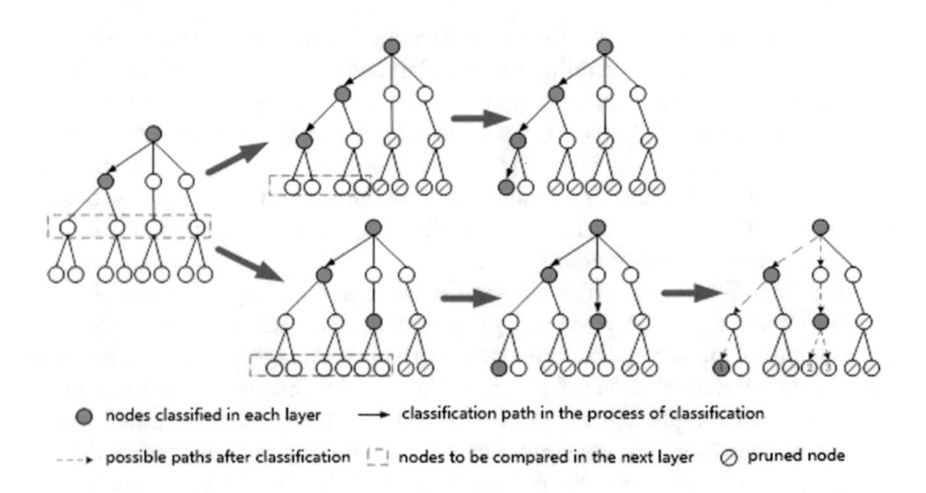

Fig. 1. Classification process of cluster center in hierarchy

3.2 Path Evaluation

In this chapter, by comparing similarity between child nodes of classification node in this layer and corresponding child nodes of upper layer node after pruning, the problem of classification error propagation in high-layer nodes can be

Algorithm 1 Top-Down Hierarchical Image Classification Algorithm(UBHIC)

Input: layers H, batches N, i batch cluster center coding sets $C_i = \{c_{i1}, \ldots, c_{ii}, \ldots, c_{in}\}$, similarity threshold θ

1: $path_j \leftarrow null, j = 1, 2, 3, \ldots$
2: **for** $i = 1 \rightarrow N$ **do**
3: **for** $j = 1 \rightarrow C_i$ **do**
4: Calculate the $dist_i^j$ between c_{ij} and all node codes in l layer
5: Get Sim_l, $Sim_l = \{sim_l^1, \ldots, sim_l^i, \ldots, sim_l^n\}$
6: **if** $min\{sim_l^1, \ldots, sim_l^i, \ldots, sim_l^n\} < \theta$ and $l = 2$ **then**
7: Generate a new node for c_{ij} under root node
8: **return** $path_1$
9: **else if** $min\{sim_l^1, \ldots, sim_l^i, \ldots, sim_l^n\} < \theta$ and $l = 3$ **then**
10: Determine whether to generate a new node for latest node in $path_1$
11: **return** $path_1$
12: **else if** $min\{sim_l^1, \ldots, sim_l^i, \ldots, sim_l^n\} < \theta$ and $l \geq 4$ and $l \leq H$ **then**
13: Get best path according to formula(3.2)
14: Judge whether a new node is generated
15: **return** $path_j$
16: **else if** $min\{sim_l^1, \ldots, sim_l^i, \ldots, sim_l^n\} \in$child of latest node in $path_1$ and $l = 3$ **then**
17: Classify to corresponding node of $min\{sim_l^1, \ldots, sim_l^i, \ldots, sim_l^n\}$
18: Add node in $path_1$ and Pruning non-sibling node
19: $l = l + 1$
20: **else if** $min\{sim_l^1, \ldots, sim_l^i, \ldots, sim_l^n\} \notin$child of latest node in $path_1$ and $l = 3$ **then**
21: Open a new path $path_j (j \geq 1)$ for all the children of latest node in $path_1$ and corresponding nodes of $min\{sim_l^1, \ldots, sim_l^i, \ldots, sim_l^n\}$ respectively
22: Pruning non-sibing node
23: $l = l + 1$
24: **else if** $min\{sim_l^1, \ldots, sim_l^i, \ldots, sim_l^n\} \in$child of latest node in $path_j$ and $l \geq 4$ and $l \leq H$ **then**
25: Classify to corresponding node of $min\{sim_l^1, \ldots, sim_l^i, \ldots, sim_l^n\}$
26: Add node in $path_j$ and Pruning non-sibling node
27: $l = l + 1$
28: **else if** $min\{sim_l^1, \ldots, sim_l^i, \ldots, sim_l^n\} \notin$child of latest node in $path_j$ and $l \geq 4$ and $l \leq H$ **then**
29: Open a new path $path_j (j \geq 1)$ for all the children of latest node in l-layer $path_1$ and corresponding nodes of $min\{sim_l^1, \ldots, sim_l^i, \ldots, sim_l^n\}$ respectively
30: Pruning non-sibing node
31: $l = l + 1$
32: **else**
33: Classify to corresponding node of $min\{sim_l^1, \ldots, sim_l^i, \ldots, sim_l^n\}$
34: Add node in $path_1$
35: $l = l + 1$
36: **end if**
37: **end for**
38: **end for**

corrected. Although the opportunity of error correction is increased, the multiple paths may be generated.

When multiple paths are generated, the nodes in the path are weighted, the scores for all paths are calculated, and the best single path is obtained by selecting smallest path scores among these paths. The weight w_j^l of the lth node in the path is calculated as shown in formula(1).

$$w_j^l = 1 - level(path_j^l)/(maxlevel + 1) \tag{1}$$

where, $path_j^l$ represents lth node in $path_j$, $level(path_j^l)$ represents conversion of l in $path_j$ to a value, and $maxlevel$ represents length of the longest path in all paths. Therefore, the closer node in path is to root node, the higher corresponding weight, and the higher score of path will be when the high-layer node is misclassified.

The score $score_j$ of $path_j$ path evaluation is calculated as shown in the Formula (2).

$$score_j = \sum_{l=1}^{n} w_j^l \times sim_l^i \tag{2}$$

where, w_j^l represents weight of lth node in $path_j$, sim_l^i represents similarity value between cluster center and node i corresponding to l-layer in the $path_j$, and n represents length of $path_j$. The smaller $score_j$, the higher degree of correct classification. Therefore, the path corresponding to minimum $score_j$ is selected as best classification path, and each node corresponding to best classification path is selected as classification node of cluster center in each layer of hierarchy.

3.3 New Node Generation

After cluster centers are classified from top-down, last node in the final classification path is judged to determine whether a new node is generated. There are two cases to generate new nodes: the first is to classify to non-leaf nodes, which do not belong to any child nodes after similarity comparison, and generate new child nodes directly under corresponding nodes; the second is to classify to leaf nodes, by judging whether cluster centers and current leaf nodes are approximate enough, if they are approximate enough, they will be directly merged with original leaf nodes, otherwise, after merging with original leaf nodes, a new sub-node is generated by clustering, as shown in Fig. 2, new nodes generated by dotted line on the far right of figure represent two new leaf nodes generated by merging original leaf nodes and re-clustering after they are classified into current leaf nodes.

To measure the approximation between sibling nodes belonging to the same parent node in any layer, the sibling nodes belonging to the same parent node in l layer are divided into a group, and the Euclidean distance between all node pairs is calculated in each group, in which the minimum value in i graph is taken as the intra graph proximity distance d_l^i , as shown in Formula (3).

$$d_l^i = min\{dist_i^1, dist_i^2, \ldots, dist_i^n\} \tag{3}$$

where, $dist_i^n$ represents the Euclidean distance of nth pair of nodes in i graph, d_l^i is the minimum Euclidean distance in l layer that belongs to all node pairs in i graph.

Similarly, in order to measure the approximation degree of the closest sibling nodes in a layer, and as a standard for judging whether to continue to branch down after being classified into leaf nodes, the calculation formula of approximate distance d_l in l layer is shown in Formula (4).

$$d_l = \min\{d_l^1, d_l^2, \ldots, dist_l^i\} \tag{4}$$

where, d_l represents the approximate minimum distance value within all groups in l layer.

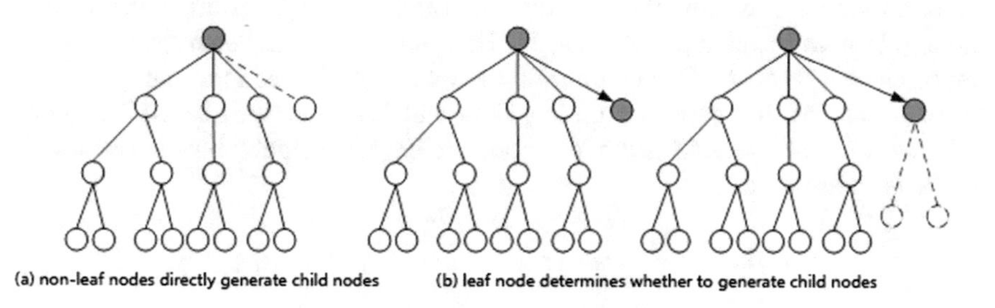

Fig. 2. Two scenario of new node generation

4 Experiments

4.1 Experimental Data and Preprocessing

This chapter evaluates proposed algorithm on two open datasets (CIFAR-10, CIFAR-100). Six kinds of CIFAR-100 images (apple, train, pickup truck, tiger, ox, tank) and ten kinds of CIFAR-10 images (airplane, car, bird, cat, deer, dog, green frog, horse, ship and truck) are selected as the experimental datasets. Six kinds of CIFAR-100 images include which are similar to the category in hierarchy and completely different from the category in hierarchy. In unlabel images dataset, 70% of them are used as training set to classify unlabel images, and 30% of them are used as testing set to test accuracy.

In the experimental preprocessing stage of this chapter, VGG model [12] is first selected as the feature extractor, and convolutional layer in front of fully connected layer is used as feature extraction layer to encode unlabel images to obtain unlabel images coding set. Batch operation of unlabel number of batch images is used, and image codes in unlabel images coding set are sequentially divided into different batches of same number. Finally, the AP clustering method [13] is used to cluster the batched batch images in sequence.

4.2 Experimental Results and Analysis

In this chapter, we choose same evaluation criteria as BUHIC algorithm, meanwhile, through experimental verification(as shown in Figs. 3 and 4), when threshold $\theta = 2$ and numbers of images in each batch is 200, BUHIC algorithm and UBHIC algorithm can achieve better classification effect.

In same data set, this chapter compares accuracy and running time of BUHIC algorithm, hierarchical classifier and dynamic hierarchical structure. As shown in Fig. 5, when the number of images is more than 2000, the running time of UBHIC algorithm is always higher than that of BUHIC algorithm.

Meanwhile, as shown in Fig. 6, from the perspective of accuracy, BUHIC algorithm has advantages. When a large number of unlabel images are labeled, there are many new images whose categories are not in hierarchy, two methods of hierarchical classifier and dynamic hierarchy need to train multiple classifiers, which will take a lot of time, BUHIC algorithm will constantly adjust the hierarchical structure, and compare all low-layer nodes, so increasing number of unlabel images, increase of running time will also increase, UBHIC algorithm adopts top-down classification method, which can achieve better results with fewer high-layer nodes.

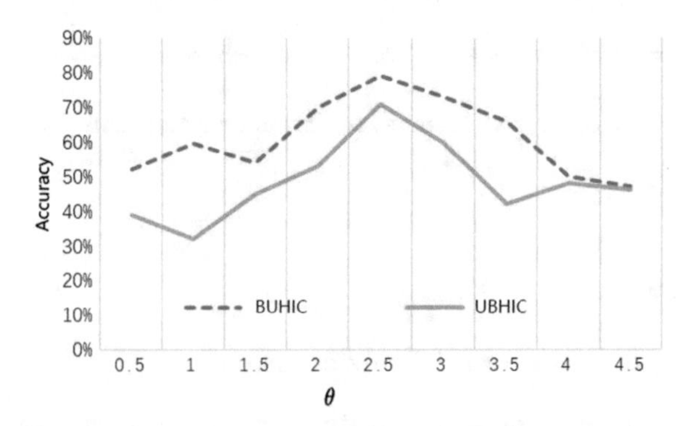

Fig. 3. Effect of threshold θ on accuracy of two algorithms

5 Summary

The priority of this chapter is efficiency. In the face of a large number of unlabel images, cluster centers are obtained by using AP clustering method, then propagation error is solved by using method of downward multiple comparison to get best classification path. Finally, a new node generation algorithm is designed to determine whether to generate child nodes. Experimental results show that the algorithm in this chapter has better accuracy and calculation efficiency in the case of large number of images. At the same time, this chapter may have problem of data imbalance on the unlabel image, which needs to carry out targeted research on the situation that there are few images in some categories later.

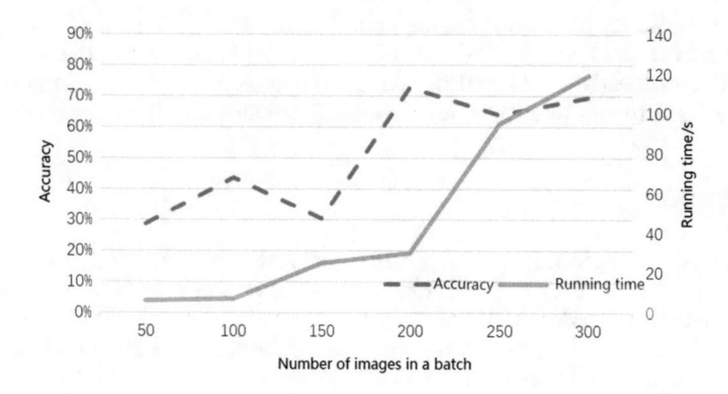

Fig. 4. Effect of numbers of images in each batch on the running of two algorithm

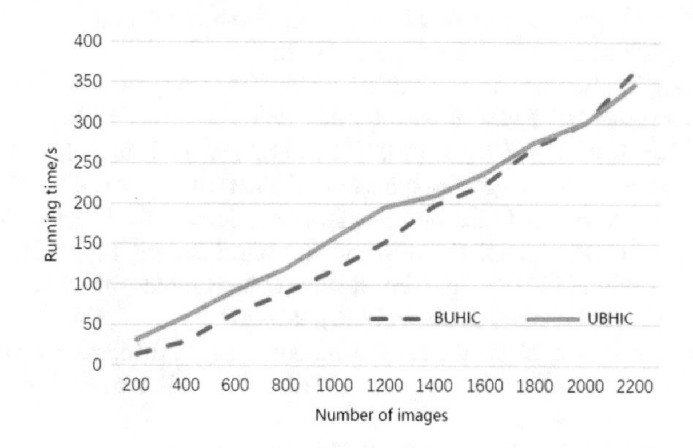

Fig. 5. Running time comparison between BUHIC algorithm and UBHIC algorithm

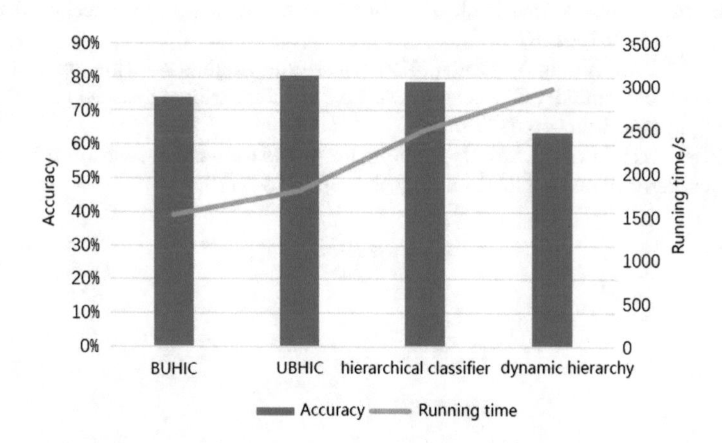

Fig. 6. Comparison of experimental accuracy and running time of four algorithms

Acknowledgment. This work was supported by the Major Projects of Technological Innovation in Hubei Province (No. 2019ABA101), the Fundamental Research for the Central Universities(No. CZT19012), and the Research Team of Key Technologies of Smart Agriculture and Intelligent Information Processing and Optimization.

References

1. Zhang X, Du S, Wang Q et al (2018) Integrating bottom-up classification and top-down feedback for improving urban land-cover and functional-zone mapping. J Remote Sens Environ 231–248
2. Hillion H, Proth J-M (2010) A top-down hierarchical classification method. J Appl Stochastic Models Business Ind 3(4):247–255
3. Wang JQ, Zhang L, Sun C et al (2020) A semi-supervised hierarchical model based on deep learning. J Data Collect Process 35(03):392–399 (in Chinese)
4. Cha J, Kim PK (2016) The automatic text summarization using semantic relevance and hierarchical structure of wordnet. In: C. broadband and wireless computing. Communication and applications, pp 215–222
5. Deng J, Dong W, Socher R et al (2009) ImageNet: a large-scale hierarchical image database. Comput Vis Pattern Recogn 248–255
6. Zheng Y, Fan J, Zhang J et al (2017) Hierarchical learning of multi-task sparse metrics for large-scale image classification. J Pattern Recogn 97–109
7. Naik A, Rangwala H (2016) Embedding feature selection for large-scale hierarchical classification. In: International conference on big data, pp 1212–1221
8. Naik A, Rangwala H (2017) HierFlat: flattened hierarchies for improving top-down hierarchical classification. J Data Sci 4(3):191–208
9. Zhao T, Zhang B, He M et al (2018) Embedding visual hierarchy with deep networks for large-scale visual recognition. J IEEE Trans Image Process 27(10):4740–4755
10. Zhang J, Zhang J, Chen S et al (2012) Constructing dynamic category hierarchies for novel visual category discovery. Intell Robots Syst 2122–2127
11. Fan J, Zhang J, Mei K et al (2012) Cost-sensitive learning of hierarchical tree classifiers for large-scale image classification and novel category detection. J Pattern Recogn 48(5):1673–1687
12. Liu X, Chi M, Zhang Y et al (2012) Classifying high resolution remote sensing images by fine-tuned VGG deep networks. In: International geoscience and remote sensing symposium, pp 7137–7140
13. Li N, Latecki LJ (2017) Affinity learning for mixed data clustering. In: International joint conference on artificial intelligence, pp 2173–2179

Topic Mining and Effectiveness Evaluation of China's Coal-Related Policy Based on LDA Model

Fang Yue[1,2], Kaimo Guo[1,2], Mingliang Yue[1,2](✉), and Wei Chen[1,2,3](✉)

[1] Wuhan Literature and Information Center, Chinese Academy of Sciences, Wuhan 430071, China
{yuef,guokm,yueml,chenw}@whlib.ac.cn
[2] Hubei Key Laboratory of Big Data in Science and Technology, Wuhan 430071, China
[3] School of Economics and Management, University of Chinese Academy of Sciences, Beijing 100190, China

Abstract. Policy text mining is widely used to evaluate the evolution of policy themes and the effects of policy interventions. In this paper, focusing on China's most abundant energy resource, latent Dirichlet allocation (LDA) model is adopted to recognize the technical themes included in the coal-related policies as well as the emphasis of policies on technical themes. Combined with publication years of the policies, the yearly strengths of the themes are calculated for evolution analysis of the technical themes. To evaluate the effects of policy interventions, technical themes included in the coal-related projects funded by the National Natural Science Foundation of China (NSFC), and their evolutions are also analyzed and cross-certificated with those acquired from policies. The analysis shows that the evolutions of policy themes agree with the nation's vision of energy transition, and the technical themes of the NSFC projects evolve in a similar trend as those of policies, with a delay of 5–6 years.

Keywords: Topic mining · LDA model · Coal · Policy theme · Policy effectiveness evaluation

1 Introduction

China is rich in coal resources, with its coal reserves ranking among the top countries in the world. The development of coal-related science and technology (S&T) plays an important role in China's energy security as well as the healthy progress of the coal industry. As one of the most important tools, S&T policies have been used to guide the direction of research and development. It is of great value to analyze the relationship between coal-related policies and S&T developments to verify the effectiveness of the policies.

Policy text mining methods have been widely used to evaluate the evolution of policy themes and the effects of policy interventions [1, 2]. Wu and Chu use classification

Q. Liang et al. (eds.), *Artificial Intelligence in China*, Lecture Notes in Electrical Engineering 653, https://doi.org/10.1007/978-981-15-8599-9_66

analysis to explore the relationships among keywords used in Macau Special Administrative Region's annual policy addresses [3]. Prior et al. use text mining strategies with features of semantic and network analysis to unravel the basic elements of UK's health policy text [4]. Talamini and Dewes carried out text mining to compare and analyze the differences between the scientific documents and the Brazilian government documents related to liquid biofuels [5]. Lee et al. quantify the Monetary Policy Board minutes of the Bank of Korea using text mining, and explain the current and future Bank of Korea monetary policy decisions [6]. In those works, text mining methods are used to extract potential information from policy texts to get a more accurate and deep understanding of the texts.

More recently, topic analysis has been further incorporated to recognize policy emphasis. Park et al. analyze the press releases of a ministry which oversees the Korean energy policy through LSA model [7]. Yukari et al. extract the topics of the financial policy in Japan after the East-Japan great earthquake disaster in 2011 using the LDA model [8]. Yang et al. conduct contrastive analysis of climate policy of China, EU, and US basing on LDA model [9]. Those works verified the effectiveness of topic modeling methods (especially LDA model and its variants) on analyzing the evolution of policy topics.

In this paper, focusing on China's most abundant energy resource, LDA model is adopted to recognize the technical themes included in the coal-related policies as well as the emphasis of policies on technical themes. To evaluate the effects of policy interventions, technical themes included in the coal-related projects funded by the National Natural Science Foundation of China (NSFC), and their evolutions are also analyzed and cross-certificated with those acquired from policies.

2 Data and Methods

2.1 Data

This article focuses on the planning policies promulgated by State Council and the central ministries of China, which represent the nation's top-level design. That is, 115 coal-related planning policy documents issued by State Council and the central ministries of China from 1995 to 2017 were collected for topic mining. In order to evaluate the effectiveness of the polices, titles of 4019 coal-related research projects funded by NSFC from 1997 to 2017 were also collected, analyzed, and cross-certificated with policies. Before LDA model was conducted, policy text and project titles were segmented using Jieba word segmentation module. During the segmentation, 22 Sogou specialized dictionaries (including energy, chemical, chemical, electric power, engineering, coal, petrochemical, mining, materials, environment, and other fields) and the headwords of *Da Ci Hai (Energy Science Volume)* were used as thesaurus to improve the segmentation quality.

The reason behind using NSFC projects to verify the effectiveness of the S&T polices is that as the most important way for researchers to get governmental support, the research themes included in NSFC projects represent the most mainstream research direction in China. The consistency among themes included in S&T polices and NSFC projects to a large extent can illustrate the effectiveness of the policies.

2.2 Analysis Method

LDA is a document-topic generation model [10]. The model presumes that the words in the topic, and the topics of the document are both subject to certain polynomial distributions. Hence, generating a document can be seen as a repeated process of selecting a topic with a certain probability and then selecting a word in the topic with a certain probability. The model can be formally represented as $\Omega = \Phi \times \Theta$, where Ω, Φ, and Θ is document-word distribution, topic-word distribution, and document-topic distribution, respectively. $\times$ presents matrix multiplication.

In this paper, LDA model is used to acquire the policy theme matrix and project theme matrix to recognize the technical themes included in the policy documents and the project titles and the distribution of the policies and projects over the themes. After multiple rounds of testing and manual interpretation, the number of topics given to LDA model for policy documents is set as 12 and for project titles is set as 30.

Since the probability of a document over a theme represents the emphasis of the document on the theme, for a particular technical theme, the strength of the theme in the policies (projects) is calculated by adding up the probabilities (i.e., weights) of the policies (projects) over the theme. By considering the publication date of the policies (the funding dates of the projects), we can then calculate the yearly distribution of the theme strengths of the policies (projects) over themes. Those yearly sequences of theme strengths are then used for theme evolution analysis, either for independent analysis (for policy only) or correlated analysis (for policy and projects).

It is to be noted that the number of published policies varies year from year. For certain years, only few numbers of policies were published. The fact to some extent makes yearly theme strengths fluctuate in a comparatively large magnitude. Hence, we calculate moving averages (window = 3) of the yearly theme strength sequences for a smoother analysis. Besides, we can get enormous theme strengths from project titles compared with those from policy documents due to the large numbers of project data used. For each yearly sequence of theme strengths, we calculate a normalized strength sequence by dividing each strength value to the sum of the strengths in the sequence before calculating moving averages, to get more intuitive comparisons of the evolution trend of the technical themes in the policies and projects.

3 Results

3.1 Technical Themes in Policies

3.1.1 Theme Strength

Through the above methods, the technical theme strength of coal-related policies was obtained, as given in Table 1 and Fig. 1. In all technical themes, strength of theme 8 (advanced coal chemistry), theme 4 (industry transformation), and theme 1 (coal mine safety and gas control) are highest. Since the 1990s, China has stepped up its support for the transformation and upgrading of the coal industry. In 1996, China issued the *Outline for the Reform and Development of the Coal Industry during the Ninth Five-Year Plan Period*. Since then, China has continued to issue a number of important planning policies to propose the development of a modern coal chemical industry. During the 13th Five-Year Plan period, China intensively released a series of important planning policies, such as *Plan of Innovative Action for the Energy Technology Revolution (2016–2030), 13th Five-Year Plan for the Development of the Coal Industry, Development Planning of the Petrochemical and Chemical Industry (2016–2020),*" *Energy Production and Consumption Revolution Strategy (2016–2030), 13th Five-Year Plan for Demonstration of Deep Coal Processing Industry*, and *Layout Plan of Innovative Development for Modern Coal Chemical Industry*. Therefore, coal chemistry and the transformation of coal industry have become the focus of coal development planning. In addition, coal mine safety, which cannot be ignored, has always been regarded as one of the important tasks of the government. Therefore, the strength of this theme ranks third. Clean and efficient coal-fired power generation is a key technology for the government to solve the pollution problem of coal-fired power. It is also an important part of China's future clean, low-carbon, safe, and efficient energy system. As a result, clean and efficient coal-fired power generation is also one of the hot themes of coal-related policies. On the other hand, themes 9–12 are relatively weakly connected to coal, so their strengths are relatively lower.

Table 1. Technical themes strength of selected coal-related policies

Theme number	Theme	Theme strength
1	Coal mine safety and gas control	15.31
2	Mining and filling of coal seams	7.24
3	Development and utilization of coalbed methane and coal gangue	7.43
4	Industry transformation	24.58
5	Clean energy generation and heating	8.08
6	Energy saving, emission reduction, and desulfurization of coal-fired power generation; combined heat and power	12.84
7	Clean and efficient coal-fired power generation	11.74
8	Advanced coal chemistry	12.09
9	Advanced gas turbine	4.33
10	Advanced control and monitoring technologies	5.06
11	Smart grid-related technologies	3.09
12	Carbon emission reduction and waste recycling	2.36

3.1.2 Theme Evolution

The 12 technical themes can be divided into three groups, that is, coal mining and industry transformation (Group I), coal utilization (Group II), and related cross-cutting

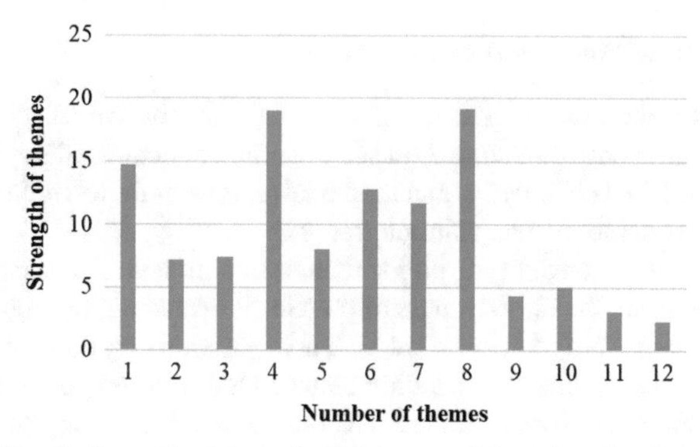

Fig. 1. Strength of the technical themes of the selected policies

technologies (Group III). The evolutions of the technical themes of three groups are shown in Fig. 2.

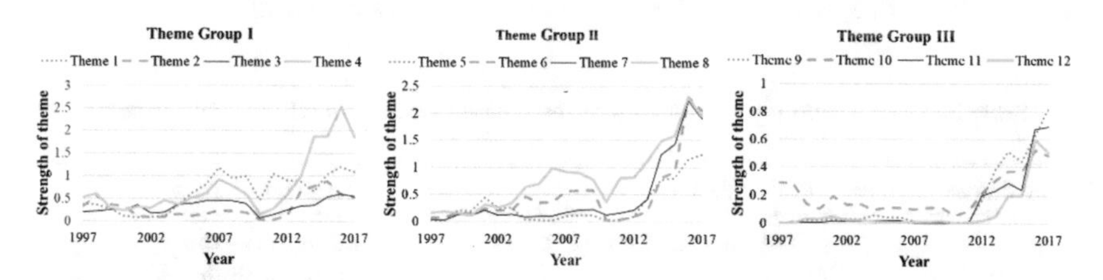

Fig. 2. Evolution of the technical themes of the selected policies

Overall, the strengths of policy themes have shown a gradual upward trend. Coal is the most abundant energy resource of china. In recent years, clean and efficient has become the principle of coal utilization. All technical themes of policies have a positive effect on promoting the clean and efficient utilization of coal, so their strengths are gradually increasing. *Action Plan of Energy Development Strategy (2014–2020)* released in June 2014 proposes to formulate and implement a plan for clean and efficient utilization of coal. In April 2015, China released the *Action Plan for Clean and Efficient Coal Utilization (2015–2020)*, which systematically deployed clean and efficient coal utilization from strategic objectives to implementation measures. During the 13th Five-Year Plan period, China implemented the *Four Revolutions and One Cooperation* strategy for the development of energy security, and issued *Plan of Innovative Action for the Energy Technology Revolution (2016–2030)*, and *Energy Production and Consumption Revolution Strategy (2016–2030)*, emphasizing the development of a clean, low-carbon, safe, and efficient modern energy system. For coal technology, these policies focus on the harmless mining of coal and technological innovation in the clean and efficient coal utilization. Therefore, the strengths of all themes increase rapidly during the 13th Five-Year Plan period. The strengths of themes in Group III are lower than those of the other themes in Group I and Group II. It may stem from the fact that these technical themes can be applied to multiple fields, so that they are not the main technical themes in the policy documents studied in this article.

3.2 Comparison of Policy and Project Themes

According to the important processes of the coal industry, we select theme 1 (coal mine safety and gas control), theme 7 (clean and efficient coal-fired power generation), theme 8 (advanced coal chemistry), and theme 12 (carbon emission reduction and waste recycling) to analyze the theme evolution trends.

By comparing the change of the policy theme strength with the change of the NSFC project theme strength, the effectiveness of policies is evaluated. Compared with policy themes, project themes are more detailed, so a policy theme may correspond to multiple project themes. We established a mapping between the selected policy themes and the NSFC project themes, as given in Table 2. According to the mapping, we accumulate the strengths of the project themes corresponding to the same policy theme and normalize the strengths for analysis. Then, the evolution of the normalized strengths of policy themes and corresponding project themes are compared, as shown in Fig. 3. Considering that the number of NSFC-funded projects has increased year by year, we have normalized the total number of NSFC-funded projects from 1997 to 2017 in the same way and used it as a reference line, which is indicated by the thin solid line in Fig. 2.

Table 2. Corresponding relationship of the policy themes and NSFC project themes

Policy theme number	Policy theme	NSFC project themes
1	Coal mine safety and gas control	Coal mine safety theory and simulation; evolution of coal seam gas; coal mine gas control; coal seam gas outburst damage; spontaneous combustion of coal; soil ecology of mines
7	Clean and efficient coal-fired power generation	Oxy-fuel combustion of coal; adsorption kinetic of coal; chemical looping combustion; circulating fluidized bed
8	Advanced coal chemistry	Catalytic conversion of CO_2; lignite pyrolysis and gasification; coal ash composite material; High temperature coal gasification; low rank coal liquefaction; catalytic conversion of syngas
12	Carbon emission reduction and waste recycling	CO_2 geological storage; CO_2 adsorption

Overall, the themes of policy and projects show a similar upward trend. Figures 3a and c show the evolutions of the two policy themes (themes 1 and 8) with relatively high total strength. It can be seen that the policy theme strengths of theme 1 and 8 increase substantially since 2003, and the growth continued until about 2007. During this period,

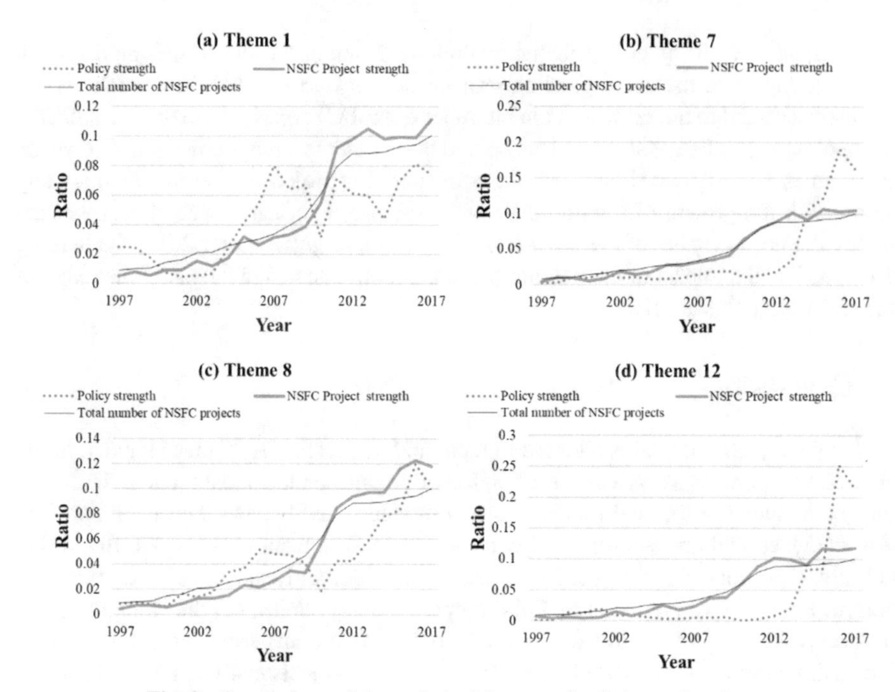

Fig. 3. Comparison of the technical themes of policies and projects

in addition to the coal-related *11th Five-Year Plan*, the central government issued a number of policies to strengthen the safety management of coal mine exploration and mining. State Council of China issued *Several Opinions on Promoting the Healthy Development of the Coal Industry* in 2005. National Development and Reform Commission and other central ministries have successively issued documents to promote the structural adjustment and sustainable development of the coal industry, which is also reflected in the policy theme strength curve of theme 8. In 2010, National Development and Reform Commission issued a notice on regulating the orderly development of the coal chemical industry, and repeatedly emphasized the acceleration of the development of modern coal chemical industry in many planning policies of coal industry and petrochemical industry during the 12th and 13th Five-Year Plan period. Therefore, the policy theme strength curve of theme 8 shows another rapid increase after 2010. The policy theme strengths of themes 7 and 12 show a similar change, which remain stable until 2012 and since then increase substantially. This may be due to the fact that China has always emphasized clean and efficient coal-fired power generation in the *12th Five-Year Plan* and *13th Five-Year Plan* of energy development and many important energy development policies since 2012, and issued policies such as *National Energy Saving Action Plan for 13th Five-Year Plan* and *Three-year Action Plan Aims for Blue Skies* to reduce air pollution of the coal. The policy theme strength trends of themes 7, 8, and 12 clearly show China's increasing emphasis on clean and efficient coal utilization in the past decade, which is consistent with the direction of the government's energy development strategy.

Compared with the policy theme evolutions of themes 1 and 8, the rapid growth of the strengths of the corresponding project themes start in 2009, in a much greater manner as regard to the growth of total number of NSFC projects. It can be seen that the change of project theme strength lags behind policy theme strength by about 5–6 years. For themes 7 and 12, the strengths of the corresponding project themes grow moderately along with the growth of total number of NSFC projects. It may due to fact that the policy theme strengths (of themes 7 and 12) start rapid growth after 2012; the lagging effect makes the impact of the policies that cannot be reflected in the data time range of the research in this article.

4 Conclusion

In this paper, LDA model is adopted to recognize the technical themes included in the coal-related policies as well as the emphasis of policies on technical themes. Technical themes included in the coal-related projects funded by NSFC and their evolutions are also analyzed and cross-certificated with those acquired from policies. We found that (1) advanced coal chemistry, industry transformation, and coal mine safety and gas control are the most emphasis of China's coal-related policies; (2) the strengths of all the policy themes grow rapidly during the 13th Five-Year Plan period, indicating China's increasing emphasis on coal; (3) themes of research projects evolve in a similar trend as those of policies, with a delay of 5–6 years, which verifies the effectiveness of policy interventions.

Acknowledgements. This work is supported by Transformational Technologies for Clean Energy and Demonstration, Strategic Priority Research Program of the Chinese Academy of Sciences (XDA21010103); Literature and information capacity building Project of the Chinese Academy of Sciences (No. E0290001); Youth Innovation Promotion Association Project of the Chinese Academy of Sciences (No. 2017221); Young Talent-Field Frontier Project of Wuhan Documentation and Information Center, Chinese Academy of Sciences (No. Y8KZ491003).

References

1. Grimmer J, Stewart BM (2013) Text as data: the promise and pitfalls of automatic content analysis methods for political texts. Polit Anal 21(3):267–297
2. Li J, Liu YH, Huang C et al (2015) Remolding the policy text data through documents quantitative research: the formation, transformation and method innovation of policy documents quantitative research. J Pub Mgt 12(2):138–159
3. Wu SH, Chu SH (2013) The text mining and classification analyses on the relationship of macau special administrative region's policy addresses from 2012 to 2013. In: International conference on engineering management science and innovation (ICEMSI 2013)
4. Prior L, Hughes D, Peckham S (2012) The discursive turn in policy analysis and the validation of policy stories. J Soc Policy 41(2):271–289
5. Talamini E, Dewes H (2012) The macro-environment for liquid biofuels in Brazilian science and public policies. Sci Pub Policy 39(1):13–29
6. Lee YJ, Kim S, Park KY (2019) Deciphering monetary policy board minutes with text mining: the case of south Korea. Korean Econ Rev 35(2):471–511

7. Park C, Yong T (2017). Prospect of Korean nuclear policy change through text mining. In: International scientific conference on environmental and climate technologies (CONECT 2017)
8. Yukari S, Takako H, Tamaki S (2014) Extraction of the financial policy topics by Latent Dirichlet allocation. In: 2014 IEEE Region 10 Conference (TENCON 2014)
9. Yang H, Yang J (2016) Quantitative analysis of policy text merged with lda model-based on the field of international climate as demonstration. J Mod Info 36(5):71–81
10. Blei DM, Ng AY, Jordan MI (2003) Latent Dirichlet allocation. J Mach Learn Res 3:993–1022

Feature Extraction and Selection in Hidden Layer of Deep Learning Based on Graph Compressive Sensing

Yifei Yuan, Lei Xu, Yiman Ma, and Wei Wang(✉)

Tianjin Key Laboratory of Wireless Mobile Communications and Power Transmission, Tianjin 300387, China
weiwang@tjnu.edu.cn

Abstract. Faced with massive high-dimensional, multi-modal, and redundant information data sets, there are many processing methods, one of which is deep learning. Although the deep learning model is effective, there is a feature redundancy phenomenon in the hidden layer after feature extraction. This phenomenon will not only cause over-fitting problems in subsequent data mining, but also cause excessive cost loss. This paper proposes a feature extraction and selection method in hidden layer of deep learning based on graph compressive sensing, which can extract low-dimensional features from complex high-dimensional features, and eliminate the redundancy of features.

Keywords: Automatic · Encoder · Graph compressed sensing · K-nearest-neighbor graphs · Convolutional neural network

1 Introduction

In the face of massive and complex structured or unstructured information and data, how to effectively distinguish or quickly process them is a challenge. Thanks to the increasing computing power of modern electronic products. Deep learning emerged as a method for scholars and experts to form a more abstract high-level representation of attribute categories or features by combining low-level features. For neural networks, the hidden layer is a collection of neurons with activation functions, and it bears an important connection between the input layer and the output layer [1]. Although the hidden layer in deep learning brings hope for solving optimization problems related to deep structures, the extracted hidden layer features still have redundancy.

At present, many articles have carried out research on eliminating the redundancy of hidden layer features. For example, eliminating redundant hidden units proposed by Jiang et al. [2, 3] is a method used on RBM. The former determines the redundancy of features by comparing synchronous changes, while the latter comparing the feature similarity with the determined threshold in fuzzy removing RBM. Some scholars add the sparse penalty factor that avoids the hidden units learning similar features to overcome

Q. Liang et al. (eds.), *Artificial Intelligence in China*, Lecture Notes in Electrical Engineering 653, https://doi.org/10.1007/978-981-15-8599-9_67

the homogenization phenomenon in training process, as stated in [4, 5]. Cross-covariance is used to regularize the feature weight vector to construct a new objective function to eliminate feature redundancy in Ref. [6]. A detailed summary of several feature selection and redundancy removal methods is described in Ref. [7]. Minimum redundancy maximum relevance(MRMR) which selects an optimal feature set according to the maximal statistical dependency criterion based on mutual information is used in Ref. [8].

Graph signal processing is often used to represent and process these irregular data. It shows that if the sampled signals are spatially or temporally related, we can construct an underlying graph, so that the signals are compressible in the corresponding transform domain to remove the redundancy in the data. Therefore, this paper proposes a method combining graph compression sensing theory to optimize the extraction and selection of hidden layer features.

The remainder of the paper is organized as follows. Section 2 details the theoretical knowledge of the models used in the experiment, including automatic encoder, K-nearest-neighbor(KNN) graphs, and graph compression sensing. Section 3 introduces the experiment process. Section 4 is the conclusion.

2 Theory

2.1 Automatic Encoder

The feature extraction method used in this paper is an automatic encoder (AE), one of the basic structures of deep learning. Each input mode is assigned a branch of the network, that is a sub-networks. In addition, there may be potential connections between different data modalities. In order to make the extracted features better retain the main information of each picture, a hidden layer named auxiliary layer is shared at the uppermost layer of each sub-network to store different data and represent relationship between different modes. The overall framework of the semi-supervised multi-modal neural network that we used is shown in Fig. 1.

The training of the entire network is divided into two parts, unsupervised pre-training and supervised joint fine-tuning (this is why the overall structure is called semi-supervised). In the pre-training stage, the objective function is the reconstruction errors of original input and the reconstructed input, and the entire training process does not require labels.

In the fine-tuning phase, the auxiliary layer is connected to the sub-networks of all modals through the weight T, defining h_m to represent the uppermost neuron of the m-th mode, and P_r to represent the labeling information of the auxiliary layer. The back propagation algorithm is used to optimize the model, and the loss function is defined as:

$$L = -\sum_{j}^{m}\sum_{i}^{N}\log(P(Y = y^{(i)}|h_m^i, T, b_{\mathrm{root}})) \tag{1}$$

m represents the number of modalities, n represents the number of training samples, y represents the class label of x, and b_{root} represents the bias. The gradient descent optimization algorithm is used for adjustment, and finally, the respective high-level abstract feature expressions are extracted from the hidden layers of each sub-network. After the network training is completed, the auxiliary layer is canceled.

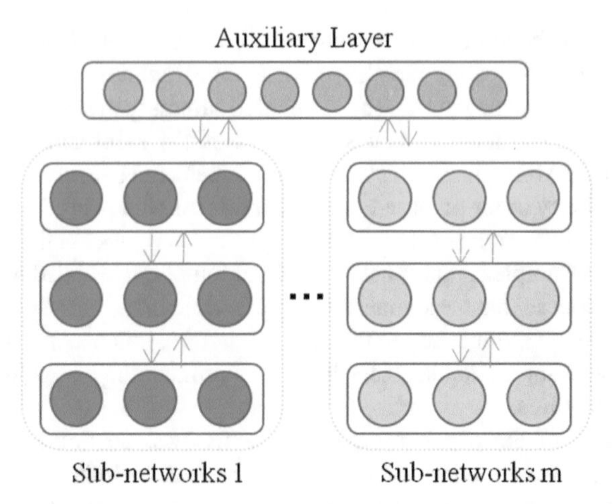

Fig. 1 Structure of semi-supervised multi-modal neural network

2.2 Graph Compressive Sensing

Graph theory plays an important role in analyzing networks. The two key tools for studying graphs are the adjacency matrix and the Laplacian matrix [9]. For an undirected, unweighted graph $G = (V, E)$, the graph Laplacian is $L = D - A$, with the adjacency matrix A and the degree matrix D. Graph compressed sensing (GSCS) that we used in this paper shows that if the sampled signals are spatially or temporally related, we can construct an underlying graph, so that the signals are compressible in the corresponding transform domain. The equation of GSCS is expressed as follows:

$$y = \Phi \Psi \mathrm{s} = \Phi \mathrm{x} \tag{2}$$

For signal supported on graph $x \in R^N$, N is the number of vertices, Φ is the measurement matrix with dimension $M \times N$.Ψ is the Laplacian eigenbasis of graph G with dimension $N \times N$.

More specifically, in order to use GSCS to achieve the purpose of eliminating the redundancy of hidden layer features, the first step required is to construct an underlying graph of the hidden layer features extracted by the automatic encoder mentioned above. From the perspective of columns, we regard each column of the low-dimensional high-level abstract feature matrix extracted by AE as a point. Then, calculate the Euclidean distance between each point (each column) and the rest, sort them from small to large. The KNN graph, used in this paper to construct an underlying graph, which is as the name implies, selects and connects the k-nearest-neighbors at each point, thereby constructing an undirected, unweighted graph G, and acquires its adjacency matrix A, the degree matrix D.

The second step is to spectral decompose the L mentioned above, the resulting feature matrix obtained is the Laplacian eigenbasis Ψ, the sparse matrix, and project the graph signal onto the substrate, which means the graph Fourier transform(GFT), to make sure the signal is compressible. A detailed theoretical analysis on why the graph Laplacian eigenbasis can be regarded as the Fourier transform of graphs is described in Ref. [10].

The third step is to determine the measurement matrix, which representing the sampling method. For general signals, independent and identically distributed Gaussian random measurement matrices can become universal compressed sensing measurement matrices. However, for the signals supported on the graph in this paper, Φ is a submatrix obtained by selecting randomly the rows of Laplacian eigenbasis Ψ. Each KNN graph generates multiple measurement matrices based on multiple random rows of the Φ. In other words, we can reconstruct the original information through only a few points represented by a few rows.

The fourth step is to solve the underdetermined equation mentioned above to obtain the sparse coefficient s based on the known measurement value y, reconstruct the original signal after obtaining the result, and compare the MSE between the original signal and the reconstructed signal. Select the Φ with the smallest MSE, which represents the optimal feature after removing redundancy.

3 Experimental Design

3.1 Database

In order to verify the effectiveness of the algorithm proposed in this paper, the database used in the experimental design is the Animal with Attributes (AWA) dataset, containing 30,475 natural pictures of animals, divided into 50 categories. It provides six high-dimensional modal features. This paper uses 8000 images in ten categories, of which 7200 are used as training sets, the rest is used as a test set, and all unlabeled data is also used in the pre-training process.

3.2 Experimental Process

The convolutional neural networks introduced as a classifier in this section use the most common two-dimensional convolutional layers. It has two spatial dimensions, high and wide, and is often used to process image data.

Block diagram of the data processing structure of the entire experiment is shown in Fig. 2

3.3 Experimental Results

This paper selects 24 best points and 32 best points from the extracted low-dimensional high-level abstract features represented by 64 points through the method of graph compression sensing, and randomly samples 24 points and 32 points, respectively. And classify these sets of data through a convolutional neural network and the experimental results are shown in Table 1 After comparison, the feature extraction and selection method proposed in this paper has a very good effect compared with randomly extracting the main features from low-dimensional isomorphic features. It can be seen from the results that the classification effect is not good when the value of K increases. When the value of K is 8, the effect is the best.

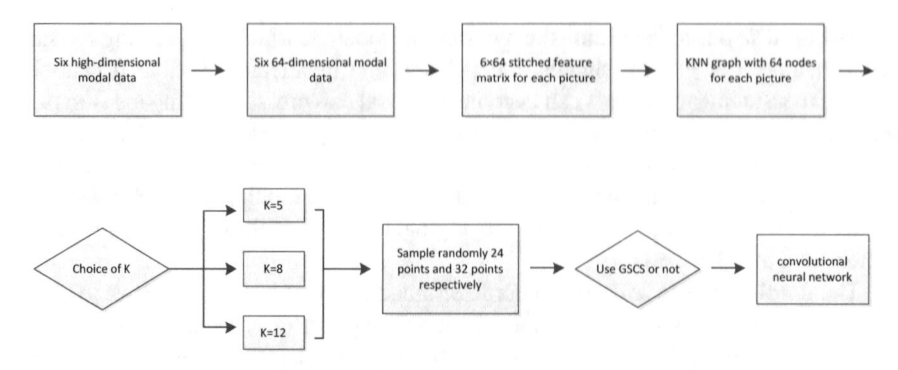

Fig. 2 Data processing structure of the entire experiment

Table 1 Comparison results of the classification accuracy

K	24 refined features		32 refined features		64 complete features		24 randomly selected		32 randomly selected	
	Train	Test	Train	Test	Train	Test	Train	Test	Train	Test
5	0.9794	0.9676	0.9944	0.9902	0.9994	0.9979	0.9114	0.2423	0.9219	0.3155
8	0.9867	0.9741	0.9935	0.9911						
12	0.9752	0.9680	0.9929	0.9899						

4 Conclusion

This paper introduces a multi-modal feature extraction and selection of deep learning based on graph signal processing structure and details the design and theory. In order to verify the effectiveness of the proposed model, a series of comparative experiments were carried out. The experimental results show that the proposed multi-modal feature extraction and selection model can effectively extract low-dimensional features from the original high-dimensional data, and remove redundant information to the greatest extent in order to obtain more refined features.

Acknowledgements. The work was supported by the Natural Science Foundation of China (61731006, 61971310)

References

1. Karsoliya S (2012) Approximating number of hidden layer neurons in multiple hidden layer BPNN architecture. Int J Eng Trends Technol 3(6):714–717
2. Jiang Y, Xiao J, Liu X, Hou J (2018) A removing redundancy restricted Boltzmann machine. In: Tenth international conference on advanced computational intelligence (ICACI), pp 57–62
3. Lü X, Meng L, Chen C, Wang P (2019) Fuzzy Removing Redundancy Restricted Boltzmann Machine: improving learning speed and classification accuracy. IEEE Trans Fuzzy Syst. https://doi.org/10.1109/TFUZZ.2019.2940415

4. Luo H, Shen R, Niu C, Ullrich C (2011) Sparse group restricted boltzmann machines. In: Twenty-Fifth AAAI conference on artificial intelligence, pp 429–434
5. Guo R, Qi H (2013) Partially-sparse restricted boltzmann machine for background modeling and subtraction. In: 12th International conference on machine learning and applications, vol 1. pp 209–214
6. Chen J, Wu Z, Zhang J, Li F, Li W, Wu Z (2018) Cross-covariance regularized autoencoders for nonredundant sparse feature representation. Neurocomputing 316:49–58
7. Mirzaei A, Pourahmadi V, Soltani M, Sheikhzadeh H (2020) Deep feature selection using a teacher-student network. Neurocomputing 383:396–408
8. Peng H, Long F, Ding C (2005) Feature selection based on mutual information criteria of max-dependency, max-relevance, and min-redundancy. IEEE Trans Pattern Anal Mach Intell 27(8):1226–1238
9. Zhu X, Rabbat M (2012). Graph spectral compressed sensing for sensor networks. In: IEEE International conference on acoustics, speech and signal processing (ICASSP), pp 2865–2868
10. Zhu X, Rabbat M (2012) Approximating signals supported on graphs. In: IEEE International conference on acoustics, speech and signal processing (ICASSP), pp 3921–3924

Vehicle Detection in Aerial Images Based on YOLOv3

Ruiheng Hu[1], Bingcai Chen[1,2(✉)], and Tiantian Tang[1]

[1] School of Computer Science and Technology, Xinjiang Normal University, No. 102 Xinyi Road, 10763 Xinjiang, China
china@dlut.edu.cn

[2] School of Computer Science and Technology, Dalian University of Technology, No. 2 Linggong Road, Liaoning, China

Abstract. In the twenty-first century, traffic problems are still troubling people. The concept of intelligent transportation was proposed to solve this problem. The main problem in intelligent traffic is to obtain information about the vehicles on the road and use this information to manage traffic problems. Most of the data now comes from the driveway monitor. The camera cannot focus on everywhere. Therefore, when some accident occurs, UAV can collect information as fast as possible. UAV is more inflexible than a fixed camera. In terms of real time, YOLOv3 is the best choice, because YOLOv3 is extremely fast. Thus, this study has important significance in promoting the development of intelligent transportation.

Keywords: UAV · Intelligent transportation · YOLOv3

1 Introduction

According to the ministry of public security's traffic management bureau, China had 319 million vehicles at the end of June 2018. People from all over the world must face the problem of urban traffic congestion. For example, the number of traffic accidents is increasing. So, the issue of traffic environmental degradation needs to be solved as soon as possible. Facing the terrible traffic environment, it is significant to consider the people, cars, and roads environment involved in the intelligent transportation system, and the primary goal of intelligent transportation system is to collect road vehicle information, and vehicle detection and identification is an important part of it.

The purpose of vehicle detection is to analyze whether there are vehicles in the detection area and use bounding boxes to represent the vehicles in the scene, so that traffic flow, road occupancy, and other related information can be obtained through vehicle detection. Vehicle detection can also be used to deal with unexpected accidents or special road conditions, parking lot utilization, urban planning, etc. [1]. So that this system will have a wide range of application prospects.

In fact, the data used to process could be collected from remote sensing or driveway monitoring systems, but there are many inconveniences. Firstly, you have to spend lots

Q. Liang et al. (eds.), *Artificial Intelligence in China*, Lecture Notes in Electrical Engineering 653, https://doi.org/10.1007/978-981-15-8599-9_68

of time waiting to access remote sensing, and the weather has a great influence on remote sensing images, such as clouds that will hide the earth's surface. As for the traffic monitoring systems that are too rigid to change the plan according to the task, unmanned aerial vehicle (UAV) becomes a better bet.

The rapid development of unmanned aerial vehicle (UAV) is accompanied by the continuous maturity of relevant technologies, such as camera platform and better battery life. Both the military and civilian communities are concerned about the development of drone technology, because of UAVs provide a complement to monitoring systems. UAV system is characterized by low cost, a wide range of use, stable pictures and video, and strong timeliness. In addition, due to the low requirements of UAV take off conditions, so that tasks could be planned according to the real-time status and further save costs.

From what has been discussed above, we present base YOLO version 3 model vehicle detection in aerial images, which the YOLO model is extremely fast and compare with other models such as R-CNN, and YOLO is more suitable for real-time detection. Besides, compared with fixed monitoring, UAV use for the task will gain a wilder view, which means there are more vehicles in one image. Hence, this project will also promote the development of "intelligent transportation system," which is of great significance and practical value.

2 Base on YOLOv3

In the intelligent transportation system, vehicle detection is the basic function which is a relatively low-level function, which provides data support for upper functions such as traffic management systems and traffic information service systems. Especially when it comes to emergency response tasks, the real-time requirement is high. The previous method was generally using hand-crafted features and a classifier within a sliding window approach [2–4]. But now some object detection algorithms, such as SPP, fast, and regional recommendation networks such as R-FCN, have continuously improved the speed and accuracy of object recognition, but still fail to meet the requirements of the industry, so this model could not complete the task efficiently. In this paper, YOLOv3 is the algorithm used to detect vehicle, and Fig. 1 shows the model of YOLOv3. This object detection treat problem as a single regression problem, from image pixels to bounding box coordinates and class probabilities. YOLO divides the input image into a 7×7 grid, and the final output is a $7 \times 7 \times 30$ tensor, which means the grid in the input corresponding to the grid in the output. In the output tensor, 30 dimensions including two confidence scores of bounding boxes and eight dimensions are to represent the places of these two bounding boxes. Each place of bounding boxes is shown as four numbers, which is x, y, w, and h. The x and y coordinates represent the center of the box. The bounding boxes' width and height are represented by w and h. And the confidence prediction of the bounding box equals the probability of the object in the bounding box times the IOU between the predicted box and ground truth box. In [5], author multiply the probabilities and the confidence predictions,

$$\text{Confidence} = \Pr(\text{Object})^{*}\text{IOU}_{\text{pred}}^{\text{truth}}$$

$$\text{IOU}_{\text{pred}}^{\text{truth}} = \frac{\text{Detection} \cap \text{GroundTruth}}{\text{Detection} \cup \text{GroundTruth}}$$

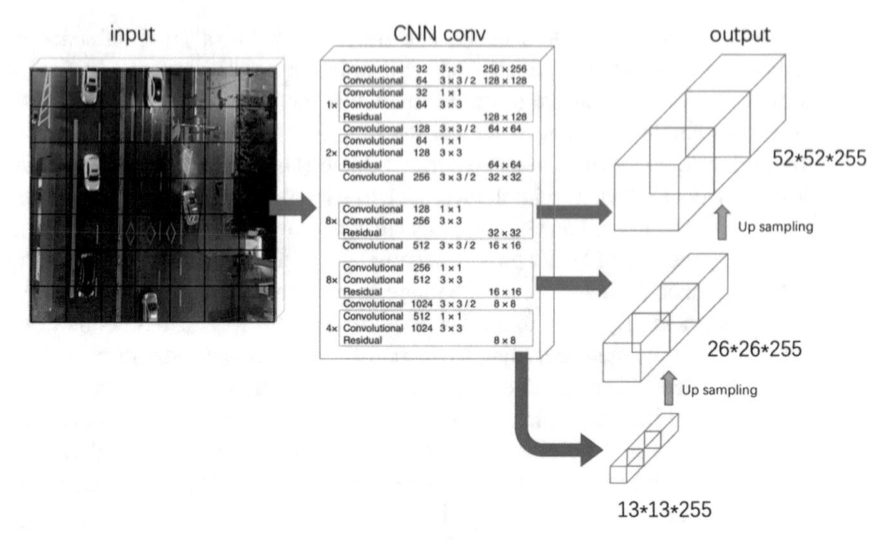

Fig. 1. Architecture of YOLOv3

The confidence scores of each bounding box mean how confident the object is in the box and how accurate the predicts is right. If a confidence score equal to zero, it means that the object is not in the box that we predict or the box we predicted is far away from the object. The confidence score to equal to the intersection over union (IOU) between the predicted box and the ground truth shows that we got the right box.

YOLO3 use the Darknet-53 network for performing feature extraction. Part of the Darknet-53 is given in Table 1. This is a huge improvement over the previous version. This Darknet-53 network contains 53 convolutional layers. And comparing with ResNet-101 or ResNet-152, Darknet has much fewer layers, and therefore, we can process faster and more efficient.

To deepen this model, YOLOv3 adds some shortcut connections between some layers to replace the one by one and three by three block in the last version. Just like residual block, and each residual block contains two convolution layers and one shortcut connection.

The third version of YOLO uses logistic loss function instead of SoftMax loss in the second version, and it shows that logistic loss could handle the problem which is one data in the dataset that has two labels. Such as a truck is also labelled as a car.

3 Experiment and Analysis

3.1 UA-DETRAC Dataset

The UA-DETRAC [6] dataset is adopted in the experiment. The dataset is mainly shot in Beijing and Tianjin. The image size is 960 * 540. Three sizes are defined: small (0–50 pixels), medium (50–150 pixels), and large (more than 150 pixels), and four conditions of cloudy, night, sunny, and rainy days are considered. Before using the dataset, we need to convert the annotation file to voc2007 format for use by YOLO.

Table 1. Part of the Darknet-53

	Type	Filters	Size	Output
	Convolutional	32	3 × 3	256 × 256
	Convolutional	64	3 × 3/ 2	128 × 128
	Convolutional	32	1 × 1	
1×	Convolutional	64	3 × 3	
	Residual			128 × 128
	Convolutional	128	3 × 3/ 2	64 × 64
	Convolutional	64	1 × 1	
2×	Convolutional	128	3 × 3	
	Residual			64 × 64
	Convolutional	256	3 × 3/ 2	32 × 32

3.2 Model Training

The initial learning rate is 0.01, and 64 samples are accumulated in Darknet for a back-propagation. Start training and saving log for generating Loss curve. The weight file is saved every 1000 times in the training of the Darknet network. The maximum number of iterations in the experiment is 500,000. Considering the influence of weather and sunlight on the aerial photography of UAV, there will be different exposures. Therefore, set the saturation parameter and exposure parameter to 1.5, that is, adjust the saturation and exposure to increase the number of samples.

When the training stops, the visual training log is generated as shown in Fig. 2 below, and the loss decreases continuously in the training until it reaches about 0.3; AVG_IOU was close to 1 in the training process, and the curve of IOU was more than 80% with the increase of batches.

3.3 Results and Analysis

As shown in Fig. 3, the cars on the road are packed in bounding boxes. The target category is shown in the upper left corner of the box.

Then, we use the test set to test the model in batches. The total number of test set samples used is 9852, which is the test set provided in dataset UA-DETRAC. The results are given in Table 2:

According to different functions, there are many evaluation indexes in the object detection task, such as recall, precision, PR curve, average precision (AP), and average

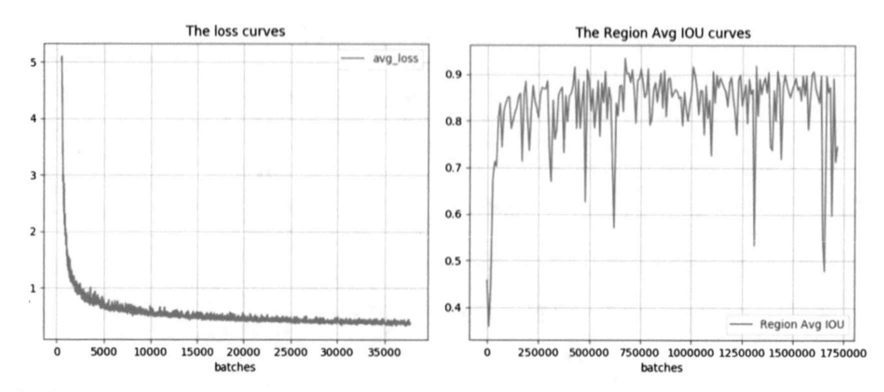

Fig. 2. Loss curve and AVG_IOU curve; left: loss function curve; right: The region AVG_IOU curves

Fig. 3. Detection results

Table 2. Results of YOLOv3 based on UA-DETRAC

TP	FP	FN	Avg_IoU	Precision
68529	4312	1990	82.17%	0.94

precision average (map).

$$\text{Precision} = \frac{\text{TP}}{\text{TP} + \text{FP}}$$

Among them, TP is the quantity predicted to be positive for the actual positive samples; FP is the quantity predicted to be positive for the actual negative samples; FN is the quantity predicted to be negative for the actual positive samples.

As can be seen from the above, most cars can be identified, including the buses which are turning. But it can also be seen from the picture that the cars far away from the main body of the picture, that is, the smaller cars, are easy to be ignored and not recognized.

4 Conclusion

In this paper, we introduced the most important component of intelligent transportation system, which is vehicle detection. We further proposed to use YOLOv3 as our detector. YOLOv3 uses some shortcut connections between some layers like what residual block does. Training with UA-DETRAC datasets and we take 80% as training set and 20% as test set, turns out the precision is 0.94. As future work, the performance could be improved by using some anchor free algorithm, since YOLO is a single-stage algorithm, the speed of this detector would be still fast.

Acknowledgements. This work was supported in part by the Tianshan Young Talent Program, Xinjiang Uygur Autonomous Region under Grant 2018Q024, in part by the Natural Science Foundation of China under Grant 61771089 and Grant 61961040, and in part by the Regional Cooperative Innovation Program of Autonomous Region (Aid Program of Science and Technology to Xinjiang) under Grant 2020E0247 and Grant 2019E0214.

References

1. Sommer LW, Schuchert T, Beyerer J (2017) Fast deep vehicle detection in aerial images. In: 2017 IEEE winter conference on applications of computer vision (WACV), IEEE, pp 311–319
2. Cheng G, Han J (2016) A survey on object detection in optical remote sensing images. ISPRS J Photogrammetry Remote Sens 117:11–28
3. Liu K, Mattyus G (2015) Fast multiclass vehicle detection on aerial images. IEEE Geosci Remote Sens Lett 12(9):1938–1942
4. Moranduzzo T, Melgani F (2014) Detecting cars in UAV images with a catalog-based approach. IEEE Trans Geosci Remote Sens 52(10):6356–6367
5. Redmon J, Divvala S, Girshick R, Farhadi A (2016) You only look once: unified, real-time object detection. In: Proceedings of the IEEE conference on computer vision and pattern recognition, pp 779–788
6. Wen L, Du D, Cai Z, et al (2015) UA-DETRAC: A New Benchmark and Protocol for Multi-Object Detection and Tracking [J]. Comput Sci

A New Node Optimization Algorithm of Wireless Sensor Network Based on Graph Signal

Lei Xu, Yifei Yuan, and Wei Wang(✉)

Tianjin Key Laboratory of Wireless Mobile Communications and Power Transmission, Tianjin 300387, China
weiwang@tjnu.edu.cn

Abstract. Graph is a great model that can efficiently deal with structural data and has excellent research value and application value in our life. Graph theory and signal processing have been perfectly combined to form graph signal processing theory, which provides a theory for studying high-dimensional and structural signals. The wireless sensor network (WSN) is a typical data transmission network based on a graph structure. Because of redundant sensor nodes in WSN, it results in large energy consumption and slow transmission rate. To solve this problem, we propose a new algorithm to optimize the number of nodes by using graph signals processing and compressed sensing (CS) theory. The new node optimization algorithm can not only reduce energy consumption but also increase the transmission rate.

Keywords: Graph signal processing · Wireless sensor network · Compressed sensing

1 Introduction

The graph structure is a good model that can intuitively and effectively represent visual data or study the large number of complex data. The graph structure is not only applied in the computer field but also has great application value in big data, machine learning, and sensor networks. In recent years, the research on WSN based on graph structure has developed rapidly [1, 2]; graph structure can be perfectly combined with signal processing to form graph signal processing theory. Graph signal theory is widely used in many fields such as transportation network [3], social relationship processing [1], brain networks [4, 5], and sensor networks [6], etc. Because processing these data, in addition to thinking of the data itself, we also need to consider the interrelated between the data.

For large-scale WSNs, research on WSN based on graph structures has been very popular in recent years, especially graph signal processing has made great breakthroughs in these fields [7]. The optimization of WSN is the choice of sensor node location. The location of the sensor node is selected to minimize the number of sensors or to optimize the performance of a given number of sensors, to achieve the best prediction of variables at undetected locations [8]. The new algorithm is to reduce the number of sensor nodes to achieve the purpose of optimizing the WSN.

Q. Liang et al. (eds.), *Artificial Intelligence in China*, Lecture Notes in Electrical Engineering 653, https://doi.org/10.1007/978-981-15-8599-9_69

CS [9, 10] is to complete the data compression during the sampling process, so that the original signal can be recovered with a small number of sampling points. In the WSN, each sensor node can be regarded as a time point, and the signal transmitted at a specific moment can be regarded as a one-dimensional signal equal to the number of nodes. Therefore, we can combine CS and graph signals to form the new algorithm theory for nodes optimization of graph-based WSNs. We can use our new algorithm to optimize the sensor nodes in the WSN.

The rest of the paper is organized as follows: In Sect. 2, we will introduce the construction of WSN and knowledge related to graph signals. In Sect. 3, we will introduce the graph compressive sensing theory. In Sect. 4, experiment. Section 5 concludes the paper.

2 Wireless Sensor Network and Graph Signal

A good WSN model plays an influential role in the optimization of sensor nodes. Therefore, we build a WSN into a two-dimensional random geometric graph model $G(n, r)$, where n is the number of sensor nodes in the WSN and r is the communication radius of each sensor. We require the communication radius and sensing radius of each sensor to be r. If the Euclidean distance between the two sensors is less than the communication radius between them, the two sensing points are considered connected. Such random geometric graphs have been widely used in model optimization of WSNs [11, 12]. In order to be more practical, we made a WSN in this way and took its appropriate sensing radius r, which required that the WSN composed of sensing radius r can cover the whole area as much as possible.

A graph is represented as $G = \{V, E\}$, where V is the number of vertices in the graph and E is the number of edges in the graph. The graph signal is defined as $f \in R^N$. The number of nodes is $N = |V|$. The adjacency matrix A is a symmetric matrix, which represents whether two nodes are connected to each other. The degree matrix D of the graph is a diagonal matrix, and the m-th diagonal element can be expressed as:

$$D(m, m) = \sum_n A(m, n) \tag{1}$$

The graph Laplacian matrix can be express as the follow:

$$L := D - A \tag{2}$$

In our new algorithm, we only select a subset of sensor nodes to participate in data transmission. Assuming S is a set of nodes of the WSN that have been selected, $S^C = V \backslash S$ is a set of nodes that have been unselected. Mutual information (MI) represents the correlation between two signals and is a criterion for measuring the similarity of two signals. In our new algorithm, the selected nodes set S can be used to reduce the uncertainty of the unselected nodes set S_C. In order to better show the similar effect, we propose an objective function in our new algorithm. The objective function is:

$$S^* = \underset{S \in V : S = |M|}{\arg\max}\ MI(S, SC) \tag{3}$$

The $MI(S, S^C)$ represents MI between two subsets S and S^C. In order to obtain the best sensor node set S and perfectly reconstruct the original signal. We need to gradually increase the number of sensor nodes in S, so that the objective function S^* can reach the maximum value as soon as possible.

3 Graph Compressive Sensing Theory

CS is a very useful theory for processing sparse signals. For a one-dimensional signal $x \in R^N$, we can express it as:

$$x = \sum_{i=1}^{N} \phi_i \theta_i \text{ or } x \phi_i \theta. \tag{4}$$

where ϕ is a sparse matrix and $\phi \in R^{N*N}$, θ is the sparse representation of the signal x under the sparse basis $\theta \in R^N$, and the equation of the CS can be expressed as:

$$y = \varphi x = \varphi \phi \theta \tag{5}$$

where φ is an sensing matrix and $\varphi \in R^{M*N}$ where M is the number of observation nodes and $M \ll N$, y is the observation vector during data transmission and $y \in R^M$.For a sparse signal x, we can use the convex optimization reconstruction algorithm to get the reconstructed signal x_1 by solving the linear equation [13].

$$\min_X \|X\|_1 \text{ s.t.} y = \phi x. \tag{6}$$

In our new algorithm, we consider a signal $x \in R^N$ can be regarded as a set of data generated by WSN with N sensor nodes at a specific time. For example, each subset x_I can find the corresponding sensor node I from the WSN. As mentioned before, our node optimization algorithm is suitable for the WSN with large number of nodes. When the WSN transmits data, we can regard the data of a specific time node as a signal $x \in R^N$. Combining with the previous CS theory, when we are collecting data on a WSN, we do not need to collect the data of all sensor nodes, but only need to collect the M ($M \ll N$) sensor nodes to transmit data to the sink nodes, and the rest N-M sensor nodes are in sleep state. In this way, we can reduce the energy consumption of the sensor nodes and improve the data transmission rate. Through the theory of Graph Fourier Transform [8] and CS, we can know that the Laplacian eigenvectors can be used as a sensing matrix [13]. However, it is different from traditional CS theory, because the WSN is based on a graph structure and the sensor nodes are not isolated but interrelated, so we cannot use the random Gaussian matrix for its sensing matrix. In our new algorithm, we need to construct a new sensing matrix. Because we only have selected M nodes as sensor nodes set from the WSN. So, we randomly and not repeatedly selected M rows from its Laplacian eigenvectors U and form a new sensing matrix $U_\Omega R^{M*N}$. The new sensing matrix U_Ω satisfies the RIP [9, 10] and orthogonality of the matrix. Then according to the theory of our algorithm, we can form a new equation of the CS:

$$y = U_\Omega x \tag{7}$$

We record the selected sensor nodes set as Ω and $y \in R^M$, where y is the measurement vector. Then, we have the sensing matrix U_Ω and the measurements y. At last, we can get the reconstructed signal x_1 by solving the l_1 convex optimization problem [13].

$$\min_X \|X\|_1 \; s.t. y = U_\Omega x. \tag{8}$$

Then we calculate the mean square error (MSE) to determine the optimal number of M and the optimal placement of sensor nodes in the WSN. The MSE can reflect the difference between the two signals and the MSE between the two signals is smaller; the two signals are more similar.

4 Experiment and Experiment Analysis

We randomly generated a WSN with $N = 500$ sensor nodes in a two-dimensional area and specify that the communication radius and transmission radius of each sensor are equal. If the Euclidean distance between two sensor nodes is less than their communication radius r, we can consider that the two sensor nodes are connected to each other. Then we generate a non-sparse signal x, and by placing the non-sparse signal x in a specified variation domain, let the original signal x becomes a sparse signal with a sparsity of $K = N/10 = 50$. Under the condition of $K = 50$ and $N = 500$, we gradually increase the number of sensor nodes in the set S where $|S| = M$, and compare the MI and MSE one by one. Finally, get the optimal sensor nodes set S and mark the selected sensor nodes. Each experiment was run for 500 times to get the best of MI and MSE. The experimental results are shown in Figs. 1, 2, and 3.

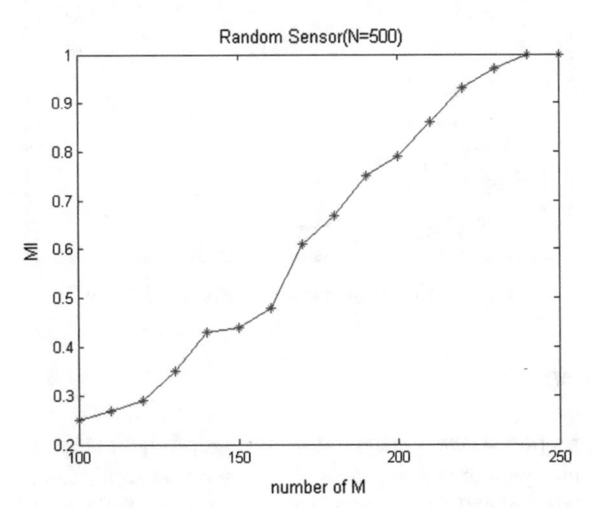

Fig. 1. The MI transformation curve between S and S^C

As can be seen from the above two figures that in the condition of sparsity $K = 50$ and $N = 500$, when $M = 220$, the original signal can be correctly reconstruct.

The red points have been selected, and the blue points have not been selected.

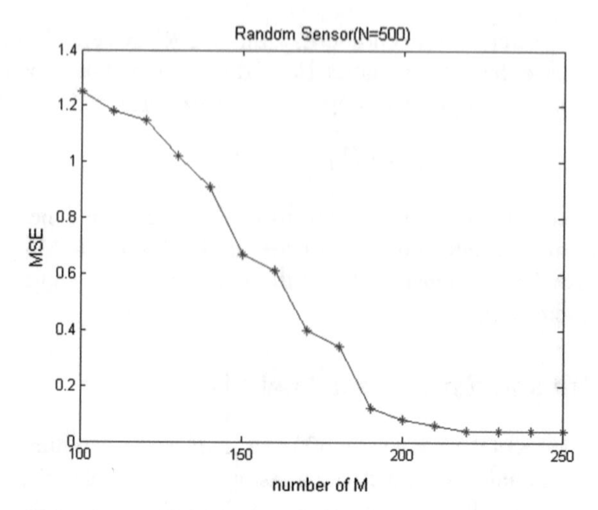

Fig. 2. The MSE transformation curve between x and x_1

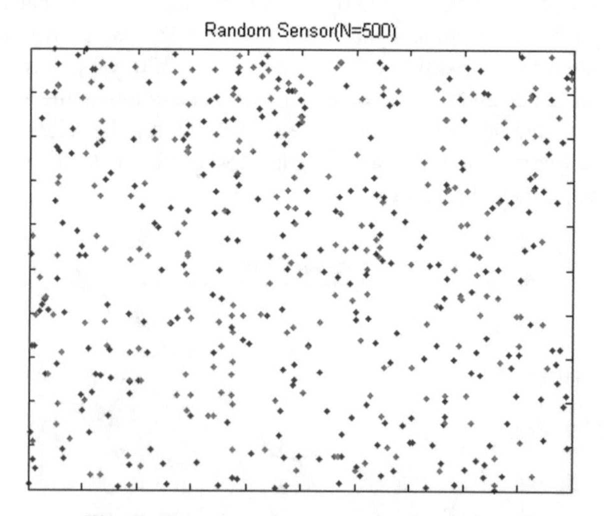

Fig. 3. The selected sensor nodes in the WSN

5 Conclusions

In this paper, we propose a new node optimization algorithm for large-scale WSNs, using the theory of graph signal processing and CS. The new node optimization algorithm that we proposed can optimize the number of sensor nodes in a WSN with large number of nodes. We can use fewer sensor nodes to transmit data when WSNs transmit data, this will not only reduce energy consumption but also increases transmission rate. However, the shortcoming of this algorithm is that the optimal node set we choose adopts the sampling idea of random sampling. Such sampling will lead to instability and randomness of reconstruction results in some condition. Therefore, our next work is to improve our algorithm that can improve stability and reduce uncertainty and randomness, to help us better achieve nodes optimization of WSN.

Acknowledgements. The work was supported by the Natural Science Foundation of China (61731006, 61971310).

References

1. Sandryhaila A, Moura JMF (2013) Discrete signal processing on graphs. IEEE Trans Signal Process 61(7):1644–1656
2. Sandryhaila A, Moura JMF (2014) Big data analysis with signal processing on graphs: Representation and processing of massive data sets with irregular structure. IEEE Signal Process Mag 31(5):80–90
3. Crovella M, Kolaczyk E (2003). Graph wavelets for spatial traffic analysis. In: Twenty-second annual joint conference of the IEEE computer and communications societies, San Francisco, CA, vol 3. pp 1848–1857
4. Higashi H, Rutkowski TM, Tanaka T, Tanaka Y (2016) Multilinear discriminant analysis with subspace constraints for single-trial classification of event-related potentials. IEEE J Sel Topics Signal Process 10(7):1295–1305
5. Huang W, Bolton TAW, Medaglia JD, Bassett DS, Ribeiro A, Van D De, Ville A (2018) Graph signal processing perspective on functional brain imaging. Proc IEEE 106(5):868–885
6. Leonardi N, Van D, De Ville (2013) Tight wavelet frames on multislice graphs. IEEE Trans Signal Process 61(13):3357–3367
7. Ortega A, Frossard P, Kovaˇcevi´c J, Moura JMF, Vandergheynst P (2018) Graph signal processing: overview, challenges, and applications. Proc IEEE 106(5):808–828
8. Sakiyama A, Tanaka Y, Ortega A (2016) Efficient sensor position selection using graph signal sampling theory. In: IEEE International conference on acoustics, speech and signal processing (ICASSP), pp 6225–6229
9. Candes E, Tao T (2006) Near-optimal signal recovery from random projections Universal encoding strategies. IEEE Trans Inf Theory 52(12):5406–5425
10. Donoho D (2006) Compressed sensing. IEEE Trans Inf Theory 52(4):1289–1306
11. Gupta P, Kumar PR (2000) The capacity of wireless networks. IEEE Trans Inf Theory 46(2):388–404
12. Boyd S, Ghosh A, Prabhakar B, Shah D (2005). Mixing times for random walks on geometric random graphs. In: Workshop on algorithm engineering and experiments and the second workshop on analytic algorithmics and combinatorics, DBLP
13. Zhu X, Rabbat M (2012) Graph spectral compressed sensing for sensor networks. In: IEEE international conference on acoustics, speech and signal processing (ICASSP), pp 2865–2868

Research and Practice of Intelligent Water Conservancy Integration Management Platform in Xinjiang

Yumeng Lin[1], Bingcai Chen[1(✉)], Zhiming Ma[1], Qian Ning[2,3], Yanting Xiao[2,3], Lun Shao[2,3], Qiang Luo[2,3], and Xinzhi Zhou[3]

[1] School of computer science and technology, Xinjiang Normal University, Urumchi 830054, China
china@dlut.edu.cn

[2] Chengdu Wanjianggangli Technology Corp. LTD, Chengdu 610041, China

[3] Joint Laboratory of Water Conservancy Informatization, School of Electronics and Information, Sichuan University, Chengdu 610065, China

Abstract. Jeminay county is the arid and poor water county with the least surface water resources in Xinjiang. There are no large rivers and lakes in the territory. Muz Taw iceberg is the only water source. The traditional water conservancy management methods are relatively backward, with high labor intensity, low efficiency and poor benefit, so it can no longer solve the current water resources management dilemma in Jeminay county, nor adapt to the current economic development situation. The modern wisdom water conservancy utilizes information technology, network technology, big data and artificial intelligence to make water conservancy management more intelligent, greatly improving efficiency and striving for sustainable development of water conservancy. Based on the smart water conservancy comprehensive management platform of Jeminay county in Altay Prefecture of Xinjiang, this chapter demonstrates the superiority of the intelligent information integrated platform in water conservancy management.

Keywords: Water management · Smart water · Artificial intelligence · Internet · Big data

1 Introduction

Water is not only a basic natural resource, but also a strategic economic asset. Water conservancy and affairs are vital guarantee for the national economy so

Bingcai Chen, professor, doctoral supervisor. Main research direction: water conservancy Internet of Things, intelligent system.

Q. Liang et al. (eds.), *Artificial Intelligence in China*, Lecture Notes in Electrical Engineering 653, https://doi.org/10.1007/978-981-15-8599-9_70

as to achieve sustainable development. In recent years, the utilization rate of surface water resources in Jeminay County has reached more than 80%, greatly exceeding the ecological warning line of 60%. The status of water conservancy in Jeminay County directly affects local food production and farmers' income. Therefore, it's necessary to strengthen the construction of water conservancy in Jeminay County and promote the continuous improvement of management capacity and service level of water conservancy in the county. Smart water conservancy aims to enhance the water governance system by applying advanced information technologies such as mobile Internet, Internet of Things, big data, cloud computing, and artificial intelligence to achieve full interconnection [1–3].

In this chapter, we will propose an intelligent water conservancy integration management platform through scientific top-level design and cutting-edge information technology applications to implement full water conservancy business. It is with one-stop integration, standardization of construction, intelligent management control, automation of early warning response, data analysis of water conservancy, and intelligent business mobility to comprehensively promote the water resources optimal allocation capability, guaranteeing sustainable use of water resources, and to boost water efficiency by setting up a modernized water conservancy development path of Smart, Intensive, Safe, and Sustainable in Jeminay County.

2 Construction Goals

We are committed to optimizing the allocation of water resources and improving the efficiency of water resources utilization and stimulating the rational development, efficient use, comprehensive conservation, effective protection and optimal allocation of water resources in the region to ensure the sustainable development of the regional economy and society.

Through the platform the regional water conservancy information system is constructed in accordance with unified technical standards and data center, which is timely and accurate for the management hydrology and water resources, project monitoring and operation. By constructing the water regime monitoring system in Jeminay county, we will strengthen the infrastructure construction of water regime monitoring, by use of scientific information systems to analyze and judge the occurrence of flood conditions in a timely manner.

Refined and water-saving agricultural irrigation is accomplished through demonstration construction of modern irrigation districts. Through the demonstration construction of modern irrigation districts in Jeminay county, a water resource utilization and management network that integrates remote monitoring, remote adjustment, and remote control with the Jeminay County Water Conservancy Bureau as the center will be erected, from water source (Hongshan Reservoir) to channel water delivery and use, with the help of the flow monitoring and remote control of the gates at each outlet of the channel, to bring out closed-loop management and precise regulation of the total canal water volume, so as to accurately grasp the channel water utilization coefficient.We will establish a project operation management system involving dam safety and engineering inspections. Through the construction of non-engineering information-based monitoring methods, together with electronic engineering inspections and automatic monitoring of dam safety, it will effectively avoid or reduce engineering safety incidents caused by hidden engineering hazards.

3 Platform Design

3.1 Overall Architecture

According to the construction goals of the intelligent water conservancy integrated management platform, as well as the top-level design of the intelligent water conservancy, the overall framework of intelligent water conservancy in Jeminay County was established. It mainly includes the Internet of Things monitoring layer, computing resource layer and business application layer. The overall framework is supported by the security system and standard system to achieve the security, unified and standardized operation of the system, as shown in Fig. 1.

The monitoring layer of the Internet of Things mainly includes remote automatic control of the gate, water condition monitoring module, water condition monitoring module, and pump station monitoring module, etc., to accomplish the collection, monitoring and transmission of water conservancy information. The computing resource layer mainly includes the virtual server resource pool of the Regional Water Resources Bureau, the central industrial control computer, the big data distributed processing center, and the GIS cloud center. The business application layer mainly includes a central application software platform, an engineering operation management module, and a mobile terminal platform, etc., and realizes the specific application of the platform in different forms. The overall operating process of the platform is that each monitoring and monitoring system transmits data to the computing resource layer including the information center of the Jeminay County Water Resources Bureau through network transmission modes such as fiber optic private network RTU BeiDou communication, and then the computing resource layer processes the data efficiently in order to make a decision, and finally deploy the central application platform software to implement various applications.

3.2 Functional Architecture

The intelligent water management platform of Jeminay County implements functional system, which are smart sensing, intelligent simulation, intelligent diagnosis, intelligent early warning, intelligent dispatch, intelligent disposal, intelligent control, and intelligent service as shown in Fig. 2.

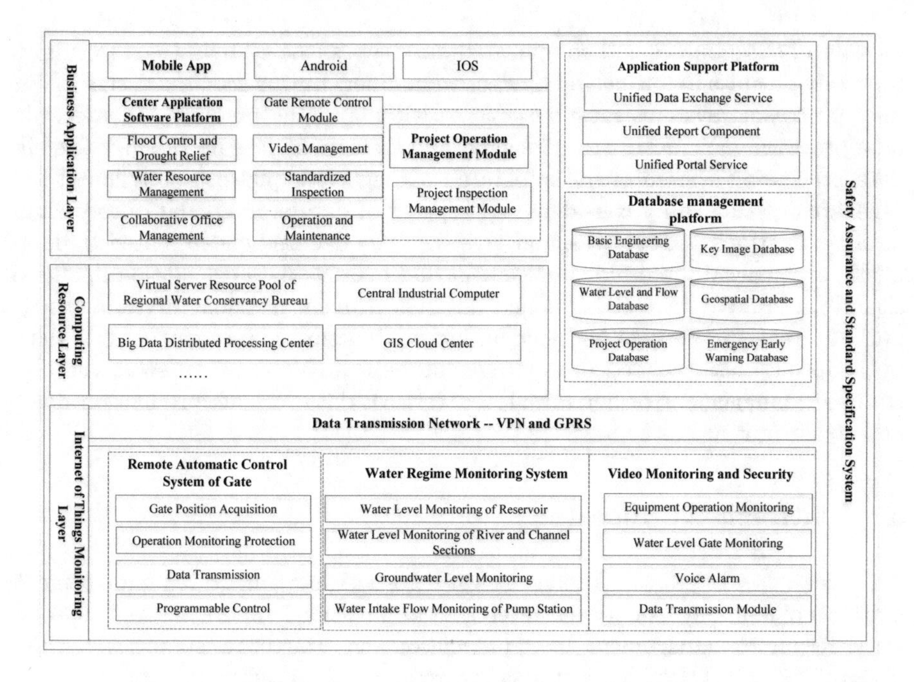

Fig. 1. Overall architecture of the intelligent water conservancy integrated management platform in Jeminay County

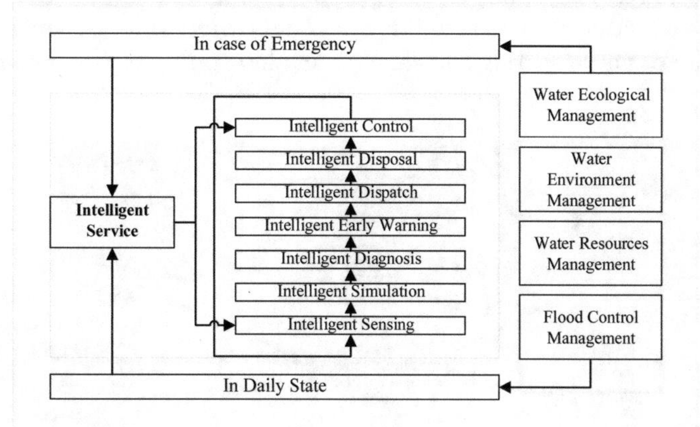

Fig. 2. Functional architecture of smart water conservancy integrated management platform in Jeminay County

3.3 Big Data Processing System

The core layer of the construction of the intelligent water conservancy integrated management platform is the computing resource layer, and the most critical of the computing resource layer is the big data distributed processing center. For massive data resources and complicated data types, the system needs to supply processing results in a timely and effective manner. In this way, the design of big data processing systems must be able to withstand the pressure of large-scale data processing and ensure the efficiency and security of data processing. The big data processing system of the integrated management platform is followed the SOA architecture, fully considered all aspects of data storage, data processing, data applications and services, and uses the extended and encapsulated Hadoop platform for distributed file management and massive data calculation in offline situations [4]. We have straightened out the problem of scalability and distribution of the full storage and calculation of structured and unstructured data by adapting well-designed distributed database architecture to solve the distributed management of horizontal and vertical data for safe production data, and concurrent access to business data.

4 Platform Application

As indicated in Fig. 3, the intelligent water conservancy integrated management platform of Jeminay County has intelligent application support in seven aspects: water resources management, flood prevention and drought resistance, water lifting projects, rural drinking water safety, reservoir dam safety, remote control of gates, and irrigation management. Through the completion of the construction of the platform, Jeminay County has accomplished comprehensive control of the county's water resources information, optimized water resources scheduling, and efficient use of water resources, building a county-wide water conservancy information system, and elevating the county's modernization of water conservancy.

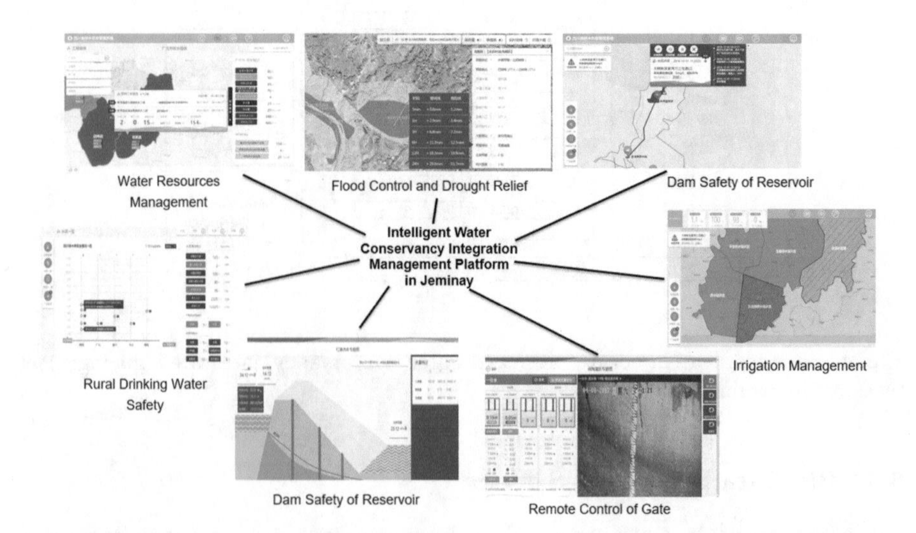

Fig. 3. Application module diagram of the intelligent water conservancy integrated management platform in Jeminay County

Reservoir dam safety management module provide real-time monitoring and recording of dam safety seepage pressure seepage data, reservoir water level and outflow data, and real-time early warning of monitoring items that exceed the threshold, and early warning push through mobile phone APP Responsible persons for dam safety can control the danger and make countermeasures at the first time.

Irrigation area water management module, consisting of irrigation area management, channel measurement, field measurement. Based on the long-term water transfer process data of the irrigation channel, the channel water transfer process data is summarized and analyzed. The water level, flow and water volume of the key cross-section of the channel are recorded in real time, and the water loss rate can be controlled.

The gates are remotely controlled and managed to centrally control 6 integrated gates in Hongshan Reservoir and downstream irrigation districts in Jeminay County. Real-time display of gate flow, gate opening, and animation of channel flow. The system provides 5 types of gate control methods to meet the gate control requirements under different water needs. We have bringed about dynamically grasping the demand and allocation of water resources in the entire irrigation district, and using intelligent systems to assist in scientific decision-making and dynamically adjust the real-time water allocation in different periods and regions during the water control process of each node (integrated gate). As shown in Fig. 4, it is the interface display of the remote control of the gate.

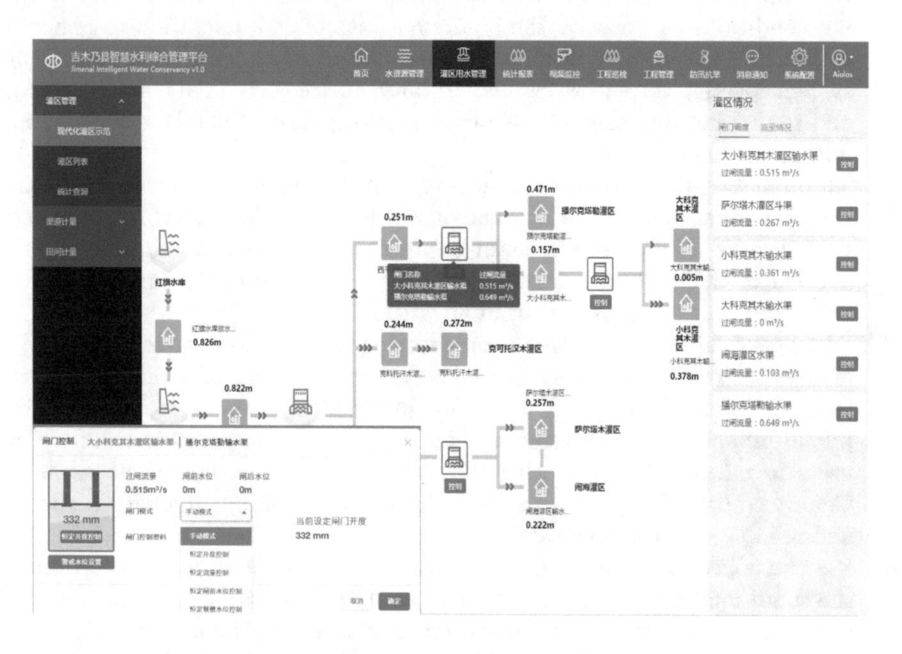

Fig. 4. Gate remote control management module

5 Conclusion

The intelligent information integrated management platform introduced in this article is an important part of the information construction of the Jeminay County Water Resources Bureau in Altay Region, which effectively elevates the management level and optimizes water resources deployment capacity of the Jeminay County Water Resources Bureau, keeps ecological balance, and bolsters up regional Water management facilities management, irrigation area water management, makes flood and drought management intelligent. The functions and benefits of smart water conservancy construction are mainly reflected in the following six aspects:

(1) It has a positive influence on improving the social and environmental benefits of water conservancy projects in Jeminay County. The results of optimal allocation of water resources, optimal scheduling, and optimal use not only improve the social benefits of water conservancy projects, but also contribute to the internal economic of water management units.
(2) It's of great significance to boost the management level of water resources in Jeminay County, that is, to achieve a comprehensive grasp of the water consumption in irrigation districts, and to control the regional water distribution situation from a macro perspective, which can be more effective optimize the allocation and dispatch of water resources.
(3) Real-time information on water conditions, moisture conditions, and working conditions can elevate the guarantee rate of safe operation of regional projects. It's needed to establish a flood and drought forecast management system covering the entire Jeminay County, grasping regional flood dynamics and improving management departments' response to floods and droughts during the flood season.
(4) Through long-term accurate monitoring of the amount of water in the reservoir, its capacity and the amount of water out of the reservoir, combined with a large number of historical data of water demand in various places, and it can efficiently predict the demand for water resources and to make a water resource scheduling plan in advance.
(5) Water users are able to grasp the engineering construction, water supply policies, water resources management and other situations of Jeminay County through information query to realize interaction in order to increase the public influence of management departments and the enthusiasm of water users to participate, promoting the water conservation bureau of Jeminay County to provide agricultural production and people's living with accurate and reliable water supply services.
(6) Via the collection of working condition information, it is possible to obtain the conditions of the project in time to apply a basis for formulating maintenance and treatment plans. Prevention and treatment in advance can avoid or reduce dam accidents and dam breaks due to hidden dangers in the project, which is conducive to ensuring the effectiveness of regional engineering.

Acknowledgments. This work was supported in part by the Regional Cooperative Innovation Program of Autonomous Region (Aid Program of Science and Technology to Xinjiang) under Grant 2020E0247 and Grant 2019E0214, in part by the Tianshan Young Talent Program, Xinjiang Uygur Autonomous Region under Grant 2018Q024, and in part by the Natural Science Foundation of China under Grant 61771089 and Grant 61961040.

References

1. Cheong SM, Choi GW, Lee HS (2016) Barriers and solutions to smart water grid development. J Environ Manage 57(3):509–515
2. Kulkarni P, Farnham T (2017) Smart city wireless connectivity considerations and cost analysis: lessons learnt from smart water case studies. J IEEE Access 4:660–672
3. Mohammed Shahanas K, Bagavathi SP (2016) Framework for a smart water management system in the context of smart city initiatives in India. J Proc Comput Sci 92:142–147
4. Ntuli N, Abu-Mahfouz A (2016) A simple security architecture for smart water management system. J Proc Comput Sci 83:1164–1169
5. Ogidan OK, Onile AE, Adegboro OG (2019) Smart Irrig Syst Water Manage Proc 10(1):25–31
6. Boulos PF (2017) Smart water network modeling for sustainable and resilient infrastructure. J Water Res Manage 31(4):1–12
7. Gao X, Nachankar V, Qiu J (2011) Experimenting lucene index on HBase in an HPC environment. In: Proceedings of the first annual workshop on high performance computing meets databases. ACM 25–28
8. Shestakov D, Moise D, Gudmundsson G et al (2013) Scalable high-dimensional indexing with Hadoop. In: IEEE 2013 the 11th international workshop on content-based multimedia indexing (CBMI), pp 207–212
9. Bu Y, Yan J Research on smart water architecture based on big data. In: Jinan C (ed) Shandong: 2017 China automation conference, vol 10, pp 766–769 (CAC 2017)
10. Ni J, Du J, Xu X et al (2018) Construction practice of intelligent river lake long system information system. J Water Conserv Inf (3):24–27
11. Xie L, Yu S, Ma Q et al (2018) Construction and development of domestic and foreign intelligent water information. J Water Supply Drain 44(11):135–139
12. Fang G (2017) Research on intelligent water production monitoring system based on cloud Internet of things. J Gansu Sci Technol 33(18):14–16
13. He X, Li Z, Guo H et al (2016) Research on the construction of Zhengzhou intelligent water system and related technologies. J Water Conserv Inf 12(6):61–67
14. Dong H (2017) Construction and key technology analysis of "smart water" platform system. J Water Conserv Plan Des (2):22–24
15. Pearce DW, Atkinson G (1993) Capital theory and the measure of sustainable development: a indictor of weak sustainability. Ecol Environ Sustain Devel (Oxford University Press)
16. Winz I, Brierley G, Trowsdale S (2009) The use of system dynamics simulation in water resources management. J Water Res Manage 23(7):1301–1323

Retraction Note to: Power Equipment Defect Detection Algorithm Based on Deep Learning

Hanwu Luo, Qirui Wu, Kai Chen, Zhonghan Peng, Peng Fan, and Jingliang Hu

Retraction Note to: Chapter "Power Equipment Defect Detection Algorithm Based on Deep Learning" in: Q. Liang et al. (eds.), *Artificial Intelligence in China*, Lecture Notesin Electrical Engineering 653, https://doi.org/10.1007/978-981-15-8599-9_26

Retraction note to: Luo H., Wu Q., Chen K., Peng Z., Fan P., Hu J. (2021) Power Equipment Defect Detection Algorithm Based on Deep Learning. In: Liang Q., Wang W., Mu J., Liu X., Na Z., Cai X. (eds) Artificial Intelligence in China. Lecture Notes in Electrical Engineering, vol 653. Springer, Singapore.

The authors have retracted this Chapter "Power Equipment Defect Detection Algorithm Based on Deep Learning" due to an error in the specific formula of average accuracy which has led to errors in the experimental results. The authors repeated their experiments and were unable to reproduce the results. All authors agree to this retraction.

The retracted online version of this chapter can be found at https://doi.org/10.1007/978-981-15-8599-9_26

Q. Liang et al. (eds.), *Artificial Intelligence in China*, Lecture Notes in Electrical Engineering 653, https://doi.org/10.1007/978-981-15-8599-9_71

Printed in the United States
by Baker & Taylor Publisher Services